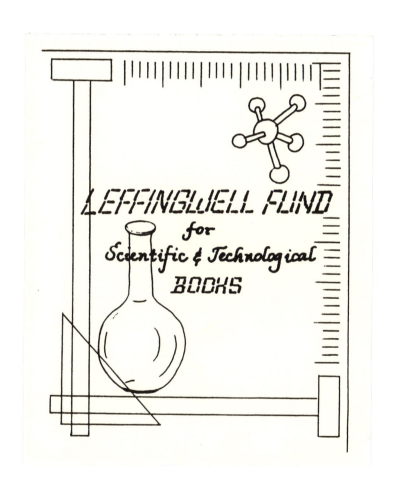

LEFFINGWELL FUND

for

Scientific & Technological

BOOKS

ENCYCLOPEDIA OF
CLIMATE CHANGE

ENCYCLOPEDIA OF

CLIMATE CHANGE

Volume 2

Energy resources and global warming
to
Organization of Petroleum Exporting Countries

Editor

Steven I. Dutch

University of Wisconsin—Green Bay

SALEM PRESS
A Division of EBSCO Information Services, Inc.
Ipswich, Massachusetts

GREY HOUSE PUBLISHING

Publisher's Cataloging-In-Publication Data
(Prepared by The Donohue Group, Inc.)

Names: Dutch, Steven I., editor.
Title: Encyclopedia of climate change / editor, Steven I. Dutch, University of Wisconsin - Green Bay.
Description: [Second edition]. | Ipswich, Massachusetts : Salem Press, a division of EBSCO Information Services, Inc. ; Amenia, NY : Grey House Publishing, [2016] | First edition published as: Encyclopedia of global warming. | Includes bibliographical references and index. | Contents: Volume 1. Abrupt climate change - Energy Policy Act of 1992 – Volume 2. Energy resources and global warming - Organization of Petroleum Exporting Countries – Volume 3. Oxygen, atmospheric - Younger Dryas, Appendixes, Index.
Identifiers: ISBN 978-1-68217-141-7 (set) | ISBN 978-1-68217-142-4 (v.1) | ISBN 978-1-68217-143-1 (v.2) | ISBN 978-1-68217-144-8 (v.3)
Subjects: LCSH: Climatic changes—Encyclopedias. | Global warming—Encyclopedias.
Classification: LCC QC981.8.G56 E46 2016 | DDC 551.603—dc23

PRINTED IN THE UNITED STATES OF AMERICA

Table of Contents

Abbreviations and Acronyms

AMO: Atlantic multidecadal oscillation
AQI: Air Quality Index
C^{14}: carbon 14
CCS: carbon capture and storage
CDM: clean development mechanism
CER: certified emissions reduction
CFCs: chlorofluorocarbons
CH_4: methane
CITES: Convention on International Trade in Endangered Species
CMP: Conference of the Parties to the United Nations Framework Convention on Climate Change, functioning as the meeting of the Parties to the Kyoto Protocol
CO: carbon monoxide
CO_2: carbon dioxide
CO_2e: carbon dioxide equivalent
COP: Conference of the Parties [to a treaty, such as the Framework Convention on Climate Change or the Convention on Biological Diversity]
COP/MOP: Conference of the Parties to the United Nations Framework Convention on Climate Change, functioning as the meeting of the Parties to the Kyoto Protocol
COP-1: First Conference of the Parties
CSD: Commission on Sustainable Development
DNA: deoxyribonucleic acid
EEZ: exclusive economic zone
ENSO: El Niño-Southern Oscillation
EPA: Environmental Protection Agency
ERU: emission reduction unit
FAO: Food and Agriculture Organization
GCM: general circulation model
GDP: gross domestic product
GHG: greenhouse gas
GWP: global warming potential
H_2: hydrogen (molecular)
HCFCs: hydroclorofluorocarbons
HFCs: hydrofluorocarbons
IAEA: International Atomic Energy Agency
IGY: International Geophysical Year
IMF: International Monetary Fund

INQUA: International Union for Quaternary Research
IPCC: Intergovernmental Panel on Climate Change
ITCZ: Inter-Tropical Convergence Zone
IUCN: International Union for Conservation of Nature
LOICZ: Land-Ocean Interactions in the Coastal Zone
MOC: meridional overturning circulation
MOP: meeting of the Parties [to a treaty, such as the Kyoto Protocol]
MOP-1: first meeting of the Parties
N_2O: nitrous oxide
NAAQS: National Ambient Air Quality Standards
NAM: Northern annular mode
NAO: North Atlantic Oscillation
NASA: National Aeronautics and Space Administration
NATO: North Atlantic Treaty Organization
NGO: nongovernmental organization
NOAA: National Oceanic and Atmospheric Administration
NO_x: nitrogen oxides
NRC: National Research Council
O^{16}: oxygen 16
O^{18}: oxygen 18
O_2: oxygen (molecular)
O_3: ozone
OECD: Organization for Economic Cooperation and Development
OPEC: Organization of Petroleum Exporting Countries
PFCs: perfluorocarbons
QELRCs: quantified emission limitation and reduction commitments
RuBisCO: Ribulose-1,5-bisphosphate carboxylase oxygenase
SAM: Southern annular mode
SCOPE: Scientific Committee on Problems of the Environment
SF_6: sulfur hexafluoride

SO$_2$: sulfur dioxide
SSTs: sea surface temperatures
SUV: sports utility vehicle
THC: Thermohaline circulation
UNCED: United Nations Conference on
  Environment and Development
UNDP: United Nations Development Programme
UNEP: United Nations Economic Programme

UNESCO: United Nations Educational, Scientific,
  and Cultural Organization
UNFCCC: United Nations Framework Convention
  on Climate Change
UV: ultraviolet
VOCs: volatile organic compounds
WHO: World Health Organization
WMO: World Meteorological Organization

Common Units of Measure

Common prefixes for metric units—which may apply in more cases than shown below—include *giga-* (1 billion times the unit), *mega-* (one million times), *kilo-* (1,000 times), *hecto-* (100 times), *deka-* (10 times), *deci-* (0.1 times, or one tenth), *centi-* (0.01, or one hundredth), *milli-* (0.001, or one thousandth), and *micro-* (0.0001, or one millionth).

Unit	Quantity	Symbol	Equivalents
Acre	Area	ac	43,560 square feet 4,840 square yards 0.405 hectare
Ampere	Electric current	A *or* amp	1.00016502722949 international ampere 0.1 biot *or* abampere
Angstrom	Length	Å	0.1 nanometer 0.0000001 millimeter 0.000000004 inch
Astronomical unit	Length	AU	92,955,807 miles 149,597,871 kilometers (mean Earth-Sun distance)
Barn	Area	b	10^{-28} meters squared (approx. cross-sectional area of 1 uranium nucleus)
Barrel (dry, for most produce)	Volume/capacity	bbl	7,056 cubic inches; 105 dry quarts; 3.281 bushels, struck measure
Barrel (liquid)	Volume/capacity	bbl	31 to 42 gallons
British thermal unit	Energy	Btu	1055.05585262 joule
Bushel (U.S., heaped)	Volume/capacity	bsh *or* bu	2,747.715 cubic inches 1.278 bushels, struck measure
Bushel (U.S., struck measure)	Volume/capacity	bsh *or* bu	2,150.42 cubic inches 35.238 liters
Candela	Luminous intensity	cd	1.09 hefner candle
Celsius	Temperature	C	1° centigrade
Centigram	Mass/weight	cg	0.15 grain
Centimeter	Length	cm	0.3937 inch
Centimeter, cubic	Volume/capacity	cm^3	0.061 cubic inch

Unit	Quantity	Symbol	Equivalents
Centimeter, square	Area	cm²	0.155 square inch
Coulomb	Electric charge	C	1 ampere second
Cup	Volume/capacity	C	250 milliliters 8 fluid ounces 0.5 liquid pint
Deciliter	Volume/capacity	dl	0.21 pint
Decimeter	Length	dm	3.937 inches
Decimeter, cubic	Volume/capacity	dm³	61.024 cubic inches
Decimeter, square	Area	dm²	15.5 square inches
Dekaliter	Volume/capacity	dal	2.642 gallons 1.135 pecks
Dekameter	Length	dam	32.808 feet
Dram	Mass/weight	dr *or* dr avdp	0.0625 ounce 27.344 grains 1.772 grams
Electron volt	Energy	eV	$1.5185847232839 \times 10^{-22}$ Btus $1.6021917 \times 10^{-19}$ joules
Fermi	Length	fm	1 femtometer 1.0×10^{-15} meters
Foot	Length	ft *or* ´	12 inches 0.3048 meter 30.48 centimeters
Foot, cubic	Volume/capacity	ft³	0.028 cubic meter 0.0370 cubic yard 1,728 cubic inches
Foot, square	Area	ft²	929.030 square centimeters
Gallon (British Imperial)	Volume/capacity	gal	277.42 cubic inches 1.201 U.S. gallons 4.546 liters 160 British fluid ounces
Gallon (U.S.)	Volume/capacity	gal	231 cubic inches 3.785 liters 0.833 British gallon 128 U.S. fluid ounces
Giga-electron volt	Energy	GeV	$1.6021917 \times 10^{-10}$ joule
Gigahertz	Frequency	GHz	—

Unit	Quantity	Symbol	Equivalents
Gill	Volume/capacity	gi	7.219 cubic inches 4 fluid ounces 0.118 liter
Grain	Mass/weight	gr	0.037 dram 0.002083 ounce 0.0648 gram
Gram	Mass/weight	g	15.432 grains 0.035 avoirdupois ounce
Hectare	Area	ha	2.471 acres
Hectoliter	Volume/capacity	hl	26.418 gallons 2.838 bushels
Hertz	Frequency	Hz	$1.08782775707767 \times 10^{-10}$ cesium atom frequency
Hour	Time	h	60 minutes 3,600 seconds
Inch	Length	in or ˝	2.54 centimeters
Inch, cubic	Volume/capacity	in³	0.554 fluid ounce 4.433 fluid drams 16.387 cubic centimeters
Inch, square	Area	in²	6.4516 square centimeters
Joule	Energy	J	$6.2414503832469 \times 10^{18}$ electron volt
Joule per kelvin	Heat capacity	J/K	$7.24311216248908 \times 10^{22}$ Boltzmann constant
Joule per second	Power	J/s	1 watt
Kelvin	Temperature	K	–272.15 Celsius
Kilo-electron volt	Energy	keV	$1.5185847232839 \times 10^{-19}$ joule
Kilogram	Mass/weight	kg	2.205 pounds
Kilogram per cubic meter	Mass/weight density	kg/m³	$5.78036672001339 \times 10^{-4}$ ounces per cubic inch
Kilohertz	Frequency	kHz	—
Kiloliter	Volume/capacity	kl	—
Kilometer	Length	km	0.621 mile
Kilometer, square	Area	km²	0.386 square mile 247.105 acres

Unit	Quantity	Symbol	Equivalents
Light-year (distance traveled by light in one Earth year)	Length/distance	lt-yr	5,878,499,814,275.88 miles 9.46×10^{12} kilometers
Liter	Volume/capacity	L	1.057 liquid quarts 0.908 dry quart 61.024 cubic inches
Mega-electron volt	Energy	MeV	—
Megahertz	Frequency	MHz	—
Meter	Length	m	39.37 inches
Meter, cubic	Volume/capacity	m^3	1.308 cubic yards
Meter per second	Velocity	m/s	2.24 miles per hour 3.60 kilometers per hour
Meter per second per second	Acceleration	m/s^2	12,960.00 kilometers per hour per hour 8,052.97 miles per hour per hour
Meter, square	Area	m^2	1.196 square yards 10.764 square feet
Metric. *See* unit name			
Microgram	Mass/weight	mcg *or* μg	0.000001 gram
Microliter	Volume/capacity	μl	0.00027 fluid ounce
Micrometer	Length	μm	0.001 millimeter 0.00003937 inch
Mile (nautical international)	Length	mi	1.852 kilometers 1.151 statute miles 0.999 U.S. nautical miles
Mile (statute or land)	Length	mi	5,280 feet 1.609 kilometers
Mile, square	Area	mi^2	258.999 hectares
Milligram	Mass/weight	mg	0.015 grain
Milliliter	Volume/capacity	ml	0.271 fluid dram 16.231 minims 0.061 cubic inch
Millimeter	Length	mm	0.03937 inch
Millimeter, square	Area	mm^2	0.002 square inch
Minute	Time	m	60 seconds

Unit	Quantity	Symbol	Equivalents
Mole	Amount of substance	mol	6.02×10^{23} atoms or molecules of a given substance
Nanometer	Length	nm	1,000,000 fermis 10 angstroms 0.001 micrometer 0.00000003937 inch
Newton	Force	N	x 0.224808943099711 pound force 0.101971621297793 kilogram force 100,000 dynes
Newton meter	Torque	N·m	0.7375621 foot-pound
Ounce (avoirdupois)	Mass/weight	oz	28.350 grams 437.5 grains 0.911 troy or apothecaries' ounce
Ounce (troy)	Mass/weight	oz	31.103 grams 480 grains 1.097 avoirdupois ounces
Ounce (U.S., fluid or liquid)	Mass/weight	oz	1.805 cubic inch 29.574 milliliters 1.041 British fluid ounces
Parsec	Length	pc	30,856,775,876,793 kilometers 19,173,511,615,163 miles
Peck	Volume/capacity	pk	8.810 liters
Pint (dry)	Volume/capacity	pt	33.600 cubic inches 0.551 liter
Pint (liquid)	Volume/capacity	pt	28.875 cubic inches 0.473 liter
Pound (avoirdupois)	Mass/weight	lb	7,000 grains 1.215 troy or apothecaries' pounds 453.59237 grams
Pound (troy)	Mass/weight	lb	5,760 grains 0.823 avoirdupois pound 373.242 grams
Quart (British)	Volume/capacity	qt	69.354 cubic inches 1.032 U.S. dry quarts 1.201 U.S. liquid quarts
Quart (U.S., dry)	Volume/capacity	qt	67.201 cubic inches 1.101 liters 0.969 British quart

Unit	*Quantity*	*Symbol*	*Equivalents*
Quart (U.S., liquid)	Volume/capacity	qt	57.75 cubic inches 0.946 liter 0.833 British quart
Rod	Length	rd	5.029 meters 5.50 yards
Rod, square	Area	rd²	25.293 square meters 30.25 square yards 0.00625 acre
Second	Time	s *or* sec	$\frac{1}{60}$ minute $\frac{1}{3600}$ hour
Tablespoon	Volume/capacity	T *or* tb	3 teaspoons 4 fluid drams
Teaspoon	Volume/capacity	t *or* tsp	0.33 tablespoon 1.33 fluid drams
Ton (gross or long)	Mass/weight	t	2,240 pounds 1.12 net tons 1.016 metric tons
Ton (metric)	Mass/weight	t	1,000 kilograms 2,204.62 pounds 0.984 gross ton 1.102 net tons
Ton (net or short)	Mass/weight	t	2,000 pounds 0.893 gross ton 0.907 metric ton
Volt	Electric potential	V	1 joule per coulomb
Watt	Power	W	1 joule per second 0.001 kilowatt $2.84345136093995 \times 10^{-4}$ ton of refrigeration
Yard	Length	yd	0.9144 meter
Yard, cubic	Volume/capacity	yd³	0.765 cubic meter
Yard, square	Area	yd²	0.836 square meter

Complete Table of Contents

Volume 1

Volume 2

Volume 3

Categorized List of Contents

Animals

Arctic and Antarctic

Astronomy

Chemistry and geochemistry

Meteorology and atmospheric sciences

Nations and peoples

Oceanography

Organizations and agencies

Physics and geophysics

Plants and vegetation

Pollution and waste

ENCYCLOPEDIA OF
CLIMATE CHANGE

Energy resources and global warming

Category: Energy

Most energy consumed by humans currently comes from fossil fuels, a source that will be exhausted within decades to centuries. Burning these fuels adds CO_2 to the atmosphere, which affects the climate.

Key concepts

extraction: removing a resource from its natural location

Hubbert's peak: the point at which the rate of production or extraction of an energy resource ceases to increase and begins to decrease

processing: changing a resource into a marketable commodity by milling, refining, and so on

transportation: movement of a resource from its place of origin to its market or processing location

Background

Many of the greenhouse gases (GHGs) in the atmosphere result from humans burning energy resources. The type of resource being burned depends on the value given to different properties of different fuels. The need for lowest cost and an ability to stockpile supplies favors coal, the need to minimize air pollution favors natural gas, and the need to be easily transported favors fuels derived from petroleum. As the need to limit emissions of GHGs becomes more pressing, it is reasonable to assume that other changes will occur.

Early American Settlers

Before agriculture, humans used fire to keep warm, cook their food, make their hunting more efficient, and promote the growth of wild plants they could eat or that would attract their preferred game. Their population density was low, and they moved frequently, so they were probably not limited by access to firewood.

As agriculture developed, it became necessary to remain in one location for extended periods. Initially, because the forests had to be cleared anyway, firewood was plentiful, and prodigious amounts were burned. A typical colonial family in New England burned 100 to 145 cubic meters of wood each year, translating to about 23 kilowatts of annual energy use. As the average colonial household contained about six people, the per capita energy use was about 4 kilowatts. For comparison, the United States today has a per capita use of about 11 kilowatts, whereas the global average is about 2.5 kilowatts per capita.

Another energy source available to early settlers was water power. In a new settlement the first major building to be constructed was usually a sawmill, often before churches or schools were built. By 1840, over sixty-five thousand water-powered mills had been operating in the United States. In particularly favorable locations such as Niagara Falls, Rochester, and Minneapolis, large flows over waterfalls permitted many large mills to coexist. Dependent on rainfall, removed by floods, and usually limited in scale by available hydraulic head and river flows, water-powered mills lost out to steam power, produced from coal.

Energy from coal did not fluctuate with the seasons, factories could grow, and they could be located where infrastructure existed to transport resources and products. Furthermore, they could become concentrated in those locations, providing jobs for thousands of city dwellers, rather than being stretched out along often unnavigable rivers.

The Industrial Revolution

By the middle of the nineteenth century the Industrial Revolution had begun, largely as a result of the steam engine. In addition to the large steam engines powering factories, smaller ones were adapted to pull trains, and with this development it became easier to transport an energy resource from where it occurred naturally to where it was needed.

To be valuable, resources need to bring in a profit when and where they are sold. The costs of bringing a resource to market, including extraction, processing, and transportation, as well as the market price, will determine whether that resource can be utilized. For example, abandoned gold dredges sit near the headwaters of dozens of streams in Alaska, intact except for

Iraq's Khawr Al Amaya Oil Platform (KAAOT) *in the North Arabian Gulf just after sunrise.* (Wikimedia Commons)

their copper wires. The costs of dismantling and transporting the other parts are greater than their market value.

Coal requires little processing, so the cost of bringing it to market entails only a trade-off between extraction and transportation. More expensive mining techniques can be used where transportation costs are lower. Coal quality (heat content, sulfur content, and so on) varies, and long-term contracts may also distort prices, resulting in remarkable discrepancies. In 2006, for example, the open market value of a ton of coal mined in Wyoming was $9.03, with a delivered price in Wyoming of $17.61; the open market value of a ton of coal mined in Pennsylvania was $37.42, with a delivered price in Massachusetts of $68.02.

The railroads and the coal industry have had a mutually beneficial relationship: Before the advent of diesel locomotives trains were a major coal user, and coal has always been a major fraction of the freight moved by trains. In addition to powering factories and trains, coal was used to heat homes. It had gradually replaced wood and by 1885 it heated more homes than wood did. Basement furnaces replaced stoves and fireplaces, and the low cost of coal meant that homes could be heated even with windows open in the middle of winter. The combined effect of hundreds of coal-burning furnaces, however, polluted the air, and the shipping, handling and storing of coal were very dirty operations.

Because it is a sedimentary rock, formed in swamps hundreds of millions of years ago, coal does not burn cleanly, but forms ash and clinker that need to be disposed of. Much less labor-intensive than wood stoves, coal furnaces still needed people to attend to them, to make sure coal was getting to the furnace and ash was being removed. As soon as oil or gas could be used, it quickly replaced coal for home heating.

Oil became an energy resource after ways were discovered to extract kerosene from it, which was used for lighting. By the beginning of the twentieth century large reliable supplies had been developed, largely in Texas, and railroads and barges transported it to markets elsewhere. The invention of the automobile and the oil burner during the first decades of the twentieth century made oil the most important energy resource in Europe and America. Crude oil must be extracted from the ground, transported to a refinery, processed, and then transported again before its useful components, gasoline and other fuels, can be used. Before this, it must be discovered.

Estimating Oil Reserves

An economic resource must be sufficiently abundant so that the costs needed to develop it can be paid off before it runs out. When a company invests in refineries and distribution systems it should know that it has sufficient reserves on hand to justify those expenses. These are called "proven reserves." To exploit these reserves, the resource must be extracted, processed, and transported and still return a profit to the company. The amount of proven reserves is a function of all of these costs as well as the market price of the final product. A company which is neither finding new resources nor selling old resources can still see change in the amount of proven reserves it owns as these costs and prices change. As the price of oil rose in 2007, the vast oil sands in Alberta, Canada, became profitable to extract and so were, in many tabulations, added to the list of proven reserves. When they are included they put Canada second, after Saudi Arabia, in total proven reserves.

In 1949, King Hubbert suggested that production of a nonrenewable resource would follow a bell-shaped curve, reaching a peak when about half of the resource had been consumed. Since then there has been considerable discussion about whether or not some region, defined geologically or politically, has reached its "Hubbert's peak" with respect to some resource. Although details vary, emotions run high, and apocalyptic predictions sprout up like weeds, a consensus is gradually developing that the world peak for conventional oil will have been, or will occur, sometime between 2005 and 2050.

As is the case with coal, natural gas requires transportation to a market before it has value. When it is encountered in oil fields with no access to pipelines, transportation costs often exceed market prices, so the natural gas is burned off, or "flared." Technological developments in the 1940's permitted the construction of extensive pipeline systems to distribute natural gas in developed countries. By 2008, there were over 480,000 kilometers of pipelines in the United States. Where pipeline infrastructure exists, much of the cost of natural gas is still in its transportation. In 2007 in the United States, for example, the average price for 28.3 cubic meters at the wellhead was $6.39, while residential users paid $13.01.

Renewables face similar distribution requirements. Solar- or wind-generated energy needs to be shipped to a market, and ideally needs to be stored until it is most needed. As each conversion step has inefficiencies, this will add to the costs of renewable energy resources.

Context

Climate change scenarios rely on estimates of the total amount of carbon-containing fuels which will be burned and the rate at which this will occur. Neither estimate is known well; however, guidance is provided by exploring resource utilization history. As reserves of petroleum are depleted, it is reasonable to expect prices to rise and alternative fossil fuel resources, such as oil sands and oil shales, to be developed. Whether or not such unconventional fossil fuels are included in climate model projections can make a substantial difference in the predicted CO_2 levels in the atmosphere.

For renewable energy resources to compete, means must be developed to store their energy and transport it to markets when and where it is needed. For example, wind energy is unlikely to be directly available during hot spells, which are usually accompanied by still, high-pressure air—yet, that is just when peak electric loads to run air conditioners are greatest. If wind energy could be stored in batteries, hydrogen, pumped-storage facilities, and other repositories, then it would be available when most needed. If it could be shipped and stockpiled safely and efficiently, then huge installations of windmills could be developed, far from populated or scenic areas, perhaps in developing countries.

Otto H. Muller

Further Reading

BP. BP Statistical Review of World Energy, June, 2008. London: Author, 2008. One of the industry standards for assessing energy resources, this publication consists of a wide variety of tables and charts, including data from all of 2007. Written by an oil company, its biases are different from those of a government agency. Charts, tables.

Campbell, C. J., and J. H. Laherrere. "The End of Cheap Oil." Scientific American 278, no. 3 (1998): 60-65. An easily understood description of Hubbert's peak, the difficulties involved in reporting proven reserves, and some of the possible effects of depleting petroleum resources. Graphs, charts.

Cloud, P. Resources and Man. San Francisco: W. H. Freeman, 1969. A report by the National Academy of Sciences; the last third concerns energy resources. Of interest is how well the predictions have held up over time. Tables, charts, graphs.

Energy Information Administration, U.S. Annual Energy Review, 2006. Washington, D.C.: Government Printing Office, 2007. One of the industry standards for assessing energy resources, this publication consists of a wide variety of tables and charts, including data from all of 2006. Written by a government agency, its biases are different from those of an oil company.

Twidell, John, and Tony Weir. *Renewable energy resources.* Third Edition, Routledge, 2015. Survey of wind, solar, tidal, biofuels and other resources, as well as discussion of their performance to date.

Wolfson, Richard. Energy, Environment, and Climate. New York: W. W. Norton, 2008. An excellent treatment of environmental energy issues; Wolfson is able to construct quantitative explanations without using intimidating math. In

addition to showing how the variables relate, he includes current values for global estimates of many of the important variables. Illustrations, figures, tables, maps, bibliography, index.

Engineers Australia

Category: Organizations and agencies
Date: Established 1919

Mission

Engineers Australia advocates a comprehensive, coordinated governmental approach to counter environmental degradation. It particularly wants to recast infrastructure, such as transportation, energy, and greenhouse gas (GHG) management, in order to ensure security, prosperity, and social harmony while fostering sustainability.

Its royal charter naming it as the Institution of Engineers, Australia, the eighty-thousand-member nonprofit organization is best known simply as Engineers Australia. It is governed by its National Council and Congress chaired by its president, has a national office in the capital (Canberra) headed by its chief executive, and maintains nine state and territorial divisions. It is affiliated as well with colleges, national committees, technical societies, and engineering interest groups.

Engineers Australia serves as a national forum for promoting all engineering disciplines. Specifically, it seeks to advance engineering science, set standards for chartering engineers, cultivate educational development among its members, instill professional ethics and integrity, and recognize achievements through awards. Its National Engineering Registration Board, established in 1994, is a consumer-protection measure to ensure that enrolled engineers use their knowledge in the public interest.

Significance for Climate Change

Engineers Australia issued a media statement in June, 2007, defining its support for a GHG

Engineering House is the national office for Engineers Australia located in the Canberra suburb of Barton, Australian Capital Territory. (Wikimedia Commons)

reduction effort that progressively decreases annual emissions. The statement quotes the group's national president, Rolfe Hartley:

The collective science and ongoing research firmly point to the need to reduce global greenhouse emissions by 2050 by between 40 and 80 percent. Any further delays in determining a target, rather than getting on with solutions, is more wasted time that we don't have. A target in this range requires only a minimal reduction in emissions of about 1 percent per year. While we need to balance the impact of meaningful action on global warming on our economy and employment opportunities, the fact remains that a slightly slower growth of the Australian economy stands in stark contrast to what may happen if the economy becomes smaller under the impacts of climate change…. While it would be naïve to expect global consensus on emission reduction programs in the short-term, it would be equally dangerous for Australia to continue to procrastinate and then opt for emissions reduction programs that lack substance and rigour. Until wider agreement on a global emissions strategy is achieved, Australia could go a long way towards meeting a meaningful emissions target through vigorously pursuing energy efficiency approaches which the International Energy Agency believes could achieve almost half of the needed emissions.

Roger Smith

Further Reading

"About Us." Engineers Australia. Engineers Australia, n.d. Web. 23 Mar. 2015.

Aliento, Willow. "Engineers Australia Commits to a Low-Carbon Future." Fifth Estate. Fifth Estate, 27 Nov. 2014. Web. 23 Mar. 2015.

Climate Change Policy. Barton: Engineers Australia, 2014. Print.

Kaspura, Andre. Australia's Energy Future: Australian Energy Policy and Climate Change. Canberra: Engineers Australia, 2007. Print.

Enhanced greenhouse effect

Category: Meteorology and atmospheric sciences

Definition

Greenhouse gases (GHGs) in the atmosphere—such as carbon dioxide (CO_2), methane, nitrous oxide, water vapor, and ozone—absorb infrared radiation from the Sun and reradiate some of it at the surface, warming Earth's atmosphere. The average temperature of the atmosphere has been estimated to be more than 30° Celsius warmer than it would be without these gases. The natural greenhouse effect occurs when this process is the result of nonhuman activities; the enhanced greenhouse effect denotes increases in the effect caused by GHGs emitted into the atmosphere by human activities.

Significance for Climate Change

The concentrations of CO_2 and, to a lesser extent, other GHGs have gradually increased in the atmosphere, especially during the twentieth century. For instance, the CO_2 content of the atmosphere in the Hawaiian Islands has increased from 313 parts per million in 1960 to 375 parts per million in 2005. Arctic ice-core samples indicate that the CO_2 content of the atmosphere has also gradually increased over longer timescales. Much of this increase in atmospheric CO_2 concentration appears to be due to human activity, although the importance of human activity relative to natural processes such as in volcanism is not clear. It is known, however, that CO_2 is released to the atmosphere by human activity, such as the burning of fossil fuels (petroleum, natural gas, and coal). Deforestation of tropical and other forests—such as the Amazon rain forest in Brazil—produces a great deal of CO_2 as the plants decay, and it also reduces an extremely important carbon sink, increasing the amount of CO_2 that remains in the atmosphere rather than being converted to biomass and oxygen. CO_2 is also liberated in cement production.

A continued increase in GHGs will likely cause a continued increase in the average temperature of the atmosphere. The greatest increase in temperature will likely be over polar landmasses. For example, a doubling of the amount of CO_2 in the atmosphere has been predicted to cause an average increase of 3° to 4° Celsius at high northern latitudes, resulting in much less snow and ice. Summer conditions might last an extra two months with a correspondingly shorter winter.

Warmer air holds more water vapor than cooler air, so global warming will likely cause the evaporation rate to increase. This increase in evaporation may result in increased drought, desertification, and water shortages in some regions. Although water availability may decrease, greater levels of atmospheric CO_2 will mean more CO_2 is available to drive photosynthesis, so some plants may benefit from this increase. Warmer temperatures at higher latitudes should also allow some crops such as wheat to be grown further to the north than at the present.

Robert L. Cullers

See also: Carbon dioxide; Climate change; Emissions standards; Greenhouse effect; Greenhouse Gas Protocol; Greenhouse gases and global warming; Methane.

Environmental economics

Category: Economics, industries, and products

Environmental economics seeks to quantify present and future losses from damaged health and depleted natural resources that are due to environmental degradation. The most efficient way to reduce these losses is determined by comparing the cost of environmental damage to the cost of mitigation options.

Key concepts

abatement cost: cost incurred when reducing a nuisance such as pollution

Coase theorem: assertion that when property rights are properly defined, people are forced to pay for the negative externalities they impose on others and market transactions will produce efficient outcomes

discount rate: percentage by which measured economic effects occurring in the future are transformed into present values

emission charges: a fixed-rate tax calculated per unit of emissions

option value: potential benefits of the environment not derived from actual use

Pareto optimum: a situation in which it is impossible to make improve any individual's condition without worsening that of another individual

willingness to accept (WTA): minimum amount of money one would accept to forgo some good or to bear some harm

willingness to pay (WTP): maximum amount of money one would give up to buy some good

Background
Environmental economics emerged as a discipline in the 1950's and 1960's, when its primary focus was on natural resource economics as concern over environmental degradation began to develop. Environmental economics borrows heavily from the theories of its precursors and is couched in many neoclassical economics assumptions. Environmental economics has evolved to address ever-emerging issues, such as pollution and fishery issues. Currently, efficiently mitigating and adapting to climate change, which is a multidiscipline issue, is a main focus.

Monetary Valuation of the Environment
Environmental economics is predicated on the concept that natural resources and environmental health can be valued monetarily, which is contested by some environmentalists. Economic value is determined by people's preferences and the intersection of supply and demand for the valued entity. Money is a convenient and familiar metric to decision makers determining "optimal" environmental policies. Monetary valuation has been established as a precedent by legally mandated economic reparations in major environmental disasters, such as the 1989 *Exxon Valdez* oil spill off Alaska.

The economic value of an environmental good is assessed subjectively, is relative to the value of other goods, and is marginal (that is, it is altered by changes in the broader state of affairs). There is no formal market for environmental goods; therefore,

economists have developed methods to elicit individuals' valuation. "Revealed preferences" methodology deduces valuation based on people's observed behaviors. Alternatively, "stated preference" methodology techniques directly ask people their valuation of environmental goods.

Monetary valuation of the environment remains an ethical question. This brand of valuation is based upon a utilitarian philosophy, which is anthropocentric and consequentialist. It can be argued that individual preferences are a poor guide for social decision making. To appropriately weigh all aspects of a decision, the environment should not be ignored; some monetary valuation is arguably better than none at all. In 1997, researchers estimated that the worldwide flow of environmental services was $33 trillion, demonstrating that the environment is a larger provider of economic goods and services than conventional economic estimates reflect.

Cost-Benefit Analysis
Economic valuation allows comparison of diverse benefits and costs associated with ecosystem conservation. In this way, the wider array of benefits and costs associated with a project can be considered in deciding which alternative produces the largest net benefit to society. Economists study how people make decisions when faced with scarcity. Scarcity implies that resources devoted to one end are not available to meet another; hence, there is opportunity cost associated with any action. For example, funds used by a municipality to retrofit a sewage treatment plant cannot also be used to improve the local community center. Pareto improvements are an important consideration in this process.

Cost-benefit analysis (CBA) is the economist's applied tool for assessing the efficiency of alternative public choices and policies. Economists favor the use of CBA because it requires transparency and encourages objectivity by the analyst. CBA provides an organizational framework for identifying, quantifying, and comparing benefits and costs of a proposed action. Generally, economists suggest policies providing greater benefits than costs; CBA allows ranking of alternatives in this manner.

Market-Based Instruments

The Coase theorem states that under ideal circumstances, when polluters and polluttees bargain, the equilibrium level of pollution is independent of the allocation of property rights. Ideal circumstances include perfect information about costs and benefits and assume no transaction costs. In the real world, externalities can lead to market failures, and the use of market-based instruments can provide more equitable outcomes. The optimal level of pollution is determined where the marginal cost of emissions and marginal benefits are equal. The target level of the policy is then calculated, taking into account the potential for gaming and principal-agent problems.

The four main economic instruments used are pollution taxes, clean technology subsidies, tradable permits, and hybrid schemes. Pollution (Pigovian) taxes and clean technology subsidies are price instruments; issuance of tradeable emissions permits is a quantity instrument.

Pigovian taxes are levied to correct negative externalities. The tax shifts the marginal private cost (MPC) curve up by the amount of the tax, thus providing incentive for producers to reduce output to the socially optimum level. Subsidies are useful when taxes are politically unrealistic, though they run contrary to the polluter pays principle and can result in efficiency losses. Emissions trading guarantees that the pollution quantity reached is optimal (unlike taxes) but does not cap the costs of achieving the target. The European Union Emission Trading Scheme (EUETS) is the largest multinational emission trading scheme in the world.

Climate Change Economics

Environmental economics is increasingly involved in climate change mitigation strategies, which present unique problems due to their global nature and the long time-horizons affected.

Estimating abatement costs can be achieved through two different approaches: resource cost estimates (RCE) and macroeconomic cost estimates (MCE). RCE rely on deriving the marginal abatement cost (MAC) curve by estimating and ranking the technologies and their emissions potential. MCE models examine behavioral responses to carbon price increases.

Economic estimation of potential global warming damages necessitates an integrated assessment (IA) approach, considering scientific findings and socioeconomic forcings. IA models differ considerably in impact representations, which are especially sensitive to the discount rate and applied elasticity of marginal utility.

Context

Environmental economists are tasked with recommending policies that reflect appropriate division of scarce resources on the societal level. However, there is controversy over the correct manner of nonmarket good and services valuation and whether environmental economics is overly anthropocentric. When dealing with environmental issues, ethical considerations are necessary. For instance, climate change economics is currently dealing with the ethical dimension of discount rate valuation. The long-term nature of climate change coupled with the complexity of climate systems complicates economic analyses by producing inherently high uncertainty as to potential outcomes.

Jennifer Freya Helgeson

Further Reading

Callan, Scott J., and Janet M. Thomas. *Environmental economics and management: Theory, policy, and applications.* Cengage Learning, 2013. Explores both theoretical and practical aspects with a strong business perspective.

Diamond, P., and J. Hausman. "Contingent Valuation: Is Some Number Better than No Number?" *Journal of Economic Perspectives* 8, no. 4 (Autumn, 1994): 45-64. Argues that monetary valuation of environmental goods can be misleading and inaccurate.

Hanneman, M. "Valuing the Environment Through Contingent Valuation." *Journal of Economic Perspectives* 8, no. 4 (Autumn, 1994): 19-43. Supports the use of monetary valuation of environmental goods.

Pizer, W. "Combining Price and Quantity Controls to Mitigate Global Climate Change." *Journal of Public Economics* 85, no. 3 (2002): 409-434. Discusses the economics of hybrid mechanisms.

Stern, N., et al. *Stern Review on the Economics of Climate Change.* New York: Cambridge University Press, 2006. This report, headed by Lord Stern for the British government, discusses the effect of climate change on the world economy. Not the first economic report on climate change, but its approach and level of detail are novel.

Tietenberg, Tom, and Lynne Lewis. *Environmental and Natural Resource Economics.* 8th ed. Boston: Pearson Addison Wesley, 2009. An introduction to environmental policy problems through an integrated discussion of economic theory and empirical evidence.

See also: Anthropogenic climate change; Damages: market and nonmarket; Deforestation; Ecological impact of global climate change; Economics of global climate change; Environmental movement; Extinctions and mass extinctions; Human behavior change; Kyoto Protocol.

Environmental law

Category: Laws, treaties, and protocols

Federal environmental laws such as the Clean Air Acts, although formulated before the modern scientific study of climate change, have become the chief vehicle for attempting to reduce GHG emissions. Several states have passed climate change initiatives, as has the international community in the Kyoto Protocol.

Key concepts

cap-and-trade agreements: transaction-based agreements that make the right to emit greenhouse gases or other pollutants a fungible commodity

greenhouse effect: increased warming of the Earth caused by absorption of heat energy by gases in the atmosphere

greenhouse gases (GHGs): gaseous constituents of the atmosphere, such as carbon dioxide, that contribute to the greenhouse effect

legal standing: eligibility to bring a civil lawsuit in a specific case, usually as a result of suffering damages

litigation: disputes in court between parties regarding the meaning and execution of laws

Background

For most of human history, the environment was to all intents and purposes unprotected by law. In Anglo-American law, the near-absolute dominion that was accorded to owners of property put little restraint on the right to consume, exhaust, or pollute resources at will. A nuisance lawsuit could be pursued if a neighbor deprived a property owner of the ability to enjoy such essentials as light, air, and water. The public trust doctrine restrained landowners from monopolizing access to navigable waters. Other than these impediments, American law left the environment open to even the most rapacious use. The preservationist movement of John Muir and Gifford Pinchot focused on conserving public lands and parks; it had little effect on private uses of land.

Environmental Law

This situation changed in the United States with the introduction of environmental laws in the twentieth century. The first environmental laws were an extension of nuisance laws and health and safety laws: protecting against environmental threats to the safety of the population, rather than the environment itself. But by the 1970s, laws to protect the environment had found broad acceptance in American political and popular culture. The National Environmental Policy Act of 1970 (NEPA) established federal guidelines for protecting the environment. In the same year, Congress established the Environmental Protection Agency (EPA) as the federal authority to oversee enforcement of the environmental laws, and the Clean Air Act Extension to control air pollution.

The Clean Water Act was enacted in 1972 to protect the purity and drinkability of water and to preserve wetlands. The Energy Policy and Conservation Act of 1975, as amended by the Alternative Fuels Act of 1988, mandated automobile fuel economy standards and encouraged the development of alternative fuels in vehicle use. The Comprehensive Environmental Response Compensation and Liability Act of 1980 provided a superfund program to clean up hazardous wastes.

Global Climate Change and the Law

In 2009, the United States contemplated legislation aimed directly at global warming; previously, legal controversies over global warming focused on the extension of existing environmental laws to greenhouse gases (GHGs). For example, the landmark case of Massachusetts v. EPA (2007) concerned the question of whether motor vehicular emissions of GHGs are air pollutants and thus subject to regulation under the Clean Air Acts (1963–90). A major barrier in all environmental law litigation is the question of legal standing—who is permitted to bring a lawsuit on behalf of the environment. Normally a plaintiff must show a concrete and immediate harm.

In Massachusetts v. EPA, the US Supreme Court allowed a coalition of environmental groups and state attorney generals to bring the lawsuit. After hearing evidence from both sides, the Court required the EPA to determine if carbon dioxide (CO_2) emissions satisfy the definitions of an air pollutant under the Clean Air Acts and are thus subject to EPA regulation. Likewise, in Border Power Plant Working Group v. Department of Energy (2003), the federal government was required to consider power plant CO_2 emissions in regulating the construction of power lines and grids. The administration of George W. Bush, however, did not believe administrative action was required as to climate change. The 2008 election of President Barack Obama signaled a change in executive enforcement. On April 17, 2009, the EPA announced that it had determined that CO_2 and other gas emissions were indeed dangers to public health and welfare and thus subject to its mandate.

Several US states took legislative action on climate change before the US Congress. California in particular passed legislation to restrict GHG emissions. As specific carbon emission legislation is passed by the states, there may be less reliance on older, more general environmental laws to address global warming concerns. The Regional Greenhouse Gas Initiative is a joint cap-and-trade program of ten northeastern states to reduce CO_2 emissions from power plants while pursuing clean energy programs. Globally, the 1992 United NationsFramework Convention on Climate Change launched the international effort to reduce carbon emissions. It found legal force in the Kyoto Protocol of 1997 establishing binding targets for reducing GHGs. However, upon taking office in 2001, President Bush renounced American participation in the protocol.

Context

Environmental legislation enacted since the 1970s has provided some basis for regulation and litigation concerning CO_2 emissions that many believe contribute to global warming. Environmentalists have made use of such historic legislation as the Clean Air Acts to try to compel the EPA to include reduction of GHGs under its regulatory mandate, a step resisted by President George W. Bush but revisited by the Obama administration. States, particularly in the northeast and California, have been more aggressive in enacting frameworks designed to reduce CO_2 emissions in a cap-and-trade system.

Given the exclusive authority that the federal government retains over national environmental questions and interstate commerce, it is hard to know how effective state legislation can be. The industrialized nations have been active in enacting frameworks, initiatives, and treaties addressing climate change, but, given the nature of international competition and national sovereignty, the future of such essentially voluntary restrictions is uncertain. One of the largest questions facing Congress is how to integrate federal policy with both state and international initiatives.

Howard Bromberg

Further Reading

Aspatore Book Staff, eds. The Legal Impact of Climate Change: Leading Lawyers on Preparing for New Environmental Legislation, Assessing Green Programs for Clients, and Working with Government Agencies on Climate Change Issues. Boston: Thomson West, 2008.

Bodansky, Daniel, Jutta Brunnée, and Ellen Hey. *The Oxford handbook of international environmental law.* Oxford University Press, 2012. At 1080 pages, the label "handbook" is a bit stretched. Treats environmental law as a field in itself rather than a variety of international law.

Gerrard, Michael, ed. Global Climate Change and U.S. Law. Chicago: American Bar Association, 2008.

Kushner, James. Global Climate Change and the Road to Extinction: The Legal and Planning Response. Durham, N.C.: Carolina Academic Press, 2009.

Salzman, James, and Barton Thompson. Environmental Law and Policy. 2d ed. New York: Foundation Press, 2007.

Environmental movement

Category: Environmentalism, conservation, and ecosystems

The organizations and individuals that make up the environmental movement have been instrumental in bringing the issue of climate change into the mainstream and to the attention of the general public. By raising public awareness, the environmental movement has helped pressure politicians and the private sector to take action.

Key concepts

direct action: nonconventional political activity, including protests and demonstrations

environmental paradox: the tendency of the environmental movement to work within the system that may be responsible for the problems the movement seeks to address

individualization of responsibility: placing responsibility for climate change on individuals rather than on businesses, corporations, or the government

precautionary principle: a policy rule that one should act to prevent severe harm, even if the likelihood of that harm occurring is unknown or potentially low

Background

The environmental movement began in the 1960's as a grassroots effort to address local and regional environmental issues. With increasing public support throughout the 1970's, it became highly professionalized and began focusing mostly on federal politics. The environmental movement became concerned with global warming in the late 1980's. Since then, the movement has employed a combination of direct action and traditional lobbying methods to prompt action by the public and private sectors to address climate change.

Promoting Awareness of Climate Change

The environmental movement has promoted both awareness and action in response to climate change through educational and motivational campaigns. These campaigns have targeted the general public, all levels of government, businesses, and international civic society. They have encouraged civic and personal responsibility, made elected officials aware of public opinion, and sought to hold them accountable for responding to it with appropriate legislation. There have been several campaigns in the early twenty-first century, including Step It Up, a grassroots effort to encourage people to lobby their congressional representatives to take action on climate change, and the Climate Change Solutions Campaign, which connects thousands of national and international members of the environmental movement. Both of these campaigns and many others rely on the Internet to reach and educate people. Another method of bringing about awareness is through films. *An Inconvenient Truth* (2006), the U.S. documentary on climate change presented by Al Gore, is a good example of this tactic.

Sponsoring Public and Private Solutions

Traditionally, the environmental movement focused on federal legislation to combat global warming. However, this strategy has proven fruitless for almost two decades. In 2005, major environmental groups began to target state and local governments, some of which are more willing than Congress to address climate change. The Cool Cities Campaign is an example of the environmental movement encouraging action at the local level. This campaign aims to persuade cities to sign the U.S. Conference of Mayors Climate Protection Agreement, a commitment started in 2005 to advance the initiatives of the Kyoto Protocol at the local level. Additionally, the movement has targeted the federal government through litigation related to the Clean Air Acts (1963-1990) in an attempt to force the Environmental Protection Agency to regulate carbon emissions.

The environmental movement has also targeted the private sector through a variety of campaigns, including efforts to discourage the production of sports-utility vehicles (SUVs). In 2004, the Apollo Alliance was established as a large-scale attempt to address climate change by forging a coalition between environmental groups and labor unions. The alliance is working on the Apollo Project, an effort to create 1 million new jobs in the renewable energy sector and reduce U.S. dependence on foreign oil.

Dealing with the Opposition

The environmental movement has used a variety of strategies to overcome opposition to action on climate change. A major argument against acting on climate change is that the lack of total scientific consensus does not warrant the economic investment needed. The environmental movement has advocated the use of the precautionary principle in making decisions about climate policy. The principle is popular in other industrial countries, particularly European Union member countries, and holds that even if the risk of climate change is uncertain, the potential for catastrophe is great enough to warrant immediate action to avoid it: Certainty of the risk of severe consequences may come only when it is too late to prevent them.

Criticisms

Despite the efforts of the environmental movement to address climate change, several critiques have emerged of the movement's dominant climate-related strategies. Kirkpatrick Sale notes that there is an "environmental paradox" that has resulted from the movement's tendency to work within the system that is causing environmental degradation rather than confront the actors and encourage changes to the system itself.

Michael Maniates provides a critique of the "individualization of responsibility," a tendency of environmental groups to encourage individuals to take on responsibility by reducing their own carbon footprints rather than confronting corporate or government entities: Environmental groups, for example, encourage people to use less fossil fuels by biking, carpooling, or driving more efficient vehicles. The argument against this tactic is that alleviating the global warming problem requires much more than individual action, and by focusing on campaigns on individuals, environmentalists implicitly absolve bigger entities that generate large amounts of greenhouse gas emissions of their responsibility. The most notable critique was presented in 2004 by Ted Nordhaus and Michael Shellenberger in their essay, "Death of Environmentalism." The essay criticizes the movement for relying on tactics that worked for issues like acid rain but are inadequate to combat a global environmental problem such as climate change. Environmental organizations responded by revamping their strategies in 2005 to include a more grassroots approach.

Context

Global warming has become a unifying issue for the environmental movement. It provides an overarching concern that covers all of the issues that environmental organizations deal with, such as biodiversity, preservation, and human health. The ability of the movement to promote the radical changes necessary to mitigate climate change has been called into question. As the organizations and individuals within the movement increase their efforts, some scientists warn that climate change may soon become irreversible. It is yet to be determined whether the sort of concerted effort of all levels of government, consumers, and individuals in nations throughout the globe will agree to implement the sorts of changes that environmentalists believe to be necessary to respond to global warming.

Katrina Darlene Taylor

Further Reading

Maniates, Michael F. "Individualization: Plant a Tree, Buy a Bike, Save the World." *Global Environmental Politics* 1, no. 3 (2001): 31-52. Reviews the emerging strategy of the environmental movement to place responsibility for solving global warming on individuals. Shows that this may be happening at the expense of seeking political solutions and policies.

Sale, Kirkpatrick. *The Green Revolution: The American Environmental Movement, 1962-1992.* New York: Hill and Wang, 1993. Presents one of the first critiques of the environmental movement's increasing neglect of grassroots organizing and solutions since the 1970's in favor of more professional lobbying and litigation tactics like those used by the for-profit community.

Schellenberger, Michael, and Ted Nordhaus. "Death of Environmentalism." n.p.: Authors, 2004. Argues that the environmental movement is failing to effect change on global warming and that a new and bolder strategy is needed. Focuses on the need to move away from traditional tactics through the reformation of liberalism, the main political position of the environmental movement.

Shabecoff, Philip. *A fierce green fire: The American environmental movement.* Island Press, 2012. Covers American environmentalism throughout history, and discusses the successes of anti-environmental forces in recent years.

See also: Biodiversity; Carson, Rachel; Catastrophist-cornucopian debate; Civilization and the environment; Conservation and preservation; Deforestation; Ecological impact of global climate change; Ecosystems; Environmental law; Friends of the Earth; Greenpeace; International Union for Conservation of Nature; Sierra Club; Sustainable development.

Estuaries

Categories: Geology and geography; water resources

Definition

Oceanographer Gerardo Perillo has crafted perhaps one of the most well-considered definitions of an estuary from the dozens that had been put forward previously:

Estuaries are flooded river valleys. Delaware Bay (top right) and Chesapeake Bay (center) are two of the best known estuaries in the United States. They were eroded when sea level was lower during the ice ages and then flooded as sea level rose. (NASA)

An estuary is a semi-enclosed coastal body of water that extends to the effective limit of tidal influence within which seawater entering from one or more free connections to the open sea, or other saline body of water, is significantly diluted with freshwater.

This definition encompasses seven geomorphic types of estuary in three categories of salinity stratification. The geomorphic types are: high relief estuaries, such as fjords; moderate-relief, winding valleys, such as rias; low-relief, drowned river valleys, or coastal plain estuaries; deltaic estuaries; tectonic estuaries; bar-built estuaries; and seasonally blocked bar-built estuaries, or blind estuaries. In terms of salinity stratification, estuaries can be salt-wedge, partially mixed, or homogeneous.

Significance for Climate Change

Coastal systems in general and estuaries in particular exert a control over biogeochemical cycles and, therefore, play an important role in regulating global climate. In the face of a changing climate, sea-level rise, sediment supply, the persistence of barrier beaches and salt marshes, water temperature changes, and freshwater supply will all alter estuaries. Changes in the magnitude of freshwater flow or in its seasonal distribution will immediately shift the stratification of existing estuaries. For instance, greater spates will drive existing estuaries toward more stratified end members. Under extreme conditions, salinity gradients could be driven out onto the shelf. At high latitudes, the reduction or elimination of permafrost and sea-ice cover will redefine the freshwater budget of local estuaries.

A transgressive shoreline creates space for estuaries, but it may be inimical to them in some situations. Bar-built estuaries, in particular, rely on the maintenance of a barrier beach. A too rapid rise in sea level could make it impossible for the barriers to maintain themselves, drowning the former estuaries. In other situations, stronger seasonality of freshwater discharge may open formerly blind estuaries or turn ephemeral estuaries into permanent, bar-built estuaries. The condition will depend on the rate of sea-level rise and on the supply of littoral sand by processes that allow barrier beaches to maintain their integrity and to migrate shoreward in the face of rising seas.

In the geological perspective, estuaries are ephemeral features, prevalent at times of high sea-level stands, especially along coasts of low relief. Rising sea levels will shift isohalines inland, possibly turning formerly true-estuarine environments into marginal seas or merely arms of the ocean. At the same time, high-energy environments will shift toward the outer part of the coast. However, because estuaries are sediment traps, their continued existence is a balance between flooding associated with rising sea level and infilling.

Estuaries disappear when they eventually fill with sediment. Climate-induced changes in the amount and seasonality of rainfall, temperature, humidity, and vegetation over wide areas will all influence the rate of infilling of estuaries newly fighting for existence. Any increased sedimentation will compete with increased sea-level rise to shorten the life of future estuaries. For example, increased rainfall and more intense spates may enhance soil erosion, and changes in both temperature and humidity may accelerate chemical weathering. Northern estuaries might be expected to be exposed to higher sediment supplies more typically associated with semitropical regions.

Human responses to climate change will exert another influence on estuaries, especially because so much human development is concentrated around estuarine shorelines. Shoreline hardening can squeeze out salt marsh development, as well as arresting shoreline transgression. Human control of flooding—such as that provided by the Thames storm-surge barriers, those at Rotterdam and St. Petersburg, and that under construction in Venice—would limit the expansion of estuaries in some regions. The construction of dams to control flooding and water supply could force some estuaries to cope with decreases in both freshwater flow and nutrient fluxes. Changes in forestation and agricultural land-use must further alter the delivery of both nutrients and sediments to estuaries.

Although the physical shifts in estuarine conditions may be as yet poorly resolved, the potential and realized impacts on marine ecosystems are well recognized. Increasing stratification, coupled with elevated temperature, exacerbates problems of hypoxia that already plague many temperate estuaries. Freshening and warming will also change nutrient delivery to the estuary, perhaps increasing total nutrient delivery while simultaneously shifting higher concentrations seaward. Pelagic species will be forced into new patterns of larval transport and recruitment and changes in foraying. Changes in one or two key, leverage species may cause broad, community-level changes. Harmful algal blooms, invasive species and the spread of diseases will alter estuarine ecology further. In Long Island Sound, for example, a crash of the American lobster population has been suggested to result from immunodepression caused by the stress of warming water temperatures.

Henry Bokuniewicz

Further Reading

Crossland, Christopher J., et al., eds. *Coastal Fluxes in the Anthropocene.* Berlin: Springer-Verlag, 2005. Synthesizes a decade of international research by the Land-Ocean Interactions in the Coastal Zone (LOICZ) project to assess the impacts on coastal ecosystems of climate change and human activity. Focuses on biogeochemical fluxes and key habitats.

Fairbridge, Rhodes W. "The Estuary: Its Definitive and Geodynamic Cycle." In *Chemistry and Biogeochemistry of Estuaries*, edited by E. Olausson and I. Cato. New York: Wiley, 1980. Classic article by a renowned scientist that lays a foundation for the global study of estuaries, including the ephemeral nature of individual estuaries in geological time.

Harley, Christopher D. G., et al. "The Impact of Climate Change in Coastal Marine Systems." *Ecology Letters* 9 (2006): 228-241. Discusses the importance of a few key, "leverage" species in maintaining marine ecosystem services and the potential impacts of temperature changes and changes in ocean chemistry.

Perillo, Geraldo M. E. "Definition and Geomorphic Classifications of Estuaries." In *Geomorphology and Sedimentology of Estuaries*, edited by G. M. E. Perillo. Amsterdam: Elsevier, 1997. Reviews physical, geochemical, and biological characteristics of estuaries and crafts a new definition in the context of multidisciplinary studies as required for coastal management.

Neilson, Bruce J., and L. Eugene Cronin. *Estuaries and nutrients.* Springer Science & Business Media, 2012. The focus is really on eutrophication, the concentration of nutrients in small bodies of water. Even in undisturbed estuaries, natural eutrophication is constant, but human activities increase it.

See also: Barrier islands; Coastal impacts of global climate change; Coastline changes; Freshwater; Ocean acidification; Ocean life; Sea sediments; Venice.

Ethanol

Category: Energy

Definition

Ethanol (ethyl alcohol, or grain alcohol) is a transportation fuel that is used as a gasoline substitute. Henry Ford's first car, the Model T Ford, was designed to run on pure ethanol. Ethanol is a colorless liquid with the chemical formula C_2H_5OH. It is produced through a biological process based on fungal or bacterial fermentation of a variety of materials. In the United States, most ethanol is produced by yeast (fungal) fermentation of sugar from cornstarch. Sugar is extracted using enzymes, then yeast cells convert the sugar into ethanol and carbon dioxide (CO_2). The ethanol is separated from the fermentation broth by distillation.

Ethanol can also be produced chemically from petroleum. Ethanol produced by fermentation is commonly referred to as bioethanol to differentiate it from chemically produced ethanol. Brazil, the second largest bioethanol producer after the United States, generates bioethanol from sugarcane. Brazilian production of ethanol from sugarcane is more efficient than is the U.S. corn-based production method. Several factors contribute to that efficiency, including climate and cheap labor. In addition, sugarcane has much a higher sugar content than corn, and no enzymes are necessary to extract cane sugar. Therefore, sugarcane yields of ethanol are twice as great as corn yields.

When energy demands and oil prices increase, ethanol becomes a valuable option as an alternative transportation fuel. In 2005, the U.S. Congress passed an energy bill that required gasoline sold in the United States to be mixed with ethanol in order to decrease the nation's dependence on oil. There is a limit to the amount of ethanol that can be mixed with gasoline in the American market, however. Nearly all modern automobiles can use E10, fuel that contains 10 percent ethanol. By contrast, E85, containing 85 percent ethanol and only 15 percent gasoline, requires specially equipped flexible-fuel engines. In the United States, only a fraction of motor vehicles use such engines. Most cars in Brazil have a flex engine, however, as ethanol use in vehicle fuels has been mandatory in that country since 1977. Brazil's ethanol policy allowed the nation to achieve energy independence in 2006.

Blending ethanol with gasoline oxygenates the fuel mixture, which then burns more completely and emits less carbon monoxide. However, ethanol has about two-thirds the energy content of gasoline by volume, so vehicles can travel less far on a gasoline-ethanol mixture than they can on pure gasoline. Ethanol also tends to be more expensive than gasoline. In addition, carcinogenic aldehydes, such as formaldehyde, are produced when ethanol is burned in internal combustion engines.

Significance for Climate Change

Burning ethanol in internal combustion engines produces carbon dioxide (CO_2), a major greenhouse gas (GHG). However, vehicles that run on pure ethanol or ethanol/gasoline mixtures can produce 20 percent less CO_2 than vehicles that burn gasoline alone. Emissions of nitrogen oxide (another GHG) are about equal for both ethanol and gasoline.

In order fully to evaluate ethanol's carbon footprint, it is necessary to determine how much CO_2 is emitting during production of the fuel. Such determinations depend on the method of ethanol manufacturing employed. For example, ethanol can be manufactured from various feedstocks, including starch (cornstarch), sugarcane, and lignocellulose. Lignocellulose is a combination of lignin, cellulose, and hemicellulose that strengthens plant cell walls. Cellulose and hemicellulose are made of sugars that can be converted into ethanol. Many energy-intensive steps are required to sustain this conversion. These steps include growing, transporting, and processing the feedstock.

Manufacturing ethanol from cornstarch requires considerable amounts of energy, usually obtained by burning fossil fuels. Burning these fossil fuels releases significant amounts of CO_2. In addition, allocating corn crops to ethanol production leads to increases in food prices, because a great number of food items in the United States include corn-based ingredients, such as corn syrup and cornstarch. Corn-based ethanol is thus not a viable long-term biofuel, but it may help smooth the transition from a petroleum-based economy into a biofuel-based economy that utilizes ethanol from other sources.

Increasing corn ethanol use will help reduce U.S. reliance on foreign oil, but it will not do much to slow global warming and will lessen the availability of corn for food. By contrast, net emissions of CO_2 during lignocellulose ethanol production can be nearly zero. Burning the lignin itself can provide enough energy to power ethanol production, alleviating the need for fossil fuels. Burning lignin does not add any net CO_2 to the atmosphere, because the plants that are used to make the ethanol absorb CO_2 during their growth. Most important, lignocellulose may be obtained from nonedible plants, such as switchgrass and poplar, or nonedible parts of other plants, such as corn stalks and wood chips. Thus, lignocellulose ethanol does not compete for food resources.

Lignocellulose is a very attractive ethanol fuel feedstock because it is in abundant supply. On a global scale, plants produce almost 90 billion metric tons of cellulose per year, making it the most abundant organic compound on Earth. In addition, cultivation of nonedible plants for ethanol production requires fewer nutrients, fertilizers, herbicides, acres of cultivated land, and energy resources.

Methods of processing the cellulosic parts of plants into simple sugars in order to ferment them into ethanol have so far been costly. The cost of ethanol generation from lignocellulose can be reduced substantially, however, by using a biorefinery-based production strategy. Similar in function to a petroleum refinery, a biorefinery utilizes every component of lignocellulose to produce useful products, thereby increasing revenues and cost-effectiveness. These products include ethanol fuel, electrical power, heat energy, animal feed, and chemicals such as succinic acid and 1,4-butanediol.

The latter can be used to manufacture plastics, paints, and other products. Major research and development efforts are under way to improve the conversion cost of lignocellulose to ethanol.

Sergei Arlenovich Markov

Further Reading

Demain, Arnold L., Michael Newcomb, and J. H. David Wu. "Cellulase, Clostridia, and Ethanol." *Microbiology and Molecular Biology Reviews* 69, no. 1 (March, 2005): 124-154. Review article about making ethanol from plant cellulose.

Geddes, Claudia C., Ismael U. Nieves, and Lonnie O. Ingram. "Advances in ethanol production." *Current opinion in biotechnology* 22, no. 3 (2011): 312-319. Recent progress has included advances in catalysts that break down cellulose. Improvements in the foreseeable future will increase yields.

Europe and the European Union

Category: Nations and peoples

Key facts

Population: European Union: 508.5 million as of 2015; the rest of Europe: 230,327,713

European Union Member States (EU-28): Austria, Belgium, Bulgaria, Croatia, Republic of Cyprus, Czech Republic, Denmark, Estonia, Finland, France, Germany, Greece, Hungary, Ireland, Italy, Latvia, Lithuania, Luxembourg, Malta, Netherlands, Poland, Portugal, Romania, Slovakia, Slovenia, Spain, Sweden, United Kingdom

Area of Europe: 10,180,000 square kilometers

Gross domestic product (GDP): European Union: $18.51 trillion (World Bank, 2014 estimate)

Greenhouse gas (GHG) emissions in millions of metric tons of carbon dioxide equivalent (CO_2e): 4,696 in 2005, 4,611 in 2013

Kyoto Protocol status: Ratified by the European Union and all member states in 2002

Historical and Political Context

After World War II, many European governments agreed on the basic understanding that there never should be a war on European territory again. Shocked by the devastation of the war and by the Holocaust, Winston Churchill soon after the war's conclusion proposed the idea of the establishment of "United States of Europe" (1946). Several institutions and organizations, including the Western Union Treaty (Brussels Treaty, 1948)—signed by Belgium, France, Luxembourg, the Netherlands, and the United Kingdom—the Organization for European Economic Cooperation, coordinating the initiatives undertaken by the Marshall plan, and the Council of Europe (1949) were established. Based on a concept by French Minister for Foreign Affairs Robert Schuman, regarding the integration of the Western European coal and steel industry, the Treaty of Paris establishing the European Coal and Steel Community (ECSC) was signed by Belgium, France, Germany, Italy, Luxembourg, and the Netherlands in 1951. The basic idea behind the ECSC was to prevent future wars by economically integrating the two main economic sectors with which a war can be fought, coal and steel. In 1953, all trade barriers regarding these two economic goods were removed within a common market.

The European Communities, the ECSC together with the European Economic Community (EEC) and the European Atomic Energy Community (Euratom), were founded in Rome in 1957. The Treaty of Rome entered into force in 1958, and it established a Parliamentary Assembly and a Court of Justice. In 1958, the Economic and Social Committee was founded. Agricultural policies were integrated from 1962 onward by means of the Common Agricultural Policy, and in 1967, a treaty was signed to establish a joint European Commission and a European Council for all three communities. While the 1970s brought much discussion but little progress toward European integration, in 1987, the Single European Act developed the EEC, including the European Monetary System (representing further progress for a system of exchange rates).

With the Maastricht Treaty of 1992 and several other commitments, the Common Market, with the free movement of goods and services, financial

Simulated view of city lights of Europe and part of the Middle East. Clouds obscure much of Britain and parts of Scandinavia. (NOAA National Geophysical Data Center)

capital, and labor, was fully established, together with the joint European currency (the euro). Many policy fields fall within the competence of the European Council, the European Commission, and the European Parliament. The majority of policy fields (for example, economic and environmental policy) are treated at the European level, with aims further to integrate tax and social policies of member states and to pass a European Constitution.

Environmental policy is an especially important topic for European integration, since environmental problems (air and water pollution, waste management, land use, climate change) do not follow national boundaries but are often truly international problems. The European Union has therefore established a dense system of environmental regulations touching upon, among other issues, air pollution, water quality standards, agricultural subsidies, transport policies, product and production standards, and the regulation and registration of chemicals.

Impact of European Policies on Climate Change

The European Union ratified the United Nations Framework Convention on Climate Change (UNFCCC) in 1993. After the adoption of the Kyoto

Protocol in 1997, the European Union signed the protocol in 1998 and ratified it in 2002, with the obligation that all EU member states had to ratify the Kyoto Protocol by no later than 2002.

The European Union and its member states have committed themselves to a reduction of greenhouse gas (GHG) emissions by 8 percent during the observation period of 2008 to 2012, compared to the base years of 1990 and 1995. The EU member states follow different national policies, including different national base years and reduction scenarios. Some of the newer EU member states do not have reduction targets according to the Kyoto Protocol, since they were transition countries at the time of the setting up of the Kyoto Protocol. However, the majority of EU member states have a reduction target varying approximately from –6 percent to –8 percent from 1990 levels. However, in 2013, the European Union issued reports showing that they have reduced carbon dioxide (CO_2) and other greenhouse gas emissions by 18 percent since 1990, far exceeding the 8 percent reduction in emissions that the European Union originally committed to.

The European Union's Climate and Energy Package, adopted in 2009, was established to work toward the goal of a 20 percent reduction of GHG emissions, 20 percent share of renewables in EU energy consumption, and a 20 percent overall energy improvement by 2020, also requirements of the second Kyoto Protocol period. This package proposes to revise the EU emissions trading system (ETS), require each member state to achieve an emissions limitation target for 2020. A major policy instrument for reducing the European Union's GHG emissions is the scheme for GHG-emission-allowance trading based on a mechanism for monitoring GHG emissions. Within the ETS, GHGs emitted by certain sources (including energy companies and major industrial emitters) can be traded in order to reduce emissions at sources where reduction is most efficient.

The EU member countries have to submit national allocation plans for the distribution of emission allowances with many recent positive results regarding the control of CO_2 emissions.

Europe and the European Union as GHG Emitters

The European Union member countries emit huge quantities of GHGs into the atmosphere. Carbon dioxide emissions amounted to 3.778 billion metric tons in 1995 and to 3.873 billion metric tons in 2005; GHG emissions were 4.762 billion metric tons of CO_2 equivalent in 1995 and 4.696 billion tons of CO_2 equivalent in 2005 in EU28. The EU15 countries, falling under the reduction scenarios according to the Kyoto Protocol, create about 3.6 trillion metric tons of CO_2 emissions per year. These emissions exhibit a major impact on global climate change, since the EU15 countries are responsible for 12 percent of global CO_2 emissions. While China (28%) and the United States (16%) have the largest share in worldwide CO_2 emissions, followed by the EU-28 (10%), while India and Russia (both 6%) emit smaller amounts of CO_2. While some small oil-producing countries such as Qatar (44.0 metric tons) and the United Arab Emirates (20.4 metric tons) have the highest per capita emissions, annual U.S. per capita emissions amount to about 17 metric tons, which is roughly twice the EU15 per capita emission. China's and India's per capita emissions are comparatively small, at 6.7 and 1.7 metric tons, respectively. However, it is estimated that the sustainable level of per capita emissions is at the most two metric tons per year, indicating that industrialized countries are largely off track from a sustainability perspective.

Regarding the importance of the different economic sectors and activities emitting GHGs, energy use, especially electricity consumption (not counting the transport sector) and the use of energy for heating buildings, is the largest emitter by far at 57.2 percent, while the transport (including international aviation) sector emits 22.2 percent of EU27's GHGs emissions; industrial processes follow with 7.8 percent.

The European and national strategies have been successful to various degrees. The EU-28 target of reducing GHG emissions to –8 percent until 2008–2012 (compared with 1990 emissions) has been quite successful as of the 2013 reporting by the EU-28 of an 18% reduction of greenhouse gases, including CO_2, far exceeding the proposed 8% reduction as per their Kyoto Protocol commitment. Out of the 28 member states of the EU-28, GHG emissions were highest (as of 2013) in Germany (21.17%), the United Kingdom (13.10%), and France (10.98%). The most significant decreases in GHG emissions among the EU-28 members were found in Lithuania (–58.2%), Latvia (–57.2%), and Romania (–56.2%). The largest increases in GHG emissions compared to 1990 among members of the EU-28 were found in Cyprus (+43.8%), Malta (+41.3%), Spain (+13.1%), and Portugal (+9.7%). Overall, eight of the twenty-eight member states showed significant decreases in GHG emissions, six member states showed significant increases in GHG emissions, and the remaining member states showed relatively modest decreases and increases in GHG emissions as compared with 1990 levels.

Summary and Foresight

GHG emissions of European Union countries are among the largest in the world. Strong action is therefore needed to decrease CO_2 emissions substantially. The European Union has issued proposals for reduction targets for the post-Kyoto period until 2020. The European Commission proposed to cut GHG emissions by 20 percent until 2020, to increase the share of renewable energy sources from currently around 12 percent to 20 percent by 2020, to increase the share of biofuels to a minimum of 10 percent of all transport fuels, and to reduce energy consumption by 20 percent based on increased energy efficiency. Additionally, the European Council discussed a reduction target of 30 percent based on 1990 emissions, provided that other developed and industrialized countries committed themselves to similar targets and efforts. Critics argue that the European Union's efforts may only slightly decelerate the increase of the CO_2 content in the atmosphere, since reductions in energy use and CO_2 emissions in Europe are—without other comparable international efforts—easily offset by increases in emissions of large countries such as the United States, China, India, and Russia.

Even if the European Union's commitments are realized and the impact on the Earth's climate is small, the economy may largely benefit. Research in new energy-efficient technologies can significantly increase the competitiveness of the European economy, and lower energy costs also increase the disposable income of households.

Regarding the instruments for further emission reduction, the emission trading system will have to be developed further, Europe-wide standards for energy efficiency will have to be set into force, and national policies with respect to energy taxation will have to be coordinated on the European level.

Michael Getzner
Updated by: Ben Riley

Further Reading

Llavador, Humberto, John Roemer, Joaquim Silvestre. "How to Allocate CO_2 Emissions." Yaleglobal, 25 June 2013. Yaleglobal Online. Web. 13 June 2016. Brief yet informative article that explores how CO_2 emissions can be allocated in a way that is environmentally and practically sustainable.

Behrens, A., Ehenhofer, C. (2008). *Energy Policy for Europe*. Retrieved from http://www.ceps.eu/system/files/book/1623pdf Comprehensive report compiled from the CEPS (Centre for European Policy Studies) task force's discussions on European energy and climate change policy.

Butorina, O. A. "Europe without the European Union." *Russia in Global Affairs* 4 (2011). An examination of the stresses on the European Union, economic and demographic. Concludes with a warning that the European Union faces stagnation if it fails to develop a dynamic sense of identity.

Commission of the European Communities. "An Energy Policy for Europe." *COM(2007)* 1, January 10, 2007. This energy strategy statement describes future EU efforts for increasing energy efficiency, increasing the share of renewables, and technological progress, and includes an "action plans" to achieving EU energy goals.

_____. "Limiting Global Climate Change to 2 Degrees Celsius: The Way Ahead for 2020 and Beyond." *COM(2007)* 2, January 10, 2007. Communication paper of the European Commission addressed to the European Council in order to discuss EU's new targets regarding combating climate change by significantly reducing GHG emissions.

_____. "Proposal for a Decision of the European Parliament and the European Council on the Effort of Member States to Reduce Their Greenhouse Gas Emissions to Meet the Community's Greenhouse Gas Emission Reduction Commitments up to 2020." *COM(2008)* 17, January 23, 2008. The European Commission proposed "effort-sharing" decision in order to implement GHG emission reduction policies in a differentiated way among EU member countries.

_____. "Proposal for a Directive of the European Parliament and the Council Amending Directive 2003/87/EC So As to Improve and Extend the Greenhouse Gas Emission Allowance Trading System of the Community." *COM(2008)* 16, January 23, 2008. A commission proposal to account for the 20 percent reduction scenario by means of developing and strengthening the EU's emission allowance trading system.

European Communities. *Europe in Figures: Eurostat Yearbook, 2008.* Luxembourg: Office for Official Publications of the European Communities, 2008. Includes statistical information (text, figures, tables) on the demography, economy, environment, and social issues of the European Union and EU member countries; it is also most valuable for details on GHG emissions and country-specific obligations and achievements.

Jordan, Andrew. *Environmental Policy in the European Union.* London: Earthscan, 2005. Describes EU's environmental policies from the viewpoint of processes, actors, and institutions, and presents case studies of EU environmental policies such as climate change; special emphasis is also laid on policy challenges due to globalization and EU enlargement.

Knill, Christoph, and Duncan Liefferink. *Environmental Politics in the European Union.* Manchester, England: Manchester University Press, 2007. Comprehensive and easy-to-read introduction to the European Union's environmental policies presenting the establishment of environmental policies such as a European issue, institutions and actors, policy processes and policy fields, and country case studies; discusses EU's capacity for environmental solutions.

Peeters, Marjan, and Kurt Deketelaere, eds. *EU Climate Change Policy: The Challenge of New Regulatory Initiatives.* Cheltenham, England: Edward Elgar, 2006. Collective and comprehensive volume presenting EU's climate change policies and frameworks. Discusses the European Union's emission allowances trading system from various perspectives, explores the links between energy policies and climate change policies, and summarizes policy principles such as good governance.

See also: Annex B of the Kyoto Protocol; Carbon dioxide; Greenhouse effect; Greenhouse gases and global warming; Industrial emission controls; Kyoto lands; Kyoto mechanisms; Kyoto Protocol; United Nations Framework Convention on Climate Change.

Evapotranspiration

Category: Meteorology and atmospheric sciences

Definition

In the hydrologic cycle, two processes—evaporation and transpiration—are responsible for returning water to the atmosphere. Evapotranspiration is the sum of these processes. Evaporation is the loss of water vapor from the surface of objects containing liquid water, such as lakes and streams, soil, and organisms. Temperature is a major factor controlling evaporation. In general, the higher the temperature, the faster the rate of evaporation. Other major factors influencing evaporation rate include the amount of water vapor already held by the air (humidity) and air movement. Low humidity and high wind speed promote evaporation.

Transpiration is the loss of water vapor primarily through the stomata of leaves and green stems of plants. All above-ground parts of plants are covered by a waxy cuticle that helps prevent water loss. As a result, the internal spaces within the plant are nearly saturated with water vapor. Stomata are microscopic pores in plants' surface layers that open or close to allow gas exchange for photosynthesis while regulating water loss to prevent desiccating the plant. Cottonwoods near Dallas, Texas, can transpire up to 120 liters of water per tree per day. In the Amazon rain forest, a single large tree can transpire up to 1,180 liters of water per tree per day.

Transpiration provides the force necessary to pull water and dissolved nutrients up from the roots to the tips of the highest branches. Within the plant, water and dissolved minerals travel in specialized xylem cells that are arranged end to end in long files. Xylem cells are programmed to die once they mature; they then form hollow tubes to transport water. When filled with water, the xylem acts like a straw, and transpiration from the top draws water up from the bottom. Transpiration provides enough suction force that water could be pulled through the xylem to the top of a tree much taller than the tallest living redwood, higher than a forty-story building.

Significance for Climate Change

Air temperature directly affects evapotranspiration; any increase in temperature will increase evapotranspiration as long as adequate water is available. In tropical rain forests, transpiration increases beginning at sunrise, and the released water vapor rises in the heated air until it condenses in cooler air at high elevation, thus forming clouds. These clouds deliver afternoon rain showers that replenish soil moisture to repeat the process the next day.

Some studies suggest that extensive clear-cutting of the forest has decreased transpiration sufficiently to decrease local rainfall. If enough water is not available during the hot daytime, stomata of the remaining trees close and transpiration is shut down. Less water vapor is returned to the atmosphere, and the daily water cycle is broken. Less water returned to the atmosphere can exacerbate drought conditions and have a snowballing effect on climate. Some scientists predict that even the Amazon rain forest could be transformed to a savanna if too many trees are cut down and there remains too little evapotranspiration to sustain regular rainfall. Similar changes can occur in temperate grasslands and in temperate forests, making them more arid.

Decreasing rates of transpiration that are due to global warming and local drying also affect photosynthesis. Some plants, classified as carbon 4 plants, are more efficient at photosynthesizing at higher temperatures and in drier conditions. These plants, which are most common in the southern United States, will spread northward as temperatures rise. By contrast, carbon 3 plants, including most trees and vegetable crops, tend to be less drought tolerant and will not cope as well with increasing temperatures. Carbon 3 plants can be expected to migrate toward cooler latitudes if their existing locations become untenable. Plant migration is generally slow, and it is questionable if a poleward migration of biomes and plant types could keep pace with a global warming trend. Even if plants could migrate fast enough, the soils of higher latitudes are not as fertile and productive as those in current temperate areas, the "breadbaskets" of the world, because they have not supported extensive plant growth for millennia.

Water loss from transpiration often can exceed evaporative water loss from a water surface of the same area. For instance, a sugarcane field in Hawaii can transpire 120 percent as much water vapor as a similar area of open water. This is due not only to the amount of leaf surface area containing stomata but also to extensive root systems that can tap soil water in excess of the flat surface area of water. Some studies predict that such natural pumping of soil water, in response to a warming environment, will deplete soil water in the same way that pumping irrigation water from underground aquifers depletes groundwater.

Marshall D. Sundberg

Further Reading

Angelo, Claudio. "Punctuated Disequilibrium." *Scientific American* 292, no. 2 (February, 2005): 22-23. Describes how deforestation and global warming in the eastern Amazon rain forest may be tipping the scale toward development of a savanna ecosystem.

Huxley, Anthony. *Green Inheritance: The World Wildlife Fund Book of Plants.* New York: Anchor Press, 1985. A botany textbook written for the general public with a concise, well-illustrated section explaining transpiration.

Pimm, Stuart L. *The World According to Pimm: A Scientist Audits the Earth.* New YorK: McGraw-Hill, 2001. Pimm uses an engaging style to describe human impacts on the Earth.

Wang, Kaicun, and Robert E. Dickinson. "A review of global terrestrial evapotranspiration: Observation, modeling, climatology, and climatic variability." *Reviews of Geophysics* 50, no. 2 (2012). A fairly technical examination of ways of surveying global transpiration from satellite data. Significant uncertainties still remain in the models used for relating raw data to transpiration.

See also: Amazon deforestation; Atmospheric dynamics; Carbon 4 plants; Carbon 3 plants; Clouds and cloud feedback; Dew point; Hydrologic cycle; Rainfall patterns; Water vapor.

Extinctions and mass extinctions

Categories: Animals; plants and vegetation

Climate change, particularly global warming, has been correlated with past extinction events. Any future short- or medium-term global warming would likely contribute to further extinctions.

Key concepts

anthropogenic: resulting from human actions

extinction: species loss; the death of the last member of a particular species

extinction event: an unusually large number of extinctions occurring in a relatively short period of time

methane hydrate: also known as methane clathrate; an extremely abundant, solid, ice-like form of methane that occurs at or beneath the ocean floor in especially deep or cold waters

Background

Although consensus about categorization is lacking, five major mass extinctions (the "Big Five") and numerous smaller extinction events are generally recognized. Mass extinctions involve global losses of at least 30 percent of species of various types and sizes occupying various habitats. Less severe extinction events involve fewer species losses (less than 25 percent) and fail to meet one or more additional criteria defining mass extinctions. Their scope may be regional rather than global, for example. Background extinctions occur between extinction events, at relatively low rates, and affect only one or a very few species at a time. Thus, they reflect the mean duration—or "life span"—of species in general as reflected in the fossil record. Although background extinctions occur infrequently, because they accumulate over very long periods of time they account for far more species losses than do mass extinctions.

Extinction, Mass Extinction, and Global Warming

It is generally impossible to identify a single cause for an extinction. By way of example, imagine that global warming allows a pathogen to extend its range from the tropics into what was formerly a temperate region. Imagine further that the pathogen's new range overlaps that of a bird that has become highly endangered by habitat loss. The pathogen causes 90 percent mortality among the birds, and the remainder dies off shortly thereafter during a series of severe storms. It would be an oversimplification to state that only habitat loss, global warming, infectious disease, or severe weather caused the extinction. Habitat loss represents the ultimate, underlying cause, although additional proximate stressors subsequently eliminated the species. Relative to single-species extinctions, extinction events typically involve far more complicated chains of causation.

Global warming is thought to have contributed to numerous extinctions—including the Permian-Triassic and Triassic-Jurassic events, as well as less severe events during the early Jurassic (around 183 million years ago) and the Paleocene-Eocene boundary (around 55 million years ago). If global warming were to occur again, it would likely contribute to future extinctions via several main mechanisms. First, warming conditions would reduce or degrade the habitats of high-latitude, high-altitude, and temperate species. For example, polar bears rely for food on the formation of winter ice in which their seal prey dens, so they would be threatened by a decrease in such ice formation. Similarly, rising sea levels resulting from melting glaciers would flood coastal areas and reduce Earth's overall island area, threatening the disproportionately high levels of biodiversity occupying these relatively small landmasses.

Although rising sea levels might increase the total volume of aquatic habitat, global warming would also place important, interrelated stresses on aquatic species. The first of these would be thermal stress. Because of water's ability to absorb and retain heat, aquatic environments experience less temperature variation than do terrestrial environments. As a consequence, aquatic species are generally adapted to a narrower range of temperature tolerances than are terrestrial species. Thus, any change in sea temperature is liable to threaten aquatic species. Because excessive heat poses a greater physiological threat to most organisms than does excessive cold, global warming would represent an especially serious stress. Contemporary coral die-offs resulting from warming oceans provide a noteworthy example.

A second threat to aquatic species relates to the fact that as water's temperature increases, it holds less dissolved oxygen, while warming waters tend to increase organismal metabolism and oxygen demand. Consequently, warmer water can support relatively fewer aerobic organisms than can colder water. Hypoxia (oxygen scarcity) and anoxia (absence of oxygen) in warming waters have been linked to a variety of extinctions. Although these factors would most directly affect aquatic species, they would also exert significant indirect effects on species dependent on aquatic organisms, such as fish-eating birds.

Warming oceans pose an additional threat to marine species, since warmer seawater would likely melt extensive seafloor deposits of methane hydrate. Melted methane hydrate reacts with the dissolved oxygen in seawater to form carbon dioxide, and the release of large quantities would greatly reduce oceanic oxygen levels, placing considerable stress on both marine and marine-dependent species by exacerbating the effects of the reduced

oxygen levels associated with increased water temperature. Such a scenario is thought to have occurred during the Paleocene-Eocene extinction, when high-latitude and deep-sea temperatures may have increased by as much as 7° Celsius. It has been predicted that contemporary global warming could cause similar effects.

Modern Versus Paleoextinctions

The fossil record's limitations make detailed characterizations of the biodiversity losses associated with extinctions difficult. Paleontologists estimate that fossils preserve less than 4-5 percent of all species that have ever lived. The fossil record's incompleteness results from numerous factors. First, species vary in composition and habitat. Hard tissue such as shell, tooth, and bone is much more prone to fossilization than is soft tissue such as muscle and skin. Since fossilization occurs only in environments where dead organisms can be preserved in sediment, species occupying habitats where sediments were routinely deposited are more common as fossils. By contrast, in areas such as mountain tops, where erosional processes dominated, fossils are rare or unknown. Thus, the fossil record reveals far more about shelled marine mollusks (which had hard parts and occupied depositional environments) than about small, delicately built uplands birds (which had fragile bones and occupied erosional environments).

Since fossilization is a rare outcome, species that were rare in life are correspondingly rare in the fossil record. As a consequence, while complete or nearly complete fossils of some marine invertebrates are common, most dinosaurs are known from one or a very few specimens, many of them fragmentary. Because marine fossils are much more common than terrestrial fossils, they are frequently used in computing extinction rates.

The final key factor limiting the fossil record's completeness is erosion, which eventually reduces fossil-bearing rock to sediment, destroying any fossils it contains. The longer fossil-bearing rock is in existence, the more opportunity there is for erosion to act on it. As a result, the fossil record of more recent extinctions is far more complete than that of older extinctions. Consequently, the details of the Cretaceous-Tertiary extinction are better understood than those of the much earlier Ordovician-Silurian extinction.

Because the fossil record is so often imprecise at the species level, paleontologists often assess extinctions on the basis of the persistence or disappearance of higher taxa, such as genera, families, or orders. The extinction of even one order, family, or genus would likely involve the loss of numerous species, although the vagaries of fossilization preclude precise estimates of species losses.

Although contemporary biologists face logistical difficulties in locating and cataloging existing biodiversity, they clearly have significant advantages over paleontologists attempting to characterize the biodiversity of vanished ecosystems. It would be highly unlikely for relatively small, rare, and delicately boned species analogous to today's black-footed ferrets and ivory-billed woodpeckers to be abundant or well-preserved in the fossil record. However, some of the same traits that make such creatures rare in the fossil record place them in danger of extinction in the modern world, where an ability to study their distant ancestors through fossil records could help scientists better understand their ecological and evolutionary roles.

Possibility of Anthropogenic Mass Extinction

Despite the difficulties paleontologists face in providing detailed characterizations of vanished life, they can render informative sketches of lost biological communities. In comparing estimates of present-day extinction rates, whose main cause is habitat loss, to those computed from the fossil record, ecologists have concluded that recent human history, including the present, probably qualifies as a period of mass extinction.

Mass extinctions are generally thought to have resulted from nonbiological causes, such as climate change, sea-level shift, intense volcanism, or meteor impact. Current extinctions, linked as they are to the activities of a single species, humans, appear unprecedented: Collectively, they represent the first mass extinction with a biotic cause. The rate of extinction might also be unusually high, at least fifty to

The Big Five Mass Extinction Events

Event	Occurrence (millions of years ago)	Percent of Marine Species Extinct
Ordovician-Silurian	c. 440	up to 85
Late Devonian	c. 360	up to 83
Permian-Triassic	c. 250	up to 95
Triassic-Jurassic	c. 200	up to 80
Cretaceous-Tertiary (K/T)	c. 65	up to 76

one hundred times greater than background extinction rates. Although often referred to as "events," mass extinctions are generally thought to have been protracted by human standards, occurring over thousands or, in many cases, millions of years. By contrast, the spike in extinction rates associated with human activities may be unprecedentedly high, occurring as it has over only several thousand years.

Context

The fossil record provides invaluable insights into extinction, but it may prove insufficient for use in predicting future extinction risks. For instance, it has been noted that few North American extinctions are associated with climate change during the roughly two-million-year duration of the Pleistocene. During periods of Pleistocene cooling and warming, species simply shifted their ranges. The assumption that species would do the same globally under near-term warming conditions ignores a complicating factor: Humans now occupy and modify a far larger percentage of the Earth's habitats. While Pleistocene species could readily shift their ranges in response to climate change, they did not face the challenge of doing so across such large areas of potentially inhospitable human-modified habitats. Vulnerability to human barriers would vary from species to species. While many birds could simply fly to preferred habitats and shift their migratory patterns in response to changing conditions, less-mobile species—particularly those with specialized habitat requirements—would be at greater risk. Just as important, anthropogenic stressors could interact both with global warming and with other abiotic stressors such as drought or volcanism to increase extinction risk.

Jeffrey V. Yule

Further Reading

Barnosky, Anthony D., Nicholas Matzke, Susumu Tomiya, Guinevere OU Wogan, Brian Swartz, Tiago B. Quental, Charles Marshall et al. "Has the Earth/'s sixth mass extinction already arrived?." *Nature* 471, no. 7336 (2011): 51-57. Mass extinctions are defined as events where 75% or more of species go extinct in a geologically short time. By that measure, we have not yet arrived at mass extinction, but for some groups the number of threatened species approaches half, and our record extends only a few centuries, not the millennia typical of geological extinction events.

Benton, Michael J., and Richard J. Twitchett. "How to Kill (Almost) All Life: The End-Permian Extinction Event." *Trends in Ecology and Evolution* 18 (2003): 358-365. Semi-technical review of potential causal factors—including global warming—and their likely interactions in Earth's largest mass extinction.

Botkin, Daniel B., et al. "Forecasting the Effects of Global Warming on Biodiversity." *BioScience* 57 (2007): 227-236. A broad survey of attempts to model global warming's effects on biodiversity.

Bond, David PG, and Paul B. Wignall. "Large igneous provinces and mass extinctions: an update." *Geological Society of America Special Papers* 505 (2014): SPE505-02. In addition to meteor impact, large volcanic events are leading contenders as causes of mass extinction. Likely environmental disruptors include toxic gases, greenhouse gases, and aerosols.

Gibbs, W. Wayt. "On the Termination of Species." *Scientific American*, November, 2001, 39-49. A nontechnical review of extinctions, mass extinctions, and extinction rates written by an able science journalist; does not focus on global warming.

Hallam, Tony. *Catastrophes and Lesser Calamities: The Causes of Mass Extinctions.* New York: Oxford University Press, 2004. An accessible overview of mass extinction phenomena that covers global warming and climate change in the context of other causal factors.

Huey, Raymond B., and Peter D. Ward. "Hypoxia, Global Warming, and Terrestrial Late Permian Extinctions." *Science* 308 (2005): 398-401. A simulation-based assessment of how two particular extinction stressors would have interacted to affect terrestrial species.

Quammen, David. *The Song of the Dodo: Island Biogeography in an Age of Extinction.* New York: Scribner, 1996. An engaging account of factors other than global warming that are contributing to species extinctions.

Raup, David M. *Extinction: Bad Genes or Bad Luck?* New York: W. W. Norton, 1991. An excellent, readable overview of extinction phenomena that emphasizes the difficulties of determining their causes; covers both current extinctions and those in the geological record.

Wilson, Edward O. *The Diversity of Life.* 2d ed. New York: W. W. Norton, 1999. An accessible assessment of biodiversity and discussion of methods for estimating contemporary extinction rates.

See also: Climate reconstruction; Dating methods; Earth history; 8.2ka event; Endangered and threatened species; Holocene climate; Paleoclimates and paleoclimate change; Pleistocene climate.

Extreme weather events

Category: Meteorology and atmospheric sciences

Scientists continue to debate the effects of global warming upon both the intensity and the frequency of extreme weather events. It is possible that one or both of these factors may increase as global temperature increases, causing increased loss of life and property.

Key concepts

cyclone: a storm system that rotates about a low pressure area *drought:* a long period of no or scarce precipitation

extratropical cyclone: a cyclone originating and subsisting outside the tropics

heat wave: an extended period of abnormally high temperatures

hurricane: a cyclone originating in the tropics

severe thunderstorms: mostly summer convective storms involving microscale rotating winds

tornado: a narrowly focused, funnel-shaped violent windstorm

wildfire: spontaneously ignited, naturally occurring fire

Background

Extreme weather events are weather systems that become abnormally severe and have high impacts on human life, property, and the environment.

In the last hundred years, especially in the latter half of the twentieth century, global surface temperature have experienced a rapid increase. This rapid warming may cause significant changes in Earth's weather patterns, including affecting the frequency and intensity of extreme weather events.

Heavy Precipitation and Floods

One of the possible consequences of global warming is that it will enhance evaporation and transpiration of water vapor from oceans, rivers, lakes, and vegetated lands. Temperature is the main factor determining the moisture-holding capacity of air. The higher the temperature, the more moisture an air parcel can hold. Although global warming may not occur uniformly across the globe, the averaged

This damage was probably caused by an intense downburst during a thunderstorm in Green Bay, Wisconsin in 1994. Downbursts are sudden downward currents of air that move horizontally when they hit the ground. They are often confused with tornadoes, but lack rotation. (© Steven I. Dutch)

temperature increase in the atmosphere should give rise to a corresponding increase of average humidity. More water content in the atmosphere will increase the probability of heavy precipitation, which can lead to floods.

Another global warming-related factor that influences heavy precipitation and flooding is a convection-related increase in the severity of storms. In addition to increased temperatures, global warming may involve a higher extension of Earth's troposphere. Both of these processes can lead to more and stronger convection, which can in turn produce more violent convective storms and heavier precipitation.

Droughts

Since global warming is not uniform, a strong warming can occur in some parts of the world while other parts of the world cool. In response to such non-uniform conditions, different patterns of atmospheric circulation can be realized in different parts of the globe. As a result, in some areas air may be enriched by moisture, and in other areas the air may lose moisture, depending on the general atmospheric circulation patterns and moisture transport in each region. Therefore, while global warming may increase the intensity and frequency of heavy precipitation and floods in some locations, it may also increase the severity and frequency of drought conditions in others.

Heat Waves

Global warming increases air temperature in both average and extreme contexts. That is, it results in an increase not only in average temperature but also in daily, monthly, and yearly maximum temperatures. The increase in temperature extremes suggests an increase in the number of hot days as well. Therefore, global warming will most likely increase both the severity and frequency of heat waves. Such heat waves, along with drought, may increase the occurrence of wildfires.

Although wildfire is not a weather phenomenon, its occurrence is closely related to weather conditions. In particular, warm temperatures and low humidity are the two necessary conditions for wildfires. Because global warming can generate warm surface temperatures and frequent

drought conditions, it will increase the likelihood of wildfires.

Tropical Storms

One of the necessary conditions for tropical-storm formation is high sea surface temperature (SST). Tropical storms typically develop over the ocean when SSTs exceed 26°-27° Celsius. Based on this criterion, recent global warming trends seem to suggest an increase in the number of tropical storms. Furthermore, tropical storms derive energy from latent heat brought by water vapor evaporated from oceans. Higher SSTs promote greater evaporation of water into the atmosphere. This factor suggests that future warm climates may also produce more powerful hurricanes and typhoons. However, the question of whether global warming would cause an increase in storm frequency is unresolved, and different studies have produced conflicting results.

Extratropical Cyclones

Unlike tropical storms, extratropical cyclones derive energy from a non-uniform temperature distribution, or a temperature contrast between locations in the northern and southern latitudes. In meteorology, such a condition is called a "temperature gradient." Strong temperature gradients will generate unstable atmospheric conditions, which will initiate large-scale cyclones. These cyclones typically occur in the cool season, and they are enforced by upper-level jet streams and also characterized by surface fronts.

Years of Greatest Accumulated Cyclone Energy (ACE)

Year	ACE	Year	ACE	Year	ACE
2005	248	1995	227	1933	213
1950	243	2004	224	1961	205
1893	231	1926	222	1955	199
				1887	182

Note: ACE is equal to the sum of the squares of the maximum sustained wind speed of each tropical storm measured in knots every six hours

A general consensus exists that global warming will decrease temperature gradients and also decrease the intensity of jet streams. The combined effect of these changes would tend to decrease the intensity and frequency of extratropical cyclones. However, some scientists argue that, because global warming tends to increase humidity, extratropical cyclones may gain extra energy from latent heat flux due to water vapor condensation.

Severe Thunderstorms and Tornadoes

Severe thunderstorms and tornadoes are convective-scale and microscale weather systems. They are different from extratropical cyclones, which are forced by large-scale temperature gradients. Thunderstorm development strongly depends on convection, which is influenced by surface radiative heating, convergence of surface flows, and topographic forcing. In a future warm climate, increase of global surface temperatures may provide favorable conditions for convection to occur. The number of both annual tornado sightings and annual tornado warning days increased over the second half of the twentieth century, but the number of the most severe tornadoes (F2-F5) exhibited a slight decrease.

Context

A 2007 report by the Intergovernmental Panel on Climate Change (IPCC) predicted that global warming would likely increase the number and frequency of extreme weather events. These increased extreme weather events would generate profound impacts on global socioeconomic development. Such impacts include, among many other things, increased risks to human life and health, increased property and infrastructure losses, increased cost and pressure on government's disaster relief and mitigation resources, and increased costs of private insurance.

Chungu Lu

Further Reading

Ahrens, C. Donald. *Essentials of Meteorology: An Invitation to the Atmosphere.* 5th ed. Belmont, Calif.: Thomson Brooks/Cole, 2008. Widely used introductory textbook on atmospheric science; covers a wide range of topics on weather and climate.

Diffenbaugh, N. S., R. J. Trapp, and H. Brooks. "Does Global Warming Influence Tornado Activity?" *Eos* 89, no. 53 (2008). Discusses the impact of global warming on tornado activity in the United States.

Intergovernmental Panel on Climate Change. *Climate Change, 2007—Synthesis Report: Contribution of Working Groups I, II, and III to the Fourth Assessment Report of the Intergovernmental Panel on Climate Change.* Edited by the Core Writing Team, Rajendra K. Pachauri, and Andy Reisinger. Geneva, Switzerland: Author, 2008. Comprehensive overview of global climate change published by a network of the world's leading climate change scientists under the auspices of the World Meteorological Organization and the United Nations Environment Programme.

Kreft, Sonke, David Eckstein, Lisa Junghans, Candice Kerestan, and Ursula Hagen. "Global climate risk index 2015: who suffers most From extreme weather events? weather-related loss events in 2013 and 1994 to 2013." (2014). Poorer countries tend to dominate statistics for extreme weather events, mostly because their low latitude makes them susceptible to heat events and tropical cyclones.

Lutgens, Frederick K., and Edward J. Tarbuck. *The Atmosphere.* 10th ed. Upper Saddle River, N.J.: Pearson Prentice Hall, 2007. Introductory textbook that covers a wide range of atmospheric sciences.

Tebaldi, C., J. M. Arblaster, K. Hayhoe, and G. A. Meehl. "Going to the Extremes: An Intercomparison of Model-Simulated Historical and Future Changes in Extreme Events." *Climate Change* 79: 185-211 (2006). Discusses model-based assessment of global warming's effects on extreme climate conditions.

See also: Average weather; Drought; Dust storms; Meteorology; Rainfall patterns; Sahel drought; Storm surges; Thunderstorms; Tornadoes; Tropical storms; Tsunamis.

Eyjafjallajökull

Category: Geology and geography

Eyjafjallajökull is a small-glacier that completely covers a subglacial volcano in southeast Iceland. The volcano has experienced four major eruptions since 1920, the most recent one lasting from late March to late May 2010. Although the eruption caused significant disruption in air travel, except for some flooding, it had no major environmental or climate effect.

Key concepts

Ash: A fine, powdery residue of lava resulting from a volcanic eruption

Eyjafjallajökull: a glacier that completely covers an active subglacial stratovolcano in southeast Iceland

Fluoride: Any compound containing the element fluorine.

Katla: A very active volcano in southeast Iceland

Olivine: An olive to brownish green mineral commonly found in igneous rocks

Tectonic Plates: The mass of rock that makes up the Earth's surface. The plates roughly correspond to the continents and surrounding oceans.

Background

Eyjafjallajökull (Eyjafjalla Glacier, in English) is a relatively small glacier that conceals an active, subglacial stratovolcano that is about ninety-three miles (150 km) southeast of Reykjavik on the coast of Iceland one to three miles (1.6 to 4.8 kilometers) from the Atlantic Ocean. Eyjafjallajökull translates as the Islands Mountain Glacier or the ice cap (jökull) of Eyjafjalla and refers to both the ice cap and its associated volcano which is part of an active chain of volcanoes in Iceland. The glacier covers thirty-nine square miles (101 square kilometers) and is the sixth largest in Iceland. The glacier spreads by extending tongues in all directions. The volcano has an elevation of 5,417 feet (1,651 meters) and has a crater about two miles (3.2 kilometers) in diameter. The vents of the volcano consist of basalt and andesite lavas derived from the magmas exposed by the diverging North American and the Eurasian tectonic plates which extend from the

southwest to the northeast nearly dividing Iceland into two halves. The volcano has experienced four major eruptions in the past 1,100 years.

Eruptions

The most recent eruption of Eyjafjallajökull started on March 20, 2010, after experiencing tremors in January and February 2010. The first eruption was effusive and nonexplosive consisting of olivine basaltic andesite magma originating from a 0.3 mile (483 meters) ice-free fissure vent. Other fissure vents subsequently opened. On April 14, 2010, the volcano erupted from the two mile (3.2 km) caldera directly under the center of ice cap. This explosive eruption caused extensive glacial water melting and flooding. When ice melt made contact with the magma, it caused a more explosive ash-laden and sulphurous gas-laden eruption spewing an ash and gas plume over six miles (9.7 km) into the atmosphere. Analysis of the ash indicated that early in the eruption cycle the ash contained a high concentration of basaltic materials whereas later eruptions contained silica ash. As the available water decreased, the ash concomitantly decreased, but fire fountains and lava flows continued. The eruption ended on May 23, 2010.

Eyjafjallajökull has been periodically active for the past 800,000 years. Previous to the 2010 eruption, the most recent eruptions were in 920, 1612 or 1613, and 1821 to 1823. The subglacial eruption of 1821 to 1823 was a major eruption that lasted for 14 months and caused extensive ice melt and flooding. Although the 2010 eruption did not involve a concomitant eruption from nearby Katla, most of the past eruptions were followed by an eruption of Katla.

Ash Content

An analysis of the ash content of early Eyjafjallajökull eruptions has demonstrated that the eruptions began with basaltic magma followed by silica-rich, small grained material (volcanic glass) which covered southern Iceland. The 2010 eruption followed this pattern. It is estimated that Eyjafjallajökull released from about 4.9 to 9.5 billion cubic feet (140 to 270 million cubic meters) of ash during the 2010 eruption. The ash emitted during the 2010 eruption contained up to 58% silica and up to 12% iron oxide. Gases emitted from the volcano included sulfur dioxide (SO_2) and carbon dioxide. Ash from the 1821 to 1823 eruption contained a substantial amount of fluoride which is harmful to plant and animal life.

Disruption of Air Travel

Fearing the microscopic, abrasive ash particles could damage aircraft and potentially lead to a loss of life, the eruption of Eyjafjallajökull resulted in a six-day (April 15 to April 21, 2010) disruption of air travel especially in northwest Europe. Airspace in many parts of Western Europe was closed as volcanic ash spewed up to six miles (9.7 km) into the atmosphere and flowed southeast into Norway, Scotland, and throughout northern Europe. It is thought that the small grained, porous and irregularly shaped ash made them less aerodynamic permitting the ash to flow farther than usual and to remain in the atmosphere for an unusually long period of time.

As a result of the disruption in air travel, the airline industry lost several billion U. S. dollars, including $1.7 billion loss by the U.S airline industry. The effects on the world economy were extensive.

Environmental Effects

The primary effect of the 2010 eruption on the environment was flooding. Much livestock was killed during the 1821 to 1823 eruption and has been attributed to the high fluoride content of the ash. Although volcanic eruptions can have an effect on the climate as has been demonstrated with the eruption of the Greek island of Santorini in ca. 1628 B. C. and Mount Pinatubo in the Philippines in 1991. Although the Mount Pinatubo eruption lowered global temperatures from about 0.7° to 0.9° F (0.4° to 0.5° C), the eruption of Eyjafjallajökull was simply too small and the ash plume did not reach far enough into the atmosphere to have any significant environmental effect.

It has been estimated that Eyjafjallajökull released up to 250,000 tons (226.8 million kg) of CO_2/day during its 6 day eruption – one thousandth of what humans produce from their various activities during the same time period. The grounding of aircraft resulted in a savings of CO_2 emissions over what the volcano released.

Charles L. Vigue

Further Reading

The Smithsonian Institution, National Museum of Natural History, Global Volcanism Program: http://volcano.si.edu/volcano.cfm?vn=372020

An excellent web site that is well referenced and chronicles the 2010 eruption of Eyjafjallajökull.

Iceland on the Web http://www.icelandontheweb.com/articles-on-iceland/nature/glaciers/eyjafjallajokull

This web site summarizes the 2010 and earlier eruptions of Eyjafjallajökull.

The British Geological Society http://www.bgs.ac.uk/research/volcanoes/icelandic_ash.html

This web site summarizes the 2010 eruption of Eyjafjallajökull.

Prager, Ellen J., editor. *Volcano: Iceland's Inferno and the Earth's Most Active Volcanoes.* Washington D. C.: National Geographic, 2010

A well-illustrated book that discusses several of Iceland's volcanoes.

The Eruption on YouTube

https://www.youtube.com/watch?v=e-TMtRh8AIs
https://www.youtube.com/watch?v=I1mAqMpbngc
https://www.youtube.com/watch?v=b4nlgDtyoU4
https://www.youtube.com/watch?v=tGPD_0SCDp4
https://www.youtube.com/watch?v=nI3HrI-Ep00

See also: Climatology, Volcanoes

Faculae

Category: Astronomy

Definition

The word facula (*plural:* faculae) derives from the Latin word for torch. Faculae are bright spots on the Sun's visible surface or photosphere. Faculae might be thought of as the opposite of sunspots. Sunspots are dark areas on the Sun's surface caused by the solar magnetic field deflecting energy coming up from within the Sun. Faculae are bright areas on the Sun's surface that are also caused by the solar magnetic field. However, for faculae, the magnetic field concentrates rather than deflects energy from the interior. Because both sunspots and faculae result from solar magnetic activity, the numbers of both increase and decrease together as they follow the Sun's activity cycle.

Faculae are most easily observed near the "edge" of the Sun. A phenomenon called limb darkening causes the edge of the Sun's disk to appear fainter from Earth, even though it emits just as much energy. Limb darkening increases the contrast between faculae and the Sun's surface, so faculae show up more clearly near the edge of the Sun.

Faculae can extend upward from the Sun's photosphere into the Sun's chromosphere, the layer directly above the photosphere. These extensions of faculae into the chromosphere are called plages. Both faculae and plages are always found around sunspots or sunspot groups. They can also occur on unspotted regions of the Sun's surface.

Near the Sun's surface, energy is transferred from the interior by convection currents similar to those that heat a room containing a radiator on one side but no fan. These convection currents make the Sun's surface very turbulent, and the turbulence causes granules, which are higher regions at the top of convection-current cells. Faculae form in the lower regions along the boundaries between granules.

Significance for Climate Change

Bright faculae and dark sunspots both increase and decrease in time with the Sun's eleven-year sunspot cycle. Accurate satellite measurements over the course of these sunspot cycles show that during maximum sunspot activity, the Sun emits about 0.15 percent more energy than during minimum sunspot activity. Thus, the total extra energy emitted by faculae is slightly greater than the total energy blocked by sunspots. The effect of the eleven-year sunspot cycles is too small and rapid noticeably to affect Earth's climate; however, there are also longer, less regular, and poorly understood cycles in the Sun's activity.

During the Maunder Minimum, from about 1645 to about 1715, there were very few sunspots and correspondingly few faculae. If the observation that, during the eleven-year cycle, the Sun is fainter during sunspot minimum holds for longer cycles, then the Sun should also have emitted less energy during the Maunder Minimum. Accurate

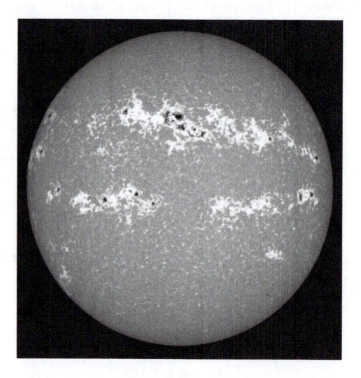

The solar disk, covered with faculae (bright areas) *and a few sunspots* (dark areas). (NASA)

instruments for measuring the Sun's total energy output did not exist in the seventeenth century, so this hypothesis could not be tested directly. However, it is known that this time period was the coldest portion of the Little Ice Age. Thus, circumstantial evidence indicates that the Sun did emit less total energy during the Maunder Minimum.

During the period from about 1000 to about 1200, the Medieval Grand Maximum, the Sun had many more sunspots and faculae than normal. If the hypothesis that the Sun emits more energy during periods of high sunspot activity is correct, then the Sun should have been more luminous than normal during these centuries. Climate research shows that, excepting the latter portion of the twentieth century, this period was the warmest of the last one thousand years. For example, a Viking colony flourished on Greenland during this period, but it was abandoned when the climate cooled again.

Climate studies combined with solar-activity studies over the last millennium show that during extended sunspot maxima Earth's climate is warmer and during extended sunspot minima Earth's climate is cooler. Faculae play an important role in these climate changes. As the bright areas on the Sun's surface, they contribute to the Sun's increased energy output during periods of high solar activity. Their lack also contributes to decreased energy output during periods of lower solar activity.

Many of the eleven-year solar-activity cycles during the late twentieth century had higher levels of maximum solar activity, so John Eddy and a few other scientists have suggested that the Sun might be entering a period of increased activity. The increased solar activity to date is probably not sufficient to have caused the current global warming. However, it might have contributed to the trend on a small scale. Understanding faculae, sunspots, and their relation to the total solar energy output is important to completely understanding long-term climate changes on Earth.

Paul A. Heckert

Further Reading

Eddy, J. A. "The Maunder Minimum." *Science* 192 (1976): 1189-1192. In this seminal article, Eddy demonstrates that the Maunder Minimum is real and argues both that the Sun's energy output varies with long-term activity cycles and that this variation causes climate changes.

Foukal, P., C. Fröhlich, H. Spruit, and T. M. L. Wigley. "Variations in Solar Luminosity and Their Effect on Earth's Climate." *Nature* 443 (2006): 161-166. Reviews the current state of knowledge of solar variability and its role in climate changes. The role of faculae is specifically discussed.

Golub, Leon, and Jay M. Pasachoff. *Nearest Star: The Surprising Science of Our Sun.* Cambridge, Mass.: Harvard University Press, 2001. This well-written book provides a detailed summary of scientific knowledge of the Sun. Includes solar surface features such as faculae and solar variability.

Hoyt, Douglas V., and Kenneth H. Schatten. *The Role of the Sun in Climate Change.* New York: Oxford University Press, 1997. Well-written and extensively documented coverage of solar variability and possible associated climate changes, as well as of the role of faculae and sunspots in causing solar variability.

Marshall Space Flight Center. Solar Physics: Photospheric Features. http://solarscience.msfc.nasa.gov/feature1.shtml (2 July 2016). Brief explanations of major features of the visible sun, including faculae, with clear photographs.

Soon, Willie Wei-Hock, and Steven H. Yaskell. *The Maunder Minimum and the Variable Sun-Earth Connection.* River Edge, N.J.: World Scientific, 2003. Explores surface features such as faculae, the Maunder Minimum, long-term solar variability, and the role this variability plays in Earth's climate changes.

See also: Little Ice Age (LIA); Medieval Warm Period (MWP); Solar cycle; Solar energy; Sun; Sunspots.

Falsifiability rule

Category: Science and technology

The falsifiability rule asserts that for a theory to count as scientific, it must be logically capable of being disproven. The falsifiability of climate change theories has been contested by some of their critics.

Key concepts

falsifiability: testability of a scientific theory by performing experiments or collecting observations that might potentially disprove the theory

falsification: disproof of a scientific theory using experiments or observations that contradict the theory

naïve falsification: erroneous claims that a scientific theory has been falsified based on isolated anomalies, poor data, or faulty observation

replicability: the ability of an experiment or observation to be repeated and confirmed

Background

Science covers such a broad range of subjects that it is very hard to come up with a simple definition of scientific proof. Many scientific ideas cannot be tested by simple experiments. Sometimes, even well-established scientific ideas require revision when new data are discovered, so scientists generally believe it is impossible to prove a theory to be absolutely true. Austrian-born British philosopher Karl Popper proposed that ideas could be regarded as legitimate science if they could be falsified—that is, if there were some test that might possibly prove the idea wrong. Falsifiability separates science from subjects such as philosophy, religion, and political ideology, for which either there is no way to disprove ideas or believers are unwilling to admit their ideas can be falsified.

Types of Falsification

The simplest form of falsification is to compare a theory with the results of an experiment or an observation in nature. If the theory does not agree with the experiment or observation, it has been falsified. Of course, the experiment or observation must be performed correctly. If there is any doubt, the test can be repeated. Therefore, replicability, the ability to repeat observations or experiments, is considered a vital part of the scientific method and is one aspect of falsification. Events outside the laboratory such as earthquakes, hurricanes, and epidemics are never exactly repeatable, but they all have features in common from one occurrence to the next.

Theories about events in the remote past or physically inaccessible events can be falsified in

Karl Popper in the 1980s. (Wikimedia Commons)

several ways. For example, one can compare ancient features in rocks with features forming now to test whether processes on Earth have changed over geologic time. One can develop theories of how stars develop and compare the results with actual stars. If the theories fail to match observations, they have been falsified.

Popper also noted that tests must be "risky." There must be a real possibility of the test failing. If one knows that a theory will always pass a given test, it is not a real test. A test in which someone dismisses or evades the falsification is also not a real test. Although many people claim their religion or ideology is scientific, very few are willing to submit to a risky test and abide by the results.

Problems with Falsification

Sometimes falsifications are wrong. A famous example is that of astronomer Simon Newcomb, who "proved" that powered flight was impossible only a few months before Wilbur Wright and Orville Wright successfully flew, because he made overly conservative assumptions about the weight and power of engines. Erroneous falsifications are common enough that author Arthur C. Clarke described them in his First Law: "When a distinguished but elderly scientist says that something is possible, he is very likely right. When he says that something is impossible, he is very probably wrong." Generally, erroneous falsifications stem from overly strict assumptions or failure to anticipate new discoveries.

One of Popper's own errors, frequently cited by antievolutionists, was to say "Darwinism is not a testable scientific theory, but a metaphysical research program." In fact, Charles Darwin himself proposed one test of evolution. He stated that evolution would be falsified if it could be shown that some change could not have occurred in a series of gradual steps. In 1978 Popper retracted his statement, saying "I have changed my mind about the testability and logical status of the theory of natural selection; and I am glad to have an opportunity to make a recantation."

Misapplications of Falsification

The more tests a theory has passed, the more confident one can be in its validity. For long-established theories, it is hard to imagine any additional test that might falsify them. However, many of the best-established ideas in science are the basis of modern technology and can be said to face the riskiest of all tests: producing practical results. If the theories ever fail to produce the expected results, they will have failed.

Many people equate falsifiability with experiment, but there are many other ways scientific ideas can be falsified. They can be shown to be contradicted by other scientific findings, or shown to be logically inconsistent. Some people believe ideas cannot be falsified simply because they are unaware of the possible tests. For example, study of light from distant galaxies shows clearly that the same laws of physics apply throughout the universe. The idea that the laws of nature are constant in space and time is falsifiable and so far has survived every test.

Many people, seeking to justify disbelief in some scientific idea, engage in naïve falsification—the use of evidence that seems to contradict established science, even if the evidence is of poor quality or supported by bad reasoning. Whenever an observation is claimed to contradict a theory, there are two possibilities: either the theory is wrong, or the observations are wrong. If a single observation is claimed to disprove a well-tested theory, almost certainly the observation is wrong. In addition, if the results are reported in popular media, the observation may be inaccurately reported or even a deliberate hoax. Finally, unsolved problems or unanswered questions do not disprove a theory.

Context

There probably is no single, all-purpose definition of science, but falsifiability is one of the most widely useful. In the case of climate change, falsifiability must be understood and applied properly. Isolated cases of unusually warm or cold weather do not prove or disprove climate change. Long-term climate trends, unprecedented changes such as large-scale melting of Arctic sea ice, and the physics of heat absorption by greenhouse gases count much more than individual weather events.

Steven I. Dutch

Further Reading

Chalmers, Alan F. *What is this thing called science?*. Hackett Publishing, 2013. Fourth edition of a work first published in 1976 and periodically revised extensively, widely regarded as the best introductory textbook on the philosophy of science.

Clarke, Arthur C. Profiles of the Future: An Inquiry into the Limits of the Possible. New York: Harper & Row, 1962.

Godfrey-Smith, Peter. Theory and Reality: An Introduction to the Philosophy of Science. Chicago: University of Chicago Press, 2003.

Popper, Karl. The Logic of Scientific Discovery. 1959. Reprint. New York: Routledge, 2002.

Popper, Karl. "Natural Selection and the Emergence of Mind." Dialectica 32, nos. 3/4 (1978): 339-355.

Tesh, Sylvia Noble. Uncertain Hazards: Environmental Activists and Scientific Proof. Ithaca, N.Y.: Cornell University Press, 2000.

Famine from global warming

Category: Diseases and health effects

There is considerable evidence that global climate change is taking place, and floods, droughts, and other extreme weather events precipitated by these changes are likely to reduce food production and distribution significantly, resulting in famine in various parts of the world.

Key concepts

climate canaries: the first victims of world climate change, who serve as early warnings to others

climate change scenario: a physically consistent set of changes in meteorological variables based on accepted projection of carbon dioxide and other trace gas levels

desertification: the gradual transformation of habitable land into desert due to climate change or destructive land use

drought: an extended period of months or years when a region experiences a deficiency in its water supply

greenhouse gas (GHG): any atmospheric gas that absorbs radiation, contributing to warming of the Earth's atmosphere

Background

Some groups deny that continuing global warming will lead to famine, arguing that the warmer climate exerts beneficial effects on food production and that the increased carbon dioxide (CO_2) production from global warming serves as a fertilizing agent for plants. The majority of research, however, paints a very different picture. Numerous researchers associated with respected organizations conclude that climate change is real and that it is possible to predict when and where the most severe famines are likely to occur.

Findings of the Hadley Centre for Climate Prediction and Research

Established in 1990, the Hadley Centre has been located in Exeter, England, since 2003 and has been recognized for the quality of the research on climate change carried out by its more than 150 scientists. In 2006, it predicted that about one-third of the Earth will become desert by 2100, as a result of drought and its consequent desertification. Those areas of the world that are already victims of drought, such as Africa, will likely experience the most severe effects. The people predicted by the Hadley Centre to be the first victims of world climate change, called "climate canaries," will be about three million pastoral nomads in northern Kenya. A way of life that has been sustained for thousands of years therefore faces eradication. Myriad herders have forsaken their traditional way of life to settle in Kenya's northeastern province after their livestock were decimated. The situation is not limited to Kenya: At least eleven million people are affected from Tanzania to Ethiopia, Eritrea, and Somalia.

Predictions of the UN Food and Agriculture Organization

The Food and Agriculture Organization (FAO) likewise predicts the most severe impact of diminished food production and resulting famine to occur in African countries below the Sahara Desert. Desertification could result in an increase of as many as 90 million hectares of arid land, an area almost four times the size of Britain. The FAO's predictions are not limited to Africa: Sixty-five developing countries, including more than half of the total population of the developing world in 1995, are expected to lose around 254 million metric tons of potential grain production because of climate change. Nor are the "extreme weather events" limited to drought and desertification. Flooding will bring devastating effects as well. During the first decade of the twenty-first century, more than six hundred floods have caused $25 billion in damage, a substantial amount of which includes the loss of some 254 million metric tons of potential cereal production. Another FAO study reported that at least ten million people in Malawi, Zimbabwe, Lesotho, and Swaziland are threatened with starvation; even at harvest time, a serious food crisis persists.

Scholze's Predictions

Mark Scholze of Bristol University has conducted research for the organization Quantifying and Understanding the Earth System (QUEST) involving world climate simulation predictions through the twenty-first century based on sixteen climate models. He poses several scenarios regarding fire, flood, and famine by the year 2100 and predicts that effects of an average of 2° Celsius in global temperature rise are inevitable and will cause deforestation of up to 30 percent in parts of Europe, Asia, Canada, Central America, and Amazonia. Freshwater shortages, likely due to drought, can be expected with a rise of between 2° and 3° Celsius in parts of West Africa, Central America, southern Europe, and the eastern United States. As trees are lost, tropical Africa and South America will be subject to flooding.

Should a 3° temperature increase occur, an even more dangerous scenario is likely: As temperatures rise, plants may begin to grow more vigorously and take up more carbon oxide from the air. When saturated, the ecosystem begins to respire more than it is taking up. Scholze's data, which are in line with findings of the Hadley Centre, indicate that this tipping point could arrive by mid-century. These phenomena would cause a decrease in worldwide cereal crop production of between 18 million and 363 million metric tons and put 400 million more people in famine conditions. Scholze insists that fossil fuel combustion must be significantly curtailed before 2040.

Context

During the past two million years, the climate on Earth has alternated between cooling and warming. Thus, one might question the concern during the latter twentieth and early twenty-first centuries over global warming. The concern arises, because the Earth is growing warmer faster than it has in the past, as more greenhouse gases (GHGs) are released into the atmosphere. Over one hundred years ago, people worldwide began using more coal and oil for homes, factories, and transportation, thereby releasing CO_2 and other GHGs into the atmosphere. Scientific data reveal that during the past century, the world's surface air temperature increased an average of 0.6° Celsius. Even one degree can affect Earth's climate.

Heavier rainfall is causing flooding in some areas, while there is extreme drought in others, resulting in famine. The first half of the twentieth century was not unusual: The period of 1900 to 1939 brought mild winters, characteristic of a high North Atlantic Oscillation (NAO) condition. However, in the 1950s, the global average temperature fell, and some thought an ice age was imminent. Then, the NAO suddenly flipped to high, and some scientists declared that the warming was a permanent phenomenon because of humans' promiscuous use of fossil fuels. If so, the likelihood of famine remains a constant.

Victoria Price

Further Reading

Bazzaz, Fakhri, and Wim Sombroek. Global Climate Change and Agricultural Production. New York: John Wiley & Sons, 1996.

Cline, William R. Global Warming and Agriculture: Impact Estimates by Country. Washington, D.C.: Center for Global Development, 2007.

McCaffrey, Paul, ed. Global Climate Change. New York: H. W. Wilson, 2006.

Roseboom, Tessa J., Rebecca C. Painter, Annet FM van Abeelen, Marjolein VE Veenendaal, and Susanne R. de Rooij. "Hungry in the womb: what are the consequences? Lessons from the Dutch famine." *Maturitas* 70, no. 2 (2011): 141-145. The Dutch famine of World War II presents a unique example of a famine in an otherwise developed country with good health care and record keeping. Infants exposed to famine early in gestation suffered increased incidence of some mental illnesses, and all infants showed increased risk of diabetes.

Fire

Categories: Meteorology and atmospheric sciences; environmentalism, conservation, and ecosystems

Forests sequester carbon, and fires return it to the atmosphere. Huge fires may account for as much as 40 percent of the anthropogenic contribution of CO_2 to Earth's

atmosphere. Early humans used fire in their agriculture. When specific human populations collapsed, the absence of these fires resulted in reforestation and the removal of CO_2 from the atmosphere.

Key concepts

carbon sequestration: removing CO_2 from the atmosphere and storing it, or the carbon within it, somewhere else

gigaton C: one billion metric tons of carbon, equivalent to 1 petagram (10^{15} grams)

hectare: an area of 10,000 square meters

natural succession: changes in the types of plants that grow in an area over time, starting with grasses and moving through shrubs to trees

slash and burn: an agricultural practice of clearing the land whereby trees are girdled, left to die, and then burned

wildland/urban interface: a region where residential housing is embedded within, or abuts against, a wild or nearly wild environment

Background

Fire is an important factor in climate, whether it is stopping or starting. Study of charcoal deposits has shown that before 1500 c.e., indigenous populations in the Amazon Basin used fire to clear forests and maintain extensive agricultural enterprises. After European plagues killed 90 percent of the Amazon population, the rain forest grew back, sequestering 10 gigatons of carbon. This may have lowered atmospheric carbon dioxide (CO_2) concentration by 2 parts per million by volume, contributing to the development of the Little Ice Age. In 1997-1998, forest fires in Indonesia are thought to have added between 0.81 and 2.57 gigatons of carbon to the atmosphere, out of the total 6.0 gigatons.

Wildfires and Public Policy

Wildfires are natural. Many habitats exist because wildfires occur and would disappear in the absence of fires. Fires occur when there is fuel, oxygen, and a source of ignition. Lightning and oxygen will always exist in the wild, so the principal variable is the fuel load. Nature limits the supply of fuel with frequent small fires, which burn through an area, removing accumulated deadwood and forest litter.

Long periods of fire suppression interfere with this process, permitting large quantities of fuel to accumulate. The subsequent fires are then larger and more intense.

CO_2 is sequestered in the wood of the forest, removing it from the atmosphere and preventing it from affecting the Earth's climate. As the natural succession results in shrubs replacing grasses and being replaced in turn by slightly larger trees, which are replaced by even larger trees, and so on, the amount of CO_2 being stored in the forest increases. When the forest reaches its climax state, the species making up the forest will stabilize. Eventually, new growth will only replace older, dying growth, and no additional CO_2 will be removed from the atmosphere. How rapidly this all occurs will vary with the climate, soil, and other factors, but it is generally thought to take fifty to one hundred years.

Early Human Burning Practices

Early humans may have observed beavers. These creatures prefer foods such as aspens, which grow fairly early in the succession. If they encounter hickories or oaks, they will frequently girdle them, chewing away a strip of bark a few centimeters wide, all the way around the tree. The trees will die, standing where they are, eventually falling over, or being burned in a lightning-caused fire, and the area in which they were standing will revert to meadow to begin the process of succession again.

If early humans wanted to create a meadow, they could emulate this behavior of beavers, using blades to girdle the trees and, after the trees had died, starting fires to speed things along. This technique is called "slash and burn," and it is still a major cause of deforestation. It causes the CO_2 that was sequestered in a forest to move into the atmosphere, where it contributes to the greenhouse effect. Some 3.5 to 4.5 million square kilometers of forest are burned in fires every year. Over 99 percent of these fires are set intentionally by humans, usually for agricultural purposes. Most are in Africa.

If agriculturalists abandon their fields, the opposite effect, reforestation, will occur. In the case of the Americas, scientists estimate that 35-90 million people died shortly after the European plagues were introduced. If each of these farmers had been

Brush fires near Shreveport, Louisiana. The size and arrangement of the fires suggests they are controlled burns. (© Steven I. Dutch)

keeping one hectare clear, and after regrowth to rain forest each hectare of land supported an additional 100 metric tons of carbon, the increase in the amount of carbon sequestered would have been on the order of 3.5-9 gigatons.

Context

Climate change will alter precipitation patterns in time and space. Some areas will become more fire prone, others less. The makeup of forests will change along with the climate, and eventually the plants growing in an area will be those that are best suited for that environment. Some will require fires and will thrive in areas that burn frequently. Sound fire management practices should permit this. As changes occur, however, those places that are drying up will have more combustible fuel available, and wildfires can be expected until this fuel is consumed. Thinning, controlled burns, and other means of reducing this fuel should be used to avoid uncontrollable conflagrations.

For the foreseeable future, fire will be used as an agricultural tool. If done on a small scale, this is unlikely to contribute substantially more CO_2 to the atmosphere than the oxidation performed by microbes. However, it should not be permitted on a large scale, particularly as part of lumbering operations, as under such circumstance the contribution of burning to atmospheric CO_2 concentrations may be large and long lasting.

Otto H. Muller

Further Reading

Bond, William J., and Brian W. Van Wilgen. *Fire and plants.* Vol. 14. Springer Science & Business Media, 2012. Although fire has been recognized as an essential feature of ecology for a long time and there are many studies of effects of fire on specific plants or areas, there have been few studies of broader implications of fire. This book attempts to create a synthesis of knowledge of fire and its effects on ecosystems and plant evolution.

Drysdale, Dougal. *An introduction to fire dynamics.* John Wiley & Sons, 2011. Updated version of a standard reference that incorporates scientific theory along with practical experience.

Nevle, Richard J., and Dennis K. Bird. "Effects of Syn-pandemic Fire Reduction and Reforestation in the Tropical Americas on Atmospheric CO_2 During European Conquest." *Palaeogeography, Palaeoclimatology, Palaeoecology* 264, nos. 1/2 (2008): 25-38. Argues that before Columbus much of the Amazon Basin was farmland, abandoned as the farmers tending it died off in the pandemics resulting from the arrival of Europeans, and that the reforestation was at least partially responsible for the Little Ice Age. Map, tables, bibiliography.

Omi, Philip N. *Forest Fires: A Reference Handbook.* Santa Barbara, Calif.: ABC-CLIO, 2005. As a handbook, this text is not intended to be read in a linear fashion, but it contains a wealth of information about wildfires and their management. It contains details about many famous forest fires in the United States and data showing how fire fighting and forest management practices have changed over time. Illustrations, figures, tables, maps, bibliography, index.

Page, Susan E., et al. "The Amount of Carbon Released from Peat and Forest Fires in Indonesia During 1997." *Nature* 420, no. 6911 (2002): 61-65. Describes how between 13 and 40 percent of the average anthropogenic contribution of CO_2 was added to the atmosphere by these fires. Maps, tables, diagrams.

Pyne, Stephen J. *Vestal Fire: An Environmental History, Told Through Fire, of Europe and Europe's Encounter with the World.* Seattle: University of Washington Press, 1997. An erudite treatment of the role of fire in the history of the world, this book is exhaustive in its scope. Illustrations, charts, tables, bibliography, index.

Ruddiman, William F. *Plows, Plagues, and Petroleum: How Humans Took Control of Climate.* Princeton, N.J.: Princeton University Press, 2005. Written for the lay public, this book provides the background and thinking behind the theory that humans have influenced the climate for the last nine thousand years, primarily through agriculture. Illustrations, figures, tables, maps, bibliography, index.

See also: Amazon deforestation; Deforestation; Forestry and forest management; Forests.

Fishing industry, fisheries, and fish farming

Categories: Animals; economics, industries, and products

Definition

Fish have been a protein source throughout history. Early fishing primarily involved individuals capturing fish near their communities for consumption or trade. Ships gave fishers access to ocean-based fisheries. Commercial fishing became industrialized by the late nineteenth century, as technological innovations helped locate, catch, and process fish. In addition to fish living in natural freshwater or saltwater fisheries, fish cultivated in fish farms' ponds or tanks represented approximately one-fourth of the fish eaten in the world. Countries benefited economically with domestic trade and by exporting valuable fish. In the early twenty-first century, fisheries generated billions of dollars globally with approximately 42 million people employed to catch fish and several hundred millions more working in related industries. Fisheries reinforced food security when climate changes caused shortages of other agricultural products.

Significance for Climate Change

For centuries, fishers realized that weather affected fish populations, but they lacked the scholarly resources to investigate their observations. During the nineteenth century, fishery researchers began

Top Ten Inland Fishing Nations

Nation	Fishery Production, 1998 (thousands of metric tons)	Percentage of Global Production*
China	2,280	28.5
India	650	8.1
Bangladesh	538	6.7
Indonesia	315	3.9
Tanzania	300	3.7
Russian Federation	271	3.4
Egypt	253	3.2
Uganda	220	2.8
Thailand	191	2.4
Brazil	180	2.3

Source: Food and Agriculture Organization.
*The top ten countries accounted for 65 percent of global inland fisheries production.

applying scientific methodology to study diverse factors affecting fish health, reproduction, and habitats. They contemplated reasons for decreased fish populations besides overfishing. U.S. Environmental Protection Agency representatives voiced concerns that climate deviations affected the quantity and quality of fisheries to the U.S. Congress in 1988. The American Fisheries Society promoted research examining how climate change might affect fisheries. Scientists consulted ships' logs and records documenting fish-catch statistics and meteorological patterns to evaluate hypotheses about the climate's possible role in fish population losses. Researchers used computer simulations to consider future climatic factors that could potentially harm fish.

Fish are exceptionally vulnerable to habitat changes. The World Wildlife Federation emphasizes that temperature increases of 2° Celsius or more are dangerous for fish. Healthy water temperatures differ according to species. Scientists link increased water temperatures to global warming, because water absorbs heat trapped by greenhouse gases (GHGs). The Intergovernmental Panel on Climate Change (IPCC) states that abnormally high ocean temperatures have occurred as deep as 3,000 meters below sea level. Water-temperature increases can interfere with fishes' oxygen supply and the physiological processes associated with maturation, digestion, and spawning. High temperatures also weaken fishes' resistance to toxic substances and pathogens that invade their habitats. Scientists evaluate fish otoliths (ear bones), which indicate growth, to study how temperatures at varying depths influence development.

Researchers have correlated temperature deviations in oceans with fish displacement due to reduced growth of food sources, such as plankton. Melting glaciers caused by global warming raise water levels and dilute the oceans, decreasing salinity. This surplus water disrupts currents crucial for transporting food and removing pollutants, including from such human sources as fishing vessels and fish-factory emissions. Fish migrate to waters with more compatible temperatures and ample food supplies, but they often fail to thrive if they are unable to adapt. Since the late 1980's, Massachusetts fishermen reported an 80 percent decline in cod catches. Fisheries managers aware of areas where wild fish have relocated can adapt management practices to minimize additional climate change damage and to replenish fisheries.

Temperature increases of almost 3° Celsius in some locations have caused freshwater in streams, rivers, and lakes to evaporate, reducing habitat size and stressing fisheries, particularly trout and salmon species in the western United States. In Montana, grayling fish in the Big Hole River decreased from ninety-six fish per kilometer of river in the 1990's to as few as eight in the early twenty-first century because of temperature changes. Some hardier species, such as smallmouth bass, sought warmer habitats, competing for resources and displacing indigenous fish. Scientists warned that loss of diverse fish genetic material diminishes fisheries and some species will become extinct. Fishery experts estimated reductions of approximately 90 percent of bull trout and 40 percent of salmon populations by 2050 if extreme heat and arid climate conditions persist.

Climate changes alter fish ecosystems in lakes, resulting in fishery populations being reduced by as much as 30 percent, as has been reported in Africa's Lake Tanganyika. In addition to temperature and precipitation deviations, climate changes slow wind velocities needed to stir nutrients from deeper lake water to the surface. Changing climates might force fish farmers to relocate stock from ponds to protected sites where hot temperatures do not threaten fish health.

Oceans become more acidic when water absorbs carbon dioxide (CO_2) in the atmosphere, bleaching coral reefs, which many fish need for habitats and nurseries. A January 16, 2009, *Science* article reported researchers' discovery that calcium carbonate in fish fecal material can control some ocean acidity. CO_2 circulating in fish blood stimulates production of calcium carbonate. Ironically, as GHG emissions increase, the amount of fish-produced calcium carbonate might too, helping counter climate change's impact on fisheries.

Elizabeth D. Schafer

Further Reading

Bavington, Dean. *Managed annihilation: an unnatural history of the Newfoundland cod collapse.* UBC press, 2011. Description of the exploitation and collapse of the Newfoundland cod fishery. Near shore fisheries became depleted in the 19th century but expanding fishing range and improved technology delayed recognition of the problem until it had reached critical levels.

Marine Resources Service, Fishery Resources Division, FAO Fisheries Department. *Review of the State of World Marine Fishery Resources.* Rome, Italy: Food and Agriculture Organization of the United Nations, 2005. Illustrated sections feature diverse regional fisheries. Climate chapter discusses scientific investigations that question whether climate changes have caused population fluctuations in fisheries.

Nelleman, Christian, Stefan Hain, and Jackie Alder, eds. *In Dead Water: Merging of Climate Change with Pollution, Over-Harvest, and Infestations in the World's Fishing Grounds.* Arendal, Norway: United Nations Environment Programme, GRID-Arendal, 2008. Comprehensive examination of global warming issues relevant to fisheries. Photographs, figures, and maps displaying past and projected climate changes supplement the text.

Robbins, Jim. "As Fight for Water Heats Up, Prized Fish Suffer." *The New York Times*, April 1, 2008, p. F4. Scientists explain how temperature changes and diversion of water for agricultural uses combined to harm fish habitats in western U.S. rivers. Includes suggestions for restoring fish populations.

Sharp, Gary D. *Future Climate Change and Regional Fisheries: A Collaborative Analysis.* Rome, Italy: Food and Agriculture Organization of the United Nations, 2003. Fisheries expert differentiates between global change and warming and how they have continually shaped ecosystems supporting fish throughout the Earth's history. Figures, glossary, bibliography, list of recommended Web sites.

Wilson, R. W., et al. "Contribution of Fish to the Marine Inorganic Carbon Cycle." *Science* 323, no. 5912 (January 16, 2009): 359-362. Highlights the contribution of fish excretions to the carbonate content of seawater; argues that, because the magnesium content of these excretions allows them to be dissolved more readily than other carbonates, they may have a disproportionate effect on ocean alkalinity.

See also: Dolphins and porpoises; Ocean acidification; Ocean disposal; Ocean life; Whales.

Flood barriers, movable

Category: Science and technology

Definition

People have innovated technology and methods to divert floodwater throughout history. Devices to block excess water from inundating land include dikes, levees, and seawalls which are primarily rigid structures. In the twentieth century, extreme floods overwhelmed many of those structures, prompting people to seek better protection to withstand future deluges. Engineers designed large movable floodgates placed at intervals in rivers and lagoons adjacent to seas. These barriers, which could be quickly maneuvered into place for specific flooding threats, were constructed from more durable materials and in greater sizes than previous flood walls. When sea levels rose, producing surges, operators raised and closed movable barriers to keep rising seawater from flooding communities. After conditions improved, barriers were opened or lowered to enable water traffic to pass unimpeded.

Significance for Climate Change

Raised seawater levels associated with global warming often results in severe flooding in coastal communities. European floods in 1953 were especially catastrophic. Government leaders in England and the Netherlands, which both suffered vast losses, sought more effective ways to control flooding. The 1953 floods resulted in engineers in the Netherlands developing the innovative Delta Project to control flooding with a complex system of heightened dikes, connected seawalls and dams, and movable floodgates which could be shut into position when storms occurred to block tidal surges from the North Sea.

The Delta Works consists of some of the Earth's biggest movable flood barriers. Beginning service in 1986, the Oosterschelde Storm Surge Barrier, built in an estuary, contains sixty-two steel gates which move on concrete piers. Saltwater necessary to sustain the estuary's ecosystem can advance through the opened gates. Built by 1997 at Rotterdam, the Maeslant Storm Surge Barrier on the Nieuwe Waterweg incorporates computer sensors which sense rising water levels from tidal surges and activates the two curved triangle-shaped gates to pivot from their home bank to meet in the river's center. The movable Hagestein Weir near Hagestein, Holland, has curved gates resembling visors which can be shut to control water flow.

In England, the Thames River represented that country's greatest flood concern, particularly in London. Officials considered reports suggesting building a river-wide barricade consisting of gates which could be moved to impede floodwater. They passed the 1972 Thames Barrier, and Flood Protection Act. Construction of a movable flood barrier started in 1974 at an east London site located in the Thames Estuary. By 1982, builders finished the Thames Barrier, which consists of ten hood-shaped steel gates mounted on concrete sills secured to the river floor. Environment Agency personnel oversee the Thames Barrier and evaluate tidal threats based on Storm Tide Forecasting Service data and computer models.

Engineers estimated the Thames Barrier can handle possible flooding conditions through 2030 and perhaps through 2070, but due to intensified global warming, that barrier has been activated more often than estimated. Since 1982, operators have shut the Thames Barrier in more than one hundred incidents, with at least fifty percent of those closings happening after the year 2000. If climate change is not slowed and sea levels rise more quickly than projected, the Thames Barrier will become insufficient before engineers envisioned, necessitating plans for construction of another barrier in the Thames Estuary.

The Italian city of Venice, located adjacent to barrier islands bordering the Adriatic Sea, often experiences flooding. Global warming contributes to rising sea levels, causing excess water to inundate Venice's lagoon and St. Mark's Square, which flooded more than one hundred times annually in the twenty-first century compared to six times yearly in the early twentieth century. Extreme floods in 1966 motivated Italian officials to consider more effective flood-control techniques. Concerned climate changes might result in tidal floods destroying Venice, engineers recommended installing movable flood barriers and by 1990 presented a prototype of the *modulo sperimentale elettromeccanico* (MOSE), also known as the Moses Project.

In 2002, government leaders decided to fund the Moses Project to build seventy-nine hollow steel gates for placement in the lagoon's Chioggia, Lido, and Malamocco inlets. Chief engineer Alberto Scotti oversaw construction starting in spring, 2003. According to plans, the gates, lowered on their hinges, are to be submerged in underwater cement structures. Operators will lift gates by pumping compressed air into them when needed. Because of the project's complexity, its completion date was shifted from 2010 to 2014.

Moses Project construction provoked controversy. Opponents stated the barriers will damage the lagoon environmentally. Continued flooding provoked by global warming, especially high waters in 2008, emphasized the need for flood control. Projections for sea level increases of a few meters by 2100 caused engineers to recognize the Moses Project might need reinforcements to block surges into Venice's lagoon.

Elizabeth D. Schafer

Further Reading

Aerts, Jeroen CJH, W. J. Botzen, Hans Moel, and Malcolm Bowman. "Cost estimates for flood resilience and protection strategies in New York City."*Annals of the New York Academy of Sciences* 1294, no. 1 (2013): 1-104. A timely analysis in the wake of Hurricane Sandy. The study asserts that hybrid green and infrastructure measures is most effective now but if threat levels increase, large storm surge barriers may be the most cost-effective solution.

Bijker, Wiebe E. "The Oosterschelde Storm Surge Barrier: A Test Case for Dutch Water Technology, Management, and Politics." *Technology and Culture* 43, no. 3 (July, 2002): 569-584. University of Maastricht professor whose civil engineer father worked on Dutch flood-control projects provides insights regarding construction, politics, public opinion, and environmental issues relevant to this movable barrier.

Donadio, Rachel. "Floodwaters, Worst Since '86, Begin to Recede in Venice." *The New York Times*, December 3, 2008, p. A12. Examines the impact of global warming on Venice and public reaction to the Moses Project, commenting that those barriers could have stopped this flood. Remarks that the barriers' opponents suggest that limiting carbon emissions can solve Venice's flooding concerns.

Hoeksema, Robert J. *Designed for Dry Feet: Flood Protection and Land Reclamation in the Netherlands.* Reston, Va.: American Society of Civil Engineers Press, 2006. Discusses historical Dutch floods, the twentieth century Delta Project, and Delta Works technology. Illustrations depict the construction and use of movable flood barriers. Glossaries.

Keahey, John. *Venice Against the Sea: A City Besieged.* New York: Thomas Dunne Books/St. Martin's Press, 2002. Incorporates climate change information in chapters focusing on efforts to protect Venice with movable flood barriers, as well as on disagreements regarding those plans. Illustrations include a Moses Project diagram.

See also: Coastal impacts of global climate change; Estuaries; Floods and flooding; Venice.

Floods and flooding

Categories: Meteorology and atmospheric sciences; oceanography

Floods along rivers and storm surges along ocean shorelines may become more severe on average as a result of higher sea levels and greater precipitation associated with global warming.

Key concepts

base flow: the normal amount of discharge in a river

discharge: the volume of water that passes a point in a river in a certain interval of time

evaporation: the absorption of liquid water into the air as water vapor

flood: unusually high levels in bodies of water, causing some water to leave those bodies and pour onto land

latitude: the distance of a point on Earth from the equator, measured in degrees of the planet's curvature

precipitation: rainfall, hail fall, or snowfall

storm surge: an abnormal rise of the ocean along a shoreline as a result of storm winds

Background

Floods along rivers occur when large amounts of precipitation, often combined with melting snow, cause the discharge of a river to increase far beyond its base flow. The floods may be even higher in cities because of rapid runoff of water from cement into storm sewers. Global warming may increase the intensity and variability of precipitation in many regions, causing more intense flooding. Floods along maritime coastlines can result from storm surges. Higher sea levels due to glacial melting during global warming and more frequent and intense hurricanes may result in more coastal flooding.

Warming of the Earth's Climate

The average temperature of the atmosphere of Earth near its surface is variable, and it has been estimated in a number of ways. Nevertheless, the average temperature before 1920 is estimated to have been about 0.2° to 0.4° Celsius lower than it was from about 1940 to 1980. The average temperature from about 1980 to the early twenty-first century was about 0.4° Celsius higher than that from 1940 to 1980. A major reason for this increase in average temperature is likely the 33 percent measured increase of the atmospheric concentration of the greenhouse gas (GHG) carbon dioxide (CO_2) between about 1958 and the early twenty-first century. Much of this increase in CO_2 was likely due to burning coal, oil, and natural gas.

Predictions of warming of the atmosphere in the future are much more difficult to make. For instance, a doubling of the CO_2 content of the atmosphere from that of the present has been predicted to cause an increase in temperature of about 2° to 3° Celsius. Warming of the atmosphere to the year 2100 has been predicted, depending on the model, to increase the temperature by as little as 2° Celsius to as much as 8° Celsius.

Predicted Influence of Climate Change on Precipitation and Floods

The gradual increase of temperature is likely to lead to more frequent warm periods, such as the extreme heat wave that occurred in Europe in the summer of 2003. Also, snow should, on the average, melt sooner in the spring and melt more rapidly and further to the northern and southern latitudes than it does now.

The increase in temperature of the atmosphere and oceans during global warming will also likely increase the amount of water that evaporates into the atmosphere. Thus, more intense precipitation should occur in some regions. On a regional scale, this effect is most likely to transport more water from the subtropics to higher latitudes. Thus, higher local rainfall may occur in some areas, causing more discharge into rivers and producing more flooding. This precipitation may be quite variable in different areas. For instance, some drainage basins in the northwestern United States have been predicted to have increased flood risk as a result of global warming. By contrast, drainage basins that receive much of their precipitation as snow have been predicted to have less flood risk in the spring as a result of there being less snow to melt.

Predicted Influence of Climate Change on Maritime Coasts

Hurricanes have been predicted to become more numerous and destructive as a result of global warming. For instance, there has been an increase in Type 4 and 5 hurricanes since 1970. This increase should result in more and greater storm surges along coastlines.

The gradual increase in temperature over time should result in more glacial melting and an increase in the volume of liquid water on the planet. Some predict that even modest increases in GHGs in the atmosphere could melt the ice sheet on Greenland and much of the Antarctic ice sheet. These effects would lead gradually to a rise in sea levels. Estimates of this rise have varied from an increase of 0.1 meter to 0.9 meter by the year 2100. This increase in sea level should result in more flooding of flat-lying areas along oceanic coastlines and an increased influence of storm surges.

This problem could be made worse in areas in which the land surface is gradually sinking, such as in southeastern England or in Venice, Italy. Venice is located at sea level, and a series of water canals inundate much of the city. The sea level has risen relative to the land surface by about 25 centimeters from about 1900 to 2000, mostly as a result of the land subsiding because of the removal of groundwater for industry. Thus, Venice risks being submerged into the ocean if the sea level rises.

Context

Some of the changes expected to result from global warming are difficult to predict. For instance, changes to the Gulf Stream associated with global warming could potentially be very important, causing vast changes in the climate. The Gulf Stream is the flow of warm, salty water from the subtropical Atlantic Ocean northward toward polar regions. The Gulf Stream loses heat to the atmosphere in cooler areas, so it becomes denser and sinks. The cooler water then flows to the south, along the eastern coast of North America. Warming of the atmosphere and ocean as a result of climate change could potentially keep the Gulf Stream warmer further to the north, making the water less dense. This lower density might keep the surface waters from sinking and partially or even completely shut down this circulation cycle in the Atlantic Ocean, drastically affecting the climate. Northern latitudes might become wetter, much of the midlatitudes could become dry, and tropical rainfall could shift. Thus, flooding and drought areas may shift drastically if the global climate continues to change.

Robert L. Cullers

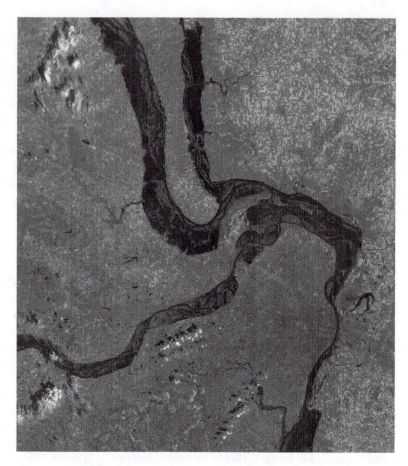

The great flood of 1993 was the worst flood ever on the upper Mississippi River. In this satellite view, the two rivers coming down from the top are the Mississppi (left) and Illinois Rivers (right). The Missouri River comes in from the left and the Mississippi exits at bottom. The city of St. Louis is between the two flooded stretches of the Mississippi and was not flooded. The flooded stretchs on the Mississippi are about 8 kilometers (5 miles) wide. (NASA)

Further Reading

Mitchell, John, Jason Lowe, Richard Wood, and Michael Vellinza. "Extreme Events Due to Human-Induced Climate Change." *Philosophical Transactions of the Royal Society* 364 (2006): 2117-2133. Describes the causes of global warming and its effects on rain and flooding, as well as other potential catastrophic effects on the Earth.

Overpeck, Jonathon T., and Julia E. Cole. "Abrupt Change in the Earth's Climate System." *Annual Reviews of Environmental Resources* 31 (2006): 1-31. Explains how the Earth's climate has changed rapidly in the past and how it may change abruptly in the future.

Ralph, F. M., and M. D. Dettinger. "Storms, floods, and the science of atmospheric rivers." *Eos* 92, no. 32 (2011): 265-266. A relatively recent insight, atmospheric rivers are streams of water vapor hundreds of kilometers wide that carry more water than the largest rivers. They transport a significant part of the water vapor in the atmosphere.and are the cause of many major floods.

Ruddiman, William F. *Earth's Climate Past and Future.* 2d ed. New York: W. H. Freeman, 2008. Ruddiman specializes in climate research, and he summarizes the background of climate science and what may happen to the climate as a result of global warming. There are many colored figures to support the discussion.

Valsson, Trausti. *How the World Will Change with Global Warming.* Reykjavik: University of Iceland Press, 2006. The causes and potential effects of global warming are described in this book. Some of the predictions are controversial. There are numerous tables and figures to support the discussion.

See also: Coastal impacts of global climate change; Coastline changes; Flood barriers, movable; Islands; Mean sea level; Sea-level change.

Florida

Category: Nations and peoples

Florida is one of the states most vulnerable to sea-level rise because of the low elevation of its land surface. The major cities along the coast, as well as Everglades National Park, are at greatest risk.

Key concepts

Category 5 storm: The most severe hurricane possible, with winds of at least 250 kilometers per hour and a storm surge of at least 5.5 meters

Everglades: A vast sawgrass marsh with occasional wooded "tree islands"

Goundwater table: the upper surface of the underground zone saturated with water. More properly termed *piezometric surface.*

St. Louis encephalitis: A form of sleeping sickness

Storm surge: an abnormally large sea-level rise occurring during a severe storm, usually driven by wind

Subtropics: A climate zone found just poleward of the tropics

Tropics: A climate zone found close to the equator that lacks a winter season

Background

Located in the southeastern corner of the United States, Florida is the nation's fourth most populous state. The state's land surface consists largely of a peninsula extending southward, bordered by the Gulf of Mexico on the west, Florida Bay on the south, and the Atlantic Ocean on the east. A panhandle that trends west from the peninsula is bordered by Georgia and Alabama on the north. The highest point in Florida is only 105 meters above sea level; the rest of the state is low-lying, with a fairly level surface.

Flooding

Flooding is a potential hazard for all of Florida's coastal areas, because the land is so flat and the ground surface is only a few meters above sea level. Heavy downpours cause serious problems. Storm drains and canals are designed to carry off excess water by gravity flow, but most of them empty into the ocean or lagoons behind the barrier islands, so a future sea-level rise would reduce the effectiveness of these drains and canals, aggravating flooding problems.

One problem is that many coastal defenses like dikes or levees are ineffective on parts of the Florida coast, because the limestone bedrock is highly porous and full of channels, so that water can leak under the dikes through the bedrock. Sealing the leaky bedrock is extremely costly and impractical.

Beach Erosion

Florida beaches experience severe erosion during storms and hurricanes, and this erosion will continue and increase if sea levels rise. Erosion along undeveloped sections of the coast is already cutting beaches back by as much as 0.5 meter per year. In developed areas, preventing beach erosion is a costly process. Property owners must armor their beachfront property with protective sea walls, or groins, while cities have to undertake dredge-and-fill projects, known as "beach renourishment" projects. Not only do the latter cost millions of dollars per kilometer, but also sand dredged from offshore

is often the same fine-grained sand that was washed away in the first place. Once it is replaced on the beaches, it is as easily washed away again, so the life expectancy of renourished beaches is just a few years or even less.

Florida's population is concentrated in the coastal zone, where the amount of residential and commercial development has steadily increased.

Owners have torn down their oceanfront cottages to replace them with two-story mini-mansions, and, where single-family houses once stood, multistory condominiums have risen. The vegetated sand dunes that once protected the coastline from beach erosion have been largely replaced by concrete structures, but many of these have proven to be just as vulnerable to wave attack as

The Florida Everglades (© Steven I. Dutch)

the dunes were, and some have even had to be torn down.

Hurricane Damage

Hurricanes are intense storms that form over warm ocean waters. Because of its southerly location, Florida may experience several of these each year. With continued warming of the oceans, Florida's hurricanes will probably occur more frequently and with greater intensity. The hurricane season lasts from June 1 until November 30, with two periods of peak activity. The first such period comes in June, when smaller storms form in the Caribbean Sea and the Gulf of Mexico, because the waters there warm first. The second activity peak comes in September, when the normally cool waters off the coast of Africa start to warm up and severe Category 5 storms may develop. The damage caused by hurricanes results from high winds as well as from the accompanying storm surges. The lower, southeastern coast of Florida has been most affected by hurricane damage over the years, but no part of Florida has gone untouched. The 2004 hurricane season was particularly stressful: Four major hurricanes hit the state during a five-week period. Following those storms, property-insurance rates for coastal dwellers skyrocketed.

Infrastructure Damage

Rising sea levels will affect Florida's highways, bridges, houses, factories, waterways, ports, airports, mass transit systems, sewage treatment plants, and water supply facilities. Unfortunately, many of these systems are already under stress as a result of long-term deferral of maintenance. In addition, the productivity of the coastal lands for agriculture, residences, tourism, recreation, and industry will be affected by rising sea levels, as will Florida's estuaries, wetlands, and coastal ground-water tables. Worse still, as salt water from the ocean intrudes along the coast, the well fields from which many Florida cities get their drinking water will become contaminated.

Environmental Changes

South Florida is classified as subtropical because of its mild winter season. Central and North Florida have more severe winters, with frequent freezing temperatures. However, global warming could cause South Florida to become tropical and cause Central Florida to develop the subtropical characteristics South Florida now has. These temperature changes would result in longer growing seasons and a poleward shift of certain plant and animal ranges. Mangroves would grow farther north, for example, and cypress, which are not tolerant of salt water, would decline and eventually die, as the rising ocean waters extended inland. Humans would be affected as well: Heat-related deaths could increase, and tropical diseases could spread. South Florida already has occasional cases of St. Louis encephalitis, and malaria, and the newly arrived Zika fever may become an issue.The numbers of these cases would probably increase as temperatures warmed, and dengue fever could appear as well.

Context

Florida is low-lying and has one of the longest coastlines of any U.S. state. As a result, it would be severely affected should sea levels rise as a result of global warming. Geologists believe that, prior to the last ice age, worldwide sea levels stood some 95-125 meters higher than present levels. As the highest point in Florida is only 105 meters above sea level, the entire state must have been submerged at that time. Evidence for this can be found in the old sand bars and sea shells scattered everywhere on the Florida land surface. A rise in sea level of only 15 meters would submerge all of South Florida, including the Everglades. If all the ice on the planet melts, almost the entire state of Florida will disappear beneath the waves.

Donald W. Lovejoy

Further Reading

Braasch, Gary. *Earth Under Fire: How Global Warming Is Changing the World.* Berkeley: University of California Press, 2007. This handsomely illustrated book surveys the changes that global warming is bringing to the planet, how these changes are caused, and what can be done about them.

Bush, D., et al. *Living with Florida's Atlantic Beaches: Coastal Hazards from Amelia Island to Key West.* Hong Kong: Duke University Press, 2004. Identifies shoreline areas that are safe or unsafe for development and the steps that coastal dwellers should take to reduce the hazards they face.

Cameron Devitt, S. E., J. R. Seavey, S. Claytor, T. Hoctor, M. Main, O. Mbuya, R. Noss, and C. Rainyn. "Florida biodiversity under a changing climate." *Florida Climate Task Force* (2012). America's tropical outpost is challenged by human disruption, sea level rise, and exotic species invasion. Changing climate may blur the distinction between native and exotic species as species adjust their ranges in response to climate change.

Warrick, R. A., E. M. Barrow, and T. M. L. Wigley, eds. *Climate and Sea Level Change: Observations, Implications, and Projections.* New York: Cambridge University Press, 1993. This classic book specifically addresses the regional effects of sea-level rise in terms of land loss, flooding, saltwater intrusion, and environmental and socioeconomic impact.

See also: Coastal impacts of global climate change; Coastline changes; Damages: market and nonmarket; Flood barriers, movable; Floods and flooding; Mean sea level; Saltwater intrusion; Sea-level change; Storm surges; Tropical storms.

Fog

Category: Meteorology and atmospheric sciences

Definition

Fog is a cloud near the ground. Fog forms as the temperature of air falls below its dew point and condensation occurs. Fog also forms and dissolves rapidly depending on whether the temperature is higher or lower than the dew point. Fogs are named based on the specific process through which the humidity has reached its saturation point, causing the fog to form.

Radiation fog forms as air temperature falls below the dew point by a rapid radiation cooling through a clear, calm night. The cool and heavy air containing this type of fog often settles in low-lying areas, such as mountain valleys. Most inland fogs are radiation fogs. Advection fog usually forms in coastal regions as warm air blows over a cold surface and loses heat to the underlying surface. Up-slope fog forms as air blows over mountain slopes, becomes cool, and reaches saturation.

Evaporation fog occurs as a cold, dry air blows over warm, moist surfaces. Thermal and humidity gradients between the two entities facilitate rapid evaporation from the surface. As the warmer, moist air from the ground mixes with cold, dry air above, it reaches saturation and forms fog. Steam fog over lakes during late fall and early winter is a typical form of evaporation fog.

Significance for Climate Change

Fog, as a surface-level cloud, has roughly the same climatic effects as low-level clouds. Fog dictates air and surface temperature patterns by controlling the amount of radiation that enters

Coastal fog creeping over the Coast Ranges, California. (© Steven I. Dutch)

and leaves the surface. High moisture content imparts fog with great thermal inertia. Thus, once a fog has formed, it will prevent air temperature from changing. In the morning, fog's high albedo allows less incoming solar radiation to reach the surface. Thus, morning fogs keep air temperatures low, allowing only gradual increases until the fog evaporates. As a result, fogs decrease daytime high temperatures.

In the night, fog traps outgoing longwave radiation and prevents air temperature from dropping as it would on a clear night. Fog also potentially warms the air near the surface, because condensation is a warming process. Overall, however, fog has a greater cooling effect than warming effect on Earth. As global warming continues, greater incidences of fog and lower clouds are expected, because warmer air can contain more water vapor. This increased fog may affect humans by reducing visibility and by trapping air pollutants to form smog. Fog can also produce precipitation, in the form of rain drizzle and light snow.

Jongnam Choi

See also: Air pollution and pollutants: anthropogenic; Air pollution and pollutants: natural; Albedo feedback; Clouds and cloud feedback; Dew point; Water vapor.

Food and Agriculture Organization

Category: Organizations and agencies
Date: Established 1945
Web address: http://www.fao.org

Mission
A United Nations agency, the Food and Agriculture Organization (FAO) gathers information about climate change that affects agricultural resources. Emphasizing food security and economic stability, the organization advises government leaders regarding policy strategies for agriculturists to adjust farming methods to continue production despite altered climatic conditions.

Food shortages and malnutrition during World War II motivated international leaders to establish the FAO on October 16, 1945. By the early twenty-first century, the FAO, headquartered in Rome, Italy, consisted of 189 member nations. The FAO supports nutritional humanitarian work by assisting agriculturists in developing and developed countries to increase cultivation of essential crops and livestock. Concerned about securing abundant food supplies worldwide, FAO consultants study agricultural conditions in diverse geographical locations, prepare publications, and present findings at meetings. They communicate with government leaders, who decide whether to incorporate FAO programs in policies.

Significance for Climate Change
FAO reports and conferences discussed climate change when addressing other issues but did not focus on that subject until 1988. At that time, FAO leaders, recognizing that global warming endangered food security, established a climate group to assess scientific activity investigating climate change to shape the FAO's objectives. By 2001, the FAO's Interdepartmental Working Group on Climate Change and integrated climate change program benefited from FAO departments representing such specific issues as animal husbandry, bioenergy, and forestry sharing climate-related work.

The FAO acquired and distributed information and funding to assist members' governments to encourage agriculturists, especially in developing countries, to utilize alternative techniques to keep farming and sustain yields despite temperature, precipitation, and other climatic deviations affecting agriculture. FAO representatives advocated conservation and recommended methods which eased climate change stresses on crops and livestock. The FAO coordinated endeavors with the United Nations Environment Programme (UNEP) and Framework Convention on Climate Change (UNFCCC).

FAO researchers utilized satellite technology to evaluate water and land resources needed for agriculture and observe climate-related disasters such

---

The FAO Mandate

The Food and Agriculture Organization provides the following statement of its institutional mandate.

Achieving food security for all is at the heart of FAO's efforts—to make sure people have regular access to enough high-quality food to lead active, healthy lives.

FAO's mandate is to raise levels of nutrition, improve agricultural productivity, better the lives of rural populations and contribute to the growth of the world economy.

FAO provides the kind of behind-the-scenes assistance that helps people and nations help themselves. If a community wants to increase crop yields but lacks the technical skills, we introduce simple, sustainable tools and techniques. When a country shifts from state to private land ownership, we provide the legal advice to smooth the way. When a drought pushes already vulnerable groups to the point of famine, we mobilize action. And in a complex world of competing needs, we provide a neutral meeting place and the background knowledge needed to reach consensus.

---

as droughts and floods. They promoted biotechnology efforts to create resilient crops that could survive such climate extremes as excess or insufficient water availability. The FAO established programs with universities to prepare experts for field assignments to assist agriculturists and show them how to adapt farming procedures to withstand climate change. FAO representatives encouraged farmers to plant trees to absorb carbons and reduce agricultural contributions to global warming by decreasing emissions from livestock and machinery used to harvest, process, and transport crops.

The FAO chaired UN Water, a program coordinating work and research concerning water resources impacted by climate change. The FAO helped governments with various water issues associated with global warming, including melting glaciers, and distributed reports such as the December, 2008, *Climate Change and Food Security in Pacific Island Countries*. The FAO stressed that solutions must be appropriate for each location and encouraged government officials from nations simultaneously using a large water resource, such as Lake Chad, to work together to address climate problems, instead of complicating situations with contradictory approaches.

The FAO studied how climate change affected indigenous populations, noting that lands they farmed often were threatened by deforestation and they were denied access to resources. FAO representatives also investigated how climate change impacted agriculturists according to gender, with males often leaving rural areas to secure consistent income.

FAO's annual World Food Day events emphasized how climate change hazards to agriculture interfered with food security. The FAO's 2006 *State of Food Insecurity in the World* report stated hunger affected 854 million people internationally. The FAO warned that climate change was the greatest threat to interfering with populations acquiring nourishment. World Food Day in October, 2008, featured a forum for participants to share ways they countered climate change.

In December, 2007, the FAO food price index indicated a forty percent price increase since 2006 due to climate changes and other factors causing shortages. The food price crisis diverted climate change discussion at a June, 2008, conference, and FAO leaders urged agriculturists to focus on crops for food, not biofuels. The 2008 FAO Initiative on Soaring Food Prices helped agriculturists in ninety countries.

In November, 2008, FAO members decided to reform that agency as a result of an independent assessment. Beginning in 2009, the FAO determined to concentrate its time and funds on several issues, including improved and expanded climate change work. FAO leaders stated they would request diplomats to incorporate food security provisions in climate change treaties. In January, 2009, FAO ideas resulted in the International Treaty on Plant Genetic Resources for Food and Agriculture, which advanced biodiversity by preserving and sharing genetic material from sixty-four crops to adapt plants impacted by climate change.

The FAO's Web site includes a climate change section, providing digital databases and publications. In 2008, the FAO initiated its electronic newsletter *Food-Climate*.

Elizabeth D. Schafer

Further Reading

Bruinsma, Jelle, ed. *World Agriculture—Towards 2015/2030: An FAO Perspective.* London: Earthscan, 2003. Chapter 13, prepared by geographer David Norse and FAO climatologist René Gommes, focuses on how agriculture both affects and is affected by climate change; projects potential food security concerns.

Food and Agriculture Organization of the United Nations. *FAO: The First Forty Years.* Rome, Italy: Author, 1985. Illustrated historical overview describes FAO programs, referring to weather issues and desertification, depicted on a map of Africa, prior to that agency initiating climate change projects.

Food and Agriculture Organization of the United Nations. *The state of food insecurity in the world 2014: strengthening the enabling environment to improve food security and nutrition.* Food and Agriculture Organization, 2015. A survey of global malnutrition. Although undernourishment is down significantly over the last decade, almost 800 million people are still chronically malnourished.

Martin, Andrew. "U.N. Food Meeting Ends With a Call for 'Urgent' Action." *The New York Times,* June 6, 2008, p. A13. Discusses why a conference initially organized to focus on climate change and biofuels instead concentrated on food prices. Participants' comments reveal diverse attitudes toward climate policies.

Rosenthal, Elisabeth. "World Food Supply Is Shrinking, U.N. Agency Warns." *The New York Times,* December 18, 2007, p. 5. Provides examples of climate change's contributions to reduced agricultural production and of FAO's adaptation strategies for farmers. Includes statistics and quotations by FAO director-general Jacques Diouf and climate scientists.

See also: Agriculture and agricultural land; Agroforestry; United Nations Environment Programme; United Nations Framework Convention on Climate Change.

Forcing mechanisms

Category: Meteorology and atmospheric sciences

Forcing mechanisms alter climate by changing the balance between incoming solar radiation and radiative energy loss. They may occur on human or geologic timescales and be anthropogenic or natural. The unpredictability of some large-scale forcing events has been used as an argument against modifying human policies to reduce the greenhouse effect.

Key concepts

astronomical forcing: climatic change triggered by changes in solar luminosity, variation in the Earth's orbit, and bolide impact

Milankovi6 cycles: variations in the eccentricity of the Earth's orbit, the tilt of the Earth's axis, and the precession of equinoxes that result in climatic variation on the scale of tens of thousands of years

planetary albedo: reflectivity of the Earth's surface to light

Background

Examining fossils, evidence from archaeological sites, and the historical record indicates that climate fluctuates markedly on timescales ranging from decades to tens of millions of years. One age's tundra may have been another age's temperate forest. Solar energy drives the Earth's climate. The amount of solar energy available depends on the balance between the amount reaching the Earth and the amount radiated back into space, both of which are variable. In addition to solar energy, Earth also receives input from gravitation and radioactive decay; these sources are dwarfed by solar input and have been essentially constant throughout the Phanerozoic era.

The Geologic Record: Evidence for Forcing

The geologic record spans of millions of years during which global climate was relatively uniform, divided by much briefer spans characterized by climatic extremes and elevated extinction rates. This record is punctuated by five recognized episodes of catastrophic extinction, believed to correlate with external climatic forcing mechanisms. The best-known of these is the end-Cretaceous event marking the end of the dinosaurs 65 million years ago. Most biologists accept that the precipitating factor was an asteroid collision that (among other effects) abruptly lowered global temperatures by ejecting massive amounts of dust into the atmosphere.

Types of Forcing Mechanisms

Climatic forcing mechanisms can be classified according to whether they are external or internal, whether their effect is to increase or decrease net energy available to the system, and the timescale on which they operate.

External Mechanisms. External forcing mechanisms include variations in the Earth's orbit, variation in solar output, impacts by comets and meteorites, and radiation from nearby supernovas. No effect attributable to the last two extremely rare, unpredictable events is evident in the Holocene. The only such event for which there is firm geological evidence is the end-Cretaceous asteroid impact that destroyed the dinosaurs. The predicted climatic effects of an asteroid impact include a short-lived global winter due to obstruction of incoming radiation, followed by a spike in temperatures from carbon dioxide (CO_2) released by burning and decaying vegetation. Oceanic currents would also be severely disrupted.

Total solar radiation varies by approximately 3 percent over the course of the eleven-year sunspot cycle. There is evidence for a ninety-year cycle in addition, and longer-term ones probably exist as well. The latter part of the Little Ice Age corresponds to the Maunder Minimum, a two-hundred-year period of reduced sunspot activity documented by historical records and proxy measurements. Whether similar extended periods of low sunspot activity contributed to global cooling at other times is uncertain. The Sun reached a low point in its cycle at the beginning of 2008.

Variations in Earth's orbit affect both total insolation and its distribution on Earth's surface. The relevant parameters are degree of eccentricity of the orbit, tilt of the Earth on its axis, and precession of the equinoxes. All of these vary in a predictable way on timescales of thousands of years, with a maximum periodicity of ninety thousand years. The entire orbital pattern is called the Milankovi6 cycles, after its discoverer. These cycles correlate well with Pleistocene glaciation episodes, and studies of Triassic-age lake sediments suggest the pattern is ancient. Earth is currently at an intermediate point in this long-term cycle.

Internal Forcing Mechanisms. Internal mechanisms include geologic events and large-scale biological changes. The timescales of such events vary from weeks, in the case of massive volcanic eruptions, to hundreds of millions of years, in the case of continental drift. The effects of changes in the Earth's magnetic field on climate are a matter of controversy. This field, generated by currents in the Earth's molten core, fluctuates on a timescale of 2,300 years and reverses polarity every 780,000 years. When the field is at its weakest, climatic variability is highest. The current weakening of Earth's magnetic field and the southward drift of the magnetic pole predict a cooling trend that Earth is not experiencing.

Plate tectonics and continental drift drive volcanic eruptions, which are major determinants of climate. Eruptions spew quantities of ash and sulfur dioxide into the atmosphere, blocking sunlight and reducing ground temperatures. There are several historical examples, most notably the Tambora eruption in 1815. Massive eruptions may also destroy enough vegetation to increase planetary albedo. On very long timescales, continental drift alters the ratio of land to water surface and distribution relative to the equator.

Oceanic Forcing. Small changes in sea surface temperature create large-scale climatic effects by altering the circulation of oceanic currents. Changes in oceanic currents may be cyclical and relatively predictable, for example those involved in the El Niño-Southern Oscillation, or they may be externally driven, unpredictable, and of longer duration. The cessation of Gulf Stream circulation in the North Atlantic during the Younger Dryas provides an example of the latter category.

Biological Forcing. Changes in the balance between photosynthesis and heterotrophic consumption alter the CO_2 content of the atmosphere. The present rise in CO_2 levels due to fossil fuel burning by humans mirrors ancient CO_2 depletion. Two episodes of massive global glaciation in the Precambrian follow periods of algal reef building and may represent global cooling due to photosynthetic carbon fixation.

Context

On a human timescale of hundreds or thousands of years, the most important forcing mechanisms are changes in atmospheric CO_2 content, the solar cycle, and volcanic eruptions. Of these, only the CO_2 content can be regulated by human activity. Sunspot cycles, volcanic eruptions, and some of the shorter Milankovi6 cycles have the potential to either reinforce or cancel out present anthropogenic global warming. Understanding and tracking forcing mechanisms is an important step in formulating rational global warming policy.

Martha A. Sherwood

Further Reading

Chambers, Frank, and Michael Ogle. *Natural Forcing* Factors *for* Climate Change *on* Time Scales *10^{-1} to 10^5 Years.* Vol. 2 in *Climate Change: Critical Concepts in the Environment.* New York: Routledge, 2002. Detailed and technical; good coverage of the research upon which conclusions are based.

Cronin, Thomas M. *Principles of Paleoclimatology.* New York: Columbia University Press, 1999. A standard textbook explaining methodology in detail, with an extensive discussion of the implications of recent discoveries for global energy policies.

Liu, Zhengyu, Zhengyao Lu, Xinyu Wen, B. L. Otto-Bliesner, Axel Timmermann, and K. M. Cobb. "Evolution and forcing mechanisms of El Niño over the past 21,000 years." *Nature* 515, no. 7528 (2014): 550-553. A study that employed modeling studies and proxy data to study behavior of El Nino cycles. Orbital parameters, meltwater discharge, carbon dioxide levels and retreat of ice sheets all played a role.

Pap, Judit M., and Peter Fox, eds. *Solar Variability and Its Effects on Climate.* Washington, D.C.: American Geophysical Union, 2004. In addition to Milankovi6 cycles, discusses sunspots and changes in solar output in geologic time.

See also: Abrupt climate change; Albedo feedback; Carbon dioxide; Earth motions; Holocene climate; Milankovi6, Milutin; Radiative forcing; Solar cycle; Sun; Volcanoes; Younger Dryas.

Forestry and forest management

Categories: Plants and vegetation; environmentalism, conservation, and ecosystems

Definition

Forests are both sources and sinks for carbon. Forest management therefore has the potential to be a means to either increase or decrease forest-based carbon storage. Forestry is a practical science concerned with creating, managing, using, and conserving forests and associated resources in a sustainable manner that meets societal goals. Forest management involves the regeneration, management, utilization, and conservation of forests to meet goals and objectives of society while maintaining the productive potential of the forest. The products, services, and values obtained from a forest include wood, water, wildlife, recreation opportunities, food products, aesthetic beauty, and many others.

A central tenet of forestry is sustainability—ensuring that the products, services, values, and inherent productivity of the resources are sustained over time. An old concept in forestry is sustained yield, meaning that the amount of resources removed from a forest is equal to the amount grown or produced in that forest. This concept has traditionally been most used to define the limits of sustainable timber production: No more timber should be removed from the forest than is grown in any given cycle. Similar concepts apply to the production of wildlife, grasses for forage or livestock, and nontimber ecosystem services such as berries, floral greenery, mushrooms, and so on.

Forest stands in which trees are roughly the same age are typically managed over rotations of twenty to over one hundred years, depending on species, site quality, and specific management objectives. Multiaged stands—which include multiple age classes of trees—are managed over cutting cycles that may range from five to over fifty years. Ideally, stands are maintained at different stages of development, so the forest as a whole may include a variety of stand ages and stand structures.

Many stands are managed to form simple structures that consist of a single tree species planted at consistent spacings and possibly with similar genotypes. Other stands are grown to have many species and multiaged structures. A regulated forest has an arrangement of stand structures that yields a constant production over time. This arrangement represents an ideal that is rarely met, because, over the long time span of forest growth, there are often changes in land ownership patterns, management directives, and regulations that affect how and which lands are managed, as well as disturbances such as fire.

Significance for Climate Change

Forestry also involves the management of fuels and their arrangement in forest stands. Forest fires occur over broad areas every year in both temperate and tropical forests and result in massive carbon emissions. Insects and pathogens represent another potential hazard to forests that may increase fuels and fire hazards, lead to large-scale carbon emissions, and reduce the ability of the forest to

Mass reforestation *programs* frequently create forests with a single tree species, all the same age, like this one in Germany. Such forests can be efficiently harvested but are aesthetically bland and not very rich in biodiversity. (© Steven I. Dutch)

meet other needs. Forests that are resistant to disturbance present an opportunity to conserve carbon on a large scale. Forests with natural functions and processes, endemic levels of insects and pathogens, and normal levels of biodiversity are said to be healthy. Regardless of specific objectives for any forest land, maintenance of forest health is a common, overriding objective for managed forests.

Forests store large amounts of carbon in living, above-ground stems, branches, and foliage; below-ground root structures and fine roots; dead, woody objects, such as logs, decomposing foliage, and twigs; and soil. Forest management activities affect these components and therefore affect a site's carbon balance. A clear-cut harvest will remove all stems but will typically not remove branches, foliage, some stemwood, or any below-ground components. A selection treatment to create a multiaged stand structure would remove a smaller number of trees. Other regeneration methods that remove intermediate amounts of trees produce intermediate results.

In addition to these direct effects on forest carbon, treatments and related soil disturbance may affect the rate of decomposition of plant materials and soil carbon. Disturbances such as fire, wind, or insect or pathogen attack can cause similar effects. Following treatment or disturbance, stands regrow, replacing trees and carbon. A forest managed on a sustainable basis would have a relatively constant level of carbon present, as well as wood fiber and other resources. Permanent removal of forests, or deforestation, generally results in reduced carbon storage. When wood and other forest products are removed and put into long-term use in a sustainable manner, however, they represent an additional source of carbon storage that supplements the forests. Likewise, wood biomass used for energy production results in temporary carbon loss in the forest but in a net carbon savings through offsets of nonrenewable fuels.

Forests represent both a source and a sink for atmospheric carbon. Under global warming climate scenarios, forests will experience greater concentrations of atmospheric carbon dioxide (CO_2), rising temperatures, and altered precipitation patterns. These changes may result in greater growth rates in some cases. Given the long life span of trees, however, they may have poor evolutionary adaptability to changing climatic conditions in their native ranges. Species can be moved to more suitable climates, but species are adapted to both climate and photoperiod.

Future forests are likely to receive greater pressure to produce wood and other ecosystem services including using wood as energy. Deforestation remains another threat to global carbon balance, as forests are replaced with agricultural lands or converted to other uses. Forests are an integral part of the global carbon equation and heavily subject to human influence. Management of forests to offset carbon losses, meet demands for wood and other forest products, cultivate ecosystem services, reduce losses to fire, provide biodiversity, and promote healthy forests all indicate the continued great importance of forestry in the future.

Kevin L. O'Hara

Further Reading

Freer-Smith, P. H., M. S. J. Broadmeadow, and J. M. Lynch, eds. *Forestry and Climate Change.* Oxfordshire, England: CABI International, 2007. An edited volume of twenty-eight chapters covering carbon sequestration, policy issues, climate change effects on forests, soil carbon, and implications for forest policy.

Kohm, K. A., and J. F. Franklin, eds. *Creating a Forestry for the Twenty-first Century.* Washington, D.C.: Island Press, 1997. Twenty-nine chapters present ecological foundations for forest ecosystem management and the concepts related to its silvicultural, landscape-level, and human dimensions.

Pukkala, Timo, ed. *Multi-objective forest planning.* Vol. 6. Springer Science & Business Media, 2013. Forest management solely for timber production is giving way to management for other purposes as well, such as hunting, recreation, and ecosystem management.

Puhe, J., and B. Ulrich. *Global Climate Change and Human Impacts on Forest Ecosystems.* Ecological Studies 13. Berlin, Germany: Springer, 2001. A fundamental treatment of climate change effects, covering a range of ecological issues and including a chapter on forest management.

See also: Amazon deforestation; Budongo Forest Project; Carbon dioxide; Carbon dioxide fertilization; Carbon 4 plants; Carbon 3 plants; Deforestation; Fire; Forests; Intergovernmental Panel on Forests; Tree-planting programs.

Forests

Category: Plants and vegetation

Forests, covering about one-third of the Earth's ground surface, are a major carbon sink. Manipulating the amount of surface area in forests arguably offers the best chance for mitigating anthropogenic climate change.

Key concepts

afforestation: creating forests on lands not previously forested

carbon sink: vegetation that incorporates carbon into its structure

dormancy: the portion of the year during which no growth occurs

growing season: the portion of the year during which photosynthesis occurs

reforestation: replacing lost forests

Background

Eight thousand years before the present, before the rise of human civilization, one-half of the Earth's surface was covered by forest. In the intervening years, 20 percent of previously forested land has lost its forest cover, resulting in the loss of a major carbon sink. One of the easiest ways for humans to prevent further carbonization of the atmosphere is to prevent further deforestation and to maintain existing forests intact.

Distribution of Forests

The forests of the world fall into three primary zones: tropical, temperate, and boreal. These zones are defined by their climate and precipitation. Trees require a minimum of 25 centimeters of rainfall per year to grow; land that, by virtue of its location, does not attract the minimum rainfall annually will not support trees. Such land will be either grassland, shrubland, or desert. Trees also require a frost-free part of the year for a growing season. Generally, the longer the growing season, the greater the annual growth.

The tropical rain forests on either side of the equator support tree growth that continues virtually all year long. About half of the world's forests are tropical forests, and they contain the greatest biological diversity of species. Because trees through photosynthesis convert atmospheric carbon to contained carbon, the half of the world's forests in tropical areas are of vital importance.

The temperate forest does not grow all year long. The falling temperatures reach a point at which growth is no longer supported, and the trees go into dormancy for part of the year. The temperate zone has a less varied collection of tree species than the tropics, but it contains a mix of coniferous and broad-leaved species, the latter of which generally lose their leaves during the dormant period.

The forest closest to the north pole is the boreal forest, composed very largely of coniferous species. It grows more slowly than the other forests, because its growing season is the shortest, and it exists in dormancy for a major part of each year. While the true polar regions may have enough precipitation to support trees, their growing season is too short for trees to grow enough to support the needles that provide respiration.

Twentieth Century Changes

The conversion of formerly forested land to other land uses was greater in the twentieth century than in any previous century. This conversion was driven in large part by the growth in human populations in all parts of the world, but especially during the latter half of the century, primarily in the tropical regions. Between 1950 and 2000, the world's population more than doubled, from 2.5 billion to 6 billion, and the developing world, largely located in the same area where tropical forests are found, contained three-fourths of that larger population. Since forests provide the largest carbon sink, the loss of many forests, especially the tropical forests, is mainly responsible for the rising proportion of carbon in the atmosphere.

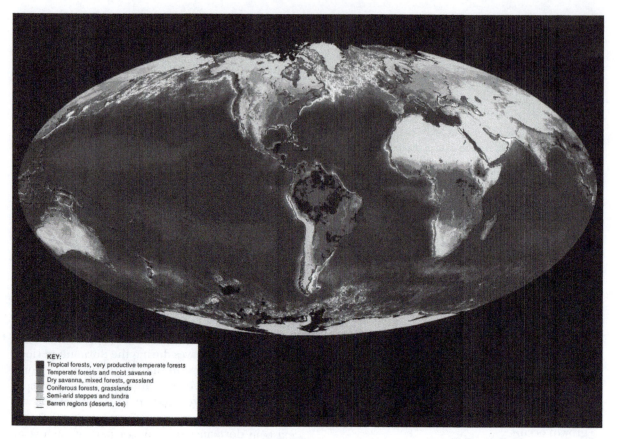

KEY:
■ Tropical forests, very productive temperate forests
■ Temperate forests and moist savanna
■ Dry savanna, mixed forests, grassland
■ Coniferous forests, grasslands
□ Semi-arid steppes and tundra
□ Barren regions (deserts, ice)

An image produced by NOAA's Polar Orbiting Environmental Satellite, showing the distribution of Earth's forests. (NASA)

The chief driver of deforestation during the twentieth century has been the demand for additional agricultural land to support the growing population. The conversion of forestland to agricultural land adds to atmospheric carbon in several ways. In many cases, the trees that are cut down to release the land for agriculture are burned, and the carbon stored in them is released into the atmosphere. Second, the carbon contained in forest soils over time is also released into the atmosphere when the trees are no longer there to prevent its release. Third, the trees that had been on the land are no longer there to capture future carbon as they grow.

Beside the need for agricultural land to grow crops to feed the growing world population, that population has experienced a change in dietary demand, particularly to include meat products. Some of the land freed up by deforestation is converted not to cropland but to pasture land. Specialized crops such as sugar or soybeans have a market price that poor populations seek to realize by converting forestland to agricultural land. In many areas, the production of specialized crops with major markets in the developed world, such as rubber, has also promoted deforestation.

Locations of Deforestation

The loss of forestland to agriculture has occurred in most parts of the world adapted to tropical forests. The Amazonian forest in Brazil and the forests in central America have been subject to important depletion since the mid-twentieth century. The deforestation has also been substantial in Southeast

Asia, notably on many of Indonesia's islands, as well as in Malaysia and Thailand. The forest in Africa has been less affected.

The temperate forest that was heavily deforested in the nineteenth century has started to bounce back as urbanization and the mechanization of agriculture have reduced the demand for agricultural land. Thus where the population is heavily urbanized, former agricultural land is being gradually reforested, as in the United States and Europe.

Much of the world's sawtimber comes from the coniferous trees in the temperate forest and from the coniferous trees that cover the boreal forest. The Russian forest constitutes one-fifth of the total forestland of the world, but in recent years much cutting to supply sawtimber to the developed world has depleted some of that forest. Combined with softwood coming from Canada's boreal forests, these two sources have supplied a major portion of the dimensional lumber used by the developed world for the construction of houses. Lumber production peaked in the United States in 1906, at 46 billion board feet, but since then lumber production supplied by U.S. forests has gone down. However, wood remains the third most productive commodity in world trade, behind petroleum and natural gas. A substantial proportion becomes fuelwood.

Context

At the Rio Conference in 1992, 158 countries agreed to try to prevent further deforestation. It has, however, proved difficult to accomplish partly because market forces are working against it, and partly because definition has proved elusive. The Kyoto Protocol of 1999 identified "reforestation" and "afforestation" as processes that could mitigate deforestation, but compensation to those who carry out such measures has not found widespread acceptance. No way has yet been found to value existing forests such that their preservation could be financially rewarded.

Nancy M. Gordon

Further Reading

Humphreys, David. *Logjam: Deforestation and the Crisis of Global Government.* London: Earthscan, 2006. Highly critical of the failure of international organizations to take on the task of preventing deforestation.

Malmesheimer, R. W., et al. "Preventing GHG Emissions Through Avoided Land-Use Change." *Journal of Forestry* 106, no. 3 (April/May, 2008). This special issue is devoted entirely to the Society of American Foresters' task force report on forest management solutions for mitigating climate change.

Pan, Yude, Richard A. Birdsey, Jingyun Fang, Richard Houghton, Pekka E. Kauppi, Werner A. Kurz, Oliver L. Phillips et al. "A large and persistent carbon sink in the world's forests." *Science* 333, no. 6045 (2011): 988-993. A survey of carbon storage in the world's forests, with extensive tables of forest biomass by country.

Stern, Nicholas. *The Economics of Climate Change: The Stern Review.* New York: Cambridge University Press, 2007. This famous report tackles the charges of those who claim that the economic costs of attempting to prevent climate change are too great.

Williams, Michael. *Deforesting the Earth: From Prehistory to Global Crisis.* Chicago: University of Chicago Press, 2003. This massive account by an author already well known for his careful evaluation of forest history is definitive.

See also: Amazon deforestation; Deforestation; Fire; Forestry and forest management; Intergovernmental Panel on Forests; Tree-planting programs.

Fossil fuel emissions

Categories: Energy; pollution and waste

With the exception of CO_2, the effluents of fossil fuel combustion devices such as automobile engines and thermoelectric power plants do not seem to have a direct effect on climate change. These pollutants are important, however, because of their deleterious effects upon the environment and human health.

Key concepts

exo: a prefix indicating 10^{18}

fossil fuel: any combustible deposit of carbon of biological origin created over millions of years of geologic history

ozone: a very reactive form of oxygen consisting of three oxygen atoms bound together loosely

particulate matter: small particles, such as fly ash and soot, emitted during the combustion of a carbon-based fuel

Background

Fossil fuels are combustible geologic deposits of carbon created from plant and animal remains subjected to high temperatures and pressures in the Earth over hundreds of millions of years. Coal, oil, and natural gas are the primary fossil fuels. When any carbon-based fuel is burned, the carbon unites with oxygen in the atmosphere to produce carbon dioxide (CO_2), the main culprit responsible for anthropogenic global warming. In addition, sulfur dioxide, nitrogen oxides, ozone, and particulate matter are often by-products of fossil fuel combustion. These pollutants detrimentally affect plants, aquatic life, and human respiratory health.

Consumption Modes

In the contemporary United States, 86 percent of all energy consumed is derived from fossil fuels, primarily oil (39 percent), natural gas (24 percent), and coal (23 percent). Some 8 percent comes from nuclear power, with the remaining 6 percent equally divided between wood and hydroelectric plants.

The energy consumed by each U.S. economic sector is as follows: residential and commercial, 35 percent; industry, 23 percent; direct transportation, 27 percent; and transportation-related uses, such as highways and other infrastructure construction, 15 percent. Some 69 percent of the petroleum consumed is for transportation, with

Annual Global CO₂ Emissions from Burning Fossil Fuels

Year	Total (mmt)*	Gas (mmt)	Liquids (mmt)	Solids (mmt)	Cement Production (mmt)	Gas Flaring (mmt)	Per Capita (metric tons)
1751	3	0	0	3	0	0	—
1775	4	0	0	4	0	0	—
1800	8	0	0	8	0	0	—
1825	17	0	0	17	0	0	—
1850	54	0	0	54	0	0	—
1875	188	0	1	187	0	0	—
1900	534	3	16	515	0	0	—
1925	975	17	116	842	0	0	—
1950	1,630	97	423	1,070	18	23	0.64
1975	4,615	623	2,132	1,673	95	92	1.13
2000	6,745	1,291	2,838	2,348	226	43	1.10
2005	7,985	1,484	3,096	3,032	315	58	1.23

Data from the Carbon Dioxide Information Analysis Center, Oak Ridge National Laboratory.
*mmt = millions of metric tons.

another 9 percent for transportation-related uses. Industry accounts for 16 percent of U.S. petroleum consumption, while the residential and commercial sectors account for only 6 percent. Of the 9 percent used for transportation, automobiles consume 40 percent, trucks 33 percent, railroads and buses 3 percent, aircaft 9 percent, water craft 6 percent, and all others 9 percent.

Environmental Impacts

Fossil fuels provide energy when carbon, the backbone of all fossil fuels, unites with oxygen in the air to produce that energy, as well as CO_2—a combustion by-product. Other elements occurring with fossil fuels, most notably sulfur, are also combusted, releasing emissions toxic to plants and animals. Nonnegligible environmental impacts also result from the extraction, processing, transportation, and waste disposal involved with fossil fuels. The two most important ecological impacts of combusting fossil fuels are the effects on climate of CO_2 emissions and the effects on health of particulate matter and the gaseous by-products of combustion.

Coal mining is accomplished through either strip mining or deep mining. Strip mining renders scores of hectares of land unusable unless they are later reclaimed and has led to mudslides when the removed overburdens are piled too high. Deep mining is prone to cave-ins and fires, and virtually all career deep miners eventually succumb to pneumoconiosis (black lung disease). Abandoned mines often leach acidic effluents into local streams, decimating the local ecology and ruining scenic vistas.

Drilling for oil leads to environmental degradation at the drill site, but even more problematic are the minor leaks and major oil spills that occur during transportation of the oil. These accidents have contaminated shorelines and estuaries, fouling beaches and killing waterfowl and aquatic life. Natural gas is prone to drilling accidents as well and is also subject to pipeline leaks during gas transportation.

Public Health Impacts

All fossil fuels emit CO_2, which is a greenhouse gas but not a direct health hazard. In addition, coal typically contains from 1 to 10 percent sulfur and many other trace elements, some of which are radioactive. When sulfur is burned with the coal, it produces sulfur dioxide, which converts to sulfuric acid in the atmosphere. Rain containing the dissolved acid (known as acid rain) will adversely affect forests, and when the acid contaminates bodies of water, fish and aquatic plants are likely to die.

Whenever a carbon-containing fuel is burned, nitrogen oxides are also created; these chemicals react with atmospheric water vapor to create nitric acid, another component of acid rain. In addition, atmospheric nitrogen oxides, as well as sulfur oxides, raise mortality rates and morbidity, particularly among those with respiratory problems. Another gaseous pollutant associated with combusting fossil fuels is ozone, a highly reactive form of oxygen, formed when nitrogen oxides combine with volatile organic compounds in automotive exhaust. Ozone, in addition to increasing morbidity in those with respiratory problems, detrimentally affects forests and reduces crop yields.

Particulate matter released when fossil fuels are burned causes respiratory illness when particles between 0.2 and 3 microns in size coat the lining deep inside the lungs. For those already burdened by respiratory ailments, increased morbidity is a likely result.

Context

In the past 150 years, the U.S. population has increased by a factor of ten, and the per capita consumption of energy has increased by a factor of five. The United States is thus consuming fifty times the energy it consumed in 1860. Over this time period, the use of wood for fuel has remained relatively constant at about 3 exojoules annually. Water was not harnessed for energy until about 1906, when Niagara Falls became the site of the first hydroelectric power plant. After World War II, the available energy from new hydroelectric plants increased to about 3 exojoules, where it has remained.

The use of coal began around 1840 and grew exponentially until 1920, when it reached 15 exojoules per year. Although the rate of increase has slowed, total annual coal use continues to increase; it is about 22 exojoules today. The use of oil, relatively minimal in the nineteenth century, reached 2 exojoules by 1900. With the twentieth century increase in automobiles, annual oil use rapidly increased to 15 exojoules in 1950, 35 exojoules in 1980, and 40 exojoules at the end of the century. Natural gas was used for lighting in the late nineteenth century at an annual rate of about 1 exojoule. As gas was increasingly used for heating, this rate increased to 5 exojoules by 1940 and 17 exojoules by 1960; it leveled off at 35 exojoules per year from 1980 through 2000.

George R. Plitnik

Further Reading

Bent, R., L. Orr, and R. Baker, eds. *Energy: Science, Policy, and the Pursuit of Sustainability.* Washington, D.C.: Island Press, 2002. An accessible introduction to crucial questions concerning energy use, consumption, and pollution.

Freese, Barbara. *Coal: A Human History.* New York: Penguin, 2004. A comprehensive survey covering in great detail the social, environmental, and political history of coal.

Hyne, Norman J. *Nontechnical Guide to Petroleum Geology, Exploration, Drilling, and Production.* 2d ed. Tulsa, Okla.: Pennwell, 2001. Details, among other items, the location, discovery, mapping, exploration, drilling, production, and reserves of oil and natural gas.

Peters, Glen P., Gregg Marland, Corinne Le Quéré, Thomas Boden, Josep G. Canadell, and Michael R. Raupach. "Rapid growth in CO_2 emissions after the 2008-2009 global financial crisis." *Nature Climate Change* 2, no. 1 (2012): 2-4. The 2008-2009 global financial crisis resulted in a sharp but transient decline in carbon emissions. Carbon dioxide emissions rapidly recovered to pre-crisis levels as the economy improved.

See also: Biofuels; Clean energy; Coal; Energy resources and global warming; Ethanol; Fossil fuel reserves; Fossil fuels and global warming; Gasoline prices; Hubbert's peak; Oil industry; Renewable energy.

Fossil fuel reserves

Categories: Fossil fuels; Energy

Definition

Fossil fuel reserves represent the total quantity of hydrocarbon fuels—such as coal, oil, and natural gas—estimated to be economically recoverable from a specified region. In order to aggregate the different physical units commonly used for each type of fuel, the reserves may be expressed in terms of energy content, using a measurement such as joules.

Calculated fossil fuel reserves are based on estimates of the physical deposits that may exist in a region. These estimates have several sources of potential inaccuracy. Even for a given oil well already in operation, geologists and engineers cannot be certain how many barrels will be extracted before it is shut down. To quantify the uncertainty of the estimates, there are three common categories of reserves. Proved reserves are those reserves that can be developed with reasonable certainty, usually taken to be 80 to 90 percent confidence, using known techniques and with prevailing market conditions. Probable reserves are those that can be developed with 50 percent confidence. Possible reserves may be developed with 10 percent confidence—for example if market conditions change or extraction technology improves.

Estimates of fossil fuel reserves rely not just on geological and engineering factors, but on economic factors as well. The operators of an oil well do not extract the entire reservoir of oil lying beneath the surface; at some point, production is discontinued because it costs more money to extract an additional barrel from the deposit than the barrel can be sold for on the market. This means that, as oil and other fossil fuel prices change, estimates of reserves will also change. The Energy Information Administration (EIA) has estimated that, as of December 2014, the United States had proved reserves of 36 billion barrels of crude oil and 354 trillion cubic feet of natural gas. As of January 2014, the EIA estimated that the United States had over 256 billion short tons of recoverable coal reserves.

Significance for Climate Change

Estimates of fossil fuel reserves at a local level are important in the oil industry. Company officials and shareholders need such estimates in order to responsibly invest in future projects. Estimates of fossil fuel reserves on a worldwide scale have many implications for the debate over climate change and energy policies.

If M. King Hubbert's theory of peak oil is correct, then the rate of worldwide oil production has reached a maximum level or will do so in the near future. According to this theory, as existing deposits are exhausted, it will become more and more difficult to locate new deposits. Over time, it will become costlier to bring a barrel of oil to market, and production will not be able to keep up with growth in demand. Those who subscribe to the peak oil theory often favor government policies that encourage the development of renewable forms of energy (such as solar and wind power), because they believe the era of conventional fossil fuel use is ending.

Other analysts dispute the peak oil theory and blame government regulations for hampering the growth in supply to meet the growing worldwide demand for energy. According to this view, production of oil (as well as other fossil fuels) would grow significantly if the US federal government lifted restrictions on energy companies for the development of the Arctic National Wildlife Refuge (ANWR) and offshore areas.

Optimists for the future of fossil fuels also argue that it only makes economic sense to look for additional reserves when the relative supply begins to dwindle. For example, in the early twentieth century, many official forecasts declared that the United States would soon run out of oil, based on known reserves at the time and the rate of oil consumption. In retrospect, these warnings were unfounded, because the oil industry explored and found new deposits as the older ones diminished. Some analysts believe the same will prove true regarding more modern warnings and that fossil fuels will continue to be an economical source of energy for decades into the future.

The role of the Organization of Petroleum Exporting Countries (OPEC) in the world oil and natural gas markets affects the treatment of fossil fuel reserve estimates. Many critics of OPEC argue that the lack of transparency applicable to national oil companies—in contrast to Western-based, private companies owned by shareholders—casts doubt upon the official reserve estimates of nations such as Saudi Arabia. For example, many analysts worry that the ministers of Saudi Arabia's vast deposits exaggerate the nation's reserves in an effort to keep the world complacent with its dependence on conventional fuels.

Robert P. Murphy

Further Reading

BP. BP Statistical Review of World Energy, June, 2007. London, England: Author, 2007.

Bradley, Robert, Jr., and Richard Fulmer. Energy: The Master Resource. Dubuque, Iowa: Kendall/Hunt, 2004.

Jakob, Michael, and Jérôme Hilaire. "Climate science: Unburnable fossil-fuel reserves." *Nature* 517, no. 7533 (2015): 150-152. A substantial portion of the earth's fossil fuels must be left in the ground if climate is to remain within the limits of a 2 degree C increase in temperature.

Simmons, Matthew. Twilight in the Desert: The Coming Saudi Oil Shock and the World Economy. Hoboken, N.J.: John Wiley & Sons, 2005.

"U.S. Crude Oil and Natural Gas Proved Reserves." U.S. Energy Information Administration. U.S. Dept. of Energy, 4 Dec. 2014. Web. 24 Mar. 2015.

Fossil fuels and global warming

Category: Energy

The coal-burning steam engine stimulated the Industrial Revolution, leading to the nineteenth century's vast commercial productivity. The invention of affordable automobiles in the twentieth century created a huge market for oil, which became America's dominant energy source. When a fossil fuel is burned, its emissions include CO_2, the main anthropogenic contributor to global warming.

Key concepts

fossil fuel: any combustible deposit of carbon of biological origin created over millions of years of geologic history

kerogen: complex compounds with large molecules consisting of carbon, hydrogen, oxygen, nitrogen, and sulfur

methane hydrates: small bubbles of natural gas trapped in a crystalline ice matrix

oil shale: a fine-grained sedimentary rock rich in kerogen

tar sands: sands containing highly viscous asphalt-like oil that can be extracted by mixing the sands with hot water or steam

Background

Fossil fuels store the chemical energy created over hundreds of millions of years as accumulated layers of plant and animal remains were subjected to heat and pressure. These organic residues transformed into coal beds, pools of oil, and pockets of gas. They include coal, oil, natural gas, oil shale, and tar sands. Since these fuels are no longer being created, they are nonrenewable resources. Equally important, when burned, the carbon unites with oxygen in the atmosphere to produce carbon dioxide (CO_2), the main culprit responsible for anthropogenic global warming.

Coal

Coal is fossilized plant material, deposited 300 million years ago when Earth was warmer and wetter. Preserved and altered by geological forces over eons of time, this material was compacted into carbon-rich fuel.

Coal mining is dirty and dangerous, because underground mines are subject to cave-ins, accumulations of carbon monoxide, and fires caused by explosive gases such as methane. In the United States alone, tens of thousands of miners have died in accidents over the past century, and even more have died or been disabled by respiratory diseases caused by the accumulation of fine dust particles in the lungs. Although strip mining is a safer and less expensive alternative, the land remaining after the overburden is removed is rendered unfit for any other use. Restoration and reclamation are now mandated by US law, but the efforts expended by mining companies are often superficial and ineffective. Coal mining also contributes to water pollution, because sulfur and other soluble minerals in mine drainage and runoff from mine tailing are acidic and highly toxic.

Coal burning releases, in addition to CO_2, many toxic metals and radioactive elements formed into gaseous compounds. Coal combustion is responsible for about 25 percent of all atmospheric mercury pollution in the United States.

Oil, Oil Shale, and Tar Sands

Oil is formed from phytoplankton, microorganisms that lived in warm, shallow seas hundreds of million of years ago. When they died, they sank to the bottom and were buried in sediments. Over eons of time heat from the Earth and the pressure of overlying layers of sedimentary rocks transformed this into kerogen deposits containing a mixture of oil, gas, and solid tarlike substances.

Drilling for oil leads to environmental degradation at the drill site, but even more problematic are water pollution due to leaks during transportation and the major oil spills caused by accidents. Such accidents have been known to contaminate shorelines and estuaries, fouling beaches and murdering waterfowl and aquatic life.

Oil shale and tar sands are unconventional resources with a large potential if they can be recovered with reasonable social, economic, and environmental costs. Western Canada has an estimated 270 billion cubic meters of tar sands from which liquid petroleum can be extracted, but the process is expensive and the environmental problems severe. A typical site yielding 125,000 barrels daily leaves 15 million cubic meters of toxic sludge and contaminates billions of liters of water each year.

Vast deposits of oil shale, rich in kerogen, are located in the United States' intermountain west. When heated to about 482° Celsius, the kerogen liquefies and can be separated from the stone. If the deposits could be extracted at a reasonable price and with acceptable environmental impact, the amount would be the equivalent of several trillion barrels of oil. The mining and extraction requires huge amounts of water (a scarcity in the west), creates air pollution, contaminates water, and leaves mountains of loose, rocky waste.

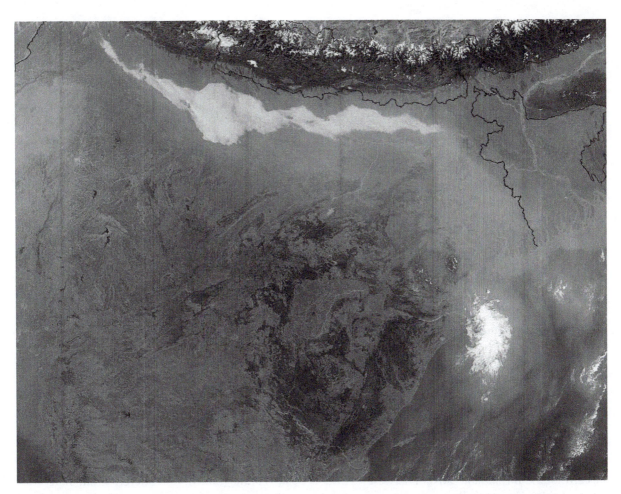

"Hazesmoke Gangeticbasin: This image, taken by Terra/MODIS instrument on December 17, 2004, shows thick haze and smoke along the Ganges Basin in northern India." (NASA)

Natural Gas

After oil and coal, natural gas is the world's third largest commercial fuel, accounting for 24 percent of global consumption. It is also the most rapidly increasing fossil fuel energy source because it is convenient, inexpensive, and cleaner burning than coal or oil. When combusted it releases half the CO_2 as an equivalent amount of coal; substituting gas for coal thus helps reduce global warming. Although it is difficult to ship across oceans, the US has an abundant easily available supply and the pipelines to transport it from source to end user. Natural gas, often released when oil is extracted, is burned off when no easy mechanism exists to deliver it to a user. Although transportation is problematic for some recoverable natural gas deposits, the world resources are estimated to be about 300 trillion cubic meters, a sixty-year supply at current rates of use.

An unconventional and as yet untapped source of natural gas is the methane hydrate deposits located in arctic permafrost and beneath deep ocean sediments. At least fifty ocean deposits and a dozen land deposits, containing about 9 trillion metric tons of methane, are known to exist. This is twice the combined amount of all coal, oil, and natural gas reserves. Although this is a possible future energy source, the complex technologies required to extract, store, and ship the methane hydrates are formidable.

Context

Fossil fuels promoted the Industrial Revolution, increased industrial productivity, contributed to capitalizing industry and farming, and still provide 85 percent of the world's energy. These resources, however, are nonrenewable; eventually the production rate will decline to the point where it is no longer economically feasible to extract the remaining fuel. When a production peak occurs and decline begins, about half of the total resource has been recovered. Assuming a modest 1 percent or 2 percent annual increase in consumption, the number of years remaining until production peaks for the three main fossil fuels is hundreds of years for coal, up to sixty years for natural gas, and no more than forty years for oil. Given the eminent end of fossil fuel dependence, the contamination of air and water caused by their extraction and combustion, and the hazards posed by global warming, alternate renewable resources must be developed and incorporated into the energy mix expeditiously.

George R. Plitnik

Further Reading

Berkowitz, Norbert. Fossil Hydrocarbons. San Diego, Calif.: Academic Press, 1997.

Hinrichs, Roger. Energy: Its Use and the Environment. 2d ed. New York: Saunders, 1996.

Jaccard, Mark. Sustainable Fossil Fuels. New York: Cambridge University Press, 2006.

McGlade, Christophe, and Paul Ekins. "The geographical distribution of fossil fuels unused when limiting global warming to 2 [deg] C." *Nature* 517, no. 7533 (2015): 187-190. To keep temperature increase within a 2 degree C range, a third of oil, half or natural gas and most coal must be left unburned. Fossil fuel production in the Arctic and unconventional energy production are incompatible with meeting that goal.

Osborn, Stephen G., Avner Vengosh, Nathaniel R. Warner, and Robert B. Jackson. "Methane contamination of drinking water accompanying gas-well drilling and hydraulic fracturing." *proceedings of the National Academy of Sciences* 108, no. 20 (2011): 8172-8176. Methane in drinking water is non-toxic and its principal danger is building up to explosive concentrations in enclosed spaces. Fluids used in fracking, however, can constitute a health risk and can contaminate ground water through the same avenues as methane.

Pfeiffer, Dale Allen. *Eating fossil fuels: oil, food and the coming crisis in agriculture.* New Society Publishers, 2013. The Green Revolution was made possible by cheap fossil fuels. In the United States, one calorie of food requires several calories of energy from fossil fuels, an ultimately unsustainable situation.

Fracking

Category: Fossil Fuels

Fracking, also called hydraulic fracturing, is a drilling process whereby fluid is pushed into the ground at high pressure to fracture shale and other poorly porous rocks. The fracturing of the shale creates fissures in the rock, which then allows oil and gas to flow out of the rock formation and into a wellbore, where oil and natural gas is then extracted.

Key concepts

Shale: A fine grained, thin layered sedimentary rock that forms from compaction of silt and mineral particles. Black shale contains organic material that breaks down to form natural gas and oil.

Shale plays: Shale formations that contain significant amounts of natural shale gas. Major shale plays in the United States include the Barnett Shale Play in Texas, the Marcellus Shale Play in the Northeastern United States, the Anadarko-Woodford Play in Oklahoma, the Granite Wash Play in Texas and Oklahoma, the Niobrara Shale Play in the Rocky Mountains, the Bakken Shale Play in Montana, North Dakota, and Canada, and the Eagle Ford Shale Play in South Texas.

Wellbore: The drilled hole, also called a borehole, which the process of fracking creates. When fracking starts, a well is drilled vertically at a surface depth of 1 to 2 miles. The vertical well is usually encased in steel or cement to ensure that well particles and oil do not leak into the groundwater.

Environmental Protection Agency (EPA): A governmental organization in the United States whose overall goal is to protect human health and the environment. Issues delegated to the EPA include issues regarding air quality, water quality, and land use.

Natural Gas: A flammable gas, consisting mostly of methane that occurs naturally underground, and can be used as fuel.

Background

Modern day hydraulic fracking began in the late 1940s, when Floyd Farris of Stanolind Oil and Gas began to study the relationship between gas and oil output in the United States. On March 17, 1949, petroleum production experts in Oklahoma performed the first commercial application of hydraulic fracking. On the same day, the Halliburton and Stanolind Oil Companies successfully fractured another oil well in Holliday, Texas.

By the 1980s technological advances in shale formation identification led to increased fracking in shale plays within the United States. Fracking also increased in the 1990s after George Mitchell created new fracking technology combining hydraulic fracturing with a horizontal drilling procedure. Fracking expanded in the early 2000s with an EPA report stating that hydraulic fracking created no threat to underground drinking water supplies. That report had numerous detractors, leading to the current EPA stance of trying to gain a deeper understanding of the potential impact of fracking on drinking water resources, as well as "factors that may influence those impacts." Presently, there is no conclusive EPA statement regarding the effect of fracking on water, communities, or the greater environment, nor is there research on the long-term impact of fracking on personal health, the environment, or water quality.

Fracking

Fracking is a way to increase natural oil and gas production. Basically, fracking is a drilling process whereby a well is drilled vertically at a surface depth of 1 to 2 miles. The vertical well is then encased in steel or cement to ensure that well particles and oil don't leak into the groundwater. Once the vertical well reaches a deep layer of rock, horizontal drilling may begin. Horizontal drilling occurs along oil bearing rock layers, as far as 1 mile (1.6 km) from the vertical wellbore. After the well has been drilled vertically and horizontally, fracking fluid is pumped down the well at extremely high pressure (approximately 9,000 pounds per square inch or 62 megapascals). The pressure fractures the surrounding rock, creating fissures and cracks through which oil and gas can flow. A pipe at the center of the horizontal portion allows oil and gas to come to the surface and be collected.

Statistics, Pros, and Cons

Fracking as a process has increased dramatically over the past decade. In the year 2000, there were over 275,000 natural gas wells drilled in the United States. In 2010, that number doubled to over 500,000; and every year, approximately 12,000 new wells are drilled within shale plays across North America. Although the process of fracking leads to high levels of domestic oil production, lower gas prices, and an increase in jobs; areas where fracking is common do pay a steep price.

Well drilling usually occurs in low income, rural areas. Land owners in these areas see the economic benefits of fracking, as they are usually paid over $2,000 an acre for permission to use land for well drilling/fracking. However, after trees are cleared and the ground is leveled, trucks and drilling equipment take over the landscape, exposing residents to increased traffic, dust, and ground tremors from artificial and natural/seismic activity.

Environmental activists also claim that fracking affects drinking water, and creates a route for potentially carcinogenic chemicals to escape into the water and air. Although proponents of fracking say fracking is a safe and economical source of clean energy, critics respond with specific examples of its dangers.

For example, in 2011, a fracking well in Pennsylvania malfunctioned, spewing thousands of gallons of contaminated fracking fluid onto the ground for half a day. In 2011, researchers from Duke University tested drinking water at sixty fracking sites throughout Pennsylvania and New York, and found that drinking water near fracking wells had dangerously high levels of methane. Fracking wells may also release carcinogenic chemical compounds such as benzene, ethylbenzene, toluene, and hexane into the air. Long-term exposure to these chemicals may cause birth defects, neurological issues, blood disorders, and certain cancers.

As fracking increases in the United States, the responsibility to ensure the safety of the land and the population around fracking sites increases as well. The EPA has called for complete transparency regarding both the benefits and long-term effects of fracking.

Although the idea of fracking is slow to spread abroad because of environmental concerns, some countries, such as Canada, India, and China, are actively pursuing research on advanced fracking techniques, so that their countries can also reap the benefits of local natural gas and oil production. In the United States, fracking continues to be a hotly debated environmental and political issue, and the EPA continues to work with individual states and stakeholders to ensure that natural gas extraction does not come at the expense of public health and quality of life.

Gina Riley

Further Reading

Energyfromshale.org. (2016). "What is Fracking." Retrieved from http://www.what-is-fracking.com/.

A pro-fracking website that contains information on fracking, as well as an in depth question and answer section. Topics addressed include the economics of fracking, environmental issues associated with fracking, land use regulations, and the effect of fracking on public health.

Environmental Protection Agency. (2016). "Natural Gas Extraction: Hydraulic Fracking." Retrieved from https://www.epa.gov/hydraulicfracturing.

Contains official EPA information regarding natural gas extraction by fracking. Discusses public health and environmental concerns, as well as the agencies' most current stance on fracking. This site also includes all official press releases by the EPA concerning Fracking and Fracking safety.

Lallanilla, M. "Facts About Fracking." *Live Science.* Web. 15 Jan 2015.

Overview of the fracking process, including sections on drilling wells, the fracking boom, fracking safety, and fracking and the law. A simple, organized, and balanced article on fracking.

See also: Gasoline prices; Fuels, alternative

France

Category: Nations and peoples

Key facts

Population: Metropolitan France: 62,150,775 (2008); 62,814,233 (2015); total population, including overseas departments: 64,057,792 (July, 2008); 66,553,766 (2015)

Area: Metropolitan France: 547,030 square kilometers; total area, including overseas departments: 643,427 square kilometers

Gross domestic product (GDP): $2.067 trillion (purchasing power parity, 2007 estimate); $2.647 trillion (purchasing power parity, 2015 estimate)

Greenhouse gas (GHG) emissions in millions of metric tons of carbon dioxide equivalent (CO_2e): 566.4 in 1990; 559.9 in 2000; 546.5 in 2006, 513 in 2010

Kyoto Protocol status: Ratified, May 31, 2002

History and Political Context

Charlemagne (742–814) united the Frankish kingdoms and ushered in a mini-renaissance, encouraging education, and the arts. After Charlemagne, France remained only nominally unified under titular kings who held little power compared to local princes such as William the Conqueror. France was eventually united under strong centralized leadership by Louis IX (1214–1270). For the next five hundred years, the country was periodically torn by internal religious conflict, as well as being threatened by foreign powers, including England, Spain, and Germany. Nevertheless, France was a major contributor to the Age of Enlightenment of the sixteenth and seventeenth centuries. France was a global power in the seventeenth and eighteenth centuries, both politically and intellectually, and vigorously guards its status as a world leader.

In the 1700s, an egalitarian and humanistic movement culminated in the French Revolution (1789–1794), which became a dominating theme in French thought and government. In 1871, the last remnants of the monarchies disappeared, and the government became a republican parliamentary democracy. The government survived many parliamentary crises from 1871 to 1958—a period that encompassed two world wars. In 1958, Charles

de Gaulle assumed power under a new constitution that included a strong presidency. This structure stabilized the government and allowed France to concentrate on its goals of fostering a strong and united Europe, as well as encouraging innovation in the arts and sciences. In 1992, France signed the Maastricht Treaty, which created the European Union. Since that time, France has worked to balance its own national interests with increased involvement in agencies and projects intended to unify and centralize Eastern and Western Europe.

Impact of French Policies on Climate Change

French environmental ideology has two fundamental roots: a devotion to intellectual innovation and a profound appreciation for the natural world. France has been the initiator or advocate of the creation of European environmental agencies beginning in 1948 with the International Union for Conservation of Nature. France was a driving force behind the Kyoto Protocol, galvanizing the 1997 conference and increasing the number of signatory nations. France has signed 130 European and worldwide agreements focused on the environment.

In February, 2005, France ratified the Charter for the Environment and added it to the preamble of the French constitution, thus assigning environmental rights and responsibilities an importance equal to that of civil liberties and economic and social rights.

The charter's ten articles include assertions that declare individuals must participate in conservation, that promote sustainable development, and that ensure the public is educated about environmental concerns. Article 5 supports the controversial precautionary principle, which states that action may be taken regarding an environmental issue even if there is disagreement in the scientific community over the severity of the problem or the best way to address it.

In addition to the charter, in 2004 France instituted a climate plan more aggressive than the Kyoto Protocol, with the goal of reducing carbon dioxide (CO_2) emissions by 54 million metric tons by 2010. This plan includes procedures to effect change at every level of French society, from large corporations to individual citizens. For both industry and consumers, those who choose lower emission technologies receive rebates, bonuses, or price reductions. Those who do not choose such technologies must pay additional fees or taxes or face punitive legal action. Examples include a measure to mitigate property taxes for energy-efficient buildings and an initiative to increase the use of biofuels. The French climate plan also includes funding for public education.

France as a GHG Emitter

According to data reported to the Climate Change Secretariat of the United Nations, the total amount of CO_2 emitted by France in 1990—the benchmark year to which levels were to be reduced—was 395.6 billion metric tons. Ten years later, the amount was 406.1 billion metric tons, an increase of 2.6 percent. In 2006, the amount was 408.7 billion metric tons, an increase from 1990 of 3.3 percent. However, all other greenhouse gas (GHG) emissions, including those of methane and hydrofluorocarbons (HFCs), have decreased 19 percent from 1990 to 2006. Thus, from 1990 to 2006, total French GHG emissions decreased by 3.5 percent.

Between 1990 and 2012, France reduced its greenhouse emissions by 13 percent. The French reliance on nuclear power—the last coal mine in France was closed in 2004—has contributed to lowering the rate of increase of CO_2 emissions, but the remaining increases, small as they are, are still troubling. One of the more difficult sectors to control is transportation, which is responsible for 26 percent of the increase in France's CO_2 emissions. The second-largest contributor to the increase is home heating, at 12 percent.

Summary and Outlook

The French desire to show leadership in environmental concerns seems to clash with the nation's ambition to also be a technologically advanced society. This has led France to attempt to integrate policies that are pro-environment with those that encourage technological competition and innovation. The French people favor measures such as green belts within industrial areas and the high-speed *train à grande vitesse* (TGV), a train that provides low-emission, energy-efficient transportation. The French citizenry seems willing to tolerate taxes and fees on ecologically unfriendly consumer goods, although occasionally there is strong opposition, as there was to a so-called "picnic tax," a tax on disposable items such as plates and tableware. The government provides incentives as well as fees, including a rebate of as much as $7,000 on cars that are particularly fuel efficient.

Although it appears that France is making great strides in improving air quality, some issues remain troubling. GHG emissions increased late in the first decade of the twenty-first century, in spite of controls and fines, and a worldwide recession generated pressure on the government to repeal or mitigate some previously established limits on GHG emissions. France was particularly concerned that some countries that were under extreme economic pressures or burdened with Soviet-era industries and power plants might rebel against EU agreements on limiting emissions. In an attempt to keep international accords from disintegrating, France has softened its stance on upholding those limits, and in 2008 it used its occupancy of the presidency of the European Union to mediate among EU member nations when conflicts arose over emissions standards and related issues.

Kathryn Rowberg and Gail Rampke

Further Reading

Bess, Michael. *The Light-Green Society: Ecology and Technological Modernity in France.* Chicago: University of Chicago Press, 2004. This award-winning book explores the conflict and eventual mediation between French beliefs in the sanctity of nature and what Bess calls "technological Darwinism"—a determination that France will not be left behind as technology advances.

European Environment Agency. SOER 2015 The European Environment: France, 2015. http://www.eea.europa.eu/soer-2015/countries/france (3 July 2016). Summary of environmental statistics and issues for France, with links to key sources (many in French).

International Energy Agency. *Energy Security and Climate Policy: Assessing Interactions.* Paris: Author, 2007. Close study of the relationship between the pursuit of national energy security and attempts to mitigate global warming in five example nations, one of which is France.

Prendiville, Brendan. *Environmental Politics in France.* Boulder, Colo.: Westview Press, 1994. Thorough examination of the influence of the "Green Movement" on French politics for the watershed years of 1970 through 1996.

See also: Europe and the European Union; International agreements and cooperation; Kyoto Protocol; Precautionary approach; United Nations Framework Convention on Climate Change.

Fraser Institute

Category: Organizations and agencies
Date: Established 1974
Web address: http://www.fraserinstitute.org

Mission

Based in Canada and grounded in conservative and libertarian philosophies, the independent economic and social research and education organization the Fraser Institute espouses free market principles. This think tank is highly critical of scientific claims that support human contributions to global warming. It focuses on advocating for freedom and competitive markets and on opposing public policy solutions such as spending, taxation, and regulation for economic problems.

Significance for Climate Change

The Fraser Institute seeks to illustrates how complex economic problems should and can be solved with innovative market solutions. The institute was established by a group of academics and business executives in order to study the impact of markets and government interventions on individuals and society. Membership includes high-profile figures and individuals from a wide range of backgrounds such as Michael Walker, an economist from the University of Western Ontario; former Reform Party leader Preston Manning; former Ontario Conservative premier Michael Harris; and Barbara Amiel, wife of columnist Conrad Black. The institute is highly critical of some widely accepted scientific evidence on climate change. Its February 1, 2007, publication titled *The Independent Summary for Policymakers: IPCC Fourth Assessment Report*, links directly to discussions included in the U.N. Intergovernmental Panel on Climate Change (IPCC) draft report.

The institute seeks to rebut popular scientific views on climate change by pointing to various studies. The basis of its claim includes data collected by weather satellites indicating that no significant warming is occurring in the tropical troposphere (the portion of the atmosphere that accounts for half of the world's atmosphere), as well as a lack of global-warming-related perturbations in long-term precipitation trends, total snow-covered area, snow depth, or arctic sea ice thickness. The institute also argues that considerable natural climatic variation exists and that greenhouse gas emissions cannot be correlated to climate change. Finally, it asserts that computer simulations tend to be inherently uncertain and inaccurate at the regional level.

The Fraser Institute has published information dedicated to specific countries. In *A Breath of Fresh Air: The State of Environmental Policy in Canada* (2008), the Fraser Institute addresses Canadians' concern for the environment and frustration with costly and intrusive environmental policies. The book outlines several market-based environmental policy solutions geared to improve the state of the environment and includes chapters dedicated to the discussion of air-pollution policy; water, forests, and fisheries management; and waste and recycling systems in Canada.

Rena Christina Tabata

See also: Canada; Conservatism; Libertarianism; Skeptics.

Freshwater

Category: Water resources

Freshwater is essential to support human life and Earth's biodiversity. Freshwater resources are under pressure from climate change and other sources, so policy makers must understand how climate affects freshwater and the importance of water resources.

Key concepts

base flow: the portion of stream flow that comes from groundwater

drainage basin: an area bounded by a continuous and topographically higher divide where water from precipitation drains downhill into a body of water

evaporation: the process by which water changes from liquid to vapor

evapotranspiration: the sum of evaporation and transpiration

groundwater: water stored beneath the land's surface

hydrologic cycle: the continuous circulation of solid, liquid, and gaseous water among the oceans, atmosphere, and continents

precipitation: the condensation of atmospheric water vapor that deposits hail, mist, rain, sleet, or snow on the Earth's surface

surface runoff: water from precipitation that flows over land surfaces to bodies of water

surface water: water found on land in such bodies as ponds, lakes, streams, rivers, wetlands, and inland seas

transpiration: the process through which water evaporates from the aerial parts of plants, especially leaves

Background

Freshwater is water that contains low concentrations of dissolved solids, such as salts. Water containing less than 1 gram per liter of dissolved solids is considered fresh, whereas water containing more than 1 gram per liter of such solids is saline. Ocean water has a salinity of approximately 35,000 parts per million. Some 97.2 percent of the total volume of the Earth's water is seawater. The majority of the Earth's freshwater is sequestered in glaciers. More than 98 percent of available freshwater is groundwater, and less than 2 percent of available freshwater is in lakes, running water, and wetlands.

Freshwater and the Hydrologic Cycle

Freshwater storage and migration are essential components of the Earth's hydrologic cycle. The ocean receives less water through precipitation than it loses through evaporation. Earth's land surface receives more water through precipitation than it loses through evapotranspiration. Surface runoff and base flow of freshwater from the land surface to the ocean are critical to maintain the balance of the hydrologic cycle, preserving sea levels and preventing permanent surface flooding.

Human Reliance on Surface Water

Even though most freshwater exists as groundwater and glacier ice, surface water—such as running water and lakes—is more important to human life, because it is readily accessible. In ancient culture, fresh surface water was mainly used as drinking water. With the development of civilization, more and more surface water was used for irrigation and industry. Water usage in modern society varies among nations. Water resources in developing countries

Sunset over Lake Superior, the largest freshwater lake in the world. (© Steven I. Dutch)

are mainly employed for irrigation and municipal uses. On the other hand, developed countries use most of their water for industry.

Water use by households, businesses, and communities is classified as municipal water use. Basic municipal water use includes drinking, washing, bathing, cooking, and sanitation. In addition, municipal water is used—especially in developed nations—for swimming pools, recreational parks, firefighting, street and car washing, and gardening. The largest use of surface water worldwide is for agriculture. Because of the increasing demand for water to grow crops, most of agricultural water use is for irrigation, or the controlled application of water to foster crop growth. As the human population grows, greater quantities and more efficient use of freshwater is necessary to meet the increasing demand for food. In addition, since the Industrial Revolution water has been used to drive almost every aspect of industry, including mining, automobile and other manufacturing, and energy production.

Impact of Climate Change on Rivers, Streams, and Lakes

Any climate change, and particularly global warming, will have a significant impact on the Earth's freshwater supply. However, the precise effect of increasing temperatures on the water supply is uncertain. Higher air temperatures can generate more precipitation, but they also increase evaporation. More precipitation and improved plant water-use efficiency due to increased atmospheric carbon dioxide (CO_2) would increase water supply. However, more precipitation and warmer temperatures could potentially extend the growing season of plants, which could result in greater transpiration and reduce water supply. The net effect of increasing temperature on water supply is thus unclear, especially on a global scale.

Studies have revealed other potential sources of dramatic climate-change-related impact on surface freshwater. Shrinking alpine glaciers would significantly reduce water supplies for many rivers and streams that originate from melting snow and glaciers while simultaneously releasing excess floods of water into glacial lakes. Even the world's largest

freshwater lake, Lake Baikal, responds strongly to the Earth's warming temperature.

Context

The availability of surface water varies among continents and countries. Today, more than 1 billion people lack access to safe drinking water. Approximately two-thirds of the world's population will be living in water-stressed areas by 2025, if no serious actions are taken to control climate change, water pollution, and water usage. Climate change has a significant impact on surface water resources. Globally, many policies have been established to reduce greenhouse gas emissions and promote the usage of renewable or alternative energy. However, a significant amount of water has been used to produce energy, and energy has been increasingly used to process water. With the pressure of population growth and limited energy and water resources, all governments and communities need to work together to develop a plan to integrate water and energy policy making. Communities and individuals need to conserve water and energy resources, and they must control pollution and mitigate climate change to ensure safe access to freshwater in the future.

Yongli Gao

Further Reading

Barnett, T. P., J. C. Adam, and D. P. Lettenmaier. "Potential Impacts of a Warming Climate on Water Availability in Snow-Dominated Regions." *Nature* 438 (2005): 303-309. Report of a research team led by Timothy Barnett at the Scripps Institution of Oceanography on the impact of climate change on snow-dominated runoff. Demonstrates that the Earth's warming temperature is likely to disrupt freshwater resources in regions that rely on snowmelt for water supply.

Bastviken, David, Lars J. Tranvik, John A. Downing, Patrick M. Crill, and Alex Enrich-Prast. "Freshwater methane emissions offset the continental carbon sink." *Science* 331, no. 6013 (2011): 50-50. Freshwater methane emissions are not generally included in climate models but this study indicates they may offset as much as a quarter of the carbon storage on land.

Cyranoski, David. "Climate Change: The Long-Range Forecast." *Nature* 438 (2005): 275-276. *Nature*'s Asia-Pacific correspondent explains how melting glaciers are threatening the livelihoods of millions of people in the Tibetan Plateau.

Hampton, S., et al. "Sixty Years of Environmental Change in the World's Largest Freshwater Lake—Lake Baikal, Siberia." *Global Change Biology* 14 (2008): 1947-1958. Details the effect of global warming on the world's largest freshwater lake.

Gleick, Peter H., and Newsha Ajami. *The world's water volume 8: The biennial report on freshwater resources*. Vol. 8. Island press, 2014. A global survey of water. Includes discussions of fracking and its effects on water, desalination, and the Syrian conflict and its connection to water.

Milly, P. C. D., K. A. Dunne, and A. V. Vecchia. "Global Pattern of Trends in Streamflow and Water Availability in a Changing Climate." *Nature* 438 (2005): 347-350. U.S. Geological Survey researchers compare the results of twelve climate models to project future annual stream flows worldwide.

Thomson, A. M., et al. "Climate Change Impacts for the Conterminous USA: An Integrated Assessment. Part 4—Water Resources." *Climatic Change* 69, no. 1 (2005): 67-88. Applies a suite of twelve potential climate change scenarios that could occur over the next century to simulate alterations to water supply in the coterminous United States.

Webber, M. E. "Catch-22: Water vs. Energy, September, 2008." *Scientific American Earth 3.0* 18, no. 4 (2008). A leading expert on the integrated study of water and energy explains the mutual dependencies between water and energy and advocates a government plan for integrated water and energy policy making.

See also: Desalination of seawater; Groundwater; Hydrologic cycle; Water quality; Water resources, global; Water resources, North American; Water rights.

Friends of Science Society

Category: Organizations and agencies
Date: Established 2002
Web address: http://www.friendsof science.org

Mission
An independent, education-based, nonprofit organization that provides regular scientific and factual commentary on the causes and impacts of climate change, the Friends of Science Society aims

> to encourage and assist the Canadian Federal Government to re-evaluate the Kyoto Protocol by engaging in a national public debate on the scientific merit of Kyoto and the global warming issue, and to educate the public through dissemination of relevant, balanced and objective technical information on this subject.

The organization, through its Internet Web site, offers "critical evidence that challenges the premises of [the] Kyoto [Protocol]" and highlights alternative causes of global climate change.

The Friends of Science Society is run by volunteers comprising mainly active and retired Earth and atmospheric scientists, engineers, and other professionals concerned about the abuse of science displayed in some of the interpretations of the Intergovernmental Panel on Climate Change and the politically inspired Kyoto Protocol. The organization uses a scientific advisory board made up of international climate scientists from around the world to "offer a critical mass of current science on global climate and climate change to policy makers, as well as any other interested parties." On its Web site, the organization declares "We do not represent any industry group, and operate on an extremely limited budget. Our operational funds are derived from membership dues and donations."

Significance for Climate Change
The stated objectives of the Friends of Science Society are to "work to educate the public through the dissemination of relevant, balanced and objective information on Climate Change, and to support real

environmental solutions" and "to spark a national and international debate on global warming." Members as well as non-members conduct literature research on scientific subjects related to global warming, the results of which are communicated to the public via its Web site and by means of newsletters.

The organization's major environmental concern is the significant shift in recent years away from the important emphasis of previous decades on continual reductions in air and water pollution, to focus almost exclusively on global warming. It claims climate fluctuations are natural phenomena and suggests that adaptation should be emphasized rather than misguided attempts at control. Substantial funds could be saved or deployed toward more urgent needs of humanity.

C R de Freitas

See also: Level of scientific understanding; Pseudoscience and junk science; Scientific credentials; Scientific proof; Skeptics.

Friends of the Earth

Categories: Organizations and agencies; environmentalism, conservation, and ecosystems
Date: Established 1969
Web address: http://www.foe.org/

Mission

Founded in San Francisco by David Brower, Friends of the Earth (FoE) is a volunteer organization that raises public awareness about activities environmentally destructive to the Earth's climate and that initiates legal action to address threats from global warming. FoE is affiliated with grassroots activist organizations in more than seventy countries that together form Friends of the Earth International.

FoE has been involved in previous initiatives against environmentally destructive activities and technologies including campaigns to stop construction of the Trans-Alaskan Pipeline and efforts to limit the growth of nuclear power. During the mid-1980's, FoE initiated a media campaign against products harmful to Earth's ozone layer such as air conditioning coolants and aerosols that contained chlorofluorocarbons. FoE has also worked to increase public awareness about human activities contributing to global warming. FoE advocates conservation, improvements in energy efficiency, and sustainable and environmentally friendly sources of alternative energy. In addition to pushing for corporate responsibility and accountability, FoE has called for changes in public policy and has taken direct action through lawsuits filed against large polluters and government regulatory agencies.

Significance for Climate Change

As a part of its campaign to raise awareness about climate change, FoE has publicized research showing evidence of a 30 percent increase in atmospheric carbon dioxide (CO_2) over the past two hundred years and predicting increases in average surface temperature of 1.4° to 5.8° Celsius by the year 2100. FoE notes that increasing global temperatures contribute to the melting of glaciers and permafrost, rising sea levels, severe flooding, heat waves, and damaging weather, all of which have disproportionate impacts on persons in developing countries. In September, 2007, FoE and thirty other organizations submitted a letter to the U.S. Congress urging its members to provide assistance to developing countries coping with global warming.

In pushing for corporate responsibility, FoE notes that many companies underreport the climate-related impacts of business ventures. To address this issue, FoE has increased pressure on automobile makers, petrochemical corporations, and electric utilities to provide information to customers and shareholders about climate-related risks associated with their activities. Under pressure from environmental groups such as FoE, the Ford Motor Corporation announced in 2005 that it would increase output of hybrid vehicles to 250,000 annually by 2010.

In addition to influencing public opinion, FoE has taken direct action to address public policies on climate change. FoE notes that industrialized countries are largely responsible for greenhouse gas (GHG) emissions that contribute to global warming. Therefore, FoE believes that these countries must take immediate and aggressive steps to reduce their carbon emissions. Among industrialized nations,

Japan, the United States, Canada, and New Zealand are most resistant to implementing policies to address the problem of climate change. With about 4 percent of the world's population, the United States is responsible for 25 percent of GHG emissions. FoE maintains that a country's production of GHGs should be tied to its population and that developing countries should not be required to sacrifice economic growth by having to institute the same reductions in emissions as industrialized countries.

In 2005, FoE filed a lawsuit in U.S. federal court to require enforcement of standards for CO_2 emissions. Working in conjunction with other environmental organizations and twelve state and local governments, FoE helped win a 2007 Supreme Court ruling that required the U.S. Environmental Protection Agency to regulate GHGs emitted by automobiles. FoE has also worked to demonstrate problems in proposed climate legislation such as the Lieberman-Warner Climate Security Bill (introduced in 2007). According to FoE, the bill fails to penalize companies that pollute and ignores recommendations made by leading scientists on ways to address global warming. In addition, FoE argues that the Lieberman-Warner Climate Security Bill's proposed targets for emissions are inadequate. Instead of endorsing the provision of pollution permits to corporations at no charge, FoE advocates the sale of such permits, with revenues used to support clean sources of energy such as wind, solar, and geothermal power.

FoE argues against increased production of biofuels and the expansion of nuclear power facilities to address the global warming crisis. It notes that growing reliance on biofuels such as corn ethanol may encourage the destruction of forests and can lead to higher food prices. Nuclear power is not viewed as a viable solution, because it remains a dangerous source of energy and produces large amounts of radioactive waste that will become a burden to future generations.

At the international level, FoE opposes a provision of the Kyoto Protocol allowing countries to meet targets for reductions in carbon emissions using a system that acknowledges benefits of carbon sinks. Carbon sinks include forests and ocean areas that absorb CO_2 produced by the burning of fossil fuels. FoE has pointed out that fast-growing monoculture plantations created as carbon sinks are harmful to biodiversity.

Thomas A. Wikle

Further Reading

Friends of the Earth. *Climate Change: Voices from Communities Affected by Climate Change.* Washington, D.C.: Author, 2007. Contains testimonials from persons in nine communities impacted by climate change.

———. *Climate Debt: Making Historical Responsibility Part of the Solution.* Washington, D.C.: Author, 2005. A call for industrialized nations to accept responsibility for past contributions to greenhouse emissions.

Porritt, Jonathan. *Friends of the Earth Hand Book.* London: Macdonald Optima, 1989. Explains how each person can limit his or her impact on the environment.

Smith, Joseph, and David Shearman. *Climate Change Litigation.* Adelaide, S.Aust.: Presidian, 2006. Examines legal issues surrounding climate change including a discussion of environmental advocacy groups.

Warhurst, Michael. "Resisting resource wars–Europe's responsibility."*Friends of the Earth* (2011). Resource use includes not just resources consumed domestically, but also those consumed externally in growing, manufacturing and transporting products

See also: Biofuels; Carbon dioxide; Environmental movement; Greenhouse gases and global warming; Sinks; U.S. legislation.

Fronts

Category: Meteorology and atmospheric sciences

Definition

A front is a band of low-pressure systems and marks the transition from one weather regime to another. It is typically formed at the boundary of two distinct air masses. In most cases, fronts are associated with a type of large-scale weather system called a midlatitude cyclone, which has a low-pressure center and causes winds to blow cyclonically (that is, in a counterclockwise direction). Midlatitude cyclones are the largest weather systems on Earth and generate most of the winter storms over the midlatitude continents. A front is a part of the midlatitude cyclone system, which trails a band of low-pressure air extending outward from the low-pressure center of the cyclone. Therefore, various weather systems, such as thunderstorms, heavy precipitation, snowstorms, and tornadoes, are also formed along the frontal band.

Different air masses are characterized by different physical properties of atmosphere, such as density, temperature, pressure, winds, and moisture. Because a front is a line that separates two different air masses, the atmosphere exhibits different physical properties on either side of a front. For example, air can change from warm to cold or from cold to warm, winds can blow from northerly to southerly or from westerly to easterly, and air can vary from dry to moist or from moist to dry across the frontal zone. Based on the movement of the frontal band and the temperature and humidity differentials, fronts can be classified as cold fronts, warm fronts, stationary fronts, and occluded fronts.

A cold front is formed when a cold and dry air mass advances and replaces a warm and moist air mass. In this case, the cold and dry air pushes and undercuts the warm and moist air ahead of it. The temperature will generally decrease in an area where a cold front is passing through. Because the cold front will lift the warm and moist air, clouds and precipitation can form at or behind the cold front.

A warm front is formed when a warm and moist air mass advances and replaces a cold and dry air mass. In this case, the warm and moist air pushes and overrides the cold and dry air, and the cold and dry air retreats. The temperature will generally rise in an area where a warm front is passing through. Because of the overriding of warm and moist air over the cold and dry air ahead of it, clouds and precipitation typically form ahead of a warm front.

An occluded front forms when a cold front catches up to and overtakes a warm front. In this case, a warm and moist air sector between the cold and warm fronts disappears, causing a complete convection of warm air in the storm center. This stage marks the full maturity of a midlatitude cyclone. Further dynamic and thermodynamic supports for the storm no longer exist, and the storm will dissipate from this time on.

A stationary front can form when a cold and a warm front move in opposite directions. When they meet, they can be locked in location. The cold and warm air mix together, so that there is no dominant overtake and apparent movement from either warm or cold air. This kind of situation often arises when fronts interact with the surface topography beneath them.

Significance for Climate Change

Fronts are important weather systems affecting people's daily lives. They mainly occur in middle latitudes, where large landmasses and dense human populations are located. A midlatitude cyclone,

A squall line marks the leading edge of a cold front. (NOAA National Weather Service Collection)

fronts, an upper-level jet stream, and specific storm tracks are all related from one to another and constitute a complete synoptic weather system (a weather system that can be analyzed on a weather map). Thus, a change in one part of the atmospheric environment will result in a change of the entire weather system. Studies show that global warming tends to widen the tropics and extend the troposphere vertically. There are many consequences of these changes. One of them is a poleward shifting of future jet streams. This shift would cause climatologic locations for midlatitude cyclones, fronts, and storm tracks to change accordingly.

The current global warming trend may also suggest a decrease of surface temperature gradient, since many observations and atmospheric model simulations indicate that a larger warming tends to occur in the colder regions. Since the horizontal temperature gradient is the key mechanism for the development of midlatitude cyclones, global warming might decrease the occurrence and intensity of midlatitude cyclones and associated fronts. On the other hand, because global warming tends to increase water content in the atmosphere, midlatitude cyclones may derive more energy from latent heat release and become more violent. There are no definite answers so far for how midlatitude cyclones and fronts are affected by these competing mechanisms.

Finally, frontal dynamics provides an important mechanism to cause convection and to form clouds. Precipitation related to fronts is a major process removing water from the midlatitude atmosphere. A potential change in frontal climatology in a future warm climate, regardless of whether it is an increase or decrease, will result in redistribution of snow and rain, changing the distribution of Earth's hydrosphere, especially in middle and high latitudes.

Chungu Lu

Further Reading

Ahrens, C. Donald. *Essentials of Meteorology: An Invitation to the Atmosphere.* 5th ed. Belmont, Calif.: Thomson Brooks/Cole, 2008. One of the most widely used introductory books on atmospheric science; covers a wide range of topics on weather and climate.

Archer, Cristina L., and Ken Caldeira. "Historical Trends in the Jet Streams." *Geophysical Research Letters* 35, no. 24 (2008). Reports an investigation of the change of location of the jet stream in response to global warming.

Berry, Gareth, Michael J. Reeder, and Christian Jakob. "A global climatology of atmospheric fronts." *Geophysical Research Letters* 38, no. 4 (2011). Despite the ubiquity of fronts in weather forecasting, there was no global climatological study of them. Fronts tend to be concentrated in mid latitudes, and over ocean basins.

Catto, J. L., Christian Jakob, Gareth Berry, and Neville Nicholls. "Relating global precipitation to atmospheric fronts." *Geophysical Research Letters* 39, no. 10 (2012). A large proportion of global precipitation is associated with fronts, cold fronts over the oceans and warm fronts over the continents.

Intergovernmental Panel on Climate Change. *Climate Change, 2007—Synthesis Report: Contribution of Working Groups I, II, and III to the Fourth Assessment Report of the Intergovernmental Panel on Climate Change.* Edited by the Core Writing Team, Rajendra K. Pachauri, and Andy Reisinger. Geneva, Switzerland: Author, 2008. Provides the most authoritative general report on climate change and global warming.

Lutgens, Frederick K., and Edward J. Tarbuck. *The Atmosphere.* 10th ed. Upper Saddle River, N.J.: Pearson Prentice Hall, 2007. Elementary introductory text that covers a wide range of atmospheric sciences.

Yin, Jeffrey H. "A Consistent Poleward Shift of the Storm Tracks in Simulation of Twenty-first Century Climate." *Geophysical Research Letters* 32 (2005). Investigates poleward shifting of storm tracks in relation to the jet stream under global warming conditions.

See also: Average weather; Extreme weather events; Hydrosphere; Meteorology; Monsoons; Rainfall patterns; Seasonal changes; Thunderstorms; Tropical storms.

Fuels, alternative

Category: Energy

Alternative fuels are sustainable, nonfossil fuels that do not duplicate the environmental and climatic damage caused by fossil fuels or nuclear power.

Key concepts

biodiesel: fuel with the chemical structure of fatty acid alkyl esters

biomass: plant and other organic material that can be used as fuel

ethanol: a colorless liquid alcohol with chemical formula C_2H_5OH

molecular hydrogen (H_2): a flammable, colorless, odorless gas

Background

The world is facing a potential energy crisis due to continuous fossil fuel energy demand and population increase. Pollution from fossil fuels, including carbon dioxide (CO_2) release, affects public health and causes global climate change. One way to solve this problem is to use alternative fuels, which provide renewable energy and generate little or no CO_2. Another reason to switch to alternative fuels is to reduce dependence on foreign oil. Some of the most commonly used alternative fuels are hydrogen, ethanol, biodiesel, and biomass.

Hydrogen

Molecular hydrogen is an ideal alternative fuel to be used for transportation, since the energy content of hydrogen is three times greater than that of gasoline by weight. It is also virtually nonpolluting and a renewable fuel. Using H_2 as an energy source produces only water; H_2 can be made from water again. A growing number of automobile manufacturers around the world are working on prototype hydrogen-powered vehicles. Only water is emitted from the tailpipe, not greenhouse gases (GHGs). These cars' motors run on electricity generated in fuel cells by chemical reactions between H_2 and oxygen (O_2). Many problems remain to be overcome before hydrogen cars become commercially viable, however.

One of the reasons for the delayed acceptance of H_2 is the difficulty of producing it on a cost-effective and climate-friendly basis. H_2 is obtained mainly from natural gas (methane and propane) via steam reforming. Although this approach is practically attractive, it is not GHG emission free. Molecular hydrogen can also be produced by electrolysis. In this case, electric energy is required to split water into H_2 and O_2. However, the process is not efficient; it requires significant expenditure of energy and purified water. One promising green method of H_2 production is biological: A great number of microorganisms produce H_2 from inorganic materials such as water or from organic materials such as sugar in reactions catalyzed by enzymes. Hydrogen produced by microorganisms is called biohydrogen. There is currently no commercial biohydrogen production process available.

The most attractive hydrogen production method for industrial applications is using photosynthetic microbes. These microorganisms, such as microscopic algae, cyanobacteria, and photosynthetic bacteria, use sunlight as an energy source and water to generate hydrogen. Hydrogen production by photosynthetic microbes holds the promise of generating a renewable hydrogen fuel, because there are large amounts of available sunlight and water. One exciting opportunity, with the potential for near-term practical applications of photosynthetic bacteria, is to use them as catalysts in the dark conversion of carbon monoxide (CO) and H_2O into H_2. This type of process is called a microbial shift reaction. The effluent gas from the experimental shift-reaction apparatus has been directly injected into small fuel cells and shown to be capable of generating enough electricity to power small motors and lamps. CO for use in the shift reaction can be produced by other microorganisms.

Ethanol

Ethanol is an alternative fuel that is mainly produced by microbial fermentation of starch crops (such as corn, wheat, and barley) or sugarcane. In the United States, most ethanol is produced by yeast (fungal) fermentation of sugar from cornstarch. With increasing energy demands and oil prices, ethanol becomes a valuable option as an alternative transportation fuel. In 2005, Congress passed an energy bill that required gasoline sold in the

United States to be mixed with ethanol. Nearly all cars can use fuel that is 10 percent ethanol (E10). Blending ethanol with gasoline oxygenates the fuel mixture, which as a result burns more completely and produces fewer harmful CO emissions. However, ethanol has about two-thirds the energy content of gasoline by volume, so vehicles running on an ethanol-gasoline mixture must be refueled more often. Ethanol is also more expensive than gasoline.

Carcinogenic aldehydes such as formaldehyde are produced when ethanol is burned in internal combustion engines. Moreover, an 85 percent/15 percent ethanol/gasoline mix (E85), which is also widely used, requires specially equipped "flexible fuel" engines. Only a fraction of the cars driven in the United States are equipped with such engines. Ethanol produced from cornstarch will help transition smoothly from an oil-based economy to an alternative-fuel-based economy. This transition will help reduce U.S. reliance on foreign oil, but it will not do much to slow global warming, will compete with food sources, and, therefore, does not represent a potential long-term energy solution. Ethanol derived from the cellulosic portion of plants offers a better alternative to cornstarch-based fuel. Cellulosic plant matter is the most plentiful biological material on Earth. Polysaccharides made of sugars can be fermented into ethanol. However, current methods of converting cellulosic material into ethanol are inefficient, so intensive research and development efforts are required.

Biodiesel

Biodiesel is a fuel obtained mainly from vegetable oil, such as soybean oil or restaurant greases. It is produced by transesterification of oils, a simple chemical reaction with alcohol (ethanol or methanol) that is catalyzed by acids or bases (such as sodium hydroxide). Transesterification produces alkyl esters of fatty acids, as well as glycerol. Biodiesel performs similarly to diesel and can be used in unmodified diesel engines of trucks, tractors, and other vehicles. Rudolf Diesel designed his first diesel engine to run on such materials, and he himself used peanut oil to power it. Biodiesel is often blended with petroleum diesel in percentages of 2, 5, or 20 percent biodiesel. It can be used as a pure fuel, but it is not suitable for such use in winter, because it thickens in cold temperatures.

Europe is the world's number one producer of biodiesel, which it generates primarily from canola oil. In the United States, biodiesel comes mainly from soybean plants. Other vegetative oils that have been used in biodiesel production are corn, sunflower, cottonseed, and rapeseed oils.

When petroleum prices increase, there is an increase in crop-based biodiesel production. Even if all crops were used as fuel, however, only a small percentage of petroleum-derived diesel fuels would be replaced. Changing from food crops to fuel crops would lead to serious food shortages and increased poverty. Moreover, production of biodiesel from crops would not significantly reduce GHG emissions so long as crop production is itself driven by burning fossil fuels. Emissions are produced by applying fertilizers to feedstock crops, transporting the feedstock to factories, processing the feedstock into biodiesel, and transporting the biodiesel to its point of use.

From the point of view of global warming, it makes the most sense to produce biodiesel from waste oil, grease, or other noncrop and nonfood sources. Waste vegetable oil is a good choice for biodiesel production, but its availability is problematic. Biodiesel from waste oil can be produced and used only locally. Another possible source for biodiesel production is microscopic algae. Research conducted by the U.S. Department of Energy Aquatic Species Program from the 1970's to the 1990's demonstrated that many species of algae produce sufficient quantities of oil to become economical feedstock for biodiesel production. The oil productivity of many microalgae species greatly exceeds the productivity of the highest-yielding oil crops. Thus, microalgal oil content can exceed 80 percent per cell dry weight, and oil levels of 20-50 percent are quite common.

Algae use oil (lipids) as a storage material. They are species of diatoms, golden algae, haptophates, and some green algae. For example, oil content in species of the green alga *Chlorella* is 28-55 percent. Compared to other crops, these green algae can grow their mass within hours. They can grow everywhere: in oceans, rivers, lakes, the snow of mountaintops, forests, and desert soils, and on rocks. They do not need much for growth, just sunlight, water, CO_2, and small quantities of mineral salts. Because—like plants—they use CO_2 to grow, algae can remove CO_2 from Earth's atmosphere, reducing the concentration of GHGs.

Cropland and potable water are not necessary for microalgae cultivation, since microalgae can grow in wastewater. Industrial production of microalgae has been achieved in open "raceway" ponds of some thousand square meters in size. These systems suffer from severe limitations, such as lack of temperature control, low attainable cellular concentrations, and difficulty in preventing contamination. Some companies concentrate on building algal photobioreactors.

Photobioreactors are named for the fact that light is the essential component for growing algae. They are closed systems comprising an array of transparent tubes or tanks in which microalgae are cultivated and monitored. The main challenge in photobioreactor design is to create a simple, inexpensive, high-cell-density, energy-efficient photobioreactor that is scalable to industrial capabilities. Photobioreactors should also provide the most efficient utilization of solar energy and allow the monitoring of culture purity and basic process parameters.

Photobioreactors can be as simple as plastic bags floating on water or thin PVC tubes allowing deep light penetration. To be used outdoors, such photobioreactors require a cooling system (such as a water sprinkler or heat exchanger) during summer and early fall, and algal cultures must be mixed to achieve uniform illumination of all cells. The future of algal biodiesel is a bright one. Therefore, development of biodiesel from algae is among the most popular energy alternatives.

Biomass

Plants and algae convert the energy of the Sun and CO_2 into energy, which they store in their biomass. Biomass burning in the form of wood is among the oldest forms of energy used by humans. Using biomass as a fuel source does not result in net CO_2 emissions, since burning biomass will only release the amount of CO_2 it absorbed during plant growth (providing its production and harvesting are sustainable). One example of biomass as a modern alternative energy source is biogas (mainly methane) production by microbial digestion of biological waste material. In India, over one million biogas plants of various capacities have been installed.

China has millions of small household digesters, which are used mainly for cooking and lighting.

Context

Alternative fuels became popular as a result of the oil embargo in the 1970's. Then, when oil prices dropped, alternative energy technologies were removed from the national agenda. As concerns about global warming have increased, however, the interest in alternative fuels has become reignited. Moreover, high oil and natural gas prices have made the need to develop alternative fuels all the more urgent. Alternative fuels under development will help protect the global climate and reduce U.S. reliance on foreign oil. These fuels still require long-term scientific, economical, and political investments, but clean, alternative fuels could become humanity's main energy source by the end of the twenty-first century.

Sergei Arlenovich Markov

Further Reading

Chisti, Yusuf. "Biodiesel from Microalgae." *Biotechnology Advances* 25 (2007): 294-306. Review on biodiesel generation from microalgae.

Nebel, Bernard J., and Richard T. Wright. *Environmental Science: Towards a Sustainable Future.* Englewood Cliffs, N.J.: Prentice Hall, 2008. Several chapters describe alternative fuels.

Service, Robert F. "The Hydrogen Backlash." *Science* 305, no. 5686 (August 13, 2004): 958-961. Good discussion about future hydrogen economy and the maturity of hydrogen power.

Wald, Matthew L. "Is Ethanol for the Long Haul?" *Scientific American* (January, 2007): 42-49. Comprehensive article about the future of ethanol fuel.

York, Richard. "Do alternative energy sources displace fossil fuels?." *Nature Climate Change* 2, no. 6 (2012): 441-443. No. Globally over the last half century, each unit of energy from non-fossil fuel sources has displaced less than one unit of fossil fuel energy.

See also: Biofuels; Clean energy; Energy from waste; Energy resources and global warming; Ethanol; Fossil fuels and global warming; Hydrogen power; Renewable energy; Solar energy; Wind power.

Gaia hypothesis

Category: Environmentalism, conservation, and ecosystems

Definition

In 1979, British atmospheric chemist James E. Lovelock and his collaborator, University of Massachusetts, Amherst, microbiologist Lynn Margulis, postulated that the chemical composition of Earth's atmosphere and oceans was actively molded and maintained by the planet's living organisms. Up to this point, scientists assumed that life adapted to the changing terrestrial environment as needed but did not mold Earth into a life-supporting planet. Lovelock's hypothesis turned this assumption around and asserted that biological organisms not only had actively transformed Earth into a life-friendly place but also continued to make it a life-friendly planet. Novelist William Golding suggested that Lovelock call his hypothesis the Gaia hypothesis in deference to the Greek goddess who gave birth to the gods.

Lovelock went further, describing Earth and its associated biosphere as one large, living, self-regulating organism. While many scientists ridiculed this characterization as a kind of neopagan, New Age religion, some environmental activists found in it a powerful metaphor. Speaking of Earth as alive and even sentient provided a strong emotional appeal for environmental causes. Describing environmental degradation as "wounding" Earth conveyed a strong moral impetus for ecological concerns.

Significance for Climate Change

The Gaia hypothesis proposes that biologically mediated, negative feedback mechanisms contribute to environmental homeostasis and make the environment suitable for life. Perhaps most controversially, the Gaia hypothesis further argues that these negative feedback mechanisms arise by means of neo-Darwinian selection.

The surface temperature of the Earth provides an example of one such mechanism. The global surface temperature has remained relatively constant, even though the energy output of the Sun

A 2005 photograph of James Lovelock, scientist and author best known for the Gaia hypothesis. (Wikimedia Commons)

has increased by 25 to 30 percent since life began on the planet. Temperature regulation results from a decrease in the atmospheric concentration of the greenhouse gas (GHG) carbon dioxide (CO_2). CO_2 reduction is due to bacteria and plants, which metabolize CO_2 into biomass. Furthermore, increased atmospheric CO_2 levels increase plant carbon sequestration and removal of CO_2 from the atmosphere. Increased carbon utilization also increases the number of marine, carbon-utilizing algae called coccolithophores. When coccolithophores die, they release dimethyl sulfide, which nucleates cloud formation. Increased cloud cover, which was initially induced by increased atmospheric CO_2, further cools Earth. Other life-dependent homeostatic processes are also thought to regulate the composition of Earth's atmosphere and the salinity of the oceans.

Some climate scientists favor the Gaia hypothesis, while others are highly critical of it. Virtually all Earth scientists recognize that living organisms have significant effects on the physical and chemical aspects of the environment. However, biologically mediated feedbacks are not intrinsically homeostatic, since some of them can destabilize the environment and are deleterious to life. For example, increased warming increases respiration rates of soil-based organisms, which increases CO_2 release from soils. Increased soil carbon release increases the greenhouse effect and global warming. This positive feedback system destabilizes the terrestrial environment. Furthermore, genomic studies of plants and animals have failed to reveal Gaian global feedback mechanisms. Likewise, natural selection affects individual organisms and is not a global process. Similarly, natural selection does not act with foresight.

With respect to global climate change, Earth, according to the Gaia hypothesis, should use a series of feedback mechanisms to regulate global temperatures. Thus, the freezing of the poles covers them with ice, which reflects the rays of the Sun and cools the planet. Likewise, increases in atmospheric CO_2 cause increased marine algal growth, which removes CO_2 and decreases greenhouse warming.

However, human intervention has short-circuited these negative feedback mechanisms. Industrial pollution of the oceans has introduced increased amounts of mercury (Hg), which prevents algal growth and destroys the primary means of regulating atmospheric CO_2 levels. Increased CO_2 levels cause increase greenhouse-based heating of the planet, which in turn melts the polar ice caps and decreases the reflectivity of Earth's surface, further heating the Earth.

In one sense, the Gaia hypothesis is strongly optimistic, since it strongly asserts that Earth and at least some of its resident life will adapt and survive. However, this optimism does not include humanity, since the environment of Earth after such dramatic climate change will probably be highly inhospitable to human life.

Michael A. Buratovich

Further Reading

Lovelock, James. The Ages of Gaia: A Biography of Our Living Earth. New York: Norton, 1995. Print.

Lovelock, James. Gaia: A New Look at Life on Earth. New ed. New York: Oxford UP, 2000. Print.

Lovelock, James. The Revenge of Gaia: Earth's Climate Crisis and the Fate of Humanity. New York: Basic Books, 2007. Print.

Margulis, Lynn. Symbiotic Planet: A New Look at Evolution. New York: Basic Books, 1999. Print.

Ruse, Michael. *The Gaia hypothesis: science on a pagan planet*. University of Chicago Press, 2013. The story of James Lovelock's Gaia hypothesis, contrasting its unfavorable reception by scientists with its favorable reception by the public.

Schneider, Stephen H., et al., eds. Scientists Debate Gaia: The Next Century. Cambridge: MITP, 2008. Print.

Gasoline prices

Categories: Fossil fuels; economics, industries, and products

It is unclear what effect climate policy will have on gasoline prices. In order to limit the atmospheric CO_2 concentration, gasoline would most likely have to be made more expensive in order to reduce consumption.

Key concepts

gasoline: a fuel, refined from oil, that is used to power automobile and other engines

oil embargo: a refusal on the part of oil-producing nations to sell oil to other nations

Organization of Petroleum Exporting Countries: group of major oil-producing nations that acts in concert to set oil prices and trade policy

peak oil: the point at which oil availability and production reaches its zenith, before the Earth's oil resources begin either to dwindle or to become prohibitively expensive to exploit

Background

In the early 1900's, when gasoline was first finding use as an automotive fuel, the inflation-adjusted cost of a gallon of gasoline was approximately $3.00. Save for a temporary rise in fuel prices during World War II, the price of gasoline generally declined from 1919 to 1973, when the price per gallon of regular gasoline stood at $1.20 per gallon.

The post-1973 history of gasoline prices has been considerably more volatile. Prices began trending upward in 1973 in response to an oil embargo by the Organization of Petroleum Exporting

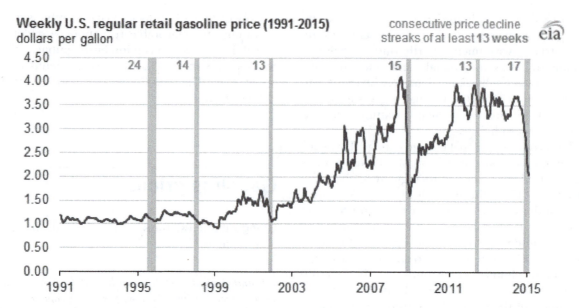

Weekly U.S. regular retail gasoline price (1991-2015)
dollars per gallon

consecutive price decline streaks of at least 13 weeks

Average gasoline prices in the U.S., 1991-2015. There was a steep drop in 2008 that was short-lived. The 2015 drop has more complex causes, including deliberate actions by Saudi Arabia to undercut rivals and extremist groups, and a rise in U.S. energy production due to increased unconventional recovery methods. (U.S. Energy Information Administration)

Countries (OPEC), with prices peaking at approximately $2.00 in 1982. After that nearly ten-year increase, there was an additional price spike corresponding to unrest in Iran, during the Iranian Revolution, when prices soared above their previous historical high of $3.00 per gallon. Subsequently, prices declined throughout the 1990's, bottoming out at $1.10 per gallon in 1998.

Prices rose sharply after 1998, nearly quadrupling to approximately $4.00 per gallon by June, 2008, then fell once again. Despite the rise in the real cost of gasoline, gasoline expenditure as a share of household income actually declined between 1950, when it was approximately 5 percent, to 4 percent in 2007.

Effects of the 1973 OPEC Embargo

In 1973, the Arab members of OPEC, as well as Egypt and Syria, stopped shipping oil to Israel and countries that were supporting Israel in the Yom Kippur War (October 6-26, 1973), including the United States, some countries in Western Europe, and Japan. In addition, OPEC reduced the production of crude oil sharply, driving up world oil prices. The world price for a barrel of oil nearly doubled in real terms from 1972 to 1974, with gasoline prices following suit.

Contrary to popular belief, long lines and gasoline shortages in the years of the OPEC embargo were not caused by a shortage of crude oil, which is a fungible commodity traded on a world market. Rather, gas lines and shortages were the result of price controls on gasoline emplaced by President Richard Nixon earlier in 1973, which prevented the rising price signal from reducing demand for refined gasoline.

With the nominal price of oil quadrupling from 1973 to 1974, oil-producing countries significantly increased their revenue, while oil-importing countries, particularly the United States, saw their economies decline. The U.S. economy shrank in 1974 and 1975, and growth in gross domestic product (GDP) also decreased significantly in Western Europe during that time, hitting negative growth rates in France, Germany, Italy, and other nations.

Causes and Effects of Price Changes

Major shifts in the price of gasoline have generally corresponded to the level of unrest in the oil-producing countries of the Middle East. Prior to OPEC's emergence in 1960 as an oil cartel, prices were generally kept low by Western-owned oil companies, particularly in Texas.

Changes in the price of gasoline lead to changes in consumer behavior, especially as it relates to automobile selection. After the 1973 oil crisis, American consumers began buying smaller cars, and Congress mandated fleet efficiency standards for American car companies in 1974 that further expanded small car offerings in the U.S. market. By contrast, the period of low gasoline prices in the late 1980's and 1990's allowed consumers to purchase larger vehicles, including pickup trucks and sport-utility vehicles.

High gasoline prices have also spurred governmental and market investment in nongasoline alternatives, such as electric vehicles and vehicles powered with natural gas, propane, and biofuels.

Peak Oil

The peak oil hypothesis is the idea that because oil is a nonrenewable resource, global oil resources will ultimately be depleted, and oil production will pass its peak and enter a permanent decline. The concept that conventional oil resources are finite is not controversial; the arguments surrounding peak oil relate not to whether but to when oil will peak and what the effects will be. Some believe that oil production has peaked or soon will, and that technology will not progress fast enough to allow people to maintain their standard of living. Others point to significant oil reserves and technological advances that allow development of nonconventional petroleum-based fuels, such as those from tar sands, which may make declining conventional resources potentially less economically damaging.

Several oil companies have commented on when they think oil resources will peak and what the implications will be. Exxon executives have argued that the oil resource base is constantly changing as resources become exploitable via new technology, and the company ran a public relations campaign in 2006 declaring that "[w]ith abundant oil resources still available peak production is nowhere in sight." Other executives and campaigns have been less optimistic. Chevron executives have said that peak oil will happen (that is, oil production will, eventually, be forced to decline permanently), but that event does not need to be a disaster. Still other executives have voiced opinions ranging from the alarmist to the sanguine.

Climate Change and Global Warming

The burning of fossil fuels is a major producer of carbon dioxide (CO_2), which most scientists believe is a contributor to global climate change. There are two major sets of proposals to mitigate CO_2 emissions: One involves limiting the amount of CO_2 emissions permissible and selling emissions rights; the other involves taxing CO_2 emissions. Both of these strategies, in order to be effective, would have to raise the price of emitting CO_2 emissions significantly, which would also raise the price of gasoline.

It is unclear at this point what effect climate-change-related policy will have on gasoline prices. In order to keep the amount of CO_2 in the atmosphere under levels that many scientists think would be dangerous, gasoline would most likely have to be made more expensive in order to reduce consumption. However, it is unknown whether policies to do so will become widespread. Also, if affordable, carbon-neutral fuels are developed, demand for gasoline (and its price) will likely decline, stimulating consumption.

Kenneth P. Green

Further Reading

Allcott, Hunt, and Nathan Wozny. "Gasoline prices, fuel economy, and the energy paradox." *Review of Economics and Statistics* 96, no. 5 (2014): 779-795. Sales data show that consumers rate the present cost of cars slightly more than the prospect of higher gasoline prices in the near future. Also, auto sales respond to changes in gasoline prices with about a six month delay.

Bryce, Robert. *Gusher of Lies: The Dangerous Delusions of Energy Independence.* New York: Public-Affairs, 2008. Bryce, an energy journalist and editor, examines whether or not calls for energy independence stand up to scrutiny. Systematically breaks down such claims, paying special attention to biofuels, which he argues are little more than a con game.

Deffeyes, Kenneth S. *Beyond Oil: The View from Hubbert's Peak.* New York: Hill and Wang, 2006. Deffeyes, a geologist, is a leading proponent of the idea that oil production is peaking and will soon enter a terminal decline as existing oil fields run dry and large new finds become increasingly scarce.

Green, Kenneth P. "Bringing Down Gas and Oil Prices." *Environmental Policy Outlook* 3 (May, 2006). Short article that discusses why gasoline and oil prices rose to new heights in 2006 and what can be done to bring such prices down.

Huber, Peter W., and Mark P. Mills. *The Bottomless Well: The Twilight of Fuel, the Virtue of Waste, and Why We Will Never Run Out of Energy.* New York: Basic Books, 2005. Argues that, contrary to those who subscribe to the peak oil theory, the world will not run out of affordable oil, since the steady move toward electrification will reduce demand for oil.

Smil, Vaclav. *Energy at the Crossroads: Global Perspectives and Uncertainties.* Cambridge, Mass.: MIT Press, 2005. Authoritative text on the history of energy use, the challenges attendant on that use, and potential solutions to the environmental impacts of energy use.

See also: Fossil fuel reserves; Fossil fuels and global warming; Fuels, alternative; Hubbert's peak; Oil industry; Organization of Petroleum Exporting Countries.

General circulation models

Categories: Meteorology and atmospheric sciences; oceanography; science and technology

Definition

The dynamics and physics that govern the atmosphere and oceans are the result of many cycles and processes interacting. A general circulation model (GCM) is a collection of mathematical equations that represent these cycles and processes using computational algorithms that can be solved on a supercomputer. In general, the components of the model, such as the atmosphere and ocean, are modeled separately and interact only at their boundaries—for instance, at the interface between the atmosphere and the ocean. GCMs include equations to solve the dynamics (motions) of the atmosphere and ocean, as well as equations and parameterizations to represent atmospheric, oceanic, sea-ice, and terrestrial physics. Processes represented by parameterizations combine equations with proportionality constants, correlations, and table lookups based on observational and experimental data.

In order to be solved on a computer, the equations and variables incorporated in a GCM must be discretized, which is typically accomplished by dividing up the atmosphere and ocean on a grid and defining the key variables such as temperature and humidity at defined points on this grid. Using variable values and the governing and parameterization equations, one can compute new variable values a relatively short time in the future. This integration process is repeated to generate longer time sequences. Validation studies that compare the results of the model computation to experimental and observational data are required to acquire confidence in the accuracy of the simulation.

Significance for Climate Change

GCMs are the fundamental tools for predicting the future evolution of Earth's climate and the impact that anthropogenic greenhouse gas (GHG) emissions or other environmental effects will have on future temperatures, rainfall, and other climatic conditions. GCMs arose from efforts at numerical weather prediction (NWP). The English mathematician Lewis Fry Richardson is credited with making the first NWP calculation when he attempted to predict the weather six hours into the future during World War I. Because he lacked computers, the calculations took six months to complete, but the basic approach was the same that would be used four decades later in NWP. The first working GCM is attributed to Normal Phillips, who completed a two-layer hemispheric atmospheric model in 1955. The late 1960's and the 1970's saw the introduction of coupled atmosphere-ocean models and the first use of a GCM to study the effects of carbon dioxide (CO_2) and pollutants in the atmosphere. From that point forward, GCMs were recognized as critical components in the study of climate change.

Four-component models (comprising atmosphere, ocean, sea ice, and land) are often referred to as atmosphere-ocean general circulation models (AOGCMs). The highest-resolution AOGCM grids have horizontal spacings of 1° (roughly 100 kilometers) and up to sixty vertical layers in the atmosphere. A typical use of these models takes an average of current conditions as the initial conditions then integrates the GCM computationally through future decades, while different parameters and boundary conditions are changed to account for different scenarios. Based on the results of these computations, the likely range of future temperatures and other future climatic conditions is determined to the extent possible given the accuracy of the model and the related inputs.

GCM simulations are also used to compute the magnitude of anthropogenic versus natural forcing of the environment in order quantitatively to isolate the human contribution to climate change. They are also used to assess the future impact of plans to reduce global GHG emissions. Key statements about global warming, such as the anticipated rise in temperature and sea level and changes in precipitation and storm patterns due to the release of anthropogenic GHGs over the next century, are the product of general circulation models.

Given their importance in the debate over global warming, GCMs are subject to much scrutiny. These models are not complete, with some physical processes generally not implemented in the models and other representations being the source of argument and uncertainty. Critical points of contention include the parameterization of cloud-radiation feedbacks and cloud microphysics, including precipitation; the impact of changes in incoming solar radiation; and the general variability of results when using different GCMs. A notable missing process in most GCMs is an explicit numerical treatment of the carbon cycle, causing most global warming studies to rely on predetermined changes in atmospheric GHG content. Ongoing work to improve modeling of physical processes, achieve higher-resolution simulations on more powerful computing platforms, and increase the use of validation studies will generate the mechanisms by which these concerns will be addressed. Even with their limitations, GCMs remain the primary tool for analyzing the future of Earth's climate.

Raymond P. LeBeau, Jr.

Further Reading

Houghton, John. *Global Warming: The Complete Briefing.* 4th ed. New York: Cambridge University Press, 2009. This overview of global warming by a leading scientist in the field includes an introductory discussion of climate modeling. Figures, references, glossary, index.

Leroux, Marcel. *Global Warming: Myth or Reality? The Erring Ways of Climatology.* New York: Springer, 2005. French climatologist Leroux, a skeptic of current global warming theory, presents criticisms of current general circulation models as part of this text. Figures, bibliography, index.

Randall, David A., ed. *General Circulation Model Development.* San Diego, Calif.: Academic Press, 2000. Collection of papers prepared for the 1998 Retirement Symposium of UCLA professor Akio Arakawa, a leading researcher in the development of GCMs. Figures, references, index.

Randall, David A., et al. "Climate Models and Their Evaluation." In *Climate Change, 2007—The Physical Science Basis: Contribution of Working Group I to the Fourth Assessment Report of the Intergovernmental Panel on Climate Change,* edited by Susan Solomon et al. New York: Cambridge University Press, 2007. IPCC assessment of the current state of general circulation models. Figures, tables, references.

Stevens, Bjorn, and Sandrine Bony. "What are climate models missing?." *Science* 340, no. 6136 (2013): 1053-1054. Adding more complexity to climate models has not increased our understanding. Better understanding of basic processes like moisture transfer from ocean to atmosphere is necessary to improve climate models.

See also: Bayesian method; Climate models and modeling; Climate prediction and projection; Ocean-atmosphere coupling; Parameterization.

Geographic Information Systems in Climatology

Categories: Geology and geography; Science and technology

Definition

A geographical information system (GIS) is, fundamentally, a map that can be queried. It is what a map becomes when computer technology enables users to make individualized choices relating to its mode of representation and the data upon which it is based. In an electronic, digital world, it becomes possible to access, represent, and analyze information in multiple, integrated ways, and geographical information systems bring these options to bear on spatial and related data. Complex relationships and processes occurring in space are thereby subject to display and investigation. GIS makes it easier to discover facts about how large-scale processes, such as weather and climate change, work and to visualize the causal roles of location (both proximity and relative distance). GIS thus serves as an interface between the basic physical sciences, such as physics, chemistry, biology, and geology, and the sciences of human social development. As Lee Chapman and John E. Thornes remark, "the assessment and monitoring of the effects of climate change is truly a multidisciplinary exercise of which GIS provides a pivotal unifying role."

GIS is usually regarded as having six principal components. First, people make and use the system and ask the questions. Second, data are identified as relevant for answering the questions. Third, computer software programs provide for display and analysis of the data. This includes not only GIS software but also databases and programs that permit imaging, drawing, statistical manipulation, and so on. Fourth, computer hardware runs the software. Hardware capabilities affect processing speed, ease of use, and the type of output available. Fifth, procedures define how the information is processed, interpreted, and used. Finally, a network links together all these elements and their real-world applications. The Internet permits GIS applications to draw upon

Example of hardware equipment for forest inventories: GPS and laser rangefinder for mapping connected to a field rugged computer. (Wikimedia Commons)

data warehouses of all kinds regardless of their physical location.

Because GIS involves the use of data drawn from many different sources, obtained and organized in many different ways, issues of quality control arise. For this reason, GIS must include information about its information—so-called metadata. Metadata provide answers to the traditional journalist's questions—who, what, where, when, why, and how—and thereby enable users to assess the relevance, reliability, and comparability of different data sets. The manner in which concepts are defined, operationalized, and measured can affect how data are recorded and what the data mean.

Significance for Climate Change

Geographical information systems play an increasing role in all fields that employ spatial data, including agriculture, ecology, forestry, health and medicine, weather forecasting, hydrology, transportation, urban planning, energy generation and policy, and climatology. By utilizing concepts such as adjacency, area, direction, coincidence, connectivity, containment, direction, length, location, and shape, the properties of geographical entities and their relationships (indeed, any data that can be mapped) can be represented in simplified models by mathematical coordinates. Geographical information systems provide ways of discovering, organizing, and presenting data about past and current climate conditions relevant to such models. These data thus comprise the basis on which competing climate models project possible future climate-change scenarios; they also provide part of the basis for assessing the accuracy of the models.

The planet as a whole can be represented as an integrated series of boxes in a stack of checkerboards, or of layers in an onion, and the data relating to each component can serve as input into mathematical algorithms representing the constraints of physical laws and the operation of natural processes. The results can then be interpreted as showing something about how these data and processes interact to produce the phenomena represented. Representations of aspects of atmosphere, land, and ocean can be combined to model meterological processes in general circulation models. Running such models repeatedly, with different assumptions about inputs and processes, results in different outcomes, and these form the basis for projections about the climate's future. Syukuro Manabe, of the National Oceanic and Atmospheric Administration (NOAA), was a pioneer of such modeling. The Goddard Institute for Space Studies (GISS) also has been a leader in the field, and its head, James E. Hansen, has played a leading role in the debate about global warming.

As geographical information systems and the remote-sensing data they so often rely on play an ever-larger role in understanding the planet's dynamic processes, it is important to remember the many ways in which data can be misleading. Joseph Farman's 1985 discovery of the Antarctic ozone hole can serve as an object lesson in this regard: As related by science journalist Fred Pearce, satellites had seen the ozone hole forming and growing over Antarctica all along, even before Farman had spotted it. But the computers on the ground that were analyzing the streams of data had been programmed to throw out any wildly abnormal readings.

Critics of GIS and other modeling technology emphasize the limitations of the climate-modeling process as a basis for reliable long-term climate forecasting. Models must include representations of natural processes that may be understood only approximately, and they must incorporate simplified parameters to stand for complex interactions in large areas of atmosphere, land, and ocean. When models that were originally developed to cover discrete phenomena are integrated, further uncertainties are introduced. Skeptics sometimes even suggest that models are designed in a way that insures desired outputs through the manipulation of parameters.

Proponents of computer modeling, in contrast, view limitations as temporary, representing natural steps in scientific progress, as models are revised in the light of researchers' improved understanding of salient factors. Such proponents often urge, however, that policy makers cannot wait patiently for the modeling process to achieve perfection, since by the time scientists can provide unassailable data, it may be too late for action. Thus, decisions, like computer models themselves, must be based on the best data and procedures available at the time.

Rebecca S. Carrasco

Further Reading

Chapman, Lee, and John E. Thornes. "The Use of Geographical Information Systems in Climatology and Meteorology." *Progress in Physical Geography* 27, no. 3 (2003): 313-330. A survey of GIS applications in climatology and meteorology; extensive bibliography.

ESRI. What is GIS? http://www.esri.com/what-is-gis (3 July 2016) Web page of one of the principal vendors of GIS software. Although obviously intended as a promotion of its products, the site includes explanations of GIS concepts and some interactive demonstrations.

Green, David, and Terry Bossomaier. Online GIS and Spatial Metadata. New York: Taylor &

Francis, 2002. Discussion of GIS, Web-based technology, and environmental informatics.

Kasianchuk, Peter, and Marnel Taggart, eds. GIS Education Solutions from ESRI: Introduction to ArcGIS I. Redlands, Calif.: ESRI Press, 2000-2004. A technical introduction to the use of GIS.

Kolbert, Elizabeth. Field Notes from a Catastrophe: Man, Nature, and Climate Change. New York: Bloomsbury, 2006. A popular treatment of the evidence for global warming; provides a good brief account of how climate models work.

Longley, Paul A., et al. Geographic Information Systems and Science. 2d ed. Hoboken, N.J.: John Wiley and Sons, 2005. An influential textbook on both GIS and its underlying mechanisms.

Maslin, Mark. Global Warming: A Very Short Introduction. New York: Oxford University Press, 2004. A concise introduction to global warming issues, particularly helpful on climate models.

Monmonier, Mark. How to Lie with Maps. 2d ed. Chicago: University of Chicago Press, 1996. A basic introduction to the complexities of data representation.

Ormsby, Tim, et al. Getting to Know ArcGIS Desktop. Redlands, Calif.: ESRI Press, 2001. An introductory workbook for the Environmental Systems Research Institute's GIS software.

Pearce, Fred. With Speed and Violence: Why Scientists Fear Tipping Points in Climate Change. Boston: Beacon Press, 2007. A leading science journalist surveys the scientific case for rapid, rather than gradual, global warming and its alarming consequences.

Geoid

Category: Geology and geography

Definition

The geoid is a mathematical model of the Earth in which the geoid surface coincides exactly with the average elevation of the surface of the planet's oceans. This model surface is extended over the globe, even through the continents, to form the geoid. This surface is smooth and is defined by mathematical calculations and numerous gravitational measurements. The geoid is a physical representation of the body of the Earth and is different from other representations of the Earth, for example the Earth's reference ellipsoid. Only since the advent of modern geomathematical computing has the geoid been calculated with high precision.

The geoid is sometimes referred to as an equipotential surface, which means that the force of gravity is everywhere the same on the geoid and that the force of gravity is everywhere perpendicular to the surface of the geoid. The form of the geoid is slightly irregular but is everywhere very smooth. The geoid has a history dating to the nineteenth century, when several scientists worked on the problem of computing a mathematical description of the Earth, or computing the shape of the geoid. Since that time, more precise measurements and better means of computing have permitted progressively more precise definitions and understandings of the geoid and its complexity.

For the average person, the geoid's most notable impact on daily life is the precision of handheld and vehicle-mounted global positioning satellite (GPS) receivers. The modern geoid allows these devices to provide extremely accurate physical locations on the Earth's surface, sometimes including elevation. However, there are more applications for the geoid than personal GPS devices, including applications that are related to environmental and climate change. The word "geoid" is the root word for the term "geodesy," which is the study of the shape of the Earth, including ways to describe that shape and practical applications incorporating it.

Significance for Climate Change

The significance of the geoid for climate change lies in its potential for very precise calculation of elevation and elevation changes on Earth. These precise calculations are made with the aid of GPS systems, which are used to compute elevations in reference to the geoid. These satellites are part of the global navigation satellite system (GNSS) operational around the world. GPS elevation readings are taken in reference to an Earth-centered reference ellipsoid, because satellites orbit about the

Bench marks like this one help define the detailed shape of the earth, or geoid. The location of bench marks is surveyed to an accuracy of millimeters. The geoid is the mathematical shape that best fits the locations and elevations of hundreds of thousands of bench marks. [© Steven I. Dutch]

center of gravity of the Earth. Using corrections, GPS elevations are computed within GPS receivers in reference to the geoid.

Knowing precise elevations is very important in studies of climate and environmental change. For example, measurements of sea-level rise or fall are important in such studies. Also, changes in elevation of ice-covered regions of Earth may help scientists understand the broader range of changes occurring there. The elevation of the surface of the ocean may vary from place to place as a result of changes in the temperature and volume of moving water masses. All these data are useful in assessing the degree, rate, and extent of climate and environmental change on Earth.

The Gravity Recovery and Climate Experiment (GRACE) is a geoid-referencing project that involves two orbiting spacecraft that were launched by the United States in 2002. In addition to detailed measurements of Earth's gravity, GRACE has conducted several environmental and climate-related studies, including studies of the new, rapid melting of the Antarctic ice sheet. The ice sheet is melting at a rate, according to GRACE, that is 75 percent faster than it was at the end of the twentieth century. These results show that Antarctic ice loss rivals the ice loss in Greenland, which was studied and documented by GRACE data in 2006.

GRACE has studied the freshwater storage patterns and human activity of many continental areas on Earth, including China's Yangtze River, Australia, and parts of North America and Africa. In particular, GRACE data show that most continental areas are drying up or losing net stored water, findings that have long-term climatic implications affecting the future of human society. The location and distribution of water on Earth has a profound effect upon the distribution of life.

The more precisely defined the geoid becomes, the more Earth parameters can be measured precisely against it. For example, a more precisely defined geoid will help scientists understand better even very small changes in the elevation of the ocean surface, which can be useful in measuring changes in circulation patterns, tides, and heat transfer in the oceans. A precisely defined geoid is useful for hydrological and glaciological modeling, which in turn provides for enhanced climate modeling and analysis. Geoid definitions are dependent upon time-variable effects, which are an integral part of the definition. Modern geoid approximations are computer models based on numerous complex variables.

David T. King, Jr.

Further Reading

Hofmann-Wellenhof, Bernhard, and Helmut Moritz. *Physical Geodesy*. 2d ed. Berlin: Springer-Verlag, 2006. Primary textbook in the field of geodesy and the geoid. Very well illustrated with problems and solutions. The second edition contains an updated discussion of the modern geoid definition.

Hooijberg, Maarten. *Geometrical Geodesy: Using Information and Computer Technology*. New York: Springer, 2008. Details both the mathematics and the computer science principles behind geodesy.

National Ocean Service. What is the Geoid? http://oceanservice.noaa.gov/facts/geoid.html (3 July

2016). Simple explanation of the geoid with links to considerably more technical sites.

National Geodetic Survey. *Geodesy: Imagine the Possibilities.* Washington, D.C.: U.S. Department of Commerce, 2000. A short book that introduces the reader to geodesy and the geoid. Can be downloaded at the National Geodetic Survey Web page at no cost. Discusses various applications now and in the future.

See also: Climate models and modeling; Climate prediction and projection; General circulation models; Mean sea level; Sea-level change; Slab-ocean model.

Geological Society of America (GSA)

Categories: Organizations and agencies; Geology and geography

Mission

The Geological Society of America is a global professional society whose members are academicians, scholars, and experts in the geosciences. The mission of the society is to be a leader in the advancement of the geosciences, to enhance the professional growth of its members, and to promote the geoscieness in the service of humankind. To implement its mission, the society holds an annual meeting, organizes numerous conferences, publishes the research of geoscientists, and fosters dialogue with the public and with decision makers on relevant geoscience issues. The society has a growing membership of more than twenty-one thousand residing in eighty-five different counties. It also has a professional staff at its headquarters in Boulder, Colorado, officers elected from among leading scientists, six regional sections that are spread throughout North America and also meet annually, numerous committees, seventeen special-interest divisions that meet in conjunction with the Annual Meeting of the Society, and

"Geological field excursion to Harpers Ferry, West Virginia, April 30, 1897, following the George Huntington Williams Memorial Lectures delivered by Sir Archibald Geikie at Johns Hopkins University. (J.S. Diller/Wikimedia Commons)

forty-five affiliated societies with which it works in partnership.

Significance for Climate Change

At their October 2006 annual meeting in Philadelphia, members of the Geological Society of America adopted the following position statement as a reflection of their institutional and individual commitment to the subject of global climate change:

The Geological Society of America (GSA) supports the scientific conclusions that Earth's climate is changing; the climate changes are due in part to human activities; and the probable consequences of the climate changes will be significant and blind to geopolitical boundaries. Furthermore, the potential implications of global climate change and the time scale over which such changes will likely occur require active, effective, long-term planning. GSA also supports statements on the global climate change issue made by the joint national academies of science (June 2005), American Geophysical Union (December 2003), and American Chemical Society (2004). GSA strongly encourages that the following efforts be undertaken internationally: (1) adequately research climate change at all time scales, (2) develop thoughtful, science-based policy appropriate for the multifaceted issues of global climate change, (3) organize global planning to recognize, prepare for, and adapt to the causes and consequences of global climate change, and (4) organize and develop comprehensive, long-term strategies for sustainable energy, particularly focused on minimizing impacts on global climate.

A public forum on the topic of climate change was also held at this meeting, allowing the GSA to engage with laypersons to discuss the implications of climate change. The major speakers at the forum were Dr. Richard B. Alley of Pennsylvania State University and Dr. Robert Jackson of Duke University.

Donald W. Lovejoy

Further Reading

"Climate Change." Geological Society of America. Geological Soc. of Amer., Mar. 2013. Web. 23 Mar. 2015.

Fletcher, Charles. Climate Change: What the Science Tells Us. Hoboken: Wiley, 2013. Print.

Kitchen, David. Global Climate Change: Turning Knowledge into Action. Boston: Prentice Hall, 2014. Print.

Maslin, Mark. Climate Change: A Very Short Introduction. 3rd ed. New York: Oxford UP, 2014. Print.

Spellman, Frank R., and Melissa L. Stoudt. The Handbook of Geoscience. Lanham: Scarecrow, 2013. Print.

George C. Marshall Institute

Category: Organizations and agencies
Date: Established 1984
Web address: http://www.marshall.org

Mission

The George C. Marshall Institute is a nonprofit organization supported by foundations, industry, and the public to improve the use of science in making public policy. Its major emphasis is on issues of the environment and national security.

The mission of the George C. Marshall Institute (GMI) is to encourage the use of science in making public policy when science and technology are major components of that policy. The institute was founded in 1984 by two respected scientists, Frederick Seitz, the former president of the National Academy of Sciences, and Robert Jastrow, an astronomer and author. The institute differs from most other think tanks, because its board of directors includes a large proportion of leading scientists and the assessments it undertakes have a strong focus on the science used in making public policy. The activities of GMI in the environmental field include civic environmentalism and climate change and, in the national defense field, bioterrorism and missile defense.

GMI carries out its mission through published reports, workshops, roundtables, and collaboration with other organizations with similar goals, such as the Hoover Institution, Doctors for Disaster Preparedness, and Physicians for Civil Defense. It has a staff that organizes workshops and roundtables and also prepares assessments of issues connected to its mission.

The Washington Roundtable on Science and Public Policy was established by GMI to bring together scientific experts and government leaders to explore policy options connected to scientific and technological questions. Its aim is to insure that policy makers are aware of the science and the uncertainties involved in these questions.

GMI has a conservative bent and has been skeptical about many mainstream scientific assessments, including those of global warming. It was a strong supporter of the Strategic Defense Initiative and emphasized the uncertainties about the negative impacts of secondhand smoke. It gives more weight to the uncertainties in the scientific basis underlying public issues than do other think tanks. Many liberal organizations and the media have attacked GMI for the stands it has taken on these controversial policies.

Significance for Climate Change

Since it was established in 1984 GMI has published more than one hundred books, reports, roundtable discussions, and other documents on the science of climate change and its policy implications.

Among these are assessments of the validity of some scientific measurements of global warming; policy implications of climate change, including arguments against the cap-and-trade policy supported by some members of Congress; and critiques of the reports of the Intergovernmental Panel on Climate Change (IPCC). Its emphasis on science and the limits of knowledge led GMI to cosponsor the Petition Project, through which more than thirty thousand scientists have signed a petition urging the United States government to reject the Kyoto Protocol on global warming.

GMI reports have been instrumental in pointing out the scientific and statistical difficulties with the hockey stick model of global temperatures used in the third IPCC report. The model indicates that the global temperatures have risen in the shape of a hockey stick since the middle of the twentieth century and are expected to continue this rapid rise. A reevaluation by GMI shows that there are serious questions about both the data and the statistics used in this model.

The cap-and-trade policy would allow Congress to set an overall cap on greenhouse gases and allow industries to auction their allotments among themselves to meet the cap. The Marshall Institute has argued against this policy, because it would distort the economy by letting Congress arbitrarily set emission values for industries, and it could be susceptible to fraud.

Because many of these reports point out difficulties and uncertainties in the science and the policies advocated by other groups, GMI has been called a denier of global warming. GMI argues that before making catastrophic and possibly irreversible policy changes, the relevant science should be improved and its uncertainties reduced. It points out that there is no consensus among scientists about either the data or the models of global warming. There could be great danger to the economy and well-being of the world posed by setting controversial policies while the uncertainties remain high.

Raymond D. Cooper

Further Reading

Ball, Timothy. *The Science Isn't Settled: The Limitations of Global Climate Models.* Washington, D.C.: George C. Marshall Institute, 2007. GMI Washington Roundtable Report in which the limitations of climate models are discussed. Points out that many of the assumptions used in the models are questionable, with biases that tend to increase predicted warming. Twenty figures.

Dunlap, Riley E., and Peter J. Jacques. "Climate change denial books and conservative think tanks: exploring the connection." *American Behavioral Scientist* (2013): 0002764213477096. Most earlier books skeptical of climate change were produced by authors having ties to conservative think tanks but in recent years, self-published books have risen sharply in number. The vast majority of climate denial books have not undergone peer review.

George C. Marshall Institute. *Climate Issues and Questions.* 3d ed. Washington, D.C.: Author, 2008. Addresses twenty-nine questions about climate change, bringing to bear the best available scientific evidence. Discusses the IPCC process and results, scientific facts about climate change that are well established, and uncertainties about these issues. Figures, tables, endnotes.

Gough, Michael. *Politicizing Science: The Alchemy of Policymaking.* Palo Alto, Calif.: Hoover Institution Press, 2003. Describes how the manipulation of science has been used to advance particular political agendas, resulting in burdens on the economy and missed opportunities. Index.

Michaels, Patrick J., ed. *Shattered Consensus: The True State of Global Warming.* Lanham, Md.: Rowman & Littlefield, 2006. Collection of ten rather technical essays by the editor and nine other experts on climate change. Reviews and highlights the differences between what has been predicted and what has actually been observed about global warming. Figures, index.

See also: Air pollution and pollutants: anthropogenic; Hockey stick graph; Offsetting; Skeptics; U.S. legislation.

Geothermal energy

Category: Energy

As an alternative to fossil fuels, geothermal energy resources may be developed where conditions warrant. This technology requires proximity to resources and the infrastructure to ship energy to market. Unlike wind and hydropower, however, availability of geothermal energy does not depend on weather or precipitation.

Key concepts

heat: thermal energy, often measured in calories or joules

heat flow: the rate at which thermal energy flows through a surface

watt: a unit of power, frequently associated with electricity, equivalent to 1 joule per second

Background

Volcanic activity heats rock, imbuing it with thermal energy. Where such activity is recent or ongoing, naturally occurring water moving through hot rock may produce natural hydrothermal systems that need only a little engineering to generate electricity. Often there is enough energy for major commercial development. Other locales may have hot rock but little water. The technology needed to produce energy commercially from these systems has not yet been developed. If the cost of producing energy from other sources gets high enough, this technology may be pursued.

Geothermal

Heat flows from hot to cold. As a result of radioactive decay, as well as some residual heat from Earth's formation, the interior of the planet is hot compared to outer space, so heat flows upward from depth through the surface of the Earth. By measuring how rapidly temperature increases with depth and determining the thermal conductivity of rocks, scientists can determine the rate of Earth's heat flow: The planet's average heat flow is about 87 milliwatts per square meter, so the total heat loss from Earth is about 44×10^{12} watts. To put this in perspective, total human energy consumption is

about 15×10^{12} watts, and total solar energy reaching the Earth is about 240 watts per square meter, or 12.65×10^{15} watts.

Heat flow varies with location, and in some places it is high enough to generate usable power. In Italy, geothermal resources have been used since 1913, when a 250-kilowatt power station was constructed. Iceland produces over one-quarter of its electricity and over 80 percent of its home heating and hot water from geothermal resources. The Geysers geothermal field in Northern California produced 2×10^9 watts at its peak in 1987, with output generally declining since then but leveling off at about 8.5×10^8 watts. In these locations, naturally occurring water migrating through hot rocks turns to steam in sufficient volumes to power a steam turbine and to heat homes. Such conditions are uncommon and are even less commonly situated in locations near significant human populations. Unlike some energy sources, however, geothermal energy at the locations where it exists is available at any time.

Hot Dry Rock

There are many areas with high geologic heat flow but no naturally occurring steam. Efforts to develop these hot dry rock (HDR) resources began in 1970 at the Los Alamos National Laboratory. The project involved drilling a 4-kilometer-deep well at Fenton Hill, New Mexico, then injecting water under pressure to produce cracks in the hot rock at that depth. The cracks were expected to be penny shaped, and another well was drilled down to where the top of the penny was expected to be. Many technological challenges were successfully met; however, the vagaries of fracture propagation within a complex geological terrain were not adequately anticipated. After thirty years of work by some of the country's top geologists and engineers and over $180 million spent, the site was closed and the wells were cemented in. Many lessons were learned, including the realization that huge investments of human and financial capital may not be sufficient to harness a "free" source of energy.

Low-Temperature Geothermal Energy

Even where heat flow is not particularly high, geothermal systems can heat and cool buildings. These

Mammoth Hot Springs, Yellowstone National Park, Montana. Yellowstone is the largest and best-known geothermal area in the world. [© Steven I. Dutch]

low-temperature systems pump fluids through the subsurface, absorbing heat in the winter and dispensing it in the summer. This is possible because the Sun and atmosphere are responsible for heating the surface of the Earth, and at a depth of a few meters the temperature is close to the median surface temperature for a given location. Electricity is needed to operate the fluid pumps and the heat pumps, but a well-designed system can be considerably more efficient than one based on fossil fuels or electricity alone.

Context

Huge quantities of energy move through the Earth's systems, providing tempting possibilities for environmentally sound, sustainable energy production. Geothermal energy is one such avenue. In places where high heat flow and naturally occurring water coexist, this resource has often already been developed. In places where there is high heat flow but no water, the risks inherent in trying to put the thermal energy to use are great, as the investments required are enormous and there is no certainty of success.

Low-temperature geothermal systems for space heating and cooling present the same problems on a smaller scale. If there were a good chance that the investment costs of such systems would be paid back quickly in fuel savings, they would be extremely popular. Design improvements, manufacturing efficiencies, and economies of scale will bring down the cost of these systems, while resource depletion and environmental costs will increase the cost of fossil fuel systems. At some point, in some areas, the former may become more economical than the latter.

Otto H. Muller

Further Reading

Brown, Donald W., David V. Duchane, Grant Heiken, and Vivi Thomas Hriscu. *Mining the Earth's heat: hot dry rock geothermal energy.* Springer Science & Business Media, 2012. The history of the hot dry rock experiments of Los Alamos National Laboratory. Although technically feasible, issues of low output and reservoir lifetime remain.

Decker, Robert W., and Barbara Decker. *Volcanoes.* New York: W. H. Freeman, 2005. An elementary-level college textbook; chapter 15 provides a good overview of geothermal energy. Illustrations, figures, tables, maps, photos, bibliography, index.

Duffield, W. A., J. H. Sass, and M. L. Sorey. *Tapping the Earth's Natural Heat.* Washington, D.C.: U.S. Geological Survey, 1994. With plenty of color photographs, this circular shows what the potential is for geothermal energy development in the United States, but it also warns of some of the threats to the environment posed by geothermal development.

Pahl, Greg. *The Citizen-Powered Energy Handbook: Community Solutions to a Global Crisis.* White River Junction, Vt.: Chelsea Green, 2007. Includes good descriptions of low-temperature systems, both municipal and residential. Photos, figures.

Tester, J., et al. "The Future of Geothermal Energy: Impact of Enhanced Geothermal Systems (EGS) on the United States in the Twenty-first Century." *Final Report to the U.S. Department of Energy Geothermal Technologies Program.* Cambridge, Mass.: MIT Press, 2006. Written in a way that a general reader will find understandable, but with sufficient technical detail to satisfy scientists and engineers, this thorough 372-page report covers everything from heat flow maps to economic forecasts based on predicted learning curves. Maps and charts.

See also: Clean energy; Energy efficiency; Energy resources and global warming; Fossil fuels and global warming; Fuels, alternative; Renewable energy; Solar energy.

Germany

Category: Nations and peoples

Key facts

Population: 82,329,758 (July, 2009, estimate); 81,770,900 (2015 estimate)

Area: 357,021 square kilometers

Gross domestic product (GDP): $2.863 trillion (purchasing power parity, 2008 estimate), $3.842 trillion (purchasing power parity, 2015 estimate)

Greenhouse gas (GHG) emissions in millions of metric tons of carbon dioxide equivalent (CO_2e): 1,226 in 1990; 981.3 in 2007, 887.2 in 2012

Kyoto Protocol status: Ratified April 26, 2002

Historical and Political Context

Because of its central location in Europe, no other country shares more borders with European countries than Germany. Germanic people were fragmented by a number of large tribes. Following the Thirty Years' War (1618-1648), which devastated German lands and killed about 30 percent of the people, the Peace of Westphalia divided the Holy Roman Empire of the German nation into numerous independent principalities, with the Austrian Habsburg monarchy and the Kingdom of Prussia as the largest players. The long road to unification included the foundation of the German Confederation in 1814, the tariff union (Zollverein), the Revolution of 1848, and the first establishment of the German nation state in 1871 amid the Franco-Prussian War. Being a latecomer, Germany pursued an aggressive path in its quest for power in Europe and abroad, leading to two lost world wars, with comprehensive devastation and unspeakable crimes committed by Nazi Germany. Being forced to learn its lesson under Allied occupations, Germany was divided in 1949 into a socialist Eastern and a capitalist Western state, both of which became the battleground (and almost the nuclear battlefield) for the Cold War antagonists the United States and Soviet Union. After the spectacular fall of the Berlin Wall in 1989, triggered by Soviet premier Mikhail Gorbachev's reforms, Germany was reunified in 1990. United Germany took a

leading role in the further establishment of the European Union, adopting the European currency, the euro in 1999.

In 2009, Germany had the largest population in the European Union and had sixteen federal states. It was home to a large number of international migrants. Germany is a federal parliamentary republic with the world's third largest economy by nominal GDP and is the world's leading exporter of goods, partly because of its scientific and technological advancements. The country has a comprehensive system of social security and the second biggest budget of development aid in the world. While military expenditure ranks sixth, Germany is an influential partner in the North Atlantic Treaty Organization and took part in military interventions in the Balkans and Afghanistan that became controversial because of a strong anti-war attitude in the population.

Beginning in 2015, large numbers of refugees from Africa and the Middle East began seeking asylum in Europe, creating a massive humanitarian crisis. Large numbers of refugees drowned when unseaworthy boats sank. Those that made landfall sought to enter the European Union since, once there, they could travel and settle freely anywhere. Germany had the most open policy toward refugees, which strained German resources and also

created friction with other European Union states who were then forced to accept refugees arriving from Germany.

Impact of German Policies on Climate Change

Concerns about the environment and sustainable development are high on the agenda in Germany. Policies have contributed to reducing air pollution and acid rain from sulfur dioxide emissions, limiting pollution from industrial effluents into water bodies, and increasing nature preservation areas. There is significant attention among the Germany public and media to the threat of global warming, including shrinking glaciers in Alpine regions, and natural hazards such as river flooding and storms.

The national government has put great emphasis on climate policy and is committed to supporting the Kyoto Protocol and other agreements for reducing greenhouse gas emissions. Germany is trying to phase out nuclear power and has taken a lead in renewable energy sources, including bioenergy and wind and solar power (such as the "100,000 roofs" solar electricity program). While the country's CO_2 emissions per capita are among the highest in the European Union (still significantly lower than those of Australia, Canada, or the United States), overall emissions are falling thanks to wide-ranging emission reduction activities.

In June, 1990, the Interministerial Working Group on CO_2 Reduction was established to develop climate protection policies and report to the federal cabinet. Germany's national climate protection strategy addresses the specifics of the economic sectors (industry, energy and transportation, commerce, trade and services) and is oriented toward achieving a sustainable energy supply by improving energy efficiency, balancing energy mix, and expanding renewable energy sources. Germany has a leading role in international climate negotiations; for example, it directed the 1995 Conference of the Parties to the UNFCCC and it hosted the U.N. Climate Secretariat in Bonn. The federal government plans to achieve its Kyoto target by implementing climate-policy instruments and measures.

The Ecological Tax Reform established a set of energy taxes (mineral oil, heating fuel, coal, electricity) and prevented 20 million metric tons of CO_2

emissions; it also created up to 250,000 jobs and additional tax revenues of 17.8 billion euros in 2005 used for the public social security system. Special regulations help to ensure the competitiveness of energy-intensive processes. The Renewable Energy Sources Act (EEG) of 2000 and the 2004 and 2009 EEG amendments granted priority to renewable energy sources. In 2006, a total of 73.8 terra-watt hours of electricity based on renewable sources was generated, corresponding to 12 percent of total German electricity consumption. At the end of 2006, 5.8 percent of primary energy requirements and 12.0 percent of electricity requirements came from renewable sources. Through 2020, the EEG is expected to reduce emissions by about 45 million metric tons of CO_2.

Since 2005, emissions trading in the European Union has created economic incentives for reducing CO_2 emissions, including large energy installations and energy-intensive industrial systems. The German Bundestag (legislature) has passed several related laws: the Greenhouse Gas Emissions Trading Act, the National Allocation Plan, and the Act on Project-based Mechanisms. The initial upper emissions limit of 499 million metric tons is to be reduced to 453.1 million metric tons by 2012, including new installations.

Germany as a GHG Emitter

When it ratified the Kyoto Protocol, the European Union (with fifteen member states at that time) committed to reducing its greenhouse gas (GHG) emissions by 8 percent by the 2008-2012 period, compared to their base-year levels (1990 and 1995). Under the EU burden-sharing agreement, Germany has agreed to reduce its emissions by 21 percent compared to 1990. German reunification was a key factor in the sharp decrease beginning in the early 1990s. Numerous fossil-fired power stations were modernized, and production in the new German state decreased considerably. According to the National Inventory Report, Germany has fulfilled a large part of its obligations. In 2007, German emissions declined by roughly 24 million metric tons compared to 2006 (–2.4 percent), bringing emissions down to below the one billion threshold for the first time, an overall reduction of 20.4 percent compared to 1990. Reasons for the steep

decline were high gasoline prices, a mild winter, a tax increase, and the increase of renewable energy sources by 15 percent.

In 2006, CO_2 accounted for 87.6 percent of all GHG emissions, mostly related to stationary and mobile combustion. Methane (CH_4) emissions from animal husbandry, fuel distribution, and landfills contributed 4.6 percent, and nitrous oxide (N_2O) from agriculture, industrial processes, and transport, 6.3 percent of greenhouse gas releases. Fluorocarbons accounted for about 1.6 percent of total emissions. Emissions changes were –14.7 percent for CO_2 by 2006, –53.8 percent for methane, and –25.3 percent for nitrous oxide.

Summary and Outlook

To achieve its Kyoto obligations and move to lower numbers beyond Kyoto, setting a target of reducing GHG emissions by 40 percent by 2020, the federal government continually reviews and refines measures already taken. There are no plans for state purchase of emissions allowances from clean development mechanism (CDM) and joint implementation (JI) projects. Instead, Germany helps to develop the EU emissions trading system by enhancing energy efficiency, renewable energy use, and additional regulatory measures. With the Project Mechanisms Act of 2005, the legal basis exists for carrying out CDM and JI projects, with the Federal Environmental Agency as the competent authority.

Jürgen Scheffran

Further Reading

Bulmer, Simon. "Germany and the Eurozone crisis: Between hegemony and domestic politics." *West European Politics* 37, no. 6 (2014): 1244-1263. Germany's leadership role within the Eurozone sometimes collides with its domestic politics.

European Environment Agency. SOER 2015 The European Environment: Germany, 2015. http://www.eea.europa.eu/soer-2015/countries/germany (3 July 2016). Summary of environmental statistics and issues for Germany, with links to key sources (many in German).

Solsten, Eric, ed. Germany: A Country Study. 3d ed.

Washington, D.C.: Federal Research Division, Library of Congress, 1996. Reviews Germany's history and dominant social, political, economic, and military aspects. Comprehensive assessment of Germany in historical and cultural context.

Umwelt Bundes Amt. *Climate Change in Germany: Vulnerability and Adaptation of Climate Sensitive Areas.* Dessau, Germany: Author, 2008. Discusses climate impacts and adaptation strategies.

——— *National Report for the German Greenhouse Gas Inventory, 1990-2006.* Dessau, Germany: Author, 2008. Official report on German GHG emissions.

——— *Progress of German Climate Change Policies Until 2020.* Dessau, Germany: Author, 2008. Puts German climate strategy in context.

See also: Clean development mechanism; Greenhouse effect; Greenhouse gases and global warming; Kyoto Protocol; United Nations Framework Convention on Climate Change.

Glacial Lake Agassiz

Categories: Geology and geography; cryology and glaciology

Definition

The ice constituting mountain glaciers and continental ice sheets is derived from ocean water. During the Pleistocene epoch, the sea level dropped 100 meters, supplying sufficient water to cover about 30 percent of the Earth with glacial ice. As climatic conditions change and deglaciation begins, ice melts; meltwaters flow into the oceans or remain on land as glacial lakes. Lake Agassiz, the largest lake in North America during Pleistocene deglaciation, was dammed to the north by the Laurentide ice sheet (LIS). This proglacial lake (a glacial lake directly in contact with glacier ice) and its drainage basin covered, at maximum, 2 million square kilometers, extending from the Rocky Mountains in Alberta to the Lake Superior basin, into South Dakota, and north to the LIS in Hudson Bay. Maximum volume was 163,000 cubic kilometers.

First recognized by W. H. Keating in 1823 and named by Warren Upham in 1880, Lake Agassiz honors Louis Agassiz, a promoter of continental ice theory. Lake Agassiz is known for its fluctuating size, shape, and depth; multiple large-volume outbursts (drawdowns); and differing meltwater outlets. Physical evidence documenting these variations include former beaches and wave-cut cliffs, lake outlets, and ancient—now dry—lake marshes.

The maximum elevation of the lake's surface was determined by a lake outlet at 70 meters above sea level. Lake outlets were controlled by the vacillating LIS. Lake Agassiz water would rise until it reached an outlet, at which point it would abruptly overflow. Runoff routes were the Mackenzie River Valley to the Arctic Ocean, Hudson Bay to the North Atlantic Ocean, the St. Lawrence River to the North Atlantic Ocean, and the Mississippi River Valley to the Gulf of Mexico. As deglaciation continued, the level of Lake Agassiz fell below the level of the eastern outlets; lands south of 53° north latitude were above the level of the lake about eighty-five hundred years ago. Approximately one thousand years later, Lake Agassiz drained into the Tyrrell Sea, an area somewhat larger than today's Hudson and James bays. The lake existed for about five thousand years.

Significance for Climate Change

Short-term, often abrupt changes of climate are driven by reorganization of ocean circulation. Studies have shown that a freshwater influx of less than 100,000 cubic meters per second may slow or shut down the North Atlantic deep water (NADW). Part of the thermohaline circulation (THC), the NADW moves cold, salty, deep-ocean water south from North America. The same studies reveal that within a glacial-interglacial transition, a freshwater outburst of less than 10,000 cubic kilometers brought significant changes to ocean circulation, as well as colder temperatures. An outburst from Lake Agassiz of 9,500 cubic kilometers through the Great Lakes-St. Lawrence drainage system initiated the Younger Dryas cooling event 12,900 to 11,600 years ago.

perturbations of the NADW and the THC, resulting in cold temperatures, especially in North America and northern Europe.

In addition, once highly reflective snow and ice have melted, the less reflective land surface will absorb more solar radiation and add more heat to Earth. Along with reduced snow cover is the loss of tundra—areas habitated by scrubby, often snow-covered vegetation located on vast regions of Europe and North America. Snow-covered tundra reflects sunlight, but, if typical tundra vegetation is replaced by taiga vegetation (the dark, evergreen forests of subarctic regions), more sunlight would be absorbed and more warming would occur.

Mariana L. Rhoades

Another cooling episode, the 8.2ka event, was preceded by an outburst of 163,000 cubic kilometers in one year from the Lake Agassiz-Lake Ojibway system. Near the end of the existence of Lake Agassiz, the waters of Lake Ojibway commingled with Lake Agassiz. The exit route for this enormous interglacial outburst was through a weakened LIS, then northward into the Hudson Strait. The Hudson Strait is 2,000 kilometers north of the Atlantic Ocean. It has been suggested that, despite the distance from the Atlantic Ocean, the rapid influx of cold freshwater brought about the 8.2ka event, the last major cooling event of the Pleistocene deglaciation.

Global warming may introduce large fluxes of cold freshwater to the North Atlantic Ocean. As the climate warms, Greenland glaciers could melt, and increased rain could fall, producing a high volume of freshwater that could slow or stop the NADW. Freshwater outbursts from Lake Agassiz during the past deglaciation document

Further Reading

Alley, R. B. *The Two-Mile Time Machine*. Princeton, N.J.: Princeton University Press, 2000. Explains climate records extracted from ice cores, as well as the human imprint on future climates.

Broecker, Wallace S., and Robert Kunzig. *Fixing Climate: What Past Climate Changes Reveal About the Current Threat—and How to Counter It*. New York: Hill and Wang, 2008. Text on climate science, glaciations, and abrupt climate change. Details consequences of global warming; concluding chapters describe remedies.

Clark, P. U., et al. "Freshwater Forcing of Abrupt Climate Change During the Last Glaciation." *Science* 293 (July 13, 2001): 283-287. Research reporting on the relationship between increased freshwater flow and the NADW, providing a mechanism for climate variability and global ice volume.

Rayburn, John A., Thomas M. Cronin, David A. Franzi, Peter LK Knuepfer, and Debra A. Willard.

"Timing and duration of North American glacial lake discharges and the Younger Dryas climate reversal." *Quaternary Research* 75, no. 3 (2011): 541-551. With the 8.2 ka cooling event widely believed to be connected to discharge from Lake Agassiz, establishing the exact location, time, and duration of discharge events to greater precision has become critically important.

Teller, J. T., and L. Clayton, eds. *Glacial Lake Agassiz.* Geological Association of Canada Special Paper 26. Toronto: University of Toronto Press, 1983. Major research detailing Lake Agassiz history, biota, geology, hydrologic basins, and post-glaciation geology.

Teller, J. T., D. W. Leverington, and J. D. Mann. "Freshwater Outbursts to the Oceans from Glacial Lake Agassiz and Their Role in Climate Change During the Last Deglaciation." *Quaternary Science Reviews* 21 (2002): 879-887. Research comparing the volumes, fluxes, routings, and timings of ten different Lake Agassiz outbursts and relating the outburst to the history of climate change.

See also: Agassiz, Louis; Cryosphere; Deglaciation; 8.2ka event; Glaciations; Glaciers; Interglacials; Last Glacial Maximum (LGM).

Glaciations

Category: Cryology and glaciology

Discovering glaciations and trying to understand them have occupied scientists for over a century. In the process, great progress has been made in understanding the drivers of climate change in the past and the likely drivers in the future.

Key concepts
glaciation: the advance of a continental ice sheet
ice age: a period of time during which major continental ice sheets advanced and retreated
interglacial: the warm period between glaciations
marine isotope stage: half of a glacial cycle, as identified in the oxygen isotope data from ocean cores

Milankovi6 cycle: period of variation in Earth's orbital parameters, including axial inclination, climatic precession, and orbital eccentricity

Background
Glaciations presented a challenge to climate science. With ice sheets today only on Greenland and Antarctica, huge continental glaciers were difficult even to imagine. As the idea took root, however, evidence for glaciations throughout geologic history became apparent. What controls the timing of glaciations is now largely understood, but how they grow or retreat—and why—continue to be important subjects of research.

Global Factors
The temperature of the surface of the Earth is a function of how much energy it receives from the Sun and how much of this energy is radiated back into space. The atmosphere plays an important role, as its clouds and aerosols limit how much light energy reaches the surface and its greenhouse gases (GHGs) limit how much infrared energy escapes. Geologic evidence suggests that average global surface temperatures have remained within a limited range over the past two billion years, and GHGs, particularly carbon dioxide (CO_2), may have been responsible for this consistency.

CO_2 is produced by volcanism and removed by weathering. Weathering is temperature dependent, so it represents a negative feedback system: Higher temperature results in more weathering, which removes more CO_2, resulting in lower temperatures, or vice versa. Although over periods measured in hundreds of millions of years this feedback loop seems to have maintained a relatively constant temperature, its effects are not instantaneous, and perturbations have occurred. When these perturbations produce colder conditions, ice ages can result.

Evidence of a number of ice ages has been identified in rocks more than one billion years old. Little is known about these ice ages, as data are sparse and difficult to interpret. Between 750 and 550 million years ago, there were several major ice ages, usually lumped together and called Snowball Earth, as there is evidence that glaciers then

During the last ice age, the Sierra Nevada of California were covered by an ice cap. This aerial view of Kings Canyon National Park shows clearly that the upper elevations were glaciated and sculpted into sharp peaks and smooth, trough-shaped valleys, while the lower elevations show the V-shaped profile typical of stream erosion. [© Steven I. Dutch]

existed at sea level near the equator. Although they are of academic interest, at the time these ice ages occurred, there were no land plants, and the atmosphere had far less oxygen, so it is not clear that efforts to understand them will help in the study of contemporary climate change.

Other ice ages, including a short one 440 million years ago and the Permo-Carboniferous ice age between 325 and 240 million years ago, provide insight into what conditions are required for ice ages. These occurred when the land on which the continental glaciers formed was over the South Pole. As continents were also over the South Pole in the time period between these two ice ages and Antarctica sat over the South Pole for 90 million years before its current glaciers formed, this location seems

to be a necessary but not a sufficient condition for ice age formation.

Pleistocene Glaciations

About 50 million years ago, the Earth began to cool. Deep-ocean temperatures gradually dropped from 13° Celsius to the present 1° Celsius. Study of ocean sediment cores has identified some minor ice ages around 40 million years ago, another ice age formed the East Antarctica Ice Sheet 34 million years ago, and a third ice age around 13 million years ago left evidence in Alaska. Orbital factors probably influenced these advances, perhaps by affecting the carbon cycle, but the details remain obscure.

Starting about three million years ago, the climate developed two different states. Since then, it has oscillated between them, causing glaciations—with continental glaciers extending down to the 40° north parallel of latitude—and interglacials such as the current period, some with temperatures even higher than those of the present. Geologists identified glacial deposits, determined that some were on top of others, and gave names to each, but the names were not standardized internationally. The name of the most recent glaciation is Valdaian on the Russian Plains, Devensian in Britain, Weichselian in Scandinavia, Würm in the Alps, and Wisconsinan in North America. The deposits of one glacier are easily removed by subsequent glaciers, so there might be some for which no evidence remains on the continents.

Oxygen isotope ratios of marine sediments, which are not removed by later glaciations, show that there have actually been around fifty glaciations. Because these ratios indicate how much water is tied up in ice, this record is now seen as the best way of delineating when an advance ended and a retreat began. Each marine isotope stage (MIS) is given a number (odd ones for interglacials, even ones for glaciations) starting with the most recent. Interglacial MIS 103 occurred 2,580,000 years ago.

Once all these additional glaciations were known, analysis showed that astronomical cycles controlled their timing, as had been suggested by James Croll in 1864 and Milutin Milankovi6 in 1941. These results are so robust that geologists now use these cycles to calibrate the more recent part of the geologic time scale.

That Milankovi6 cycles control the timing of glaciations is beyond dispute. How they do so, however, is not well understood. Feedback systems in the oceans, the atmosphere, the biosphere, or perhaps elsewhere are needed to amplify the tiny signal produced astronomically. Identifying and understanding these systems will help scientists understand current climate change.

Context

Skeptics who wonder if the climate is changing should consider glaciations. Doomsayers, who fear the human race is threatened by global warming, should also consider glaciations. Much of geologic history concerns things that happened so long ago that it is easy to dismiss or ignore them, but glaciations are recent history. *Homo erectus* was on the planet and using fire twenty or so glaciations ago. All of human evolution has taken place as glaciers ebbed and flowed. Human migration to North America occurred 14,600 years ago, just after the peak of the last glaciation. Ice sheets then, although they were getting smaller, still covered most of Maine and northern parts of New York, Vermont, and New Hampsire. They are not there now. Climate changes.

Otto H. Muller

Further Reading

Ehlers, Jurgen, Philip Leonard Gibbard, and Philip D. Hughes. *Quaternary glaciations-extent and chronology: a closer look.* Elsevier, 2011. A massive, thousand-page summary of the latest data on Quaternary glaciation, including areas not often thought of as glaciated, like Hawaii, Central America and Southeast Asia.

Imbrie, J., and K. P. Imbrie. *Ice Ages: Solving the Mystery.* Short Hills, N.J.: Enslow, 1979. An excellent history of the efforts to understand what caused the ice ages. Full of personal anecdotes, it gives a sense of the excitement and frustrations experienced by cryologists. Many diagrams, including some now classic ones.

Macdougall, J. D. *Frozen Earth: The Once and Future Story of Ice Ages.* Berkeley: University of California Press, 2004. Easily understood by a general reader, this book explains how glaciations were discovered and understood. Chapter 10 focuses on the role glaciations played in evolution in general, and in human evolution in particular. Photos, diagrams, maps, an annotated bibliography of general-interest books.

Ruddiman, William F. *Earth's Climate Past and Future.* 2d ed. New York: W. H. Freeman, 2008. This elementary college textbook has several sections concerning glaciations and their causes, with particular attention paid to how they can elucidate climate change. Illustrations, figures, tables, maps, bibliography, index.

Glaciers

Category: Cryology and glaciology

Glaciers are very sensitive to global temperature changes, retreating and growing as the Earth warms and cools. Because of this sensitivity, geologists have made careful measurements of the changes in benchmark glaciers in order to monitor the effects of global climate change and have studied the history of glaciers as a means of reconstructing the sequence of temperature fluctuations in Earth's past.

Key Concepts

alpine glaciers: large masses of ice found in valleys, on plateaus, and attached to mountains

cryosphere: the portion of the Earth's surface that is composed of frozen water

firn: the intermediary stage between snow and ice

glacial ice: ice created by the compression of snow, sometimes saturated with meltwater that is refrozen

ice caps: masses of ice covering areas smaller than 50,000 square kilometers

ice sheets: masses of ice covering large landmasses

Background

Of the freshwater ice of the world, 99 percent is located in Antarctica and in Greenland: 90 percent of the ice is located in Antarctica (an area of 14 million square kilometers containing 27.6 million cubic kilometers of ice); 9 percent is in Greenland (an area of 1.726 million square kilometers containing 2.85 million cubic kilometers of ice), and 1 percent exists in the glaciers and ice caps scattered throughout the world. The two ice sheets of Antarctica and Greenland represent about 10 percent of the Earth's land area and contain more than three-quarters of its freshwater.

Glacial Landscape

Geologists can provide useful information about the history of glaciation by studying the geographic features left by glaciers that have long since melted away. When alpine glaciers form, they produce an amphitheater-like depression called a "cirque" at their highest point. When the compression of snow produces ice in a depth of about 20 meters, the ice begins to flow downward along the valleys of former streams. The ice deepens and widens the stream valley to produce a glacial trough. Valleys of tributary streams are filled with ice as well, but the shorter length of the tributary glaciers and the lesser discharge produce less deepening of the channel. Typically, these tributary glaciers erode their channels only down to the current ice surface of the main glacier, resulting in distinctive "hanging valleys" that end in a precipice at their juncture with the valley produced by the main body of the glacier. The height of these hanging valleys indicates the depth of the ice in the main glacier.

Materials the glacier erodes are pushed to its side, creating lateral moraines, or to its lower end, creating a frontal or end moraine, or accumulate under the glacier, creating a ground moraine, or till. These features identify the greatest breadth (lateral moraines) and furthest extent (terminal moraine) of ancient glaciers. The sediments deposited in glacial lakes—called varves—can be used to determine the length of the deposition process. Distinctive small hills called drumlins are formed under an advancing glacier and, because of their teardrop shape, indicate the direction of the ice and water flow. Meltwater accumulates in lakes found in the cirques; these lakes are called tarns. Other lakes occupy the depressions in the glacial troughs. When the glacier retreats, it uncovers a U-shaped valley instead of the V-shaped valley it had prior to the glaciation.

Continental ice sheets transform the landscape differently. When they retreat, they leave a multitude of lakes, many of which are round (kettle lakes). Others are elongated in the direction of the ice flow—such as the Finger Lakes in New York. By reconstructing the history of glaciation from these

A large glacier on the flanks of Mount Rainier, Washington. (© Steven I. Dutch)

alterations made to the landscape, glaciologists and geologists have been able to contribute greatly to the understanding of the nature of Earth's climatic changes.

Ice Formation and Location

Glaciers are masses of ice that are produced where the summer temperatures fail to melt the snow that fell during the preceding winter. Over time, this snow is compressed by overlying layers of more recent snow, forcing out some of the air that exists around snow crystals. This air escapes toward the surface; the density of this older snow steadily increases, until the snow turns into ice. At that point, the density of the ice varies from 0.85 to 0.91. Ice therefore floats on liquid water, whose density is 1. The transformation of snow into ice is slower in polar areas, where compression is the major

mechanism at work, because air temperatures remain low all year round, producing very little, if any, meltwater to be refrozen. In temperate climates, ice forms more rapidly than in the polar areas because there are periods of melting when the temperature is above the freezing point of water. The meltwater soaks the snow and refreezes during the next colder period. This process is much faster than compression to achieve the density of ice.

Glaciers can exist at any latitude, even along the equator. They are present on every continent except Australia. The necessary condition for glaciers is that the air temperature remains low enough to prevent melting of the last winter snow. Because temperature decreases with increasing elevation, glaciers are found on high mountains or volcanoes. In Africa, Mounts Kenya, Kilimanjaro, and Ruwenzori have small glaciers (though they are retreating rapidly). In South America, there are many small tropical glaciers located in the Andes as well. In North America, Europe, and Asia, small glaciers dot the summits of high mountain ranges and volcanoes. Glaciers gain mass—called "accumulation"—by the deposition of snow in the highest elevation, called the cirque—a large amphitheater at the summit of glaciers.

Glaciers lose mass (a process called "ablation") by melting, sublimation, and calving. Melting takes place in a glacier where the temperature is greater than 0° Celsius. This can occur where the air temperature reaches this value or at the underside of the glacier, where friction of the ice on the ground beneath the glacier causes the temperature to increase, producing meltwater. This water lubricates the underside of the glacier, resulting in faster movement.

The second component of ablation is sublimation. sublimation is the transformation of ice directly into water vapor, without an intermediate liquid stage. In Antarctica, sublimation is a major contributor to the ablation of ice because the ice sheet is affected by very strong winds which enhance the process. The third form of ablation is calving. It involves the breaking of the end of the glacier when it reaches an ocean or a lake. calving produces icebergs—masses of ice that float because of the lower density of ice (0.85-0.91) compared to that of water (1). Calving is responsible for about 40 percent of ablation in Greenland and 80 to 90 percent in Antarctica.

The term "mass balance" refers to the difference in mass between accumulation and ablation. A glacier will grow longer if accumulation is greater than ablation over a period of time. This is called a glacial advance. A glacier will get shorter if the amount of ice removed by ablation exceeds the amount of snow that accumulates in the coldest part, the cirque. When glaciers become shorter they are said to retreat. This does not mean that the ice stops moving downward from the cirque to the lower end of the glacier, called the front. Glacial retreat indicates only that the front of the glacier will be found closer to the cirque.

Today, most glaciers in the world are retreating. Although there are a few exceptions where glaciers are advancing, the worldwide trend is a steady retreat in response to the general raising of the Earth's temperatures.

Glaciology and Paleoclimates

The science that studies the cryosphere is called glaciology. One of the first important glaciologists was Louis Agassiz, a native of Switzerland, who studied the glaciers of the Alps in the nineteenth century. Building on Agassiz's work, modern glaciologists discovered that glaciers not only are a good source of information about the global impact of recent environmental changes but also provide valuable data about the long history of climate change on this planet. Modern study of ice cores, cylinders of ice retrieved from glaciers, has shown that ice records the temperatures of the Earth atmosphere at the time the snow fell. The paleotemperatures are inferred from the composition of the water that makes up the ice.

In nature, water is made of two atoms of hydrogen and one atom of oxygen; however, oxygen has three isotopes (elements that have the same number of protons but different numbers of neutrons). The lightest and most abundant of the three is O16. O18 is the heaviest but exists in much smaller quantities. Higher Earth temperatures make it easier for the heavier O18 molecule to evaporate, resulting in snow—and therefore glacial ice—that has an increased proportion of it. During glaciations, the lower temperatures lead to a depletion of O18 in the ice of glaciers. The same principle is applied to ice when two isotopes of hydrogen are measured. Because the ice sheets of Antarctica and Greenland are very thick, ice cores obtained by drilling into these ice sheets are very long. They therefore provide a very long record of the climates of the past (known as "paleoclimates"). Based on the cores retrieved so far, scientists have been able to identify a sequence of glacial and interglacial cycles that covers the last 800,000 years.

The Earth has had many ice ages in its 4.5-billion-year history. Most recently, beginning about one million years ago, the Great ice age occurred during a time period called the Pleistocene. This glacial period formed an ice sheet in North America, Northern Europe, Northern Asia, and Antarctica that expanded until it reached its maximum extent about twenty thousand years ago.

The Pleistocene Glaciation was not uniformly cold; short interglacial periods of warming occurred several times. Finally, about ten thousand years ago, the warming trend continued, melting the ice sheets and uncovering the northern continents. The northern part of Canada became free of ice about six thousand years ago. The mountain glaciers attached to the high mountain ranges of the American West are remnants of the Pleistocene period. In more recent centuries, the Earth experienced a shorter period of cold temperatures, called the Little Ice Age, beginning about 1650 and ending approximately in 1850, during which the Earth cooled by about 1° Celsius. (The term "Little Ice Age" is used differently by different writers. Many use it to refer to the climate cooling from about 1300 to 1850, while others use it for the latter half of that interval, when cooling was greatest, beginning around 1550 or 1600.) The mountain glaciers that advanced during this period are currently retreating in response to today's higher temperatures.

Context

At their present rate of melting and retreating, glaciers are having a major impact on the populations living in their vicinity. The increase in meltwater can increase the production of hydroelectricity for a short while but at the same time may impact the population's well-being in the very near future by decreasing the amount of available water for irrigation or electricity production when the glaciers will have completely disappeared. Monitoring glacier changes and drawing scientific conclusions about their retreat is not something new. As early as 1894 scientists began cataloging glaciers and their changes. These findings were published by the World Glacier Monitoring Service. Maximum extents of glaciers were computed by using the position of their terminal moraines, and their volumes were estimated by measuring the height of their lower end since it corresponds to the height of ice that used to occupy the glacial trough.

In the 1970s, during the International Hydrological Decade declared by United Nations Educational, Scientific, and Cultural Organization, the Temporal Technical Secretariat for the World Glacier Inventory was created and began making a comprehensive inventory of more than 100,000 glaciers worldwide. Since then, with the help of satellite instruments, it has been determined that there are almost 200,000 glaciers, thousands of whose outlines, retreats, and advances are readily mappable. These measurements allow scientists to rapidly assess the impact of the warming of the Earth on the cryosphere, and their study has proven a valuable tool in monitoring their reaction to the warming of the Earth's atmosphere.

The work of the Intergovernmental Panel on Climate Change and the research conducted for the Fourth International Polar Year (2007-2008) organized by the International Council for Science in conjunction with the World Meteorological Organization focused on understanding the extremely complex relationships between glaciers and climates. One of the goals of the United States National Committee for the International Polar Year was the creation of a network of observation platforms to monitor glaciers in order to provide reliable data by which scientists can able to assess the impact of global warming both on the glaciers themselves and upon the global ecosystem of which they are an essential part.

Denyse Lemaire and David Kasserman

Further Reading

Benn, Douglas I., and David J. A. Evans. Glaciers and Glaciation. New York: Arnold, 1998. Print.

Bennet, Matthew, and Neil Glasser. Glacial Geology: Ice Sheets and Landforms. New York: Cambridge UP, 1997. Print.

Davies, Bethan. "Mapping the World's Glaciers." Antarcticglaciers.org, 25 Nov. 2014. Web. 23 Mar. 2014.

Hambrey, Michael, and Jurg Alean. Glaciers. New York: Cambridge UP, 2006. Print.

Krüger, Tobias. Discovering the Ice Ages: International Reception and Consequences for a Historical Understanding of Climate. Leiden: Brill, 2013. Print.

Pfeffer, W.T., Arendt, A.A., Bliss, A., Bolch, T., Cogley, J.G., Gardner, A.S., Hagen, J.O., Hock, R., Kaser, G., Kienholz, C. and Miles, E.S., 2014. The Randolph Glacier Inventory: a globally complete inventory of glaciers. *Journal of*

Glaciology, 60(221), pp.537-552. Summarizes the statistics of almost 200,000 glaciers in a global inventory. The article includes information on accessing the data, but utilizing it will require the ability to use geographic information system programs.

Trewby, Mary. Antarctica: An Encyclopedia from Abbott Ice Shelf to Zooplankton. Toronto: Firefly, 2002. Print.

Glacken, Clarence J.
American geographer and historian

Born: March 30, 1909; Sacramento, California
Died: August 20, 1989; Sacramento, California

An environmental historian who specialized in studying the relationships between culture and nature, Glacken was one of the first twentieth century scientists to recognize the connection between human activities and the environment.

Life

For most of his career, Clarence J. Glacken was a faculty member at the University of California, Berkeley, in the Geography Department, although his academic post followed many varied types of employment. He was a native of California and attended Berkeley as an undergraduate student. There he was profoundly influenced by Frederick Teggart, author of *The Theory of History* (1925) and other works, experiences that first introduced Glacken to broad interdisciplinary scholarship which he later cultivated. He graduated during the Great Depression, however, and spent most of the 1930's and 1940's working not as an academic but in public service, for various government agencies, to provide relief to farmworkers entering California after fleeing the Dust Bowl. These experiences significantly influenced his life and increased his awareness of the interactions between humans and the environment.

In 1937, Glacken took almost a year off to travel the world, including Europe, the Mediterranean, and East Asia. When he returned to the United States, he continued to work with displaced farmworkers until 1941, when he was drafted into the U.S. Army during World War II. During his service, he was trained as a specialist in Japanese languages and culture and was posted to Korea. On his discharge, Glacken served in Korea as deputy director of the Bureau of Health and Welfare in the military government, where he studied deforestation. Later, the Pacific Science Board invited him to do an ethnographic study of three villages in Okinawa, which resulted in a highly regarded book, *The Great Loochoo: A Study of Okinawan Village Life*, published in 1955.

Glacken's varied experiences in different worlds fueled his interest in studying the history of ideas about nature and culture. Consequently, in his forties, Glacken began work on a Ph.D. in geography at The Johns Hopkins University, where he completed a thesis titled *The Idea of the Habitable World*. After he graduated in 1952, he was appointed to the faculty in the geography department at the University of California, Berkeley, where he joined Carl Sauer. He was a faculty member there for thirty-seven years—the duration of his career.

Climate Work

Glacken was an environmental historian who specialized in the relationships between culture and nature. Climate scientists have only recently begun to rely on environmental historians to enhance their understanding of the dynamics of climate change impacts. Although the effects of expanding civilization and industrialization on climate was intensely debated worldwide during the nineteenth century, Glacken was probably the first twentieth century scientist to be concerned with these issues. As a faculty member and through his writings, he inspired many generations of students of the history of geographical ideas and contributed to historical and intellectual foundations for the environmental and conservation movement. The current understanding of how scientists throughout the ages have considered the interactions between nature and humankind was greatly enhanced by Glacken's scholarship.

Glacken wrote many books and papers about the history of ideas about nature and man. In 1955, he presented a paper titled "Changing Ideas of a Habitable World" at the conference Man's Role in Changing the Face of the Earth, a symposium that brought together the historical and current knowledge of these issues that was available at the time. This symposium represented a changing awareness among American scientists and indicated the beginning of a new era in which people began to realize that their actions affected nature and the physical world.

Glacken's most influential and prominent contribution was a book on nature and culture in Western thought from ancient times to the end of the nineteenth century, titled *Traces on the Rhodian Shore: Nature and Culture in Western Thought from Ancient Times to the End of the Eighteenth Century*. This widely appreciated work brought together ideas in a manner that transcended geography as a discipline. Published in 1967, this text predated the resurgence of environmentalism in America. It originally was meant to be only an introductory chapter to a larger work. Glacken's primary purpose in this book was to uncover the variety and complexity of thinking on nature and culture in the Western tradition and to show how contemporary interests in conservation have long histories. Although he had accumulated a significant amount of material for a sequel that covered the modern period and included art, science, and philosophy, he was unable to finalize it for publication.

C. J. Walsh

Further Reading

Agnew, John A., David N. Livingstone, and Alisdair Rogers. *Human Geography: An Essential Anthology*. Hoboken, N.J.: Blackwell, 1996. Part 3 of this five-part book is concerned with interactions between nature, culture, and landscape; includes a chapter on Glacken.

Glacken, Clarence J. *Traces on the Rhodian Shore: Nature and Culture in Western Thought from Ancient Times to the End of the Eighteenth Century*. Berkeley: University of California Press, 1967. Glacken's most influential work, which helped redefine the discipline of geography.

Hughes, J. Donald. *What is environmental history?*. John Wiley & Sons, 2015. Attempts to define environmental history and trace its evolution, and describes Glacken's role in founding the discipline

Oakes, Timothy S., and Patricia L. Price, eds. *The Cultural Geography Reader*. New York: Routledge, 2008. Suitable for students, this reader includes discussions of history and its relationship to cultural geography, as well as a chapter on the works of Clarence Glacken.

See also: Anthropogenic climate change; Human behavior change.

Global Cooling Event of AD 535-547

Category: Climatic Events and Epochs

A series of massive volcanic eruptions in 535, 540, and 547 CE caused a marked drop in global temperatures and the onset of the Late Antique Little Ice Age (LALIA), which persisted for more than a century. The ensuing famine, social disruption, and plague had profound effects on the subsequent history of Europe, China, and Mesoamerica.

Key Concepts

Volcanic veiling: A drop in solar radiation reaching the earth due to ash and particularly sulfuric acid aerosols from massive explosive volcanic eruptions.

Tree ring analysis: Examining cores from living and dead trees for distinctive patterns of broad and narrow rings that document unusual growing conditions and provide a precise date for their occurrence.

Bolide: A comet, asteroid, or large meteorite that impacts the earth. The effect of large bolide impact on climate can be dramatic.

Sea-ice ocean feedback: A process whereby a cooling trend is prolonged and intensified as increased sea ice in the arctic increases planetary albedo.

Late Antique Little Ice Age (LALIA): A period of low global temperatures from 535 to around 750 CE.

The Byzantine (Eastern Roman) Emperor ruled in an era of catastrophes, including a civil uprising that destroyed much of Constantinople (now Istanbul), a global cooling event, probably due to volcanic activity, and a plague possibly caused by the cooling event. Shown here is the former Church of Holy Wisdom (Hagia Sophia), which was built in 532-537 to replace one destroyed in the uprising. It has survived an amazing number of earthquakes and invasions. (© Steven I. Dutch)

Background

Europe at the height of Roman power experienced a warm, benign climate favoring high agricultural productivity and trade. A gradual cooling trend ensued after 200 CE, but even after the "official" fall of the Roman Empire in 476 CE, the institutions of the classical world persisted to a considerable degree. Then, within a period of a few years beginning in 535 CE, both Europe and China were hit with plummeting summer temperatures, extreme weather events, crop failures, famine, pestilence, and massive social disruption, at a level that greatly exceeded the 1816 year without a summer and also exceeded the 1257–1325 cooling in terms of its proportional impact. Low temperatures persisted for more than a century. The Byzantine historian Procopius described a darkening of the sun that persisted for more than a year and coincided with cold temperatures and crop failures. This and other sources suggest massive volcanism as the cause. Some historians now refer to the period 535–700 CE as a literal Dark Age and emphasize climate change, not internal social phenomena, as the primary determinant of the final fall of the Roman Empire.

Natural causes of LALIA

Investigating the precise timing and causes of LA-LIA requires extensive use of proxy measurements. Record keeping and objective observation in the sixth and seventh centuries were inferior to those of Roman times and the later Middle Ages. For much of the world, including those portions of it where catastrophic volcanic eruptions occur, the sources are mainly mythological.

Volcanic eruptions leave traces in ice cores in the form of elevated sulfur content and ash particles. Layers from the sixth century are present in both Greenland and Antarctica, and can be dated with some precision. Ice core data pinpoints a large sulfur-rich eruption near the equator in 535 or 536, another event in 540, and one in 547. A plausible source for the 535 eruption is Ilopango Caldera in El Salvador, with a VEI6 eruption variously estimated at 40–80 km² in tephra volume. Some of the decline in Mayan civilization in the sixth century may be due to direct effects (ash, pyroclastic flows) of the Ilopango eruption. The eruption may have been sulfur rich, like the relatively small 1982 eruption of El Chichón in Mexico, which had a significant climatic impact. El Chichón itself is a possible source for one of the other eruptions seen in ice cores. Krakatoa, which erupted explosively in the sixth century, has also been proposed. In the absence of dated chronicles, connecting ice core data to specific eruptions is difficult. Radiocarbon dates of buried wood have an uncertainty of ca. 100 years for the first millennium, volcanoes in Latin America, Polynesia, and Southeast Asia are not well studied, and any eruption remnants now underwater are likely to be overlooked.

Tree ring data from the Alps, Scandinavia, California, and the Altai mountains all show an abrupt transition to narrower rings, indicating colder and drier conditions, in 535 or 536. Growth suppression persisted for decades. Growth rates in marginal habitats in the northern hemisphere did not return to pre-535 levels until after 750 A.D. There is a drop in 680 which corresponds to elevated sulfate levels in ice cores.

Volcanism alone does not explain the severity and duration of the Late Antique Little Ice Age. The sixth century also corresponded to a period of low sunspot activity, as estimated by radioisotopes in ice cores and also by models based on more recent solar minima. A recent and plausible model points to sea-ice-ocean feedback as a possible factor. In this model, sea ice failed to melt during cold summers, increasing the reflectivity of the earth's surface. A series of eruptions a few years apart renewed the veiling aerosol, creating more cooling and more extensive ice. Eventually production of cold dense water off Greenland increased the flow of warm water into the North Atlantic, ultimately cancelling out cooling due to albedo. Sinking of cold water would draw warm surface waters northward.

One somewhat controversial theory, articulated by Michael Baillie and based heavily on Celtic oral tradition, postulates bolide impact as one of the causes of LALIA. Halley's Comet passed close to the earth in 530 and debris from its tail has been suggested as a source for atmospheric veiling. The reasoning is not particularly sound, but some object large enough to produce atmospheric effects could have struck the Pacific Ocean. There is some evidence of bolide impact in the Gulf of Carpenteria, north of Australia, from the right time frame.

Effects on the human population

Many traditional prophesies, notably the *Book of Revelations*, associate darkening of the sun and moon with famine, war and pestilence. The A.D. 535 event caused cold and drought induced crop failures in China, most of Europe, the Middle East, and central and South America. It was followed closely in 540 by the Plague of Justinian, which killed a third of the population of Constantinople and ravaged seaport towns as far north as the Baltic, returning in waves over the course of the next century. It is thought that famine weakened the human population, increased movement facilitated transmission, and drought drove infected rodents from their normal habitats. This also occurred in the fourteenth century. In a controversial but plausible hypothesis in his book *Catastrophe*, David Keys traces all of the large population dislocations of the sixth and seventh centuries to a single massive volcanic eruption, which he identifies as Krakatoa. The population crash in Europe led to a contraction of trade and the human institutions sustained by it. His calling the book "An investigation into the origins of the modern world" is an exaggeration, but the human impact was certainly profound

Martha A. Sherwood

Further Reading

Barras, Colin. The Year of Darkness. *New Scientist* 221 (2952): 34-38. Discussion of comet hypothesis.

Büntgen, Ulf et al. Cooling and Societal Change during the Late Antique Little Ice Age from 536 to around 660 A.D. *Nature Geoscience* 9: 231-236. 2016.

Focus on tree ring data; discussion of population dislocations more cautious and better supported than Keys (2000).

Keys, David. *Catastrophe! An Investigation into the Origins of the Modern World.* New York, Random House, 2000. Popular and somewhat sensationalist.

Miller, Gifford H. et al. Abrupt onset of the Little Ice Age triggered by volcanism and sustained by sea-ice/ocean feedbacks. *Geophysical Research Letters* 2012, Vol. 39 (2). Article and research refer to the 14th century.

See also: Composting; Arctic; Volcanoes

Global climate

Category: Meteorology and atmospheric sciences

The global climate system is complex and dynamic, greatly complicating attempts to evaluate or predict long-term alterations to Earth's climate. However, physical measurements of specific parameters can be made, rendering climate change quantifiable.

Key concepts

chinook/foehn wind: a warm, dry wind on the eastern side of the Rocky Mountains or the Alps

climate controls: the relatively permanent factors that govern the general nature of the climate of a region

evaporation: the process by which a liquid changes into a gas

Hadley circulation: an atmospheric circulation pattern in which a warm, moist air ascends near the equator, flows poleward, descends as dry air in subtropical regions, and returns toward the equator

Inter-Tropical Convergence Zone (ITCZ): a low-pressure belt, located near the equator, where deep convection and heavy rains occur

Köppen climate classification system: a system for classifying climate based mainly on average temperature and precipitation

monsoon: a seasonal climate system characterized by wind and precipitation patterns

precipitation: liquid or solid water particles that fall from the atmosphere to the ground

rain shadow: the region on the lee side of a mountain where precipitation is noticeably less than on the windward side

subtropical high belt: a high-pressure belt where warm, dry air sinks closer to the surface

transpiration: the process by which water in plants is transferred as water vapor into the atmosphere

Background

Climate is a general characterization of long-term weather and environment conditions for a specific location. Several major factors influence climate in a given region, including latitudinal position, the distribution of land and water, and elevation. Ocean currents, prevailing winds, and the positions of high- and low-pressure areas also have significant climatic effects.

Heterogeneous distributions of heat and water result in rich and varied climates. In particular, the tropical regions receive more energy from solar radiation than they emit in the form of infrared heat. The polar regions, by contrast, receive less energy from the Sun than they emit as heat. As a result, the tropics are regions of heat surplus, while polar regions are deficient in heat. Moreover, because tropical regions include large expanses of ocean, there is more water stored in the tropical atmosphere than is stored in the atmosphere at high latitudes.

The imbalance in the heat and water budget in the tropical and polar regions often leads to circulations that transport heat and water from and to these regions. These transports are typically carried out by both atmosphere and oceans. The weakening or strengthening of heat and water transports is an important signal for climate change.

Several methods have been developed to classify global climate. The most widely used method is based on the Köppen classification system. Designed by the German climatologist Wladimir Köppen (1846-1940), this method uses the average annual and monthly temperature and precipitation to describe a global climate for various climate zones. In this method, global climate is divided into the following six major groups, and each group is divided into subgroups.

Tropical Moist Climate (Group A)

Tropical moist climate is typical of most of Earth's tropical regions (from the equator to about 20° latitude into each hemisphere). The climate of these regions is characterized by year-round warm temperatures and abundant rainfall. In the tropics, the annual mean temperature is typically above 18° Celsius, and typical annual average rainfall exceeds 150 centimeters. Tropical moist climate is divided based on rainfall characteristics into three subtypes of climate: tropical wet, or tropical rain-forest climate (Af); tropical monsoon climate (Am); and tropical wet-and-dry climate (Aw).

The tropical rain-forest climate exhibits constant high temperatures and abundant year-round rainfall. As a result, it is marked by dense vegetation, typically composed of broadleaf trees, jungles, and evergreen forests. A large number of diversified plants, insects, birds, and animals inhabit the tropical rain forests. Many lowlands near the equator are in this type of climate, which includes the Amazon River Basin of South America, the Congo River Basin of Africa, and the East Indies, from Sumatra to New Guinea.

Unlike the tropical wet climate, the tropical wet-and-dry climate has distinctive wet and dry seasons. Although the annual precipitation usually exceeds 100 centimeters, during the dry season the average monthly rainfall can be less than 6 centimeters. The dry season lasts more than two months. The tropical wet-and-dry climate dominates most of tropical Africa, tropical South America, and South Asia. The variations of dry and wet seasons in these regions are closely associated with the migration of the Inter-Tropical Convergence Zone (ITCZ) in the tropics.

The tropical monsoon climate exists between the tropical rain-forest and wet-dry climates: It has abundant rainfall, in excess of 150 centimeters per year, but the rains do stop briefly, typically for one or two months. Tropical monsoon climate can be seen along the coasts of Southeast Asia and India and in northeastern South America. In contrast to the wet-dry climate, the rain and the pause of rain in these areas are related to monsoonal circulation.

Dry Climate (Group B)

Just outside the tropics, most of the continental land located between approximately 20° and 30° latitude in both the Northern and the Southern Hemispheres is in arid or semiarid climates. Precipitation in these areas is scarce most of the year, and evaporation and transpiration exceed precipitation.

The arid climate (BW) is the true desert climate and can be found in the Sahara Desert in Africa, a large portion of the Middle East, much of the interior of Australia, Central Asia, and the west coasts of South America and Africa. These areas are located in the subtropical high belt, which is caused by descending air from the Hadley circulation.

Around the margins of the arid regions, semiarid (BS) areas enjoy a slightly greater rainfall. The light rains of semiarid climes support the growth of short bunch grass, scattered low bushes, trees, and sagebrush. This climate can be found in the western United States, southern Africa, and the Sahel.

The Moist Subtropical Midlatitude Climate (Group C)

Most subtropical midlatitude regions are farther poleward from the major dry-climate latitudes. These areas extend approximately from 25° to 40° latitudes in both the Northern and the Southern Hemispheres. This climate has distinct summer and winter seasons. Winter is mild, with average temperatures for the coldest month of between -3°

and 18° Celsius. The regions in this climate belt are typically humid and have ample precipitation.

There are three major subtypes in the group C climate: humid subtropical (Cfa); west coast marine (Cfb); and dry-summer subtropical, or Mediterranean (Cs). The humid subtropical climate typically presents hot and muggy summers, but mild winters. Summers experience heavy rains, while winters are slightly drier. This climate type can be found principally along the east coasts of continents, such as the southeastern United States, eastern China, southeastern South America, and the southeastern coasts of Africa and Australia.

The west coast marine climate has cool summers and mild winters and produces more precipitation in winter than in summer. The largest area with this climate is Europe. Finally, the dry-summer, or Mediterranean, climate is distinctively characterized by extreme summer aridity and heavy rains in winter. Countries surrounding the Mediterranean Sea and the U.S. West Coast, including Northern California and Oregon, are in this type of climate.

Moist Continental Climate (Group D)

The moist continental climate is located farther north of the moist subtropical midlatitude climate zone, from 40° to 50° north latitudes. This climate mostly occurs in North America and Eurasia. The general characteristics of the moist continental climate are warm-to-cool summers and cold winters. The average temperature of the warmest month exceeds 10° Celsius, and the coldest month's average temperature generally drops below -3° Celsius. Winters are severe, with snowstorms, blustery winds, and bitter cold. The climate is controlled by a large landmass.

The group D climate is further divided by summer temperature into three major subtypes: humid continental with hot summers (Dfa), humid continental with cool summers (Dfb), and subpolar (Dfc). Both winter and summer temperatures in the Dfa climate are relatively severe. That is, winter is cold and summer is hot. Farther north is the Dfb climate, which experiences long, cool summers and long, cold, windy winters. The subpolar climate presents severely cold winters and short summers. In the subpolar region, moisture supply is limited. Therefore, precipitation is low.

Polar Climate (Group E)

The polar climate exists over the northern coastal areas of North America and Eurasia, Greenland, the Arctic, and Antarctica. It is characterized by low temperatures year-round. Even during the warmest month, the temperature is below 10° Celsius. Precipitation is scarce in these parts of the Earth.

The polar climate can be divided into two subtypes: the polar tundra (ET) climate and the polar ice-cap (EF) climate. The tundra climate occupies the coastal fringes of the Arctic Ocean, many Arctic islands, and the ice-free shores of northern Iceland and southern Greenland. In these regions, the ground is permanently frozen to depths of hundreds of meters, a condition known as permafrost. In summer, the temperature can remain above freezing, allowing tundra vegetation to grow. The monthly mean temperature under the ice-cap climate is mostly below 0° Celsius. The ice cap occupies the interior ice sheets of Greenland and Antarctica. The growth of plants is prohibited, and the landscape is perpetually covered with snow and ice. Many studies show that these regions are most sensitive to global warming and have experienced rapid snow and ice melting in recent decades.

Highland Climates (Group H)

The distribution of global mountain ranges and plateaus creates another type of climate. Climate in highland regions is unique. Highland climates are characterized by a great diversity of conditions. Because air temperature decreases with altitude, climatic changes corresponding to those from group B to group E will be experienced when ascending mountain slopes. In general, every 300 meters of mountain elevation will correspond to a change of climate type.

In addition to the drop of temperature with increased altitude, orography modifies precipitation and wind patterns in many ways. For example, a mountain's windward slopes typically receive more precipitation than its leeward slopes. Therefore, more dense vegetation grows on the windward slopes of large mountains, such as the western slope of the Rockies, than on the leeward slopes, such as the eastern slope of the Rockies. Often, the leeward foot of a mountain receives very little precipitation. These areas are often called "rain shadows." Leeward mountain foots are also subject to downslope mountain winds from time to time, especially during winters. These winds are called "chinook wind" in North America, or "foehn wind" in Europe.

The most prominent highland climate occurs over the Tibetan Plateau, where the average elevation is over 4,000 meters. In North America, highland climates characterize the Rockies, Sierra Nevada, and Cascades. In South America, the Andes create a continuous band of highland climate. Many of these mountains and highlands play central roles in monsoonal circulation, an important global climate system in various parts of the world.

Context

Global climate is a complex system that involves interactions among Earth's atmosphere, hydrosphere, cryosphere, lithosphere, and biosphere. In an even larger context, the global climate is just a part of the Sun-Earth system. For a particular place on Earth, the formation of the local climate pattern is dependent upon a set of climate controls. Despite its complexity, the global climate can be classified according to two basic physical parameters: mean temperature and precipitation. Future climate change can be measured and quantified by closely monitoring the change of these parameters in various parts of the world.

Chungu Lu

Further Reading

Ahrens, C. Donald. *Essentials of Meteorology: An Invitation to the Atmosphere*. 5th ed. Belmont, Calif.: Thomson Brooks/Cole, 2008. One of the most widely used introductory books on atmospheric science; covers a wide range of topics on weather and climate.

Geer, Ira W., ed. *Glossary of Weather and Climate*. Boston: American Meteorological Society, 1996. Collection of terms, concepts, and definitions related to weather and climate.

Hsiang, Solomon M., Kyle C. Meng, and Mark A. Cane. "Civil conflicts are associated with the global climate." *Nature* 476, no. 7361 (2011): 438-441. Although many authors have speculated on links between climate and civilization, there has been little rigorous study of this hypothesis. The authors find a significant correlation between the start of civil conflicts and El Nino events.

Lutgens, Frederick K., and Edward J. Tarbuck. *The Atmosphere*. 10th ed. Upper Saddle River, N.J.: Pearson Prentice Hall, 2007. Elementary introductory text that covers a wide range of atmospheric sciences.

U.S. Climate Change Science Program. *Our Changing Planet: A Report by the Climate Change Science Program and the Subcommittee on Global Change Research for Fiscal Year 2007*. Washington, D.C.: Author, 2006. A report of the Climate Change Science Program, established in 2002 to empower the U.S. national and global communities with sufficient science-based knowledge to manage risks and opportunities for change in the climate and related environmental systems.

See also: Climate and the climate system; Climate models and modeling; Continental climate; Maritime climate; Mediterranean climate; Monsoons; Ocean-atmosphere coupling; Polar climate; Weather vs. climate.

Global Climate Coalition

Category: Organizations and agencies
Date: Established 1989; disbanded 2002
Web address: http://www.globalclimate.org/ (deactivated)

Mission

The Global Climate Coalition (GCC) is a defunct industry group that worked to set the climate change agenda of the United States. This organization was influential and successful in its endeavors. The GCC dissolved amid increased understanding in climate science and the rise of public opinion in support of public policy to address climate change. The coalition met many of its goals prior to its dissolution.

The GCC was formed by representatives of multinational corporations reliant on the use of fossil fuels, particularly in the energy and transportation sectors. Prominent members of the GCC included BP, Royal Dutch Shell, Dupont, Ford Motor Company, Daimler Chrysler, Texaco, General Motors, and Exxon (now ExxonMobil). The mission of the GCC was to influence U.S. public policy regarding climate change.

The GCC was organized in reaction to the findings of the Intergovernmental Panel on Climate Change (IPCC). A GCC representative argued at the libertarian think tank the Cato Institute that the IPCC report was flawed and represented the views of pro-climate-change activists. The GCC rejected all government-mandated reductions of greenhouse gas (GHG) emissions in the United States. The GCC was against U.S. participation in the Kyoto Protocol. The coalition argued that climate science and the human impact on climate change were both too uncertain to justify coercive government policy intended to lessen GHG emissions. The GCC conducted lobbying and public relations campaigns to spread the skeptical view of climate change to the public.

Significance for Climate Change

The GCC was an influential factor in setting the U.S. climate change agenda. The GCC pressured the Bill Clinton administration to reject command-and-control environmental regulations intended to address climate change. The coalition argued that burdensome regulations designed to address climate change would decrease economic growth and hinder the economic competitiveness of U.S. companies. The GCC was an influential actor in preventing U.S. ratification of the Kyoto Protocol. In 1998, the Senate voted to reject the Kyoto Protocol 95-0.

The demise of the GCC began with the withdrawal of BP, a major multinational oil company. BP broke ranks with the GCC in May, 1997. The GCC lost members as increasing scientific evidence and rising consensus alarmed governments, and people across the world demanded action on climate change. In 2002, the GCC disbanded, stating, "At this point, both Congress and the Administration (Bush II) agree that the United States should not accept mandatory cuts in emissions required by the Protocol." The GCC claimed its mission had been accomplished, so there was no reason to maintain an active presence to shape the climate change agenda.

The skeptical view of climate change found representation in the George W. Bush administration. The administration took the view that climate science was uncertain and asked the National Academy of Sciences (NAS) for assistance in determining if IPCC reports had been tampered with by pro-climate-change-policy activists.

Although the GCC has dissolved as an organization, its activities made a lasting impact in setting the climate policy agenda in the United States. Many in the business sector continue to insist that there is no need to transform the global economy to address climate change. In June, 2008, National Aeronautics and Space Administration climatologist James E. Hansen argued that corporate executives who promote contrarian arguments against climate change theory should be held liable for their actions. Hansen contends such activities are "crimes against humanity and nature" and that executives from many fossil-fuel-intensive corporations consciously attempt to confuse the public on the state of understanding and agreement among climate scientists.

Justin Ervin

Further Reading

Bellamy Foster, John. *Ecology Against Capitalism.* New York: Monthly Review Press, 2002. Argues that climate change is an issue ill addressed by capitalist institutions and is caused by the drive for unending economic growth. Sees organizations such as the GCC as unsurprising given that global capitalism is reliant on the combustion of fossil fuels.

Brown, Lester R. *Eco-Economy: Building an Economy for the Earth.* New York: W. W. Norton, 2001. Documents the decline of the GCC as members began to define their corporate interest in terms of the need to address climate change.

Dunlap, Riley E., and Peter J. Jacques. "Climate change denial books and conservative think tanks: exploring the connection." *American Behavioral Scientist* (2013): 0002764213477096. Most earlier books skeptical of climate change were produced by authors having ties to conservative think tanks but in recent years, self-published books have risen sharply in number. The vast majority of climate denial books have not undergone peer review.

Lewis, Paul. "U.S. Industries Oppose Emission Proposals." *The New York Times*, August 22, 1995. Documents GCC resistance to U.S. participation in climate change negotiations. The GCC called on the U.S. government to study the impact that reductions in greenhouse gas emissions would have on the U.S. economy.

Pilkington, Ed. "Put Oil Firm Chiefs on Trial, Says Leading Climate Change Scientist." *The Guardian*, Monday, June 23, 2008. Discusses James E. Hansen's return to Congress after he first warned of climate change in 1988. Hansen claims industry groups have been consciously spreading misinformation on climate change that has slowed government reaction to address climate change.

See also: American Enterprise Institute; Brundtland Commission and Report; Catastrophist-cornucopian debate; Cato Institute; Competitive Enterprise Institute; Conservatism; Economics of global climate change; Skeptics.

Global dimming

Category: Physics and geophysics

Definition
The first widespread effort to measure solar radiation began during the 1957-1958 International Geophysical Year (IGY), when a global network of radiometer stations was established to measure the Earth's radiation budget. In the following years, scientists discovered that the amount of solar radiation reaching the Earth's surface was not constant. From 1960 to 1990, measurements of solar radiation decreased worldwide by 4 to 6 percent. The greatest rates of decline were in the heavily populated, industrialized regions of the Northern Hemisphere; smaller decreases were measured over the Arctic and Antarctic. The term "global dimming" was coined during the 1990's to describe this observed reduction in surface solar radiation.

Many scientists contend that global dimming results from the release of anthropogenic aerosols, or air pollution. Aerosols are solid particles or liquid droplets, typically 0.01 to 10 micrometers in size, that are produced by natural or anthropogenic sources and remain suspended in the atmosphere. Anthropogenic sources include the release of sulfur dioxide (SO_2) and black carbon from the combustion of fossil fuels. Some scientists contend that aerosols cause dimming by absorbing solar radiation or scattering it back into space and by modifying the optical characteristics and coverage of clouds. In clouds, anthropogenic aerosols act as cloud condensation nuclei. Their presence significantly increases the number of sites on which clouds droplets can form, leading to the formation of clouds with many more small droplets than would occur naturally. This process leads to the formation of clouds that reflect greater amounts of solar radiation because of their higher albedos, expanded coverage, and longer lifetimes.

Significance for Climate Change
Global dimming is important to the broader discussion of climate change for many reasons. For one, it has fostered a more focused discussion of the global impacts of anthropogenic aerosols. While scientists have long observed the large-scale impacts of aerosols, such as the global cooling that followed the eruption of Mount Pinatubo in 1991, air pollution was often regarded as a small-scale problem confined to urban and industrialized regions. Observations of global dimming have fostered a broader consideration of how aerosols impact the global climate system.

Some scientists hypothesize that global dimming may have masked the effects of global warming from the 1960's to the present. They suggest that observations of increased global temperature, often attributed to GHG emissions, may be more significant than observed, because global dimming was simultaneously acting to cool surface temperatures. This theory provides a basis to explain why surface temperatures rose by 0.3° Celsius from 1960 to 1990, while the amount of solar radiation reaching the Earth's surface declined by 4 to 6 percent.

Since 1990, the hypothesized impact of global dimming has raised concerns among global warming experts, because measurements of solar radiation suggest that dimming has waned in many parts of the world, a phenomenon that has earned the label "global brightening." Some scientists hypothesize that focused efforts to reduce air pollution, such as the United States' 1990 Clean Air Act Amendments, along with changes in the global economy, have fostered a worldwide reduction in aerosol emissions. Since aerosols have a much shorter residence time in the atmosphere than GHG emissions, there is a concern that continued efforts to reduce anthropogenic aerosols will cause global warming to proceed more rapidly over the next century than had been predicted.

In addition to its impact on global temperature, some scientists hypothesize that global dimming affects the hydrologic cycle by reducing the rate of evaporation and precipitation worldwide. Evidence for this hypothesis was originally uncovered from pan evaporation data collected over the latter half of the twentieth century, when scientists observed an annual decrease in the amount of water evaporated at many sites in the Northern Hemisphere. This result was unexpected, as it was generally thought that higher temperatures associated with global warming would increase rates of evaporation. Subsequent research, however, suggested that solar radiation may be a more important factor than originally believed in determining evaporation rates.

Provided this context, the reduction in solar radiation observed since the 1950's is viewed by many scientists as additional evidence for global dimming. It provides a basis to explain why evaporation rates decreased from 1960 to 1990 while global temperatures increased on average. In addition to the impacts outlined above, atmospheric scientists are investigating the broader impacts of global dimming on global circulation patterns, the distribution of rainfall worldwide, the vertical temperature profile, and the ability of plants to conduct photosynthesis under changing solar conditions.

Jeffrey C. Brunskill

Further Reading

Hansen, James, et al. "Earth's Energy Imbalance: Confirmation and Implications." *Science* 308, no. 5727 (June 3, 2005): 1431-1435. Presents a detailed discussion of how anthropogenic emissions of GHGs and aerosols affect the global climate system. Provides an important basis for reasoning about the response of the global climate system to changes in surface solar radiation and the ability of atmospheric models to predict future trends.

Roderick, Michael L., and Graham D. Farquhar. "The Cause of Decreased Pan Evaporation over the Past Fifty Years." *Science* 298, no. 5597 (November 15, 2002): 1410-1411. Discusses the pan evaporation paradox, the unexpected observation that evaporation rates decreased worldwide from 1950 to 2000 while global temperatures increased. The results of this study are regarded by many as evidence of global dimming.

Stanhill, Gerald. "Global Dimming: A New Aspect of Climate Change." *Weather* 60, no. 1 (January, 2005): 11-14. Stanhill was one of the first scientists to publish observations of decreasing surface solar radiation. This article summarizes the evidence for global dimming since the 1960's and discusses many of the causes and consequences of the phenomenon.

Wild, Martin. "Enlightening global dimming and brightening." *Bulletin of the American Meteorological Society* 93, no. 1 (2012): 27. Sunlight received at the earth's surface decined from the 1950's to the 1980's ("dimming") but has been increasing since then ("brightening"). These changes appear due to aerosol emissions into the atmosphere.

See also: Aerosols; Air pollution and pollutants: anthropogenic; Albedo feedback; Clouds and cloud feedback; Evapotranspiration; Hydrologic cycle; Rainfall patterns.

Global economy and climate change

Category: Popular culture and society

The global economy and climate change are two interrelated processes of globalization. As the economy grows, so do factors associated with climate change, which can both drive and impede economic growth.

Key concepts

capitalism: an economic system in which the means of production are privately owned and operated with the goal of increasing wealth

commodity: anything that has commensurable value and can be exchanged

economic interdependency: a state of affairs in which the economic processes of a group of nations are mutually dependent

globalization: the worldwide expansion and consequent transformation of socioeconomic interrelationships

industrialization: the process of transformation from an agrarian society based on animal and human labor to an industrial society based on machines and fossil fuels

liberal institutionalism: a school of thought that focuses on cooperation between countries derived from agreements and organizations

neoliberalism: a school of economic thought that stresses the importance of free markets and minimal government intervention in economic matters

Background

The global economy is the result of a process of increasingly global economic integration that began in the sixteenth century. European powers spread capitalism to colonies that provided cheap labor (including slaves), abundant natural resources, and vital new markets for goods and services. The expansion of European capitalism led to the export of industrialization in the nineteenth and twentieth centuries. The United States worked to reorganize the global economy following the instability of the early twentieth century that culminated with World War II. This reorganization was negotiated at Bretton Woods, New Hampshire, in 1944. The International Monetary Fund (IMF) and the World Bank were created at the Bretton Woods Conference. The General Agreement on Tariffs and Trade (GATT) was signed in 1947. Together, these institutions and agreements were designed to include as much of the world's people and territory as possible under a global capitalist economy. The Soviet Union and China limited this vision at the time. Members of the former Soviet Union and China have since become vital actors in the global economy.

Economic Globalization

The global economy represents a central process of globalization. As such, it increases economic interdependency. Economic globalization is made possible through advances in information and transportation technology. With the rise of technology, many of the world's countries have become interconnected through complex networks of economic production and consumption.

Economic globalization is composed of many actors. Countries legalize rules regarding international economic transactions. Multinational corporations dominate international economic production and the global trade in goods and services. International financial institutions finance the global flow of goods and services and provide money capital for foreign investment.

The Global Economy and Climate Change

The spread and growth of the global economy increases the energy intensity of the world's countries. The increase in the volume of exchange of goods and services, over greater distances, has increased the use of fossil fuels. More people drive more kilometers in automobiles and fly longer distances in airplanes. More people use increasing amounts of electricity, most of which is produced by fossil fuels. The expansion of global economic activity requires a continuing increase in the consumption of fossil fuel. The consumption of fossil fuel is the major factor causing climate change.

Economic globalization increases the human transformation of the environment. Deforestation results from the human need for lumber for construction and land for agriculture. Rising living standards lead to increased construction and the increase of the consumption of meat. Deforestation decreases the Earth's ability to remove carbon dioxide from the atmosphere. Increased meat consumption increases the concentration of the greenhouse gas (GHG) methane in the atmosphere.

Problems associated with the global economy and climate change have gained the attention of powerful actors within the global economy. The IMF and the World Economic Forum (WEF) agree that climate change, if left unchecked, is likely to

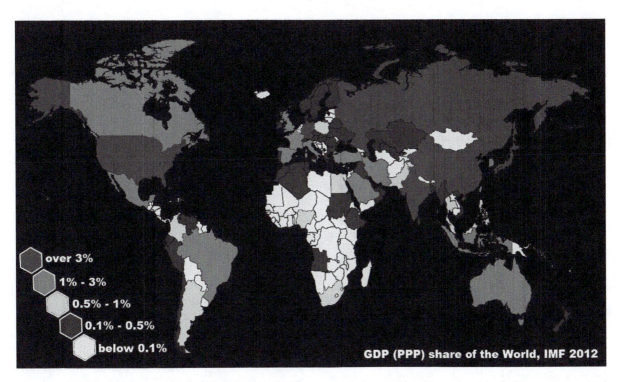

Gross domestic product based on purchasing-power-parity (PPP) share of world total in 2012. (Quandapanda/ Wikimedia Commons)

destabilize the global economy. In a report, produced by the White House in 2014, announced that climate change would likely cause a 40 percent decrease in global GDP. This could lead to a global economic depression and violent conflict. Climate change could dissolve global economic networks, creating shortages of vital economic inputs, leading to global economic decline.

Political Economic Theory and Climate Change

The impact the global economy has on climate change is addressed by theories of political economy. One's adherence to a particular theory greatly impacts the way one interprets the relationship between the global economy and climate change. For example, neoliberal economic theorists argue that global markets will distribute the technologies needed to address climate change. New technologies, such as wind generators, photovoltaic solar panels, hybrid automobiles, and fuel cells will circulate across the globe under free market capitalism.

Liberal institutionalist theorists agree with neoliberals about technological transfer. They argue, however, that the global economy requires active public management to address climate change. Liberal institutionalists cite the importance of cooperation among countries to address climate change. This cooperation is best realized in the form of international governmental organizations and agreements, such as the Kyoto Protocol.

Some theorists argue the global economy is unsustainable. These theorists propose dramatic transformations for the economy. Ecological economists argue environmental problems such as climate change are symptoms of the Earth no longer being able to assimilate human economic activity. Ecological economists argue that the global economy is unsustainably depleting Earth's natural capital at an ever-increasing rate. This condition cannot last indefinitely, because the Earth is a finite system. Ecological economists argue the global economy must attain an optimal scale or face devastating consequences.

Context

The global economy has created unprecedented opportunities and problems for humanity. The global economy has created unprecedented wealth, but it has also increased social instability while contributing to environmental problems such as climate change. Climate change transcends the ability of individual countries to create solutions. To confront climate change, countries of the world will have to cooperate at unprecedented levels. Rich countries will have to promote policies to help poor countries address climate change. Those living in rich countries and enjoying energy-intensive lifestyles have little right to demand economic sacrifices from the poor.

Humanity faces a serious economic paradox with climate change. In order to remain stable, the global economy must grow. However, to address climate change, this economic growth must be achieved while diminishing the factors responsible for climate change. Given a global economy that is based on the combustion of fossil fuels, this will be no easy feat. The global economy and climate change are interconnected but contradictory functions. As the global economy grows, the dangers of climate change increase. As the dangers of climate change increase, global economic growth is threatened. This relationship is critically important and offers no easy solutions.

Justin Ervin

Further Reading

Brown, Lester R. Plan B 3.0: Mobilizing to Save Civilization. 3d ed. New York: W. W. Norton, 2008.

Daly, Herman E. "Sustainable Growth: An Impossibility Theorem." In Valuing the Earth: Economics, Ecology, and Ethics, edited by Herman E. Daly and Kenneth N. Townsend. Cambridge, Mass.: MIT Press, 1996.

Ervin, Justin, and Zachary A. Smith. Globalization: A Reference Handbook. Santa Barbara, Calif.: ABC-CLIO, 2008.

Manyika, James, Michael Chui, Jacques Bughin, Richard Dobbs, Peter Bisson, and Alex Marrs. *Disruptive technologies: Advances that will transform life, business, and the global economy.* Vol. 12. San Francisco, CA: McKinsey Global Institute, 2013. Describes twelve future transformative technologies. Four are varieties of information technology, three are applications, three are energy technologies and genomics and advanced materials complete the list.

"Strengthening Community Forest Rights Is Critical Tool to Find Climate Change."World Resources Institute. World Resources Institute, 23 July 2014. Web. 20 Mar. 2015.

Wallerstein, Immanuel The Modern World System. New York: Academic Press, 1974.

Global energy balance

Category: Astronomy

Definition

The global energy balance is the overall inflow and outflow of energy to and from Earth. The different processes that control this balance are atmospheric reflectance, absorption by the atmosphere and Earth's surface, and reradiation of energy to space by Earth itself. Electromagnetic energy arrives in the form of solar radiation that passes through the atmosphere and interacts with the surface.

When electromagnetic energy hits an object, such as Earth, three processes can occur. Transmission occurs when the energy passes through an object without affecting it. Reflection occurs when the energy bounces off the object, and absorption occurs when the energy is taken up by the object. Incoming solar radiation exists at all wavelengths, but the dominant intensity is within the visible band (400-700 nanometers), at 550 nanometers. Of this energy, 30 percent is reflected from the atmosphere back into space, 25 percent is absorbed by the atmosphere, and the remaining 45 percent is absorbed by Earth's surface.

The energy absorbed by Earth's surface is converted to thermal energy and raises the surface temperature. Because the Earth itself is a warm object, it also emits electromagnetic energy to space. The energy emitted by Earth occurs in longer wavelengths than incoming solar radiation, with the maximum intensity at 10 millimeters. Of this energy, 70 percent passes through the atmosphere and is transmitted to space. However, 30 percent is reabsorbed by atmospheric greenhouse gases (GHGs), including carbon dioxide (CO_2), methane, nitrous oxide, and water vapor. This reabsorbed energy contributes to further warming of Earth. The balance between the amount of incoming solar radiation and the amount of reflected, absorbed, and transmitted energy controls global wind patterns and oceanic circulation. Shifts in this balance determine the overall warming and cooling of the planet.

Significance for Climate Change

Changes to the global energy balance, which is affected by both natural and anthropogenic processes, are a fundamental cause of climate change. Processes that affect the global energy balance are often intertwined with one another, resulting in various feedback mechanisms that can enhance or reduce the effects of the original change. Natural changes in the global energy balance can be caused by variations in Earth's orbit or in the output of solar radiation—both of which affect the amount of energy coming into Earth.

The shape of Earth's orbit and the tilt of Earth's axis change slightly over timescales ranging from 11,000 to 100,000 years. These changes can cause the Earth to be somewhat closer to or further away from the Sun, leading to a relative increase or decrease in the amount of solar radiation that reaches Earth. In the geologic past, these types of changes are manifested as climate shifts on timescales of millions of years, known as greenhouse and icehouse conditions. During greenhouse times, Earth is ice-free because of high concentrations of GHGs. During icehouse times, Earth is covered by variable proportions of ice, and the climate shifts between interglacial times (less ice) and glacial times (larger continental ice sheets) on timescales of tens of thousands of years. The most recent ice age peaked at the Last Glacial Maximum, approximately 20,000 years ago. Approximately 850-630 million years ago, Earth may have been completely covered by ice, creating a so-called Snowball Earth. This hypothesis, however, remains highly controversial among scientists. Blocking some proportion of incoming solar radiation remains a goal of certain geoengineering proposals as a means to prevent future global warming.

Changes in GHG concentrations also affect the global energy balance—and therefore, global temperature—because of changes in the amount of reradiated energy trapped on Earth. Volcanoes are natural sources of GHGs, and geologic evidence suggests that there is a correlation between widespread volcanism and greenhouse times in the past. Anthropogenic changes in the global energy balance are thought to be caused by the addition of GHGs to the atmosphere. The most notable of these is the production of CO_2 through the burning of fossil fuels, although the production of methane should not be discounted. Increased concentrations of GHGs in the atmosphere lead to increased absorption of energy radiated from Earth, which, in turn, leads to further warming of the surface. The Intergovernmental Panel on Climate Change (IPCC) estimates that anthropogenic changes in atmospheric CO_2 concentrations have resulted in a global temperature increase of 0.74° Celsius between 1905 and 2005. Removing CO_2 from the atmosphere and sequestering it in reservoirs such as the deep ocean, ocean sediments, or rock formations constitutes a second geoengineering approach to reduce future global warming.

Anna M. Cruse

Further Reading

Intergovernmental Panel on Climate Change. *Climate Change, 2007—The Physical Science Basis: Contribution of Working Group I to the Fourth Assessment Report of the Intergovernmental Panel on Climate Change.* Edited by Susan Solomon et al. New York: Cambridge University Press, 2007. Comprehensive treatment of the causes of climate change, written for a wide audience. Figures, illustrations, glossary, index, references.

Kunzig, Robert. "A Sunshade for Planet Earth." *Scientific American* 299, no. 5 (November, 2008): 46-55. Explains the basic premises of geoengineering and discusses the potential pros and cons of a variety of geoengineering approaches. Figures.

Rapp, Donald. *Assessing Climate Change: Temperatures, Solar Radiation, and Heat Balance.* New

York: Springer, 2008. Presents evidence for climate change—both natural and anthropogenic. Rapp is a systems engineer who attempts to consider the Earth system as a whole. Figures, tables, index, references.

Stephens, Graeme L., Juilin Li, Martin Wild, Carol Anne Clayson, Norman Loeb, Seiji Kato, Tristan L'Ecuyer, Paul W. Stackhouse Jr, Matthew Lebsock, and Timothy Andrews. "An update on Earth's energy balance in light of the latest global observations." *Nature Geoscience* 5, no. 10 (2012): 691-696. Despite their importance in climate modeling basic parameters of earth's energy balance remain poorly known. These include the amount of long wave radiation reaching the surface and the total amount of precipitation.

Walker, Gabrielle. *Snowball Earth: The Story of the Great Global Catastrophe That Spawned Life as We Know It.* New York: Crown, 2003. A presentation of the Snowball Earth hypothesis, as developed by Paul Hoffman, for nongeologists. Combines geological theory and data with stories of Hoffman's travels to remote locations to conduct his research. Index, references.

See also: Climate and the climate system; Climate engineering; Earth history; Earth motions; Glaciations; Greenhouse effect; Greenhouse gases and global warming; Interglacials.

Global Environment Facility (GEF)

Category: Organizations and agencies
Date: Established 1991

Mission

The Global Environment Facility (GEF) provides grants and concessional funds to recipient countries for projects and activities that protect the global environment. GEF programs focus on environmental problems involving climate change, biological diversity, international waters, and depletion of the ozone layer.

Jointly implemented by the United Nations Development Programme (UNDP), the United Nations Environment Programme (UNEP), and the World Bank, the GEF was launched as a pilot program in 1991. The GEF secretariat, which is functionally independent from the three implementing agencies, reports to the Council and Assembly of the GEF. The council is the main governing body and consists of representatives from thirty-two countries, eighteen of which receive GEF funds and fourteen of which are nonrecipient nations. Since 1994, any country that is a member of the United Nations or one of its specialized agencies has been able to become a GEF participant by filing a notification of participation with the GEF secretariat. By 2007, 178 countries were participating in the GEF program; Somalia became the 178th participant nation in that year.

Countries may be eligible for GEF project funds in one of two ways: by meeting the eligibility requirements to borrow from the World Bank or to receive technical assistance funds from the UNDP or by meeting the eligibility requirements for financial assistance through either the Convention on Biological Diversity or the UN Framework Convention on Climate Change (UNFCCC). In either case, the country must also be a participant in the Convention on Biological Diversity or the UNFCC to qualify for GEF funds.

The organizing principle of the GEF is that no new bureaucracy will be created to support it. The UNDP is responsible for technical assistance activities and helps identify projects and activities consistent with the GEF's Small Grants Program, which awards Nongovernmental organizations (NGOs) and community groups around the world. Responsibility for initiating the development of scientific and technical analysis and for advancing environmental management of GEF-financed activities rests with the UNEP, as does the management of the Scientific and Technical Advisory Panel, which provides scientific and technical guidance to the GEF. The World Bank serves as the repository of the GEF trust fund and is responsible for investment projects. It also seeks resources from the private sector in accordance with GEF objectives.

GEF resources aim to facilitate projects related to biodiversity, climate change, international waters, land degradation, the ozone layer, and persistent organic pollutants. Projects and activities must be approved by the GEF Council. If at least four council members request a review of the final project documents, the council will conduct the review prior to granting approval of GEF funds.

Significance for Climate Change

As of 2008, the GEF was supporting more than six hundred climate-change-related projects and programs in scores of developing nations around the world. These projects help those nations to contribute to the objective of the UNFCCC by supporting measures to reduce the risk or minimize the adverse effects of climate change. The climate-related projects GEF supports center on lowering greenhouse gas (GHG) emissions through adopting renewable energy sources, energy-efficient technologies, sustainable transportation methods, and clean energy sources.

The GEF also supports adaptation to climate change through programs that attempt to cope with the negative impacts of climate change faced by countries, regions, and communities at particular risk. In addition to allocating approximately $250 million per year for such projects, the GEF oversees several funds under the UNFCCC: the Least Developed Countries Fund (LDCF), the Special Climate Change Fund (SCCF), and the Special Priority on Adaptation (SPA). The latter supports programs and projects that may be integrated into national policies and sustainable development planning. The GEF also assists member nations in their communications with and reporting to the UNFCCC, including reports on GHG inventories.

Finally, the GEF sponsors outreach to educate the public. In December 2007, for example, the GEF joined forces with UNEP, the World Bank, Global Initiatives, and Artists' Project Earth to create Fragile Planet, a video about human impact on climate change.

Alvin K. Benson

Further Reading

Axelrod, Regina S., David Leonard Downie, and Norman J. Vig, eds. The Global Environment: Institutions, Law, and Policy. 2d ed. Washington, D.C.: CQ Press, 2004.

Global Environment Facility. Valuing the Global Environment: Actions and Investments for a Twenty-first Century. Washington, D.C.: Author, 1998.

Horta, Korinna, and Robin Round. The Global Environment Facility—The First Ten Years: Growing Pains or Inherent Flaws? Sterling, Va.: Pluto Press, 2002.

World Bank. The World Bank Group's Global Environment Facility Program (May 21, 2014). http://www.worldbank.org/en/topic/climatechange/brief/gef [accessed 3 July 2016]. Summarizes the aims of the Global Environment Facility, which funds environmental projects. Since 1991, the GEF has funded 790 projects in 120 countries.

Young, Zoe. A New Green Order? The World Bank and the Politics of the Global Environment Facility. Sterling, Va.: Pluto Press, 2002.

Global monitoring

Category: Meteorology and atmospheric sciences

To know what is happening with global climate, researchers must be able to determine exactly what has happened in the past, which requires good data taken at the same place over a period of several decades. Global monitoring is a collective term used in relation to the measurement of a wide variety of atmospheric and surface properties on a global or regional scale at frequent intervals.

Key concepts

global: relating to the whole Earth

homogeneous climate data: a sequence of values of a climate variable, such as precipitation, which have been observed under the same or similar conditions and with the same or similar measuring equipment; the combination of climate data from two localities that are near each other is often made when considering climate change *monitoring:* the systematic observation of an element such as precipitation, sea surface temperatures, or wind speed; such observations are usually made every six hours and sometimes every three hours, hourly, or (conversely) only once daily

observing systems: systems of collectively gathering and analyzing temperature and other atmospheric observations, particularly (in modern times) through the World Meteorological Organization

regional: relating to areas such as the Pacific, the Atlantic, the tropics, and large land areas, such as North America or Australia

Background

The monitoring of the climate and related variables on global and regional scales is essential if correct analyses of what has happened, and what

is happening, to the atmospheric climate are to be made. For example, it is often mentioned, especially by the media, that the climate is getting warmer, or cooler, or drier, or wetter, or more humid, or windier, and so on. What the true situation is, however, cannot be determined without rigorous and ongoing collection and examination of data. For example, are the polar ice areas increasing or deceasing, and are glaciers advancing or retreating? Are these trends different from what happened a few decades ago? The correct monitoring of the global climate and related variables is clearly necessary to advance our knowledge of the true situation.

Global Climate Monitoring System

Temperature and other atmospheric observations have been made in many parts of the world for

---

Global Monitoring Principles

According to the GCOS Web site, "effective monitoring systems for climate should adhere to the following principles":

1. The impact of new systems or changes to existing systems should be assessed prior to implementation.

2. A suitable period of overlap for new and old observing systems is required.

3. The details and history of local conditions, instruments, operating procedures, data processing algorithms and other factors pertinent to interpreting data (i.e., metadata) should be documented and treated with the same care as the data themselves.

4. The quality and homogeneity of data should be regularly assessed as a part of routine operations.

5. Consideration of the needs for environmental and climate-monitoring products and assessments, such as IPCC assessments, should be integrated into national, regional and global observing priorities.

6. Operation of historically-uninterrupted stations and observing systems should be maintained.

7. High priority for additional observations should be focused on data-poor regions, poorly-observed parameters, regions sensitive to change, and key measurements with inadequate temporal resolution.

8. Long-term requirements, including appropriate sampling frequencies, should be specified to network designers, operators and instrument engineers at the outset of system design and implementation.

9. The conversion of research observing systems to long-term operations in a carefully-planned manner should be promoted.

10. Data management systems that facilitate access, use and interpretation of data and products should be included as essential elements of climate monitoring systems. Furthermore, operators of satellite systems for monitoring climate need to:

   (a) Take steps to make radiance calibration, calibration-monitoring and satellite-to-satellite crosscalibration of the full operational constellation a part of the operational satellite system; and

   (b) Take steps to sample the Earth system in such a way that climate-relevant (diurnal, seasonal, and long-term interannual) changes can be resolved.

(continued)

(continued)

Thus satellite systems for climate monitoring should adhere to the following specific principles:

11. Constant sampling within the diurnal cycle (minimizing the effects of orbital decay and orbit drift) should be maintained.

12. A suitable period of overlap for new and old satellite systems should be ensured for a period adequate to determine inter-satellite biases and maintain the homogeneity and consistency of time-series observations.

13. Continuity of satellite measurements (i.e. elimination of gaps in the long-term record) through appropriate launch and orbital strategies should be ensured.

14. Rigorous pre-launch instrument characterization and calibration, including radiance confirmation against an international radiance scale provided by a national metrology institute, should be ensured.

15. On-board calibration adequate for climate system observations should be ensured and associated instrument characteristics monitored.

16. Operational production of priority climate products should be sustained and peer-reviewed new products should be introduced as appropriate.

17. Data systems needed to facilitate user access to climate products, metadata and raw data, including key data for delayed-mode analysis, should be established and maintained.

18. Use of functioning baseline instruments that meet the calibration and stability requirements stated above should be maintained for as long as possible, even when these exist on de-commissioned satellites.

19. Complementary in situ baseline observations for satellite measurements should be maintained through appropriate activities and cooperation.

20. Random errors and time-dependent biases in satellite observations and derived products should be identified.

Source: Global Climate Observing System.

more than a century, and in some places for more than two hundred years. Initially these observations were made by single entities, but during the last hundred years, particularly during the last fifty years, there has been an effort to centralize and consolidate these data.

The World Meteorological Organization (WMO) and its forerunner, the International Meteorological Organization (IMO), have, for more than one hundred years, been at the forefront of organizing research on and monitoring the world's climate. In particular, since 1992 the Global Climate Observing System (GCOS) has supported this research according to several key principles: operation of historically uninterrupted stations and observing systems should be maintained; high priority for additional observations should be

focused on data-poor regions, poorly observed parameters, and regions sensitive to change; and operators of satellite systems for monitoring climate need to sample the Earth system in such a way that climate-relevant (diurnal, seasonal, and long-term interannual) changes can be resolved.

Regional Monitoring

A good example of "regional monitoring" is the South Pacific Sea Level and Climate Monitoring Project (SPSLCMP), which was developed in 1991 as the Australian government's response to concerns raised by member countries of the South Pacific Forum over the potential impacts of human-induced global warming on climate and sea levels in the Pacific region. The first three phases of the project established a network of twelve high-resolution Sea Level Fine Resolution Acoustic Measuring Equipment (SEAFRAME) sea-level and climate-monitoring stations throughout the Pacific.

U.S. National Climate Center

Another example of "regional monitoring" is the Climate Prediction Center (CPC) of the National Weather Service (part of the National Oceanographic and Atmospheric Administration, NOAA). The CPC collects and produces daily and monthly data, time series, and maps for various climate parameters, such as precipitation, temperature, snow cover, and degree days for the United States, the Pacific islands, and other parts of the world. The CPC also compiles data on historical and current atmospheric and oceanic conditions, the El Niño-Southern Oscillation (ENSO) Index, and other climate patterns, such as the North Atlantic and Pacific Decadal Oscillations, as well as stratospheric ozone and temperatures.

U.K. Climate Research Unit

A significant global center for compiling temperature and other climate data sets is the Climate Research Unit (CRU) of the University of East Anglia in the United Kingdom. Some of the data produced are available online, and other sets are available on request. CRU endeavors to update the majority of the data pages at timely intervals. Data sets are available in the following categories: temperature, precipitation, atmospheric pressure and circulation indices, climate indices for the United Kingdom, data for the Mediterranean and alpine areas, and high-resolution gridded data sets.

World Data Center System

A good example of a "global center" is the World Data Center (WDC) system, which was created to archive and distribute data collected from the observational programs of the 1957-1958 International Geophysical Year (IGY). Originally established in the United States, Europe, Russia, and Japan, the WDC system has since expanded to fifty-two centers in twelve countries. Its holdings include a wide range of solar, geophysical, environmental, and human dimensions data. The WDC is maintained by Model and Data (M&D), which is hosted at the Max Planck Institute for Meteorology, located in Germany.

Context

In terms of climate change, the monitoring of the climate is extremely important. A number of internationally recognized groups collect, analyze, and publish—mainly through Web sites—climate and other data for various areas of the world, including the globe as a whole, the land as a whole, the ocean as a whole, and regions such as the tropics and polar areas. In addition to the climate variables, a wide variety of other variables are monitored on a global or regional scale, such as the extent of sea ice, sea surface temperatures, carbon dioxide, and methane. All of these values, when correctly analyzed, provide important indicators of what changes have occurred during the past (weeks to many decades) in the broader atmospheric environment. Such information is essential if correct answers to questions relating to the changing climate are to be obtained.

W. J. Maunder

Further Reading

Duren, Riley M., and Charles E. Miller. "Towards robust global greenhouse gas monitoring." *Greenhouse Gas Measurement & Management* 1, no. 2 (2011): 80-84. A robust system would include not just direct gas measurements but integration with models and statistical data like inventories to assess the effectiveness of mitigation strategies.

Kininmonth, William. *Climate Change: A Natural Hazard*. Brentwood, Essex, England: Multiscience, 2004. Kininmonth has a career in meteorological and climatological science and policy spanning more than forty-five years. His suspicions that the science and predictions of anthropogenic global warming extend beyond sound theory and evidence were crystallized following the release of the 2001 IPCC assessment report. His book gives information about global and regional monitoring on various scales.

National Aeronautics and Space Administration. "Solar Physics." Available at http://solarscience.msfc.nasa.gov/SunspotCycle.shtml. Presents data on the solar cycle, the Maunder Minimum, and other Earth-Sun interactions.

Singer, S. Fred, and Dennis T. Avery. *Unstoppable Global Warming: Every Fifteen Hundred Years*. Lanham, Md.: Rowman and Littlefield, 2008. This book is dedicated to those thousands of research scientists who have documented evidence of a fifteen-hundred-year climate cycle over the Earth. Refers throughout to various aspects of monitoring of the global climate.

University of Colorado at Boulder. "Sea Level Change." Available at http://sealevel.colorado.edu/results.php. Presents tables, maps, time series, and other data on global sea level.

University of East Anglia. "Climatic Research Unit." Available at http://www.cru.uea.ac.uk/. The CRU presents data, information sheets, and the online journal *Climate Monitor*.

University of Illinois. "The Cryosphere Today." Available at http://arctic.atmos.uiuc.edu/cryosphere/. Offers frequently updated data on the current state of Earth's cryosphere.

U.S. National Space Science and Technology Center. Available at http://www.nsstc.org/. The mission of the NSSTC, an arm of NASA, is "to conduct and communicate research and development critical to NASA's mission in support of the national interest, to educate the next generation of scientists and engineers for space-based research, and to use the platform of space to better understand our Earth and space environment and increase our knowledge of materials and processes."

See also: Air quality standards and measurement; Geographic Information Systems in Climatology; Ice cores; Industrial emission controls; Mauna Loa Record; Tree rings.

Global surface temperature

Category: Meteorology and atmospheric sciences

Definition

Global surface temperature is an estimate of global mean air temperature at the Earth's surface, based on thermometer measurements made at land-based weather stations. Because about 70 percent of the Earth's surface is covered by water, land-based data are supplemented by sea-surface-temperature (SST) measurements. Estimates of air temperature are based on the assumption that there is a simple link between the SST and that of the air above. Usually, SST measurements are based on measurements of the temperature of seawater that is taken aboard ships for use as an engine coolant. In the past, SST measurements were made of water taken from buckets tethered to ropes and thrown overboard. Supplementary SST data are gathered from data buoys, small island stations, and shipboard, nighttime measurements of marine air temperatures.

Significance for Climate Change

Three agencies have taken responsibility for the combined global surface temperature record: the Climate Research Unit of the University of East Anglia; the U.S. National Aeronautics and Space Administration's Goddard Institute for Space Studies; and the Global Historical Climate Network of the U.S. National Oceanographic and Atmospheric Administration. To determine global surface temperature changes over time, these agencies locate and analyze anomalous departures from thirty-year temperature averages. These analyses are most commonly based on the area-weighted

---

Deviations in Mean Global Surface Temperature

The following table lists deviations in average global surface temperature from the baseline temperature average set during the period between 1951 and 1980.

Year	Deviation (in 0.01° Celsius)	Year	Deviation (in 0.01° Celsius)
1880	−12	1960	−2
1890	−21	1970	+4
1900	−6	1980	+28
1910	−21	1990	+48
1920	−17	2000	+42
1930	−4	2005	+76
1940	+14	2008	+54
1950	−17		

Data from Goddard Institute for Space Studies, National Aeronautics and Space Administration.

global average of sea-surface-temperature anomalies and land-surface-air-temperature anomalies. Based on these analyses, the mean global surface temperature of the Earth shows a warming in the range of 0.3°-0.7° Celsius over the past century, or a statistical average of about 0.003°-0.007° Celsius per year. Global data sets from the various agencies show slightly dissimilar trends, as the data are processed in different ways.

Questions arise as to the representative nature of the data on which global surface temperature calculations have been based. These data come from weather stations unevenly distributed over the Earth's surface, mostly on land, close to towns and cities, and predominantly in the Northern Hemisphere. Land use can have significant effects on local climate. The best-documented examples of such effects are urban heat islands, which are significantly warmer than their rural surroundings. Cities replace natural vegetation with surfaces such as concrete and asphalt that can warm them by several degrees Celsius. Thus, a disproportionate number of weather stations being located in urban environments may skew data regarding global averages. Many weather stations are located at airports, which were originally located in rural areas that have since been developed. Thus, while the data collected at such stations remain reliable measurements of the local, urban environment, they may no longer be equally reliable indicators of global trends.

C R de Freitas

See also: Albedo feedback; Atmospheric dynamics; Climate and the climate system; Climate change; Heat vs. temperature; Sea surface temperatures.

Global Warming Hiatus

Category: Climactic events and epochs

The term Global Warming Hiatus refers to the global mean temperature either staying at levels similar to past levels or increase of the global mean temperature at a significantly

lower rate than the long-term average. Various environmental factors such as natural fluctuations in ocean temperatures and cloud and sun variations can have a cooling effect on the earth's surface temperature and thus, contribute to the occurrence of a global warming hiatus.

Key Concepts

Global Warming: Increase in the average global surface temperatures due to human use of greenhouse gases.

Global Warming Hiatus: Refers to a slowing down or complete stagnation of the average (mean) global temperature increase. Sometimes can be caused by decreases in greenhouse gas and fossil fuel emissions as well as factors attributed to natural variability.

Climate Change: Significant and long-term change in climate patterns on the earth, often caused by global warming, emissions of greenhouse gases, burning of fossil fuels, and other human as well as naturally occurring factors that have an effect on the overall climate.

Mean Global Surface Temperature: The average temperature of the earth measured from the level of the earth's surface.

Greenhouse Gases: Gases that absorb infrared radiation.

Positive and Negative Anomaly: A positive anomaly is a term used primarily by scientists measuring earth temperatures to designate that a temperature is higher than the long-term average. In contrast, a negative anomaly designates that a temperature is cooler than the long-term average.

Background

In 2001, measurements of the earth's global surface temperature taken by the four major agencies that attempt to create a complete record of global temperatures discovered something unusual. All four records showed that while carbon dioxide emissions continued to increase, the mean (average) global surface temperature did not appear to significantly increase. The reasons for this occurrence were debated among scientists. Some claimed that this development was at least partially due to natural fluctuations in ocean temperatures, while others thought that the term "global warming hiatus" was wrong or at best misleading because of the fact that carbon dioxide emissions continued to increase despite surface temperature increases significantly slowing down. Whatever the case, this slowdown in global surface temperatures is unusual, as the global surface temperature on earth has been steadily and consistently increasing since 1972 according to the agencies that specialize in recording and measuring global temperatures. While global temperatures do fluctuate to some degree due to natural variability factors such as cloud and sun patterns as well as ocean temperatures, findings from the early 2000s show a slowdown of global surface temperature increases that surpasses common naturally occurring fluctuations in global temperature.

Factors and Causes of Hiatus

While the beginning of the most recent global warming hiatus is said to have started in 2001, records of global temperatures show a slowing down of temperature increases starting in the late 1990s. However, it is not clear as to how much of this trend was due to the events of El Nino in 1998 and La Niña in 2000. As global temperatures fluctuate to some degree, the cooling effect of ocean temperature fluctuations (especially from the Pacific Ocean) does contribute to the warming slowdown. Despite the significant slowing of global surface warming, sea levels and atmospheric greenhouse gas levels have steadily continued to rise during the period of the hiatus, showing that overall global warming did not stop. These factors seem to suggest that the term "hiatus" is somewhat misleading, as it is clear that the temperature increase never completely stopped and even global surface temperatures (the measurement that showed the most drastic warming slowdown) slightly went up throughout the early 2000s .

However, starting in 2014, global temperatures have increased steadily again, suggesting that the global warming slowdown of the early 2000's may be ending. In fact, many leading scientists say that 2015 was the hottest year ever recorded as per global surface temperature measurements. While global surface temperatures are generally the prevailing standard by which global warming is currently measured, oceanic and atmospheric data also show a steady and clear warming trend (this trend did not stop during the time period of the supposed hiatus). This recent finding further supports the widely held belief among scientists and environmentalists that global warming continued throughout the period of the global warming slowdown of the early 2000s but perhaps showed up in different and less obvious ways. The increasingly erratic and inconsistent weather patterns of recent years also support the evidence that the global warming slowdown has already ended or is soon coming to an end.

The Big Global Warming Hiatus

From roughly 1940 to1972, a cooling phenomenon occurred in which there was a near zero increase in global temperatures. This is often referred to as the "big" global warming hiatus, due to the longer period of occurrence than the recent global warming hiatus as well as to the more dramatic cooling observed during this period. The big global warming hiatus is largely attributed to solar dimming caused by increased airborne pollution. After World War II ended, there was a boom in industry and no major pollution control measures were in place, resulting in a steady increase of aerosols released into the troposphere. This, combined with more frequent volcanic eruptions that released a significant amount of aerosols into the stratosphere, contributed to the cooling trend as aerosols reduce the amount of incoming solar energy by reflecting the incoming energy back into space. However, it is important to note that methods of measuring ocean temperatures were changed in 1945, which means that the records from this period may not be completely accurate compared with current temperature measuring standards.

In 1955, the Air Pollution Control Act was passed in the United States to combat increasing air pollution. The Clean Air Act was enacted in 1970, further reinforcing and improving upon the 1955 act. The Clean Air Act required air quality standards to be set by the United States Environmental Protection Agency. This development also required individual states to submit plans for adhering to and maintaining the standards of air quality specified in the act. This is important to note as the big global warming hiatus ended in 1972, two years after the act was passed, showing the act's effectiveness in lowering aerosol emissions. However, greenhouse gas emissions continued to increase in the 1970's, and without the high levels of aerosols to offset the warming due to greenhouse gases and widespread deforestation, global warming resumed.

Measurement of Global Temperatures

Scientists measure and ultimately compute global temperature levels by a combination of measurements taken from the air, earth surface, and ocean surface. Ships and buoys are used as stations where scientists take comprehensive measurements of ocean temperature levels, while earth surface levels are measured on land stations and air levels are often taken by satellite. Air, surface, and ocean temperatures are taken every day and are constantly compared to "normal" levels for the location and time the measurement was taken. The term "normal" in this context refers to the 30 year long-term average for that time and location. Day to day levels are averaged among all of the different stations that exist throughout the world on a monthly, seasonal, and yearly basis. When measurements deviate from the long-term average for the time and location, the specific differences from normal levels are studied to attempt to determine the potential causes of the discrepancy. Differences from the long-term average are known as anomalies. As discussed in the "key concepts" section, a positive anomaly means that the temperature is warmer than the long-term average and a negative anomaly refers to the temperature being colder than the long-term average.

The data currently used and averaged to come up with the yearly global temperatures is compiled by four major datasets located in the United Kingdom, United States, and Japan. The United Kingdom Met Office Hadley Centre and University of East Anglia Climatic Research Unit collaborate to form the HadCRUT4 dataset. In the United States, the NASA Goddard Institute for Space Sciences produces the GISTEMP dataset, and the National Oceanic and Atmospheric Administration (NOAA) create the MLOST dataset. In Japan, the Japan Meteorological Agency produces the JMA record. The yearly records of these datasets are averaged to compute the global temperature. While the results of these four major datasets are consistently similar, slight differences in the reports do sometimes occur and are studied, taken into account and debated among scientists. Attempts at producing a complete record of global temperatures by agencies began in 1850 in the United Kingdom, 1880 in the United States, and 1891 in Japan, and air temperature measurements from the troposphere date back to 1979 due to satellites being needed to take temperature measurements. Recent data are compared with older data typically in 15, 30, and 50 year chunks to better pinpoint recurring and/or unusual trends and fluctuations in global temperatures.

Conclusion

The so-called global warming hiatus that started in 2001 seems to be currently ending based on the evidence of recent findings that show that global temperatures have resumed their previous steady increases. The term "hiatus" was commonly used in the early 2000s to refer to the then current trend of a significant slowing down of global surface temperature increases. However, most scientists today agree that the hiatus was overstated at the time and would have been more accurately described as a "warming slowdown," especially considering that the global surface temperature continued to rise throughout the period of the hiatus, only at a significantly less drastic level than had been the case before the hiatus occurred. This slowdown in global surface temperature increases has been partially attributed to ocean temperature fluctuations having a cooling effect on the earth, though atmospheric and oceanic temperature measurements continued to show steady increases, suggesting that global warming never stopped; rather it manifested in different areas of the earth.

Benjamin Riley

Further Reading

Hawkins, Ed. "Making Sense of the Early 2000s Warming Slowdown." *Climate Lab Book*. 24 February 2016. Web. 26 May 2016. This article explores the validity of the global warming hiatus of the early 2000s and provides in depth information about global temperature trends and fluctuations. The article makes use of relevant data to prove that there has been a significant global warming slowdown and to inform the reader about recent global warming levels using a well researched and facts based approach.

Pidcock, Roz. "How Do Scientists Measure Global Temperature." *Carbon Brief.* 16 January 2015. Web. 26 May 2016. An excellent and comprehensive article that explains how global temperatures are measured. The article also discusses the four major datasets that collect and track global surface temperatures to compute the yearly global temperature, as well as explaining how and why discrepancies occur between datasets and how they are resolved. Terrific article for those who are curious about how global temperatures are measured.

Kosaka, Yu, Shang-Ping Xie. "Recent Global Warming Hiatus tied to Equatorial Pacific Surface Cooling." Research study, 2013. Very comprehensive scholarly article that discusses the connection between a recent surface cooling trend in the Pacific Ocean and the global warming hiatus. Results, conclusions, and methods used in Kosaka and Xie's simulation study are discussed, as well as in depth information about global warming trends, factors relating to the recent global warming hiatus.

Gillis, Justin. "Global Warming 'Hiatus' Challenged by NOAA Research." *New York Times* 4 June 2015. Print. Discusses how new research conducted by the NOAA challenges the claim of a global warming hiatus in the early 2000s due to the claim being made based on data that may have been inaccurate or outdated. The article also briefly explains global temperature measurement methods and interviews scientists both affiliated and not affiliated with the NOAA.

See also: Global warming potential, Ocean-atmosphere coupling, Greenhouse gases and global warming; El Niño-Southern Oscillation and global warming

Global warming potential

Category: Pollution and waste

Definition
Global warming potential (GWP) is an index based upon radiative (infrared-absorbing) properties of well-mixed greenhouse gases (GHGs). It can be used to estimate the relative potential future impacts of GHG emissions on the global climate. Specifically, the GWP of a given GHG is the time-integrated global mean radiative forcing from the instantaneous release of 1 kilogram of that GHG, relative to that of a reference GHG, usually carbon dioxide (CO_2). The GWP is a function of a gas's lifetime, concentration, and effectiveness at absorbing thermal infrared radiation.

Significance for Climate Change
GWP is the index used by parties to the United Nations Framework Convention on Climate Change (UNFCCC) to quantify the greenhouse-enhancing potential of trace atmospheric gases. The GWP is used to evaluate the effects of anthropogenic interference in the climate system and of GHG emission reductions. GWP is a purely physical index and does not take into consideration the costs and benefits of climate policy initiatives, cost discounting, or the location of emissions or regional climates, among other factors. Thus, GWP does not

Global Warming Potentials of Major Greenhouse Gases

Greenhouse Gas	Global Warming Potential*
CO_2 (carbon dioxide)	1 (reference)
CH_4 (methane)	25
N_2O (nitrous oxide)	298
HFC-23 (hydrofluorocarbon-23)	14,800
SF_6 (sulfur hexafluoride)	22,800

One-hundred-year time horizon.

measure economic, cultural, or regional factors useful in determining climate- and GHG-related policies.

Other limitations of the GWP index exist. The choice of time horizon for integration ranges from short-term changes (such as responses of cloud cover to surface temperature changes) to long-term effects (such as changes in ocean level). For purposes of the first commitment period under the Kyoto Protocol, the parties to the UNFCCC agreed to a one-hundred-year time horizon for integration of GWPs as a mid-term balance between long-term (five-hundred-year) and short-term (twenty-year) climate effects. The current one-hundred-year GWPs for select GHGs are listed in *Climate Change, 2007*, a report prepared by the Intergovernmental Panel on Climate Change (IPCC).

The mid-term time horizon favors reductions in GHGs whose GWP is greatest over one-hundred-year periods, while diminishing the importance of reducing short-lived GHG emissions in the near future and long-lived GHGs over the long term. For example, GWPs for methane over twenty-year, one-hundred-year, and five-hundred-year time horizons are 72, 25, and 7, respectively. The value of reducing methane under analyses based on the short-term time horizon is nearly three times that of reducing it according to analyses employing the mid-term time horizon. Short-term analyses, however, have motivated the United States to create the Methane to Markets initiative, which seeks to capture or reduce methane emissions to address near-term climate change cost-effectively.

Uncertainties in the GWP arise from assuming a linear radiative forcing function under a small emission pulse (1 kilogram). That is, GWP calculations assume that 100 kilograms of a GHG will have a GWP one hundred times that of 1 kilogram of that GHG. This is not necessarily the case, as emissions of a GHG may build on or interfere with each other, cause positive or negative feedback loops, or otherwise increase or decrease their effects on climate in a nonlinear manner. As more information on radiative forcing is gathered, confidence in GWP values will increase.

Kathryn Rowberg

See also: Annex B of the Kyoto Protocol; Carbon dioxide; Carbon dioxide equivalent; Certified emissions reduction; Emissions standards; Greenhouse Gas Protocol; Greenhouse gases and global warming; Kyoto Protocol; United Nations Framework Convention on Climate Change.

Gore, Al
American statesman and environmental activist

Born: March 31, 1948; Washington, D.C.

A longtime congressional and administrative governmental advocate for environmental issues, Gore was awarded the 2007 Nobel Peace Prize for his work in raising awareness about such issues, particularly global warming.

Life

Al Gore was born on March 31, 1948, in Washington, D.C., the son of then-congressman, later U.S. senator, Albert Arnold Gore, Sr., of Carthage, Tennessee, and his wife, Pauline LaFon Gore. Gore attended Harvard University from 1965 to 1969, earning a bachelor of arts in government. He saw military service in Vietnam with the Twentieth Engineers from 1970 to 1971 and spent a year in the divinity school at Vanderbilt University before landing a position as a reporter for *The Tennessean*, Nashville's largest daily newspaper. After two years at Vanderbilt Law School, Gore was elected to the U.S. House of Representatives in 1976.

He was re-elected three times; in 1984, he ran successfully for the Senate, where he served until 1993. In 1988, he announced his intention to enter the race for the Democratic Party's presidential nomination. Failing to overtake front-runner and eventual nominee Michael Dukakis of Massachusetts, Gore dropped from the race after losing heavily in the New York State primary election. Apart from his environmental efforts, Gore's most notable legislative activity lay in the field of informational technology. As chair of the House Committee on Science and Technology, he was the most active legislative advocate for promoting the Internet. While in the Senate, he sponsored the High Performance Computing and Communication Act of 1991. His contribution in the crafting of this piece of legislation was so pervasive that it was dubbed the Gore Act.

Gore initially decided against involvement in the 1992 presidential campaign, citing family concerns arising from his son's near-fatal traffic accident and lengthy recuperation. However, when Democratic nominee Bill Clinton offered him the vice presidential slot on the party ticket, the Tennessee senator accepted. His strong performance in the television debate with Republican vice presidential candidate Dan Quayle and independent candidate Admiral James B. Stockdale is seen by some as a factor in the Democratic electoral victory in November, 1992. He served as vice president for the duration of the Clinton administration, 1993-2001.

Breaking somewhat with Clinton in the wake of the Monica Lewinsky scandal, Gore easily secured the Democratic presidential nomination in 2000, selecting U.S. senator Joseph Lieberman of Connecticut as his running mate. Though Gore won the popular vote over George W. Bush, it was the Republican candidate who took the presidency by tallying a margin of 271-266 votes in the Electoral College in a campaign marred by a ballot recount controversy in Florida.

Climate Work

Gore himself attributes the genesis of his interest in the environment to listening to his mother reading from the groundbreaking book *Silent Spring* (1962) by Rachel Carson. Then, while studying at Harvard, he was much influenced by the lectures and research of Roger Revelle, one of the world's first academics to monitor atmospheric carbon dioxide levels scientifically and to tie them to global temperature increases. So active was Gore at taking the lead in advocating green causes during his congressional and

senatorial tenures that he earned the nickname Mr. Ozone. In 1992, Gore published *Earth in the Balance*, wherein he pressed for what he termed a "global Marshall Plan" comprising a united effort by the more technologically advanced nations to set aside specific funding in order to assist underdeveloped countries in coping with their own ecological crises. However, apart from raising awareness about the global warming issue through committee hearings and publications, Gore was able to push through little in the way of far-reaching ecological legislation.

It would not be until his vice presidential term that some of his ideas would effectively permeate

---

Things Fall Apart

In Earth in the Balance, *Al Gore describes the societal breakdown that occurred in the early nineteenth century in response to climate anomalies caused by a volcanic eruption. He suggests that such breakdowns will be repeated and worsen as Earth's climate systematically alters.*

Beginning in 1816, "the year without a summer," widespread crop failures led to food riots in nearly every country of Europe, producing a revolutionary fervor that swept the continent for three years.

In France, for example, the existing government fell and the conservative Duc de Richelieu was asked to form a new one. Everywhere governments struggled to maintain social order as an unprecedented crime epidemic surged in the cities. The Swiss were stunned by the wave of criminal activity. Even the number of suicides increased dramatically, along with executions of women for infanticide. . . .

Although no one realized it at the time, the proximate cause of this suffering and social unrest was a change in the composition of the global atmosphere following an unusually large series of eruptions of the Tambora volcano, on the island of Sumbawa, Indonesia, in the spring of 1815. Scientists estimate that 10,000 people were killed in the initial eruption and approximately 82,000 more died of starvation and disease in the following months. However, the worst effects on the rest of the world were not felt until a year later, by which time the dust ejected into the sky had spread throughout the atmosphere and had begun to dramatically reduce the amount of sunlight reaching the surface of the earth and to force temperatures down.

through higher governmental circles, generating some interest from the Clinton administration. Even at that, efforts in that direction would be hampered by conservative Republican electoral gains that resulted in Republican control of Congress. Increasingly focused on the 2000 presidential campaign, Gore began drawing criticism from environmental groups for his apparent inactivity. Even his greatest effort in the cause of alleviating the threat to global warming posed by greenhouse gases—his intervention to save the Kyoto Protocol on Climate Change—elicited mixed reactions. Negotiations almost broke down over the question of whether developing countries could be held to the same gas emission standards as the more technologically advanced nations. The U.S. Senate threatened to sabotage the treaty if Third World nations were not equally accountable. In December, 1997, Gore traveled personally to Kyoto and, probably putting his political career on the line, was able to devise a compromise calling for accountability, but also for lower emission standards.

During the course of his post-vice presidential years, Gore has worked for environmental policy change through the Alliance for Climate Protection, a nonprofit educational and lobbying organization dedicated to reversing global warming, and Generation Investment Management, a business corporation he founded in 2004 to encourage investment in firms that demonstrate environmental consciousness. In 2006, Gore gained his greatest international fame through the simultaneous release of a film and book titled *An Inconvenient Truth*. The DVD recording of the film, especially, in which Gore himself adroitly lectured upon and explained global warming at the popular level, was so acclaimed that it earned the former vice president the 2007 Nobel Peace Prize, which he shared with the Intergovernmental Panel on Climate Change, and a share of the credit for the 2006 Academy Award for Best Documentary Feature.

Raymond Pierre Hylton

Further Reading

Al Gore, The Future: Six Drivers of Global Change, New York, Random House, 2013. Gore focuses on economic globalization, digital communications, shifting balances of power, a "flawed economic compass," biotechnology, neuroscience, and changes in the relationship between humans and the earth.

Gore, Al. *Earth in the Balance: Ecology and the Human Spirit.* New York: Rodale Books, 2006. The first major expression of Gore's assessment of the environmental crisis and his earliest "blueprint" for action focused on solutions.

_____. *An Inconvenient Truth: The Planetary Emergency of Global Warming and What We Can Do About It.* Reprint. New York: Rodale Books, 2006. The fine tuning, expansion, and elaboration of Gore's ideas as initially expressed in *Earth in the Balance*. This book—in conjunction with the film of the same title—brings the issues to a more popular level and is generally considered to depict Gore at his most engaging and articulate.

Maraniss, David, and Ellen Nakashima. *The Prince of Tennessee: The Rise of Al Gore.* New York: Simon & Schuster, 2000. Reveals Gore's considerable prior involvement in environmental matters, its extent and nature, and its role in his formation as a political figure.

Turque, Bill. *Inventing Al Gore: A Biography.* New York: Houghton Mifflin, 2000. Takes the sweeping view that Gore's environmental focus derives from a religious zeal born out of formative childhood experiences.

Zelnick, Bob. *Gore: A Political Life.* Washington, D.C.: Regnery, 1999. Categorizes Gore's commitment to global warming issues as not only long-standing but also quasi-religious in nature.

See also: Carson, Rachel; Environmental movement; *Inconvenient Truth, An*; U.S. legislation.

Greenhouse effect

Category: Meteorology and atmospheric sciences

Definition

The greenhouse effect warms the lower portion of a planet's atmosphere when heat is trapped there by gases—such as water vapor, carbon dioxide (CO_2), methane, and nitrous oxides—that prevent it from escaping into space. As a result of their molecular structure, these gases are dominant absorbers and emitters of infrared radiation. They absorb infrared energy from a planet's surface and reemit it in all directions. A significant fraction of this reradiated energy is directed back to the planet's surface, resulting in an increase in average temperatures. This effect is somewhat analogous to the trapping of heat by a greenhouse, but the retention of heat in a greenhouse is due mostly to reduced cooling caused by the prevention of convection: Only a small amount is due to trapped infrared radiation.

Significance for Climate Change

The greenhouse effect is a natural phenomenon that has been occurring on Earth and other planets for millions of years. It allows Earth to support life. If heat were not trapped in Earth's atmosphere, the planet would be approximately 33° Celsius cooler than it is now. A large percentage of Earth's natural greenhouse warming is caused by water vapor. If the greenhouse effect were enhanced, the Earth would become warmer, which could cause problems for humans, plants, and animals.

In the mid-1950's, an enhanced greenhouse effect was recognized as a concern. As a result of anthropogenic (human-induced) activities, atmospheric concentrations of the greenhouse gases (GHGs) were on the rise. This trend was associated with an increasing global atmospheric temperature. Industrialization resulted in an increase in the use of fossil fuels, which increased GHG emissions. The global mean annual temperature rose by approximately 0.5° Celsius between 1890 and 2000. Most of that increase occurred after 1970.

In addition to rising global temperature, observations of glaciers indicate that more of them are retreating than are growing. For example, eight glaciers that were advancing on Mount Baker in the northern Cascades in 1976 were all melting back at their termini by 1990. Observed climate changes associated with rising global temperatures, retreating glaciers, and increased polar ice melting are consistent with the effects that might be produced by an enhanced greenhouse effect accelerated by burning more fossil fuels. Verification of this connection depends primarily on computer-generated climate models. Existing models must be improved by addressing the coupled, nonlinear, dynamic nature of climate and by expanding the climatological database to include additional factors that influence the greenhouse effect, such as volcanic production of GHGs and the emission of methane from landfills and melting tundra.

Alvin K. Benson

See also: Atmosphere; Atmospheric chemistry; Atmospheric dynamics; Atmospheric structure and evolution; Carbon dioxide; Climate and the climate system; Climate change; Enhanced greenhouse effect; Greenhouse gases and global warming; Industrial greenhouse emissions; Methane.

Greenhouse Gas Protocol

Category: Economics, industries, and products
Date: Published September, 2001; revised March, 2004

Definition

Prior to 2001, there was not a tool for corporations to assess greenhouse gas (GHG) emissions. To remedy this void, the World Business Council for Sustainable Development (WBCSD) and the World Resources Institute (WRI) began work on a method for corporations to account for and report their GHG emissions. The result was the Greenhouse Gas Protocol. WBCSD is a coalition of 170 international companies whose shared interest is sustainable development guided by growth economically balanced with ecology and social progress. Members come from more than thirty-five countries. WRI is a nonprofit organization of over one hundred people from many different occupations who work toward protecting the Earth while improving people's lives. There were five ideals controlling the production of the protocol: relevance (that the program provide accurate information to those needing the information, both inside and external to the company), completeness (that the program account for all sources of emissions), consistency (that the program produce meaningful and useful data year after year), transparency (that the procedures be apparent to anyone interested), and accuracy (that emissions are measured correctly with the smallest possible degree of uncertainty). The protocol was published in September, 2001, and in March, 2004, a revised version was published.

Significance for Climate Change

The accounting tool consists of three modules: The first one is for corporations to use in calculating and reporting GHG emissions, the second is for calculating indirect emissions, and the third is used to calculate the effect of reduction projects. The GHGs covered by this protocol are the gases covered by the Kyoto Protocol: carbon dioxide (CO_2), methane (CH_4), nitrous oxide (N_2O), hydrofluorocarbons (HFCs), perfluorocarbons (PFCs), and sulfur hexafluoride (SF_6).

There were five objectives used in the design of this accounting tool. First, it would be a tool that would aid companies in making a true and accurate measure of their GHG emissions. Second, the procedure would be simpler and less expensive than current practices. Third, it would provide the means to acquire the information needed to control and reduce GHG emissions. Fourth, the protocol would provide the information to allow companies to join GHG programs, both voluntary and required. Fifth, the protocol is to increase communication between companies and provide a standard method to report emissions. The procedures are divided into three scopes. Scope 1 is derived from emissions that are produced by equipment or machinery owned or controlled by the company. Most of these emissions are direct emissions; some activity of the company causes the emission. Scope 2 is the electricity indirect GHG emissions. These are the emissions generated in producing the electricity used by the company in its equipment or machinery. This includes electricity that is lost during transport or distribution. Scope 3 is the emissions due to the exploration, production, and transportation of the fuel used to generate electricity.

The 2004 revised edition added more guidelines to the procedure, more case studies, and appendixes on indirect emissions. Information in the revised edition is also designed to aid companies in setting a target for future emissions now that they know their current values. The protocol is also set up to allow the companies to more easily evaluate and trade GHG offsets. Offsets increase the amounts of GHGs that can be emitted without breaking governmental rules. A company with low rates of emissions may sell offsets to a company with high output to allow the second company to stay within the rules.

The Greenhouse Gas Protocol is designed to allow companies to evaluate their current emissions and set targets for reduction of GHGs. It even allows a company to assess different reduction procedures to determine which might be best. Countries can now set standards for companies in an effort to control the emission of GHGs. By the use of this standard accounting system, one company will not have an advantage over another company when meeting the standards. It will also be easier for the government to monitor companies' emissions with the new accounting system.

C. Alton Hassell

Further Reading

Franks, Jeremy R., and Ben Hadingham. "Reducing greenhouse gas emissions from agriculture: avoiding trivial solutions to a global problem." *Land Use Policy* 29, no. 4 (2012): 727-736. Mitigating greenhouse emissions from agriculture faces many difficulties including lack of defined units and cost-effectiveness data. In addition, mitigation in one location may inadvertently increase overall global emissions.

Gerrard, Michael B., ed. *Global Climate Change and U.S. Law.* Chicago: American Bar Association, Section of Environment, Energy, and Resources, 2007. Discusses from a legal viewpoint the different initiatives—regional, state and local—undertaken to control GHGs. Illustrations, bibliography, index.

World Business Council for Sustainable Development, World Resources Institute. *The Greenhouse Gas Protocol: A Corporate Accounting and Reporting Standard.* Geneva, Switzerland: Author, 2001.

The first edition of the protocol to assess greenhouse gases. Illustrations, bibliography.

_____. *The Greenhouse Gas Protocol: The GHG Protocol for Project Accounting.* Geneva, Switzerland: Author, 2005. The 2005 edition of the protocol to assess GHGs. Illustrations, bibliography.

See also: Carbon dioxide; Environmental economics; Greenhouse effect; Greenhouse gases and global warming; Hydrofluorocarbons; Kyoto Protocol; Methane; Nitrous oxide; Perfluorocarbons; Sulfur hexafluoride.

Greenhouse gases and global warming

Categories: Meteorology and atmospheric sciences; Pollution and waste

GHGs are trace atmospheric gases that trap heat in the lower atmosphere, causing global warming. Such warming has been associated with droughts, tornadoes, ice melting, sea-level rise, saltwater intrusion, evaporation, and other climatic changes and effects.

Key concepts

aerosols: small particles suspended in the atmosphere

anthropogenic: deriving from human sources or activities

fossil fuels: fuels (coal, oil, and natural gas) formed by the chemical alteration of plant and animal matter under geologic pressure over long periods of time

global dimming: reduction of the amount of sunlight reaching Earth's surface

global warming: an overall increase in Earth's average temperature

greenhouse effect: absorption and emission of radiation by atmospheric gases, trapping heat energy within the atmosphere rather than allowing it to escape into space

greenhouse gases (GHGs): atmospheric trace gases that contribute to the greenhouse effect

Background

Greenhouse gases (GHGs) have both natural and anthropogenic sources. They allow sunlight to pass through them and reach Earth's surface, but they trap the infrared radiation released by Earth's surface, preventing it from escaping into space. These trace atmospheric gases thus play an important role in the regulation of Earth's energy balance, raising the temperature of the lower atmosphere. GHG concentrations in the atmosphere have historically varied as a result of natural processes, such as volcanic activity. They have always been a small fraction of the overall atmosphere, however, exhibiting significant effects on the climate despite their low concentrations. Thus, small variations in GHG concentration may have disproportionate effects on Earth's climate. Since the Industrial Revolution, humans have added a significant amount of GHGs to the atmosphere by burning fossil fuels and cutting down trees. Scientists estimate that the Earth's average temperature has already increased by $0.3°$ to $0.6°$ Celsius since the beginning of the twentieth century.

Some heat from the Sun is reflected back into space (squiggled arrows), but some becomes trapped by Earth's atmosphere and reradiates toward Earth (straight arrows), thus heating the planet.

GHG Sources and Atmospheric Physics

The atmosphere comprises constant components and variable components. It is composed primarily of nitrogen (78 percent) and oxygen (21 percent). Its other constant components include argon, neon, krypton, and helium. Its variable components include carbon dioxide (CO_2), water vapor (H_2O), methane (CH_4), sulfur dioxide (SO_2), ozone (O_3), and nitrous oxide (N_2O). The variable components affect the weather and climate because they absorb heat emitted by Earth and thereby warm the atmosphere. In addition to the variable natural atmospheric GHGs, anthropogenic halocarbons, other chlorine- and bromine-containing substances, sulfur hexafluoride, hydrofluorocarbons, and perfluorocarbons contribute to the greenhouse effect.

CO_2, composed of two oxygen atoms and one carbon atom, is a colorless, odorless gas deriving from carbon burning in the presence of sufficient oxygen. It is released to the atmosphere by forest fires, fossil fuel combustion, volcanic eruptions, plant and animal decomposition, oceanic evaporation, and respiration. It is removed from the atmosphere by CO_2 sinks, seawater absorption, and photosynthesis.

Methane is a colorless, odorless, nontoxic gas consisting of four hydrogen atoms and one carbon atom. It is a constituent of natural gas and fossil fuel. It is released into the atmosphere when organic matter decomposes in oxygen-deficient environments. Natural sources include wetlands, swamps, marshes, termites, and oceans. Other sources are the mining and burning of fossil fuels, digestive processes in ruminant animals, and landfills. Methane reacts with hydroxyl radicals in the atmosphere, which break it down in the presence of sunlight, shortening its lifetime.

Nitrous oxide is a colorless, nonflammable gas with a sweetish odor. It is naturally produced by oceans and rain forests. Anthropogenic sources include nylon and nitric acid production, fertilizers, cars with catalytic converters, and the burning of organic matter. Nitrous oxide gas is consumed by microbial respiration in specific anoxic environments.

Sulfur dioxide is released during volcanic activities, combustion of fossil fuel, transportation, and industrial metal processing. This gas is more reactive than is CO_2, and it rapidly oxidizes to sulfate. It produces acidic gases and acid rain when it reacts with water and oxygen.

Ozone (triatomic oxygen) is a highly reactive, gaseous constituent of the atmosphere. A powerfully oxidizing, poisonous, blue gas with an unpleasant smell, it helps create smog. It is produced in chemical reactions of volatile organic compounds or nitrogen oxide with other atmospheric gases in the presence of sunlight. Oxygen and ozone absorb a critical range of the ultraviolet spectrum, preventing this dangerous radiation from reaching Earth's surface and making possible life on Earth.

Halocarbons have global warming potentials (GWPs) from three thousand to thirteen thousand times that of CO_2; they remain in the atmosphere for hundreds of years. These compounds were commonly used in refrigeration, air conditioning,

and electrical systems, but their use has been regulated as a result of their environmental and climatic effects.

Effect on Climate Change

The Working Groups of the Intergovernmental Panel on Climate Change (IPCC) presented a synthesis report in 2007, providing an integrated view of climate change from multiple perspectives. The report observed an increase of global air and ocean temperatures, melting of snows, and rising sea levels. The report estimated the one-hundred-year linear trend of Earth's average temperature between 1906 and 2005 at an increase of 0.74° Celsius, significantly greater than the trend from 1901 to 2000 (0.6° Celsius). The increase of temperature contributed to changes in wind patterns, affecting extra-tropical storm tracks and temperature patterns.

Global average sea level has risen between 1961 and 2001 at an average rate of 1.8 millimeters per year and between 1993 and 2008 at an average rate of 3.1 millimeters per year. The increase is due largely to melting glaciers and polar ice sheets. Satellite data between 1978 and 2008 show that average annual extent of Arctic sea ice shrank by an average of 2.7 percent per decade. The average summertime extent shrank far more, an average of 7.4 percent per decade.

Increases have been reported in the number and size of glacial lakes and the rate of change in some Arctic and Antarctic ecosystems. Runoff and earlier spring peak discharge in many glacier- and snow-fed rivers have also increased. These increases have in turn had effects on the thermal structure and water quality of the rivers and lakes fed by this runoff. Both marine and freshwater systems have been associated with rising water temperatures and with changes in ice cover, salinity, oxygen levels, and circulation patterns. These ecological changes have affected algal, plankton, and fish abundance.

Precipitation has increased in the eastern parts of North and South America, northern Europe, and northern and central Asia. It has decreased in the Mediterranean and southern Africa. These patterns also have affected algal, plankton, and fish abundance. Globally, since 1970, a greater area of Earth's surface has been affected by drought.

Changes in atmospheric GHG and aerosol concentration, as well as solar radiation levels, affect the energy balance of Earth's climate system. Global GHG emissions increased by 70 percent over preindustrial levels between 1970 and 2004. CO_2 emissions increased by 80 percent, but they began to decline after 2000. The global increase in CO_2 and methane emissions is due to fossil fuel and land use, particularly agriculture.

Coastlines are particularly vulnerable to the consequences of climate change, such as sea-level rise and extreme weather. Around 120 million people on Earth are exposed to tropical cyclone hazards. During the twentieth century, global sea-level rises contributed to increased coastal inundation, erosion, ecosystem losses, loss of sea ice, thawing of permafrost, coastal retreat, and more frequent coral bleaching.

Anticipated future climate-related changes include a rise in sea level of up to 0.6 meter by 2100, a rise in sea surface temperatures by up to 3° Celsius, an intensification of tropical cyclones, larger waves and storms, changes in precipitation and runoff patterns, and ocean acidification. These phenomena will vary on regional and local scales. Increased flooding and the degradation of freshwater, fisheries, and other resources could impact hundreds of millions of people, with significant socioeconomic costs. Degradation of coastal ecosystems, especially wetlands and coral reefs, affects the well-being of societies dependent on coastal ecosystems for goods and services.

Context

In response to global warming, changes are being implemented to reduce GHG emissions. The United Nations Framework Convention on Climate Change prepared the 1997 Kyoto Protocol. Under the protocol, thirty-six states, including highly industrialized countries and countries undergoing transitions to a market economy, entered into legally binding agreements to limit and reduce GHG emissions. Developing countries assumed nonbinding obligations to limit their emissions as well.

In the energy sector, fuel use is slowly transitioning from coal to natural gas and renewable energy (hydropower, solar, wind, geothermal, tidal, wave, and bioenergy). In the transport sector, fuel-efficient, hybrid, and fully electric vehicles are being designed and marketed, and governments are attempting to motivate commuters to use mass-transit systems. More efficient uses of energy, including low-energy lightbulbs, day lighting, and efficient electrical, heating, and cooling appliances are being developed and deployed.

Industrial manufacturers have implemented electrical efficiency measures as well, and they have begun recycling, as well as capturing and storing CO_2. Crop and land management techniques have also improved, leading to an increase in soil carbon storage and the restoration of peaty soils and degraded land. Rice cultivation techniques have been improved, and livestock management techniques are being developed to reduce methane and nitrogen emissions. More controversially, dedicated energy crops are being grown to replace fossil fuels.

Afforestation, reforestation, forest management, reduced deforestation, and harvested wood product management are also being geared toward reducing GHG emissions. Forestry products are in use for bioenenergy to replace fossil fuels. Improvements are being made in tree species, remote sensing for analyses of vegetation and soil carbon, and mapping of land use. In the waste industry, methane is being recovered from landfills and energy is being recovered from waste incineration. Organic waste is more widely used for composting, wastewater is minimized, and the wastewater produced is treated and recycled. Biocovers and biofilters are being developed to optimize methane oxidation.

Ewa M. Burchard

Further Reading

Dlugokencky, E. J., et al. "Continuing Decline in the Growth Rate of the Atmospheric Methane Burden." Nature 393 (1998): 447-450.

Gore, Al. An Inconvenient Truth: The Planetary Emergency of Global Warming and What We Can Do About It. Emmaus, Pa.: Rodale, 2006.

Le Treut, H., et al. "Historical Overview of Climate Change." In Climate Change, 2007—The Physical Science Basis: Contribution of Working Group I to the Fourth Assessment Report of the Intergovernmental Panel on Climate Change, edited by Susan Solomon et al. New York: Cambridge University Press, 2007.

Montzka, Stephen A., Edward J. Dlugokencky, and James H. Butler. "Non-CO_2 greenhouse gases and climate change." *Nature* 476, no. 7358 (2011): 43-50. Carbon dioxide is not the only villain. Gases like methane and nitrous oxides also have significant effects. Since gases like methane have short lifetimes, reducing their emission can offer a quick reduction in global warming.

Walker, Gabrielle. An Ocean of Air: Why the Wind Blows and Other Mysteries of the Atmosphere. Orlando, Fla.: Harcourt, 2007.

Greening Earth Society

Category: Organizations and agencies
Date: Established 1998
Web address: http://www.greening earthsociety.org/ (deactivated)

Mission

The Greening Earth Society (GES), created by the Western Fuels Association, is a public relations organization promoting the idea that higher levels of greenhouse gases (GHGs), particularly of carbon dioxide (CO_2), are beneficial because they promote plant growth. GES was founded on Earth Day, 1998, by Frederick Palmer, chief executive officer of the Western Fuels Association, a group of coal-based rural electric cooperatives and municipal utilities. The purpose of GES is to promote the benefits of "moderate" global warming. The group is based in Arlington, Virginia, where both Palmer and Ned Leonard, director of communications and government affairs of GES and of the Western Fuels Association, worked as registered lobbyists. After 2000, Bob Norrgard, general manager of Western Fuels, assumed the directorship of GES. Its scientific advisers include Patrick J. Michael, professor of environmental sciences at the University of Virginia; Robert E. Davis, professor of climatology at the University of Virginia's Department of Environmental Sciences; Robert C. Balling, director of the Laboratory of Climatology at Arizona State University; and Sallie Baliunas and Willie Soon of the Harvard-Smithsonian Center for Astrophysics.

GES was founded in response to attempts by the Environmental Protection Agency to regulate CO_2 as a pollutant. Higher concentrations of CO_2, GES argues, not only are not harmful, but also at moderate levels actually "increase plant productivity, water use efficiency, and their resistance to a variety of environmental stresses including heat, drought, cold, pests, deficient nutrients, and air pollution." In a 1999 ad in *The Washington Post*, GES articulated its position that CO_2 is a nutrient, not a pollutant, explaining,

Carbon Dioxide (CO_2) levels have risen in direct proportion to human population since the 15th century. The good news is that today's higher CO_2 levels are producing more abundant plant life and greater agricultural yields. Unfortunately, good news like this doesn't command the same attention as news about catastrophic climate change.

According to GES, there is no scientific consensus about the anthropogenic causes of climate change, contrary to media reports. Environmentalists who argue that global warming represents a looming catastrophe, it argues, are merely alarmists who either misunderstand or intentionally misrepresent scientific data.

Significance for Climate Change

Until 2003, GES published the *World Climate Report*, a biweekly newsletter intended to point out "the weaknesses and outright fallacies in the science that is being touted as 'proof' of disastrous warming." The group's Web site described the newsletter as "the perfect antidote against those who argue for proposed changes to the Rio Climate Treaty, such as the Kyoto Protocol, which are aimed at limiting carbon emissions from the United States." Since 2004, the newsletter has been published by New Hope Environmental Services, still under the direction of editors Michael, Davis, and Balling. The international science journal *Nature* recognized *World* Climate Report as presenting a "mainstream skeptic" view of global warming.

From 2000 to 2003, GES also issued Virtual Climate Alerts, brief press releases whose titles included "Emissions Trend Doesn't Support Claim of Ecological and Economic Disaster," "More Cold Truth Dispels Hot Air," and "How Popular Coverage of Melting Arctic Sea Ice Overlooks Relevant Long-Term Research." In addition, GES issued annual State of the Climate Reports; scientific reports, including "In Defense of Carbon Dioxide: A Comprehensive Review of Carbon Dioxide's Effect on Human Health, Welfare, and the Environment" (1998) and "The Internet Begins with Coal: A Preliminary Exploration of the Impact of the Internet on Electricity Consumption" (1999); and the videos *The Greening of Planet Earth* (1992) and *The Greening of Planet Earth Continues* (1998). The group is named as a sponsor of the Web site www.co2and climate.org, an annotated collection of links, and it has funded research through the Arizona State University Climate Data Task Force. By the end of 2008, the Greening Earth Society Web site had been suspended, many of its publications were available only on other sites, and the Western Fuels Association Web site no longer provided a link to GES.

Cynthia A. Bily

Further Reading

Dunlap, Riley E., and Peter J. Jacques. "Climate change denial books and conservative think tanks: exploring the connection." *American Behavioral Scientist* (2013): 0002764213477096. Most earlier books skeptical of climate change were produced by authors having ties to conservative think tanks but in recent years, self-published books have risen sharply in number. The vast majority of climate denial books have not undergone peer review.

Goodell, Jeff. *Big Coal: The Dirty Secret Behind America's Energy Future*. Boston: Mariner, 2007. Goodell, an editor of *Rolling Stone*, presents a lively analysis of America's dangerous reliance on coal and its hidden costs in the forms of global warming and lung disease.

Michael, Patrick J., and Robert C. Balling. *The Satanic Gases: Clearing the Air About Global Warming*. Washington, D.C.: Cato Institute, 2000. Two Greening Earth Society members argue, in accessible prose, that environmentalists and politicians have exaggerated the causes and dangers of global warming.

Poole, Steven. *Unspeak: How Words Become Weapons, How Weapons Become a Message, and How That Message Becomes Reality*. New York: Grove Press, 2007. Poole's witty analysis of the politics of language includes a discussion of the loaded term "global warming" and of the language of industry-funded propaganda groups.

Rampton, Sheldon, and John Stauber. *Trust Us, We're Experts! How Industry Manipulates Science and Gambles with Your Future*. New York: Jeremy P. Tarcher/Putnam, 2001. Explores the deceptive practices used by industry, including employment of public relations front organizations and the misuse of data, to influence public opinion on issues including global warming.

See also: Carbon dioxide; Carbon dioxide fertilization; Skeptics.

Greenland ice cap

Category: Cryology and glaciology

The largest mass of ice in the Northern Hemisphere resides atop Greenland, which is home to 10 percent of the world's ice. This ice has been melting at an accelerating rate, especially in coastal areas.

Key concepts

albedo: the fraction of radiation reflected by a surface

ice cores: cylindrical samples of ice taken from glaciers or ice caps that are used by scientists to determine atmospheric composition in the past

Jakobshavn glacier: a major outlet glacier in Greenland whose movement toward the sea has been accelerating

tipping point: point at which feedbacks take control and propel a climatic forcing past a point where control or mitigation is possible

upwelling: a rising of warm water to the surface, displacing colder water and potentially melting coastal ice

Background

Greenland's ice is only a fraction of Antarctica's, but it is melting more rapidly, in part because summers are warmer, allowing for more rapid runoff. The amount of ice lost by Greenland during 2007 was equivalent to two times all the ice in the Alps, or a layer of water more than 0.8 kilometer deep covering Washington, D.C., according to Konrad Steffen, a professor of geography at the University of Colorado, who also directs the Cooperative Institute for Research in Environmental Sciences.

Six Meters of Sea-Level Rise

Philippe Huybrechts, a glaciologist and ice-sheet modeler at the Free University of Brussels, has modeled the behavior of Greenland's ice sheet, finding that, with an anticipated annual temperature increase of 8° Celsius, the ice sheet would shrink to a small glaciated area far inland and the worldwide sea level would rise by 6 meters. Erosion of the Greenland ice sheet due to prolonged warming would be irreversible.

Upon close examination of paleoclimatic proxies, scientists examining ice cores from Greenland have found that its climate can change very quickly. The island's climate altered to a different state within one to three years at the onset of the present interglacial period, for example. Other scientists have found evidence that Greenland's surface has been covered with pine forests within a few hundred thousand years of the present, with the ice mass a fraction of today's size.

Ice melt in Greenland has accelerated significantly since 1990, according to a report in the

View of the Greenland Ice Cap [© Steven I. Dutch]

Journal of Climate coauthored by Steffen. A scientific team surveyed the rate of summer melting there between 1958 and 2006, and found that the five largest melting years occurred after 1995. The year 1998 was the biggest (454 cubic kilometers), followed by 2003, 2006, 1995, and 2002. Melting in 2007 exceeded that of all previous years. Scientists have been keeping close watch on many glaciers, one of the most notable being the Jakobshavn glacier (the probable source of the iceberg that sank the *Titanic*), which has been accelerating toward the ocean.

Melting Acceleration

Greenland is largely an artifact of the last ice age, held in place by the albedo (reflectivity) of its massive ice sheet. Once the ice melts, present-day climatic conditions will not allow its re-creation. Greenland's coastal ice often melts with a boost from the sea, warm water upwelling, flowing from below. Scientists have been debating the amount of additional warming that could cause Greenland's ice cap to cross a "tipping point," beyond which continued melting will be inevitable. During the winters between 2003 and 2007, Greenland lost two to three times as much ice in summer melt as it regained during winter snows.

Greenland's ice loss is accelerating irregularly year by year as well. During the 2007 melting season, with temperatures 4° to 6° Celsius higher than the previous thirty years' average, 450 billion metric tons of ice vanished, 30 percent more than the previous year and 4 percent more than the previous record, in 2005. During years of record melting, such as 2005 and 2007, high-pressure systems over Greenland keep storms away, clearing skies, allowing the Sun to shine for extended hours. Greenland's ice-mass loss between 2004 and 2006 was two and a half times the loss between 2002 and 2004.

While increases in the rate of ice loss to date have been greatest in southern Greenland, melting has been spreading northward, especially along the coastline. By the summer of 2007, the ice cap was studded by more than one thousand shallow meltwater lakes, some as wide as 5 kilometers. Tens of millions of cubic meters of water swirl from these lakes to the base of the ice sheet within a matter of days, opening huge waterfalls where none had previously existed.

Context

Melting of the Greenland ice sheet's western reaches increased by about 30 percent from 1979 to 2006. Ice is melting most rapidly at the edges of the ice sheet. Although Greenland's ice sheet has been thickening at higher elevations due to increases in snowfall (warmer air holds more moisture, thus more snow), the gain is more than offset by an accelerating mass loss, primarily from rapidly thinning and accelerating outlet glaciers.

Scientists using pollen records from marine sediment off southwest Greenland have deduced that much of the Greenland ice sheet has melted relatively quickly during sharp natural warming in Earth's climate, some warming as recent as 400,000 years ago, when carbon dioxide levels were lower than they are today. During these spells, boreal coniferous forest covered much of the island. These spells of abrupt warming (abrupt, that is, in geologic time) reduced the ice sheet to about one-quarter of its present size and by themselves have raised world sea levels 4 to 6 meters.

Bruce E. Johansen

Further Reading

Alley, Richard B. *The Two-Mile Time Machine: Ice Cores, Abrupt Climate Change, and Our Future.* Princeton, N.J.: Princeton University Press, 2000. A popular version of Alley's scientific work on climate change in the Arctic, with special attention to rapid climate change in Greenland.

Appenzeller, Tim. "The Big Thaw." *National Geographic*, June, 2007, 56-71. Survey of ice melt in the Arctic, Antarctic, and mountain glaciers, with explanations describing why the process has been speeding up.

Hanna, Edward, et al. "Increased Runoff from Melt from the Greenland Ice Sheet: A Response to Global Warming." *Journal of Climate* 21, no. 2 (January, 2008): 331-341. The article, co-authored by Konrad Stephen, that details the acceleration of Greenlandic ice melt between 1990 and 2007.

Murray, Tavi. "Greenland's Ice on the Scales." *Nature* 443 (September 21, 2006): 277-278. Provides ice-mass measurements for Greenland and discusses how they have changed in the years surrounding the turn of the century.

Nghiem, S. V., D. K. Hall, T. L. Mote, Marco Tedesco, M. R. Albert, K. Keegan, C. A. Shuman, N. E. DiGirolamo, and G. Neumann. "The extreme melt across the Greenland ice sheet in 2012." *Geophysical Research Letters* 39, no. 20 (2012). Virtyally the entire Greenland ice cap experienced melting in 2012. The only other known everts of that magnitude happened in 1889 and during the Medieval Warm period.

Paterson, W. S. B., and N. Reeh. "Thinning of the Ice Sheet in Northwest Greenland over the Past Forty Years." *Nature* 414 (November 1, 2001): 60-62. Greenland ice losses are presented in historical context.

Steffensen, Jørgen Peder, et al. "High-Resolution Greenland Ice Core Data Show Abrupt Climate Change Happens in Few Years." *Science* 321 (August 1, 2008): 654-657. A description of how quickly Greenland's climate has changed in the past.

See also: Abrupt climate change; Antarctica: threats and responses; Arctic; Cryosphere; Deglaciation; Glaciations; Glaciers; Ice shelves; Polar climate.

Greenpeace

Categories: Organizations and agencies; Environmentalism, conservation, and ecosystems
Date: Established 1971

Mission

An independent international activist and research organization, Greenpeace is funded almost entirely by contributions from its members. It works to produce and distribute educational and advocacy materials, support grassroots activism, encourage pressure on legislatures and other national and international bodies, and conduct original research on global warming, clean energy, and other issues.

Greenpeace's first action in 1971 was to protest US nuclear testing off the Alaskan coast at the island of Amchitka, attempting to stop a test by steering a fishing boat named Greenpeace I into the path of military vessels. In 1974, the organization launched its first environmental effort with the Save the Whales campaign, capturing the attention and imagination of people worldwide. These first activities illustrated what would become Greenpeace trademarks: well-publicized actions of nonviolent resistance based on solid science, attracting widespread public attention to pacifist and environmental causes.

As of 2008, Greenpeace had expanded to become an international organization with a presence in forty-two countries and a wide-ranging, six-part mission. Its website identifies climate change as the single most serious global threat, and it lists working to support new forms of energy as the first action in the mission statement. The other goals of the organization include defending oceans and forests, reducing the use of hazardous chemicals by finding alternatives, encouraging more responsible agriculture, and the original goal of stopping nuclear proliferation and working for peaceful disarmament.

To combat global warming, Greenpeace takes a multipronged approach. Its Project Hot Seat campaign targets individual members of the US Congress, encouraging supporters to pressure legislators to support global warming bills. Greenpeace also conducts and publishes research projects, engages in legal action against large emitters of greenhouse gases (GHGs), and campaigns actively to sway opinions of participants in United Nations climate conferences. Its website and publications attempt to reach a broad audience with information about global warming, refutation of arguments made by skeptics, and opportunities for activism.

Significance for Climate Change

Greenpeace has an international headquarters in Amsterdam, as well as regional offices in more than thirty countries. It has more than 250,000 contributing members in the United States and 2.8 million

The Greenpeace ship Rainbow Warrior visiting the port of London before going on to Edinburgh (Leith). (Glen/ Wikimedia Commons)

worldwide. With a small paid staff and thousands of volunteers, Greenpeace works to increase awareness of global warming among the general population, and it targets particular experts and officials at opportune moments, as it did when it lobbied participants in the United Nations Climate Conference in Bali in 2007.

Greenpeace publishes research reports with titles such as False Hope: Why Carbon Capture and Storage Won't Save the Planet (2008) and Impacts of Climate Change on Glaciers Around the World (2004), legislative and press briefings including Kyoto and the Bali Mandate: What the World Needs to Do to Combat Climate Change (2007), and a 2004 activists' guide for students interested in organizing to help their college campuses switch to clean energy, as well as brochures, videos, and blogs. Two of its most significant climate change campaigns are the research project ExxonSecrets, which targets scientists and lobbyists employed by the Exxon Corporation to downplay the importance of global warming, and Energy [R]evolution, whose name is a combination of the words "revolution" and "evolution," suggesting the need for a massive overhaul of the way the world produces and consumes energy.

Through the project Greenpeace Solutions, the organization collaborates with business, technology, financial, and government organizations to help individual consumers make good purchasing decisions. The group developed and marketed the GreenFreeze refrigerator, which does not use dangerous hydrofluorocarbons (HFCs), a GHG. Though the product is not widely known in North America, more than 150 million units have sold in Europe and Asia. Greenpeace has also cooperated with European and United Nations groups to develop the SolarChill battery-free solar refrigerator, and it has worked in the United States to develop and promote wind power.

In August, 2002, Greenpeace filed suit, along with the environmental organization Friends of the Earth and the city of Boulder, Colorado, against the US Export Import Bank and the US Overseas Private Investment Corporation, accusing them of helping to fund dangerous fossil fuel projects with taxpayer money. The suit claims that these government agencies should have studied the potential climate change effects of these projects before granting funding.

Cynthia A. Bily

Further Reading

Bohlen, Jim. *Making Waves: The Origins and Future of Greenpeace*. Montreal: Black Rose Books, 2001.

Ferguson, Christy, Elizabeth A. Nelson, and Geoff G. Sherman. *Turning up the Heat: Global Warming and the Degradation of Canada's Boreal Forest*. Washington, D.C.: Greenpeace, 2008.

García, María M. "Perception is truth: How US newspapers framed the "Go Green" conflict between BP and Greenpeace." *Public Relations Review* 37, no. 1 (2011): 57-59. An analysis of United States media coverage found that BP (formerly British Petroleum) was generally assigned blame and placed in a negative light while Greenpeace was treated positively. However, the study also found evidence of media hegemony, suggesting that future conflicts may not be framed so positively for environmentalism.

Weyler, Rex. *Greenpeace: How a Group of Journalists, Ecologists, and Visionaries Changed the World*. Emmaus, Pa.: Rodale, 2004.

Ground ice

Category: Cryology and glaciology

Definition

Ground ice is found in cavities, voids, pores, and other openings in frozen or freezing ground. It is a feature of a permafrost region, but the two terms are not interchangeable. A permafrost region is defined as an area where the soil and rocks making up the land remain below the freezing point for two or more consecutive years and can include areas with little or no water content. Permafrost regions account for nearly one-quarter of the landmass of the Northern Hemisphere, and some of these areas have been frozen since the Pleistocene ice age. Ground ice differs from glacier ice as well. Glaciers are created when fallen snow compresses into thick ice masses over long periods of time, and they are capable of riverlike movement, albeit at a pace so slow as to be imperceptible to the human eye.

Ground ice is formed once the temperature drops below 0° Celsius, when most of the moisture in the soil freezes. There are two different types of ground ice: structure-forming ice and pure ice. Structure-forming ice, which holds sediment together, comes in multiple varieties, including ice crystals, intrusive ice, reticulate vein ice, segregated ice, and icy coatings on soil particles. Pure, or massive, ice exists primarily toward the surface and is found as ice wedges, massive icy beds, and pingo cores. Massive ice beds are those with a minimum ice-to-soil ratio of 2.5 to 1.

Ancient ground ice, known as fossil ice, is a valuable source of historical information for geologists, paleontologists, and climatologists. Fossil ice preserves organic material and provides a measurable history of sediment and air quality covering hundreds of thousands of years. Cores drilled from fossil ice are studied to learn about the climates of the distant past.

In 2002, the National Aeronautics and Space Administration reported that the Odyssey orbiter had identified ground ice on Mars, a find that was confirmed by the Phoenix lander in June, 2008. The extreme cold and thin atmosphere on Mars would cause surface water to vaporize as dry ice does on Earth, but ground ice buried under the surface remains frozen and should prove invaluable to studies of the planet's history.

Significance for Climate Change

Ground ice and permafrost play crucial roles in the industrial development of a region's energy and mining industries, as well as in the infrastructure needed to support such enterprises. It is possible to build in the presence of ground ice in its structure-forming state, so long as its properties are taken into consideration.

The presence of ground ice is closely linked to a region's climate, especially to the temperature at ground level. Its impact varies based on a variety of factors, including drainage, snow cover, soil composition, and vegetation. Ground ice affects both topography and vegetation, as well as a region's response to changes in climate and population.

In regions where the temperature remains close to freezing, a change of only a few degrees can be enough to cause ground ice to start melting. Changes can occur naturally, with normal temperature fluctuations, or as a result of ground clearing through construction or forest fire. When structure-forming ground ice melts, the surrounding terrain is significantly weakened, creating slope instabilities and thaw settlement.

Scientists predict that as much as 90 percent of the permafrost in the Northern Hemisphere could melt by the end of the twenty-first century, with most of the thaw taking place in the top 3 meters of terrain. Some regions in Alaska and Siberia have already experienced collapsed infrastructure and increased rock fall in the high elevations. The resulting meltwater would eventually reach the oceans, causing sea levels to rise around the world. Since the 1930's, Arctic water runoff has increased by 7 percent, and it is projected to reach a 28 percent increase by 2100.

Permafrost traps and holds approximately 30 percent of Earth's carbon, sequestering it from the atmosphere. Melting ground ice will release the trapped carbon into the atmosphere, contributing to the greenhouse effect, which in turn will cause even more ice to melt.

Coastlines in permafrost regions become more vulnerable to erosion as the ice retreats. The Alaskan island of Shishmaref, home to a native population whose ancestors have lived there for more than four thousand years, lost 7 meters of coastline annually between 2001 and 2006.

P. S. Ramsey

Further Reading

Alley, Richard B. *The Two-Mile Time Machine: Ice Cores, Abrupt Climate Change, and Our Future.* Princeton, N.J.: Princeton University Press, 2000. Overview of the study of ice cores and how scientists use them to determine historical weather and geological data.

Davis, Neil. *Permafrost: A Guide to Frozen Ground in Transition.* Fairbanks: University of Alaska Press, 2001. Study of permafrost and the various ways in which humans have lived with it and changed it throughout history.

Gosnell, Mariana. *Ice: The Nature, the History, and the Uses of an Astonishing Substance.* Chicago: University of Chicago Press, 2007. Examines in detail ice and its impact on Earth and its inhabitants.

Grosse, Guido, Benjamin Jones, and Christopher Arp. "Thermokarst lakes, drainage, and drained basins." (2013): 325-353. Thermokarst lakes form by melting of permafrost and tend to accelerate permafrost degradation, but lake drainage allows permafrost to form again.

MacDougall, Doug. *Frozen Earth: The Once and Future Story of Ice Ages.* Berkeley: University of California Press, 2004. A scientific look at the ice ages and their geological impact.

Pielou, E. C. *After the Ice Age: The Return of Life to Glaciated North America.* Chicago: University of Chicago Press, 1991. Surveys the impact of the last ice age and North America's evolution after the ice retreated.

See also: Cryosphere; Glaciers; Greenhouse effect; Groundwater; Permafrost; Sea-level change; Sequestration.

Groundwater

Category: Water resources

Definition

There are approximately 1.385 billion cubic kilometers of water on, above, and in the Earth. However, the vast majority of the Earth's water (around 97 percent) is found in its oceans. Groundwater, though constituting less than 2 percent of the world's water, accounts for more than 98 percent of its available freshwater resources. In the United States, more than half of the population depends on groundwater, while in semiarid to arid regions the percentage of people using groundwater is almost 100 percent.

Groundwater is water found in pore spaces (pores) in soil and rocks. It differs from soil water in that groundwater is found below a water table—the demarcation separating saturated pores from unsaturated pores. Soil or rock that is saturated with water and is capable of supporting a well is

called an aquifer. An aquifer can comprise one or multiple rock or sediment layers, ranging in size from a few meters to several hundred kilometers long, several meters thick, and meters to kilometers in width. An aquifer could be unconsolidated material, such as sand, or consolidated material, such as sandstone and other rocks. Limestone also makes good aquifers.

Aquifers may be divided into two main types. Open aquifers enjoy direct access to surface water. Closed aquifers sit underneath confining layers of semi- or nonpermeable material, which prevent such access. Water recharges aquifers by infiltration or seepage through permeable materials, and the recharge rate depends on the type of surface material, land cover, slope, and amount and intensity of precipitation melt.

Significance for Climate Change

The effect of global warming on groundwater varies from place to place, with differing effects on water quality and quantity. In semiarid to arid regions, the higher rate of evaporation increases the

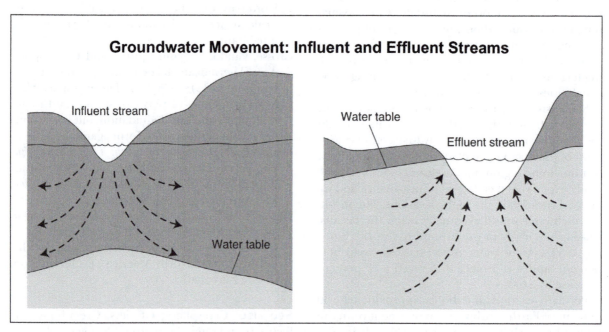

Groundwater Movement: Influent and Effluent Streams

Influent stream

Water table

Water table

Effluent stream

Two examples of groundwater movement: Influent streams, found in dry climates, cause an upward bulge in the water table as water from the stream migrates downward. Effluent streams, found in wet climates, are produced by a high water table that migrates water into channels in the ground's surface.

Center-pivot irrigation systems like this one typically tap a well. The sets of wheels are independently driven and linkages turn the drive motors on and off to keep the pipe in a straight line. Center-pivot irrigation is the cause of the circles seen on aerial views of farmland. But they have also made it possible to farm areas otherwise too dry to farm, and their reliance on groundwater has seriously depleted many sources of groundwater. (© Steven I. Dutch)

concentration of dissolved salts in the groundwater, as is the case in Lake Chad, in Africa, where the surface water in some closed areas is becoming more saline. This saline water may recharge the African groundwater system, making the water saline.

With global warming, snow and ice caps will melt, resulting in rapid sea-level rise. This rise may force saline waters into aquifers, reducing the water's suitability for human and ecological use. Already, such seawater intrusion has occurred along some coastal aquifers, including one in El Paso, Texas.

Global temperature rise will lead to an increase in precipitation in the Northern Hemisphere. As snow and glaciers melt, the groundwater table will rise in most places. Sinkhole numbers may increase in limestone terrains with higher rates of solution weathering resulting from acid rain. These sinkholes could become conduits through which contaminants from the land surface could reach the groundwater.

Groundwater resident time (the length of time water stays in an aquifer) varies from a few days to several thousand years. Generally, the shorter the resident time, the greater the infiltration rate or the shallower the aquifer. In areas such as Libya, where resident time can stretch to hundreds or thousands of years, drought-induced demands for groundwater can cause faster depletion of that water. In areas where the land surface is highly permeable and where rainfall or snowmelt is heavy, the resident time may be low; such areas may experience a dramatic rise in groundwater level. Assuming the anthropogenic effect upon groundwater is minimized or maintained at its current rate, a reduction in rainfall would lead to the lowering of water tables. Such lower water tables could cause ground compaction, loss of permeability of the land surface, and an increased runoff that could lead to flooding. Groundwater withdrawal could also lead to ground subsidence, as has occurred in Texas and California. Heavy cracks, rills, or gullies may develop on the surface, as in Lake Chad, where the soil is clayey.

Global warming is likely to increase flood and drought conditions in different parts of the world. This in turn would affect the rate of evaporation, lowering water levels in rivers and lakes during the summer months. Some of these water bodies could become groundwater recharging zones. Near-surface water tables will lose increasing amounts of water to evaporation as surface temperatures increase. In addition, surface water infiltration will lead to a reduction of groundwater flows to lakes.

Solomon A. Isiorho

Further Reading

Anderson, M. P. "Introducing Groundwater Physics." *Physics Today* 60, no. 5 (May, 2007): 42. Good, succinct summary of the principles of groundwater.

Fetter, C. W. *Applied Hydrogeology.* 4th ed. Upper Saddle River, N.J.: Prentice Hall, 2001. An easy-to-read textbook on hydrogeology. Presents several case studies and includes student versions of modflow, flownet, and aqtesolv software.

Hudak, Paul F. *Principles of Hydrogeology.* 3d ed. Boca Raton, Fla.: CRC Press, 2005. A simple book on hydrogeology that assumes readers have some geology and algebra skills.

Jackson, R. E., A. W. Gorody, B. Mayer, J. W. Roy, M. C. Ryan, and D. R. Van Stempvoort. "Groundwater protection and unconventional gas extraction: the critical need for field-based hydrogeological research." *Groundwater* 51, no. 4 (2013): 488-510. The principal risk from methane in groundwater is the buildup of potentially explosive concentrations. The authors recommend geochemical sampling and field testing of possible avenues for contamination.

Todds, David Keith, and Larry W. Mays. *Groundwater Hydrology.* Hoboken, N.J.: Wiley, 2005. Covers the basics of groundwater hydrology and includes modeling methods.

Younger, P. L. *Groundwater in the Environment: An Introduction.* Malden, Mass.: Blackwell, 2007. This is a good book for those interested in groundwater that does not require a geology background.

See also: Freshwater; Hydrologic cycle; Hydrosphere; Saltwater intrusion; Water quality; Water resources, global; Water resources, North American; Water rights.

Group of 77 and China

Categories: Organizations and agencies; nations and peoples
Date: Established June 15, 1964
Web address: http://www.G7.org

Mission

In the early 1960's, at the height of the Cold War, the agenda at the United Nations reflected a world divided between East and West. As part of the United Nations mission to offer a forum for the representation of the global community, efforts began as early as 1962 to organize the developing countries of the nonaligned Third World into a coalition within the United Nations. The original protocol for establishing the network of less influential nations was to address questions of trade. The interests of the United States and its allies and the Soviet Union and its satellites threatened smaller countries, which demanded relaxed trade barriers and stabilized commodity prices. "Solidarity" was seen as the only way to promote the economic interests of this diverse international community of nations.

The coalition was officially chartered with the Joint Declaration of the Developing Countries in October, 1963, and first convened on June 15, 1964, at the close of the first session of the United Nations Conference on Trade and Development in Geneva. Originally, the charter included seventy-seven nations that represented six continents. The group endorsed the central vitality of the United Nations as the forum most appropriate for presenting—and addressing—trade problems. Given the wide range of cultural and religious elements among the group's membership, the Group of 77 became a paradigm for the mission of the United Nations itself: forging community from diversity.

The first meeting of the Group of 77 was held in Algiers in September, 1967. That assembly adopted a superstructure for the organization, a protocol for annual meetings to be held in conjunction with the opening of the General Assembly in New York. Currently member nations number 130, although the name of the organization was retained for its historical importance. The Republic of China has been recognized as a conditional invitee since 1981. In 1971, a separate chapter-entity, the Group of 24, was established to work specifically on international monetary issues.

Significance for Climate Change

After more than three decades of monitoring the economic development, trade growth, health care, and technological development of developing countries, the Group of 77 in the mid-1990's shifted its agenda to consider the ramifications of the growing alarming data that projected dire consequences for global climate patterns should the emission of greenhouse gases (GHGs) fail to be significantly curbed. Global warming accelerated by the burning of oil, gas, and coal—the primary energy resource of the powerful dominant world economies—created a frustrating problem for those less industrial countries. Many of these countries were largely agricultural and most lacked the developed industrial grid that produced GHGs; hence, these countries were not part of the GHG emission problem. Yet, as climatologists began to gather data, it was increasingly clear that impoverished countries would bear the initial brunt of any catastrophic change in the Earth's climate patterns.

The ambitious and controversial 1997 Kyoto Protocol called for thirty-eight developed countries to reduce fossil fuel emissions by an average of 5.2 percent by 2012. Since that agreement, the Group of 77 has become aggressive in its efforts to represent the impact of global warming on developing countries. Each successive chair of the powerful caucus has endeavored to stress that the economies of developing countries are most vulnerable to the earliest effects of global warming. The Group of 77, in a series of communiqués issued through its agencies, has argued that the strategy proposed by the Kyoto Protocol to help smaller nations—transferring to these nations cutting-edge technology as a way to minimize the production of GHGs from their obsolete facilities—would not sufficiently address the magnitude of the fast-approaching problems.

The Group of 77 pointed out the need for an international water management strategy to handle diminishing water supplies; the need to chart the catastrophic impact of irregular weather patterns (most notably the increase in droughts, heat waves, and cyclones); the need to assess the impact of dropping of sea levels on trade and navigation routes and the economies that rely on them; the need for strategies to handle the sharp increase in food riots in areas stricken by droughts; and the impact on previously stable governments as countries struggle to address these concerns. Given the reluctance of entrenched dominant world economies to reduce fossil fuel emissions, the Group of 77 aggressively pursued its role as agitator, arguing that the crisis of global warming will demand the cooperation of developing nations as the only way

to make their interests part of the international dialogue on addressing the crisis.

Joseph Dewey

Further Reading

Aldy, Joseph E., and Robert N. Stavins. *Architectures for Agreement: Addressing Global Climate Change in a Post-Kyoto World.* New York: Cambridge University Press, 2007. Assesses the relationship between dominant economies and developing countries necessary to address the threat of global warming. Argues the central role of the United Nations in developing a workable framework for enforcing the Kyoto Protocol.

Mingst, Karen A., and Margaret P. Karns. *The United Nations in the Twenty-first Century.* Boulder, Colo.: Westview Press, 2006. Informed analysis of the role of the United Nations in a variety of pressing issues, including global warming.

Northcutt, Michael. *A Moral Climate: The Ethics of Global Warming.* Phoenix, Ariz.: Orbis, 2007. Fascinating and accessible argument that examines the moral dimension of the conflicting interests of industrialized and developing nations in addressing global warming.

Sauvant, Karl P. *Group of 77: Evolution, Structure, Organization.* New York: Oxford University Press, 1980. Still the most comprehensive and reliable history of the organization's mission; emphasizes trade inequities and the need for solidarity.

Zhang, ZhongXiang. "In what format and under what timeframe would China take on climate commitments? A roadmap to 2050." *International Environmental Agreements: Politics, Law and Economics* 11, no. 3 (2011): 245-259. As one of the world's largest and fastest growing economies, as well as one of the largest emitters of greenhouse gases, China looms especially large in global climate change scenarios. Simply altering the course of such a massive entity is difficult and China's reliance on coal further complicates matters.

See also: Annex B of the Kyoto Protocol; China, People's Republic of; International agreements and cooperation; Kyoto Protocol; Poverty; United Nations Framework Convention on Climate Change.

Gulf Stream

Category: Oceanography

The Gulf Stream moves huge amounts of water, salt, and heat around the Atlantic Ocean. Many scientists have argued that the position of the Gulf Stream has shifted at times in the geologic past, causing significant changes in the climate of Greenland and northern Europe.

Key concepts

Coriolis effect: the deflection of moving objects caused by Earth's rotation

geostrophic current: a current driven by horizontal gradients in pressure that are balanced by the Coriolis effect

North Atlantic drift: a current that diverges from the Gulf Stream, moving warm water northward to areas where it will cool and sink

subtropical gyre: a gigantic lens of warm water, usually occupying a substantial portion of an ocean basin, that rotates because of geostrophic currents

thermohaline circulation: circulation driven by differences in density

westerlies: the prevailing winds in the midlatitudes, resulting from a combination of vertical circulation patterns in the atmosphere and the rotation of the Earth

Background

The Gulf Stream has a peak flow of more than a hundred times that of all the rivers and streams on Earth, combined. Much of this is warm water from the Gulf of Mexico or the Caribbean, being transported to the North Atlantic, transferring a great amount of heat. A strong belief has developed that this ocean current is responsible for the mild climate of Europe. Recent data and global climate models indicate, however, that it is only one of several factors, as the heat transported by winds is several times greater.

Geostrophic Flow and the North Atlantic Gyre

The Sun warms the surface waters of the oceans, expanding them and reducing their density. The prevailing wind patterns in the subtropics (near 20° latitude) pile up these surface waters into hills, which are actually lens-shaped water masses floating on denser, cooler water, and bounded in part by continents. These hills have "peaks" no more than two meters higher than their edges, thousands of kilometers away. As the water at the surface tries to flow down this slope, the Coriolis effect, which is a consequence of the Earth's rotation, causes it to be deflected. In the Northern Hemisphere this deflection is to the right, so the circulation actually follows the hill contours, forming a rotating gyre. In the North Atlantic and North Pacific, subtropical gyres rotate in a clockwise sense. Constraints of flow on a rotating planet require that gyres are asymmetric, with a very steep slope on the western side. The rate at which these geostrophic currents flow is proportional to the slope. The Gulf Stream is a result of this steep slope, produced by a combination of solar heating, basin geometry, prevailing wind directions, and the rotation of the Earth.

The magnitude of this flow is prodigious. Peak flows are about 150×10^6 cubic meters per second. Put another way, a mass of ocean water, equivalent to the mass of the entire human population, flows by in less than four seconds. In a year the Gulf Stream moves 4.71 million cubic kilometers of water, which is huge compared to the annual runoff from all the continents (40,000 cubic kilometers) or even the total water moving between the surface of the Earth and the atmosphere (489,000 cubic kilometers).

Heat Transport

Along with all the water, the Gulf Stream moves a great deal of heat to the north. At the northern end of its range (near 50° north latitude), the North Atlantic Drift, with a flow of 14×10^6 cubic meters per second, diverges from it, moving this water farther north, where it cools and sinks. If it loses 20° Celsius in the process, or 20 calories per gram, about 1×10^{15} watts are released. This is fifty times the total human energy use of 0.018×10^{15} watts, and about half of the total poleward ocean heat transport on the planet. However, the heat transported by the atmosphere is about four times greater.

The cause of this disparity lies in the heat of vaporization of water. Every gram that evaporates moves at least 540 calories of heat from the ocean to the atmosphere. Of the 413,000 cubic kilometers of water evaporated from the sea every year,

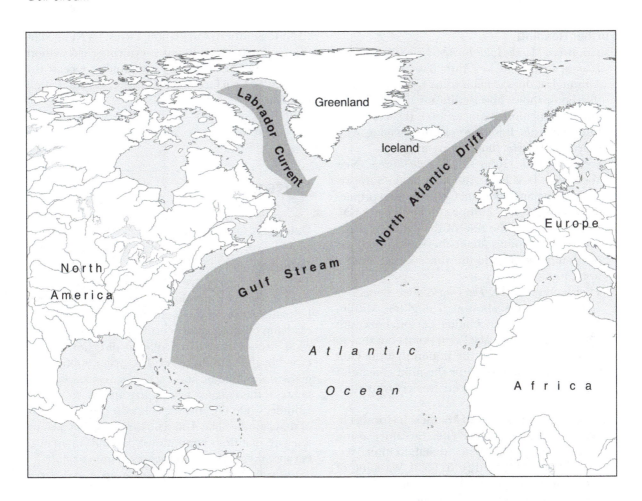

perhaps 56,000 cubic kilometers, evaporate near the equator in the Atlantic and precipitate farther north. That would release 4×10^{15} watts.

The Gulf Stream may not be directly responsible for the mild climate of Europe, but it probably is responsible for the warm temperatures of the North Atlantic. If these temperatures fall, the ocean surface may become covered with ice, the wind will not be warmed, and Europe will surely chill. Furthermore, the strength of the North Atlantic Oscillation (NAO) may well depend on these warm temperatures, and hence on the Gulf Stream.

Context

Intuitively, it would seem that a current that brings about 1×10^{15} watts of heat into the North Atlantic should be a dominant player in the climate of Europe. Since its discovery, many people have thought so and have written about it. Recent studies indicate, however, that the heat brought by the Gulf Stream is only a tenth of the heat transported, it is responsible for only a few degrees of warming, and it does nothing to explain why Europe is 15°-20° Celsius warmer, in winter, than places at similar latitudes on the East Coast of North America. Most of the heat, and all of the heat producing the contrast between Europe and North America, comes from atmospheric transport and the heating of the surface waters during the summer. Because the myth is so entrenched in the collective understanding of climate science, changing the way people think about it has proven to be extremely difficult. The strength of the NAO may be more important to Earth's climate than the strength of the Gulf Stream; however, the Gulf Stream may control the NAO.

Otto H. Muller

Further Reading

Fagan, Brian M. *The Little Ice Age: How Climate Made History, 1300-1850.* New York: Basic Books, 2000. Although it does contain some scientific errors, this book shows how significant the North Atlantic Oscillation has been over the last seven hundred years. It is written in an engaging, easily accessible style. Illustrations, maps.

Gaskell, Thomas Frohock. *The Gulf Stream.* New York: John Day, 1973. Although he does not repudiate the myth that the Gulf Stream is primarily responsible for Europe's mild climate, Gaskell emphasizes the role of the wind. An early source of warnings about the warming effects of a buildup of CO_2 in the atmosphere. Illustrations, photographs, maps.

MacLeish, William H. *The Gulf Stream: Encounters with the Blue God.* Boston: Houghton Mifflin, 1989. Provides a fascinating personal account of the Gulf Stream. MacLeish points out that to most oceanographers, the notion that the Gulf Stream was responsible for the mild climate of Europe has generally been treated with derision. Illustrations, maps.

Phrampus, Benjamin J., and Matthew J. Hornbach. "Recent changes to the Gulf Stream causing widespread gas hydrate destabilization." *Nature* 490, no. 7421 (2012): 527-530. Warming of deep waters can potentially cause breakdown of frozen gas hydrates and possible slope failure along the edge of the continental shelf.

Seager, R. "The Source of Europe's Mild Climate: The Notion That the Gulf Stream Is Responsible for Keeping Europe Anonymously Warm Turns out to Be a Myth." *American Scientist* 94 (2002): 340-31. An accessible treatment of the research conducted by Seager and others to debunk the myth. Illustrations, photographs, maps.

Seager, R., et al. "Is the Gulf Stream Responsible for Europe's Mild Winters?" *Quarterly Journal of the Royal Meteorological Society* 128, no. 586 (2002): 2563-2586. Somewhat technical, this work describes the details of the work done by Seager and others to debunk the myth.

See also: Atlantic heat conveyor; Atlantic multidecadal oscillation; Ekman transport and pumping; El Niño-Southern Oscillation; Gyres; La Niña; Maritime climate; Meridional overturning circulation (MOC); Ocean-atmosphere coupling; Ocean dynamics; Sea surface temperatures; Thermohaline circulation.

Gyres

Category: Oceanography

Definition

The oceans' gyres are basin-scale circulation patterns in which the net flow of water occurs in a circular pattern around the basin. Each gyre is generally made up of four distinct currents that are driven by wind stresses, and its circulation direction is governed by the Coriolis effect. The major, subtropical gyres in the Atlantic and Pacific Oceans are located between the equator and approximately 45° north latitude, while the smaller, subpolar gyres lie north of that latitude. The Antarctic Circumpolar Current is a gyre that flows continuously around Antarctica, because there are no landmasses to impede it.

Individual currents that make up a gyre have different characteristics. Because of the Coriolis effect and resulting differences in sea surface elevation, the western boundary currents (currents that flow on the western side of the ocean basins and flow northward from the equator) are narrow, are relatively deep (up to 1,200 meters), and have velocities of up to 178 kilometers per day. Examples of western boundary currents include the Gulf Stream in the Atlantic Ocean and the Kuroshio Current in the Pacific Ocean.

Eastern boundary currents have widths of up to almost 1,000 kilometers but are only 500 meters deep. Examples of eastern boundary currents include the California Current in the Pacific Ocean and the Canary Current in the Atlantic Ocean. Offshore Ekman transport associated with the eastern boundary currents leads to upwelling and high levels of productivity in these surface waters. Circulation of the gyres is completed by transverse

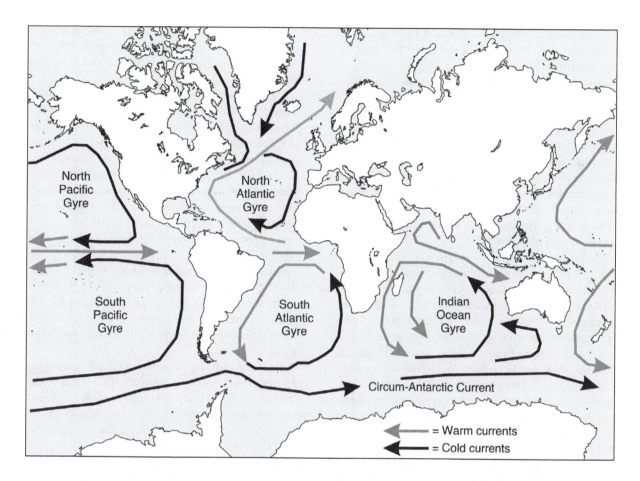

currents that flow east and west across the ocean basins, connecting the western and eastern boundary currents.

The subtropical gyres are associated with persistent high-pressure regions in the atmosphere, which leads to a net motion of surface water to the center of the gyre—a process known as convergence. These regions of high pressure are associated with low annual rainfall totals, so the salinity of water in the center of the gyres is somewhat elevated. Elevated salinity and convergence lead to downwelling, so the central gyres are regions of low productivity.

Significance for Climate Change

The gyres transfer large amounts of heat from the equator to the poles. For example, the Gulf Stream carries heat north from the Caribbean Ocean, travels along the East Coast of the United States, and then curves eastward toward Europe. The heat released from the Gulf Stream may lead to warmer average temperatures in Europe than those found at similar latitudes in North America. It has been hypothesized that global warming will lead to an increased influx of freshwater to the Arctic Ocean, which could block the northward flow of the warm, salty water of the Gulf Stream. In turn, this could prevent heat transport from the equator and could lead to cooling of the Northern Hemisphere.

The strength of the gyral currents can vary on decadal timescales, such as is observed in the Atlantic Multidecadal Oscillation (AMO), which manifests as cycles in average sea surface temperatures (SSTs), as lesser or greater amounts of warm water are transported from equatorial regions. The intensity of hurricanes is strongly dependent on SST, with stronger, more frequent hurricanes occurring when SSTs are higher. Thus, if global temperatures increase, the amplitude of the AMO will increase, leading to the circulation of warmer water and increased hurricane intensity.

Decadal variations in gyral flow can also affect the extent of coastal upwelling, salinity, and nutrient concentrations. Scientists from the Georgia Institute of Technology have discovered the existence of the North Pacific Gyre Oscillation. Variations in the mode of the oscillation are thought to cause historical variations in fish populations that are critical to Pacific fisheries. While such oscillations are part of the natural climate cycle, evidence indicates that the amplitude of the oscillations may increase with global warming.

Water in the center of the gyres is isolated from the rest of the world's oceans. An important consequence of this isolation is the low levels of phytoplankton productivity of the water—a situation described as "oligotrophic." The major source of nutrients to support productivity in the central gyres is upwelling of deep waters. However, downwelling, not upwelling, occurs in the central gyres, making these waters the equivalent of large, open-ocean deserts. As global temperatures have increased, the size of these ocean deserts has increased. Scientists from the National Oceanic and Atmospheric Administration and the University of Hawaii have examined a nine-year series of remotely sensed ocean color data from the SeaWiFS satellite and concluded that these open-ocean deserts have expanded by up to 15 percent. The expansion of the oligotrophic regions is consistent with computer models of oceanic vertical stratification in the gyres, but the rate of expansion exceeds all model predictions.

Anna M. Cruse

Further Reading

Di Lorenzo, Emanuele, et al. "North Pacific Gyre Oscillation Links Ocean Climate and Ecosystem Change." *Geophysical Research Letters* 35, no. 108607 (April, 2008). Overview of the discovery of the North Pacific Gyre Oscillation and its consequences for Pacific ecosystems and fisheries industries.

Open University. *Ocean Circulation.* 2d ed. Boston, Mass.: Elsevier Butterworth-Heinemann, 2005. This introductory oceanography textbook covers atmospheric circulation, surface currents, and thermohaline circulation. The theory of fluid flow throughout the ocean is developed with a minimum of mathematics (knowledge of algebra and trigonometry is assumed). Illustrations, figures, tables, maps, references, index.

Polovna, Jeffrey J., Evan A. Howell, and Melanie Abecassis. "Ocean's Least Productive Waters Are Expanding." *Geophysical Research Letters* 35, no. 103618 (February, 2008). Presents the SeaWiFS data, showing that the ocean's oligotrophic regions in the subtropical gyres are expanding at rates exceeding those predicted by computer models.

Trenberth, Kevin E. "Warmer Oceans, Stronger Hurricanes." *Scientific American* 297, no. 1 (July, 2007): 44-51. Describes how warmer oceans and variations in gyral currents such as the Gulf Stream affect decade-scale circulation patterns and the formation of hurricanes.

van Franeker, Jan A., and Kara Lavender Law. "Seabirds, gyres and global trends in plastic pollution." *Environmental Pollution* 203 (2015): 89-96. Studies of stomach contents of dead seabirds can be used to measure plastic concentrations far out to sea.

See also: Atlantic heat conveyor; Atlantic multidecadal oscillation; Ekman transport and pumping; El Niño-Southern Oscillation; Gulf Stream; La Niña; Mean sea level; Meridional overturning circulation (MOC); Ocean dynamics; Plankton; Sea surface temperatures; Thermocline; Thermohaline circulation.

Hadley circulation

Category: Meteorology and atmospheric sciences

Definition

Named for eighteenth century meteorologist George Hadley, who explained how it worked, the Hadley circulation is a loop of air that starts near the equator. The hot air in the equator region rises to the troposphere level and is carried toward the poles. At about 30° latitude, the air drops back to the surface, creating a high-pressure area. The air is pulled back toward the equator, completing the loop and generating the trade winds. Because the Earth is spinning, the air traveling back to the equator goes toward the southwest in the Northern Hemisphere and is called the Northeast Trades. In the Southern Hemisphere, the air movement generates the Southeast Trades. The low-pressure area where the Northeast Trades meet the Southeast Trades and where the hot air rises is called the Inter-Tropical Convergence Zone (ITCZ). The ITCZ does shift with the seasons.

The Hadley circulation is caused by solar heating. The intensity of the Hadley circulation is related to the sea surface temperature. The hot air will carry moisture aloft, but as the air rises, it cools and is able to contain less moisture, causing the large amount of rainfall in the equatorial region. Still, it is moist hot air that travels toward the poles and drier cool air that blows back toward the equator.

The Hadley circulation is one of the major regulators of the Earth's energy budget. It spreads heat collected at the equator to the northern and southern subtropical areas. It also carries heat into the troposphere, where it can radiate into space. Hadley's ascending limb controls the rainfall in the tropical areas, where large amounts of rain occur. The descending limb controls the dryness of the subtropical area. Although the Hadley circulation covers only part of the Earth, it covers the area where a large percentage of the people of the world live.

Significance for Climate Change

A difference in the patterns of the sea surface temperature would force a change in precipitation and cause a change in the Hadley circulation. A change in that system would cause a change in the flow of heat, momentum, and humidity along the meridians. The Earth's overall radiative balance, along with the monsoon systems and the ocean circulation, are also affected by a change in Hadley circulation. Different climate models diverge when they project the effect of an increase in greenhouse gases (GHGs) on the Hadley circulation. Some models indicate an increase of the intensity of the Hadley circulation, causing a more arid subtropical region to develop as the rising GHG concentration causes an increase in sea surface temperature. Other models indicate a weakening of the Hadley circulation.

Since 1950, there has been a more intense Hadley circulation, with a consequential increase in rainfall in the equatorial oceanic region and a drier tropical and subtropical landmass. This increase has accompanied a stronger westerly stratospheric flow and an increase in cyclones in the middle latitudes. The driving force behind these changes has been identified as the warming of the Indo-West Pacific tropical waters. The increased sea surface temperature difference between the winter and summer hemisphere tropics causes a stronger Hadley circulation.

Models indicate that the solar forcing of the increased Hadley circulation is more intense with GHGs than without. It is not clear from the models proposed whether the more intense Hadley circulation is due to a natural fluctuation, is anthropogenic, or is the combined result of natural and human factors. Solar forcing models indicate more evaporation and thus more moisture carried aloft. This causes less cloud cover and more solar heating. Thus, solar forcing seems to represent a positive feedback loop. The more solar heating, the less cloud cover and more solar heating.

Models of past data indicate that the ITCZ may have shifted over time by more than just the annual summer-to-winter position shift. Another shift would cause different areas of the Earth to be dry or to be rain-soaked. The extent of the Hadley circulation is influenced by several different factors according to models. One factor that was shown not to be related is the mean global temperature. Just increasing the Earth's temperature will not change the area covered by the Hadley circulation. The Hadley circulation can be likened to El Niño-Southern Oscillation events. Since 1976, the increase in number and strength of El Niño events has caused an increase in the strength of the winter Hadley circulation.

C. Alton Hassell

Further Reading

Diaz, Henry F., and Raymond S. Bradley, eds. *The Hadley Circulation: Present, Past, and Future.* Dordrecht, the Netherlands: Kluwer Academic, 2004. Essay collection that grew out of a conference examining the Hadley circulation from several different viewpoints. Illustrations, maps, references.

Hadley, G. "Concerning the Cause of the General Trade-Winds." *Philosophical Transactions of the Royal Society of London* 39 (1735): 58-62. George Hadley takes Halley's concept and expands on it.

Halley, E. "An Historical Account of the Trade Winds, and Monsoons, Observable in the Seas Between the Tropicks, with an Attempt to Assign the Physical Cause of the Said Winds." *Philosophical Transactions of the Royal Society of London* 16 (1686): 153-168. The original article explaining the Hadley circulation. Illustrations.

Hasegawa, H., R. Tada, X. Jiang, Y. Suganuma, S. Imsamut, P. Charusiri, N. Ichinnorov, and Y. Khand. "Drastic shrinking of the Hadley circulation during the mid-Cretaceous Supergreenhouse." *Climate of the Past* 8, no. 4 (2012): 1323-1337. The Hadley circulation (Trade Wind belts) can be mapped in the geologic past by studying wind deposition patterns in fossil desert deposits. This study suggests that the belt underwent drastic shrinkage toward the Equator during the height of the Cretaceous warm period.

O'Hare, Greg, John Sweeney, and R. L. Wilby. *Weather, Climate, and Climate Change: Human Perspectives.* Harlow, England: Pearson Prentice Hall, 2005. Comprehensive review of weather and its causes, including the Hadley cell. Illustrations, references, index.

Watterson, Ian Godfrey, and Edwin K. Schneider. "The Effect of the Hadley Circulation on the Meridional Propagation of Stationary Waves." *Quarterly Journal of the Royal Meteorological Society* 113, no. 477 (July, 1987): 779-813. Based on Watterson's doctoral dissertation of the same name, this article distills one of the only extended studies of Hadley circulation, presenting its most salient insights and conclusions.

See also: Atmospheric dynamics; Climate and the climate system; Climate models and modeling; Climate zones; Evapotranspiration; Inter-Tropical Convergence Zone; North Atlantic Oscillation (NAO); Ocean-atmosphere coupling.

Halocarbons

Category: Chemistry and geochemistry

Definition

The halocarbons are a group of partially halogenated organic compounds, in which carbon atoms link with halogen atoms by a covalent bond. (The halogen atoms include bromine, chlorine, fluorine, and iodine.) The most common type of halocarbon contains chlorine and belongs to the subclass of chlorocarbons, which includes substances such as carbon tetrachloride and tetrachloroethylene. The other common subclass of halocarbon, the fluorocarbons, contains fluorine and includes polytetrafluoroethylene (Teflon) and perfluorocarbons (PFCs). Some of the Freons, though not all, are halocarbons. Halocarbons are produced and valued because of their nonflammability, low chemical reactivity, and low toxicity.

The halocarbons include several compounds that affect the environment, compounds such as chlorofluorocarbons (CFCs), hydrochlorofluorocarbons (HCFCs), hydrofluorocarbons (HFCs), halons, methyl chloride, and methyl bromide. Though it is possible for halocarbons to form naturally, from volcanic activity or by the interaction of halogen and plant matter, most halocarbons are produced by humans for industrial and chemical use and include substances used as refrigerants and fire extinguishants.

Significance for Climate Change

Many halocarbon compounds have a high global warming potential (GWP). Of these compounds, three (CFC-11, CFC-12, and CFC-113, all created by human activity) are the worst offenders. These gases are not reactive, so they are able to remain in the stratosphere for hundreds of years.

How Halocarbons Affect the Earth's Heat Retention. As light from the Sun enters the atmosphere, some of that light is scattered by molecules in the air or is reflected from clouds back into space. Some of the light that reaches Earth's surface is also reflected back into space, such as light that hits snow or ice. However, much of the light that reaches Earth is absorbed and retained as heat. The Earth's surface warms and emits infrared photons, which make several passes between the Earth and the atmosphere, warming the atmosphere and the Earth as they go back and forth. Eventually, these infrared photons return to space.

The greenhouse gases (GHGs), which include many halocarbons, are able to absorb infrared photons, transferring their energy from the photons to gas molecules and thereby trapping thermal energy that would otherwise be released into space. Eventually, this absorption of energy causes a net change in the Earth's energy balance, increasing the overall amount of thermal energy held within the system. Different GHGs have different GWPs and lifetimes, measurements of the amount they contribute to global warming and the duration of that contribution, respectively. Both the GWP and the lifetime of halocarbons are unusually high. They may remain in the atmosphere for up to four hundred years, for example, continuing to affect the global climate long after their initial release.

How Halocarbons Deplete the Ozone Layer. Halocarbons containing chlorine and bromine also deplete the ozone layer. Until the 1970's, CFCs, perhaps the best known of the halocarbons, were used as propellants, solvents, and cleaners. Realizing that these compounds were depleting the ozone layer, many nations agreed to control halocarbon emissions when they signed the Montreal Protocol.

Ozone is a form of oxygen; it is formed of three atoms of oxygen bound together. Ozone in the stratosphere blocks harmful solar radiation. Halocarbons help speed the natural destruction of ozone. For example, CFCs are able to break down ozone when they combine with it in high-frequency ultraviolet (UV) light. During this breakdown, ozone molecules are destroyed, but chlorine reforms and is able to take part in the breakdown again. In this way, just one chlorine atom in the stratosphere is thought to be able to destroy about 100,000 ozone molecules.

Ozone depletion can have serious effects. The less ozone is available to shield the Earth, the more UV-B rays are able to penetrate the atmosphere and reach the planet's surface, where they cause skin cancer and cataracts in humans. Moreover, the disruption of the normal wavelengths of light reaching Earth's surface affects plants and their growth rates. It is also possible that the change in atmospheric radiation affects global wind patterns and therefore climate.

Halocarbons do not disintegrate easily and may become corrosive agents when burned, which makes them difficult to dispose of safely. Halocarbons are also implicated in liver disease, eye cataracts, and skin cancer, and they may affect the human immune system.

Marianne M. Madsen

Further Reading

Forster, P., and M. Joshi. "The Role of Halocarbons in the Climate Change of the Troposphere and Stratosphere." *Climatic Change* 71, no. 1 (July, 2005): 249-266. Discusses how halocarbons containing chlorine and bromine affect atmospheric temperature and ozone-layer depletion.

Hodnebrog, Ø., M. Etminan, J. S. Fuglestvedt, George Marston, G. Myhre, C. J. Nielsen, Keith P. Shine, and T. J. Wallington. "Global warming potentials and radiative efficiencies of halocarbons and related compounds: A comprehensive review." *Reviews of Geophysics* 51, no. 2 (2013): 300-378. A tabulation of greenhouse potential for a large number of mostly synthetic industrial compounds.

Middlebrook, Ann M., and Margaret A. Tolbert. *Stratospheric Ozone Depletion.* Sausalito, Calif.:

University Science Books, 2000. Describes how human activities affect ozone destruction.

Ozone Depletion, Greenhouse Gases, and Climate Change. Washington, D.C.: National Academy Press, 1989. Proceedings of a 1988 conference on atmospheric contributors to climate change. Includes a chapter entitled "The Role of Halocarbons in Stratospheric Ozone Depletion."

World Meteorological Organization. *Scientific Assessment of Ozone Depletion, 1994.* Global Ozone Research and Monitoring Project Report 37. Geneva, Switzerland: Author, 1995. Overview of the state of the ozone layer and factors contributing to its depletion.

See also: Aerosols; Chemical industry; Chlorofluorocarbons and related compounds; Cosmic rays; Greenhouse effect; Halons; Hydrofluorocarbons; Ozone; Perfluorocarbons.

Halons

Category: Chemistry and geochemistry

Definition

Halons are gaseous compounds of carbon, bromine, and another halogen (such as hydrogen, chlorine, or fluorine). These gases, also known as bromofluorocarbons, include the chlorofluorocarbon (CFC) family. They are classified by the U.S. Environmental Protection Agency as Group II of the Class I ozone-depleting substances.

Halons were first used during World War II as fire-extinguishing agents. They are particularly useful in extinguishing electrical fires, because they are covalently bonded and thus do not react with electrical equipment. They continued to be used through 1994, when they were banned by the Montreal Protocol as ozone-depleting substances.

Significance for Climate Change

Halons are members of the class of halogenated hydrocarbons, which have been shown to deplete the ozone layer. By some estimates, halons are three to ten times more destructive of the ozone layer

than are other substances in this family, including CFCs, because they contain bromine. The bromine radicals in halons react with ozone particles and remove them from the stratosphere. They sometimes react with atmospheric chlorine as well, magnifying their effects on ozone. The bromine-containing halons are much more destructive of atmospheric ozone than are other compounds, such as those containing chlorine, because bromine compounds are much more likely to disassociate in ultraviolet light, so many more ozone molecules are destroyed before the bromine molecules diffuse.

Halon use worldwide has historically been less than CFC use, so halons are not as widely known as ozone-destruction agents. However, halons are more destructive than some other ozone-depleting compounds, because their concentration in the atmosphere is still rising as a result of their long lifetimes.

Less environmentally destructive fire-extinguishing agents have been found to replace halons. These include halocarbon-based agents that absorb heat and inert gas agents that deplete oxygen, smothering fires. Both of these agents are less threatening to the ozone layer than are halons.

Marianne M. Madsen

Further Reading

Miziolek, Andrzej W., and Wing Tsang, eds. *Halon Replacements: Technology and Science.* Washington, D.C.: American Chemical Society, 1995. A seminal source on research, development, and testing of substances to replace halons. Arose out of a symposium on the subject and provides a summary of that symposium in addition to collected papers. Includes author index and detailed topical index.

Schwarzenbach, René P., Philip M. Gschwend, and Dieter M. Imboden. *Environmental Organic Chemistry.* 2d ed. Hoboken, N.J.: Wiley, 2003. Discusses environmental factors and how they affect organic compounds. Includes problems and case studies.

Taylor, Gary. *Eliminating Dependency on Halons: Case Studies.* Paris: United Nations Environment Programme, 2000. Case studies describing successful removal of halons according to the Montreal Protocol.

See also: Aerosols; Chemical industry; Chlorofluorocarbons and related compounds; Cosmic rays; Greenhouse effect; Halocarbons; Hydrofluorocarbons; Ozone; Perfluorocarbons.

Health impacts of global warming

Category: Diseases and health effects

Ambient temperature in general, and systematic temperature changes in particular, have both direct and indirect impacts on human health. Direct effects include increases or decreases in heatstroke, hypothermia, and metabolic and physiological disorders. Indirect effects include increases or decreases in drought, famine, severe weather events, and disease.

Key concepts

cardiac diseases: heart diseases, including those of relevant blood vessels (cardiovascular) and those caused by decreased blood supply (ischemic)

cerebrovascular disease: illness caused by abnormality in blood vessels supplying the brain

hemoconcentration: increased concentration or thickening of the blood resulting from dehydration or cooling of the body

mortality and morbidity: statistics on death and illness in population groups

pulmonary disease: illness of the lungs or affecting the breathing process

Background

The threat of possible severe climate change, including predictions of global warming from politicians, media leaders, and scientists, has aroused worldwide concern about the potential effects of such change on human health. The United Nations' Intergovernmental Panel on Climate Change (IPCC) reported in 2007 that multiple computer-model predictions were consistent with significant climate warming occurring by 2050 and potentially disastrous warming occurring by 2100. These predictions depended on various scenarios of increasing greenhouse gases (GHGs), especially carbon dioxide (CO_2), in the atmosphere and were based on the anthropogenic hypothesis, which holds that the most important variable in climate predictions is the human contribution of atmospheric CO_2 from burning fossil fuels. Less significant contributing factors include methane production by farm animals.

There is significant scientific dissent on the causation of climate change, since human historical records show temperature variation profoundly affecting agriculture and worldwide human activity for centuries before the Industrial Revolution. Furthermore, ice cores and other paleontological records show violent variations of temperature, CO_2, and methane for hundreds of thousands of years before human civilization.

The IPCC computer models predict increases in average global surface temperatures of 1.0 to 1.9° Celsius by 2050. These models calculate temperature rises of 2.0 to 4.5° Celsius by 2100, depending on whether CO_2 emissions stabilize downward at 5 gigatons per year or increase to as much as 28 gigatons per year by the end of the century.

Prior temperature records show that humans have been exposed to temperature shifts of similar magnitudes. The cooling that occurred from 1940 to 1970 was estimated at about 0.2° Celsius, and the warming from 1970 to 1998 is estimated at 0.5° to 0.7° Celsius. The average surface temperature in the decade following the 1998 El Niño did not increase significantly, and physicists who study solar effects on climate have predicted possible cooling because of the decrease in solar activity (such as sunspots).

S. Fred Singer and Dennis Avery have compiled extensive reports on prior temperature fluctuations, with examples of past climate cycles showing increased temperatures of more than 1° Celsius during the Medieval Warm Period and decreases of 0.3° Celsius during the Little Ice Age. South African data from stalagmite temperature proxies indicate temperature increases of up to 4° Celsius in the Medieval Warm Period.

Urban Heat Island Effects

As Earth's population continues to migrate from rural areas to more densely concentrated suburbs and cities, more people become exposed to urban heat islands. Cities are warmer than surrounding countryside, because traditional roofs and paving surfaces absorb more solar heat than do dirt and vegetation, and significant heat is generated by industry, power plants, residential heating, and air conditioning. Large cities have shown average temperature increases of as much as 3° Celsius (Tokyo, 1876-2004) to 4° Celsius (New York City, 1822-2000). These data are relevant, because potential health effects may be evaluated by comparing the effects of these localized temperature changes to projected future temperature changes.

Historical Context

Scientists have been speculating about the health effects of climate change since the concept of global warming became widespread. In 1992 and 1995, IPCC members expressed concern that increases in the number and severity of heat waves could cause a rise in deaths. The 1992 report found that temperature increases were more prevalent in the winter and at night, diminishing the health effects of extreme cold weather; summer temperature increases have fallen, which diminishes deaths from heat waves. This could explain the IPCC's 1995 statement that global warming could result in fewer cold-related deaths.

In 1995, Thomas Gale Moore published the first of his pioneering efforts, "Why Global Warming Would Be Good for You," followed in 1998 by "Health and Amenity Effects of Global Warming." He estimated that a temperature increase of 2.5° Celsius in the United States would cause a decrease of forty thousand deaths per year from respiratory and circulatory disease, based on U.S. mortality statistics as a function of monthly climate change.

In 1997, the Eurowinter Group published "Cold Exposure and Winter Mortality from Ischaemic Heart Disease, Cerebrovascular Diseases, Respiratory Diseases, and All Causes in Warm and Cold Regions of Europe." This was a landmark study that elucidated the mechanisms of serious illness from cold, which are dominated by hemoconcentration, which increases blood viscosity ("sludging"). Hemoconcentration can cause death from blockage of vessels serving the heart and brain tissue,

Global Deaths Due to Climate-Related Disasters, 1900-1999

Decade	Deaths (thousands)	Death Rate (per million people)
1900's	128	79
1910's	25	14
1920's	485	242
1930's	446	209
1940's	180	76
1950's	211	71
1960's	167	49
1970's	74	19
1980's	66	14
1990's	32	6

Data from: Indur M. Goklany, *The Improving State of the World.* Washington, D.C.: Cato Institute, 2007.

and it accounts for half of all excess cold-related mortality.

The 1997 Eurowinter Group study was followed by "Heart Related Mortality in Warm and Cold Regions of Europe: Observational Study," which was published in the *British Medical Journal* in 2000. These two studies provided data on mortality rates as a function of mean daily temperature in Athens, Greece; London, England; and Helsinki, Finland, providing the most comprehensive collection of evidence that mortality decreases as temperature increases, over most of the current climate range in Europe.

In 2005, Robert E. Davis furnished a survey on climate change and human health, published in *Shattered Consensus: The True State of Global Warming.* He predicted that human adaptation "will be key in determining the ultimate impacts of climate change." He demonstrated that some adaptations are already taking place that effectively mitigate negative impacts of global warming. His data, for example, show that excess mortality due to heat waves in many U.S. cities dropped to essentially zero in the three decades following 1964. This decline in heat mortality was especially evident in

Southern cities, where high heat and humidity are common, but also spread northward. This happy trend could be attributed to air conditioning, better health care, architectural changes, and public health measures such as shelters. The salutary result, however, is prevention of major death events that were previously associated with heat waves and diminution of negative effects of climate warming.

In 2006, A. J. McMichael and his colleagues published "Climate Change and Human Health: Present and Future Risks." This was an attempt at a comprehensive evaluation of the direct and indirect health risks associated with warming, including infections and vector-borne diseases such as malaria. The evaluation assumes that the maximum daily mortality in higher-temperature periods will be equal to or greater than the maximum mortality in colder periods, resulting in heat-related deaths increasing far more than the lives saved by warming of the cold periods. This hypothesis does not stand up to previous data from the United States that showed that mortality in winter due to cardiac, vascular, and respiratory disease is seven times greater than summer. This ratio is about 9 to 10 in Europe.

The most comprehensive data on daily mortality, from all causes, as a function of the day of the year, show a clear relationship, with maximum mortality in January and minimum mortality in the warmest months of July and August. These data strongly indicate that warming of average daily temperatures would cause a decrease in mortality in winter far greater than the slight increase of mortality from summer heat.

Current Findings

In early 2008, the British Department of Health released "Health Effects of Climate Change in the U.K., 2008," an update of previous reports from 2001-2002 edited by Sari Kovats. It used IPCC models that predicted an increase of mean annual temperatures in the United Kingdom between 2.5 and 3° Celsius by 2100. They found that there was no increase in heat-related deaths from 1971-2002, despite warming in summers, suggesting that the British population is adapting to warmer conditions. Cold-related mortality fell by more than one-third in all regions. The overall trend in mortality for

the warming (from 1971-2002) was beneficial. The report states, in summary, that "winter deaths will continue to decline as the climate warms."

Quantitative Estimates of Benefit

The data from the Eurowinter Group on mortality versus temperature can be used to estimate mortality benefits from climate warming. The authors actually drew "straight-line" fits to the slope of the data. The slopes for Athens, Helsinki, and London vary between 1 and 2 percent decreased mortality per degree Celsius increased temperature. This would lead to an estimated decrease of twenty-five to fifty thousand deaths per year in the United States for a 1° Celsius temperature rise. This can be compared to thirty thousand deaths per year from breast cancer, thirty thousand for prostate cancer, and about forty thousand from motor-vehicle accidents.

Bjørn Lomborg, a distinguished Danish environmental economist, has estimated 1.7 million fewer deaths in the world per year from moderate warming, or 17 million by year 2100. (He also notes that deaths from cold are nine times greater than deaths from warmth.) Note that the converse is also true; that is, cooling would cause similar increases in death rates. Heat deaths often represent "displacement"; that is, weakened people die a few days or weeks before prior expectation. Deaths due to cold, on the other hand, usually result in months to years of life lost. Thus, the benefits in life expectancy from warming in cold periods may be much more than nine times greater than lifespan lost in warm periods.

More Severe Climate Change

The slopes of the data on mortality versus temperature are fairly linear over temperature variations of more than 20° Celsius. Thus, the benefit of warming (and the risk of cooling) should be fairly proportional to the temperature change, for climate shifts of more than 2-4° Celsius. Increasingly urbanized populations have already been exposed, and presumably adapted, to warming of 2-4° Celsius, because of the urban heat island effect. A major drop in climatic temperatures could be more devastating, especially in rural and less developed societies.

The Developing World

Unfortunately, there is a dearth of data on mortality versus temperature in the developing world, especially by comparison to Europe and the United States. Lower standards of living and health care could decrease technological response to climate change, but less developed societies have already evolved adaptation techniques and behaviors that promote survival under potentially adverse conditions such as daily and seasonal climate changes.

The most prominent variable affecting health and survival of less advanced societies is the amount and variety of food available. Food crop production is enhanced by increased CO_2 in the atmosphere, which promotes photosynthesis. Increased CO_2 also allows plants to flourish with less water, giving them greater resistance to drought. It is estimated that the current increase in CO_2 has already caused increased food production by about 10 percent, independent of other factors. If CO_2 continues to increase as expected, significant improvements in nutrition may benefit populations in developing nations, no matter how the climate changes.

There are data on global death rates on climate-related disasters, however. Goklany has shown that death rates fell between 1930 and 2004. It is not possible to separate the decreasing mortality from storms, floods, drought, and so on from deaths due to heat and cold, but the overall trend is extremely beneficial and is taking place during a period of multiple climate changes and increasing GHG emissions.

Context

The direct effects of possible climate warming on mortality are likely to be beneficial and of substantial magnitude by comparison to mortality from disease and accidents. The impact on human life expectancy may be proportionally more significant because of the different characteristics of death due to cold versus heat.

Quality of life and health may also improve. Large populations have migrated from the northern to the southern United States, experiencing an increase in average temperature of more than 5° Celsius, resulting in improved health and life expectancy. There are similar effects beginning in Europe.

It is reasonable to expect better health and better health statistics in the industrialized world, with moderate climate warming. Better nutrition and human adaptation can be expected in the developing world. The consequences of widespread climatic cooling are likely to be much more threatening.

Howard Maccabee

Further Reading

Baer, Hans A., and Merrill Singer. *Global Warming and the Political Ecology of Health: Emerging Crises and Systemic Solutions*. Walnut Creek, Calif.: Left Coast Press, 2009. A reading of global warming as a condemnation of capitalism that includes significant discussion of the health effects of climate change.

Goklany, Indur M. *The Improving State of the World*. Washington, D.C.: Cato Institute, 2007. Well-researched and graphically documented book on economic development, technological change, and environmental effects on the human condition. Concludes optimistically that reconciling human well-being with climate change, sustainable development, and population trends is possible.

Lomborg, Bjørn. *The Skeptical Environmentalist: Measuring the Real State of the World*. New York: Cambridge University Press, 2001. Lomborg is a world-leading pioneer in environmental economics, with path-breaking arguments, backed with data, on improvements in human welfare and prosperity despite perceptions of global warming, pollution, and other environmental changes. This is a profound and comprehensive masterwork.

McMichael, Anthony J. "Globalization, climate change, and human health." *New England Journal of Medicine* 368, no. 14 (2013): 1335-1343. Globalization affects human health in a large number of ways, including changes in dietary habits and food production, rapid spread of diseases, and externalization of environmental impacts.

Musil, Robert K. *Hope for a Heated Planet: How Americans Are Fighting Global Warming and Building a Better Future*. New Brunswick, N.J.: Rutgers University Press, 2009. Comprehensive approach to global warming through the lens of public health policy.

Singer, S. Fred, and Dennis T. Avery. *Unstoppable Global Warming: Every Fifteen Hundred Years.* Rev. ed. Blue Ridge Summit, Pa.: Rowman & Littlefield, 2008. Explains multiple natural cycles that affect climate, especially the variations of the solar and cosmic-ray effects. Shows how data are developed to measure climate variation in the distant past and how historical events such as the Medieval Warm Period are not consistent with the anthropogenic hypothesis.

See also: Asthma; Carbon dioxide fertilization; Coastal impacts of global climate change; Diseases; Extreme weather events; Famine from global warming; Floods and flooding; Poverty; Skin cancer; World Health Organization.

Heartland Institute

Category: Organizations and agencies
Date: Established 1984
Web address: http://www.globalwarming heartland.org

Mission

An American, libertarian, free-market-oriented, tax-exempt 501(c)(3) nonprofit organization, the Heartland Institute researches and develops free market solutions to social and economic problems, including environmental problems. The institute was established in 1984 and is based in Chicago, Illinois. Its activities are directed by a fifteen-member board of directors that meets quarterly. As of 2008, thirty full-time staff, including editors and senior fellows, oversee the organization's day-to-day activities. The think tank focuses on issues such as government spending, taxation, education, health care, free market environmentalism, and global warming.

Significance for Climate Change

The Heartland Institute asserts that there are no reliable data supporting the notion that global warming mechanisms or trends have ever taken place or are taking place presently. Furthermore, it claims that a moderate degree of global warming is beneficial to the environment and humans worldwide. The institute has partnered with other global warming skeptic organizations such as the Cooler Heads Coalition, which has itself been widely criticized for its work against penalizing big carbon dioxide (CO_2) emitters.

At the Heartland Institute's March, 2008, conference, global warming skeptics from around the world were brought together in New York. At the meeting, participants collectively criticized the Intergovernmental Panel on Climate Change, as well as proponents of any scientific studies that reported positive correlation between human activities and global warming. The organization's numerous publications and conferences have all shared a similar theme and tone.

The Heartland Institute received extensive scrutiny of its publication procedures in April, 2008, after environmental journalist Richard Littlemore revealed that it engaged in questionable practices. In compiling a list of "Five Hundred Scientists with Documented Doubts of Man-Made Global Warming Scares," the organization included over forty-five scientists as coauthors on various articles with which they had no affiliation, that they did not agree to coauthor, or that made claims with which they disagreed. Following this scandal, when scientists came forward to demand removal of their names from the list, the Heartland Institute claimed that they had no legal or ethical grounds to remove or amend the original list of names.

The institute has also been criticized for its policies on appointing and recruiting members of its board of directors. In the past, executives from such corporations such as ExxonMobil, an oil company, and Philip Morris, a tobacco company, have served on the Heartland Institute's steering committee.

Rena Christina Tabata

See also: American Enterprise Institute; Catastrophist-cornucopian debate; Cato Institute; Competitive Enterprise Institute; Cooler Heads Coalition; Friends of Science Society; Heritage Foundation; Journalism and journalistic ethics; Libertarianism; Nongovernmental International Panel on Climate Change; Pseudoscience and junk science; Reason Public Policy Institute; Skeptics.

Heat capacity

Category: Physics and geophysics

Definition

The total heat capacity of an object is the amount of energy needed to raise its temperature one degree Celsius. Heat capacity is usually measured in joules per Celsius degree. Thus, if an object has a heat capacity of 500 joules per Celsius degree, adding 500 joules of thermal energy will increase its temperature by 1° Celsius. Adding 1,500 joules will increase its temperature by 3° Celsius, and so forth. Removing the same amount of thermal energy will decrease the object's temperature by the same amount.

An object's heat capacity is related to both its composition and its mass. Scientists divide heat capacity by mass to determine an object's specific heat capacity, or specific heat, which is a property of composition alone. Heat capacity and specific heat are thus closely related, but heat capacity is a property of a particular object, while specific heat is a more general property of a type of material.

Particularly for gases or liquids, the heat capacity and specific heat can differ under different conditions, such as pressure or volume. Hence, specific heat capacity tables often list specific heats at constant pressure and at constant volume.

Significance for Climate Change

Using various climate change computer models, climate researchers try to predict changes in global surface temperature. One of the many input parameters for these calculations is the change in the energy content of Earth's climate system. Energy sources can include incoming solar radiation, energy trapped by greenhouse gases, waste energy from machines or industrial processes, and so forth. After scientists determine the net energy change in Earth's climate system, they need to calculate the total heat capacity of the climate system, including Earth's atmosphere, oceans, and surface.

Using the calculated change in thermal energy contained in Earth's climate system and its total heat capacity, scientists can calculate the net change in Earth's temperature. The idea is relatively simple, but it is very difficult to calculate both the total thermal energy and the heat capacity for Earth's climate system. Therefore, it is difficult to make accurate predictions about changes in Earth's average temperature.

These complexities make global warming models very uncertain. Different global warming models will make different predictions as a result of these uncertainties.

Paul A. Heckert

See also: Bayesian method; Climate models and modeling; Climate prediction and projection; General circulation models; Heat content.

Heat content

Category: Physics and geophysics

Definition

Touch a fluorescent light bulb. It may feel pleasantly warm, but it will not feel extremely hot. However, the gas inside a working fluorescent bulb is usually at a temperature of more than 10,000° Kelvin. For comparison, the Sun's surface temperature is 5,800° Kelvin. If the gas inside a fluorescent light bulb is at such a high temperature, why is the surface of the bulb relatively cool?

The key is heat content. The gas inside the bulb is at a high temperature, but, because the gas is so thin and there are relatively few gas atoms inside the bulb, relatively little total heat energy is contained within the gas. Heat content is the total thermal energy an object contains. The gas inside the bulb has a high temperature but a low heat content, so there is not enough thermal energy to burn one's hand.

A key related concept is heat capacity, which is the amount of heat energy needed to raise the temperature of an object. The gas in a fluorescent bulb has a low heat capacity, so the total amount of thermal energy needed to raise its temperature is small. Thus, it required relatively little heat energy to raise the temperature of the gas, which also contributes to explaining why the system contains relatively little energy.

Significance for Climate Change

The total heat content of Earth's climate system, including its atmosphere, oceans, and surface, can change through a variety of processes. For example, the amount of direct solar radiation striking Earth's surface can change, either by the Sun's energy output changing or by the atmosphere's transparency changing. Changes in the concentration of greenhouse gases in Earth's atmosphere change the amount of thermal energy trapped within the atmosphere rather than radiated back into space. Industrial processes and waste heat from all types of engines pump thermal energy into Earth's climate system. These and many other processes change the total heat content of the planetary climate system.

As this total heat content changes, Earth's average temperature changes. Predictions of changes in Earth's temperature depend upon knowledge of the planet's current total heat content, as well as of all processes affecting Earth's climate system. Such comprehensive knowledge is elusive, so predicting future temperature changes is extremely difficult, and vastly divergent predictions are common.

Paul A. Heckert

See also: Bayesian method; Climate models and modeling; Climate prediction and projection; General circulation models; Heat capacity.

Heat vs. temperature

Category: Physics and geophysics

Definition

The concepts of heat and temperature are closely related, so these quantities are often confused, and the terms are sometimes used interchangeably (but incorrectly) by laypersons. Temperature is a measurement of the average kinetic energy—that is, physical motion—of the atoms in a substance. The kinetic energy of an object is calculated according to the following formula:

$$E_K = \tfrac{1}{2}MV^2$$

where E_K is kinetic energy, M is mass, and V is velocity. The kinetic energy of atoms is generally due to a combination of their directional velocity and vibrational motion. This energy may be measured by a thermometer.

The two most common temperature scales in use are the Fahrenheit and Celsius (formerly centigrade) scales. In the Fahrenheit scale, the freezing point of water is defined as 32° and the boiling point is defined as 212°. In the Celsius scale, the freezing point of water is defined as 0° and the boiling point is defined as 100°. In science and engineering, the Rankine and Kelvin temperature scales are used. These are called absolute temperature scales, because their zero reference level is absolute zero, the lowest temperature that can theoretically exist. At absolute zero, the total kinetic energy of all the atoms in a substance is zero.

Heat is the form of energy transferred across the boundary of an object as a result of a temperature difference across that boundary. Consider what happens across a pane of window glass on a cold winter day. The air inside the heated house is warmer than the air outside. This temperature difference across the window pane causes an energy flow through the glass from the warm air inside to the cold air outside: The outer air next to the glass is warmed, while the air inside the house cools, requiring a heating system to maintain a constant, comfortable temperature. There are three types of heat transfer: conduction, convection, and radiation. All result from a temperature difference between an object and its surroundings.

Conduction is the primary mechanism of heat transfer in solids. It occurs because of molecular activity in the solid. Convection is the primary mechanism of heat transfer through fluids and results from bulk mixing between fluid layers. Radiation, the third type of heat transfer, is the only mechanism that can transfer heat through a vacuum. When the Sun heats the Earth, there is no solid between them through which heat can be conducted, nor is there is any fluid through which heat can be convected. The temperature difference between the Sun and the Earth still causes heat transfer to the Earth by means of electromagnetic waves. Electromagnetic waves can also transfer energy in the form of light, X rays, or radio waves.

Heat is measured as a function of the temperature change of a substance when heat transfer occurs. The commonly used units for heat are the calorie, the joule, and the British Thermal Unit (BTU). A calorie is the amount of heat required to raise the temperature of one gram of water by one degree Celsius. A BTU is the amount of heat required to raise the temperature of one pound of water by one degree Fahrenheit. One joule is equal to 0.239 calorie.

Significance for Climate Change

Heat transfer to the Earth's oceans and atmosphere is the first step in, and at the very heart of, climate change. Most of the industrial or mechanical processes in which humans engage release heat as a waste by-product. For example, all fossil fuel and nuclear power plants generate waste heat as a by-product of producing power. A typical power plant may have an efficiency of between 35 percent and 50 percent, meaning that this percentage of the stored energy in a fuel is converted to useful energy, while the remaining energy is discharged to the atmosphere or some body of water as waste heat. This heat transfer to the environment causes a temperature increase of the environment. This temperature increase is a contributor to global warming. Even a degree or two increase in the average temperature of the oceans or the atmosphere can cause profound effects upon Earth's biosphere. All combustion engines also dissipate waste heat to the environment, as do heating and cooking appliances. Only power produced by wind, water, or geothermal energy does not directly contribute to global warming, although these types of power production can have other effects.

Eugene E. Niemi, Jr.

Further Reading

Cengel, Yunus A., and Michael Boles. *Thermodynamics: An Engineering Approach.* 5th ed. New York: McGraw-Hill, 2006. Sophomore- or junior-level college textbook dealing with heat and energy. Includes a simplified discussion of global warming and the greenhouse effect.

Holman, Jack P. *Heat Transfer.* 9th ed. New York: McGraw-Hill, 2002. College engineering textbook dealing with the various mechanisms of heat transfer.

Thurman, Harold V., and Elizabeth Burton. *Introductory Oceanography.* 9th ed. Upper Saddle River, N.J.: Prentice Hall, 2001. Includes a section on air-sea interaction and its relationship to global warming.

See also: Atlantic heat conveyor; Global surface temperature; Heat capacity; Heat content; Latent heat flux; Sea surface temperatures; Urban heat island.

Heritage Foundation

Category: Organizations and agencies
Date: Established 1973

Mission

The Heritage Foundation is a conservative public policy research organization, or think tank, based in Washington, DC. It is widely supported, with more than 390,000 individual, foundation, and corporate donors, and it promotes conservative ideas and principles as solutions to current problems. The foundation maintains a database of policy experts, including those with expertise in global warming issues, and publishes summary statements and analyses of proposed climate change legislation. Its stated mission is to "formulate and promote conservative public policies based on the principles of free enterprise, limited government, individual freedom, traditional values, and a strong national defense."

Using an executive summary format more likely to be read by government officials than are other formats, the Heritage Foundation distills complicated topics into shorter policy papers. Historically, the foundation has had considerable political

WASHINGTON (May 13, 2010) *Chief of Naval Operations* (CNO) *Adm. Gary Roughead speaks at the Heritage Foundation's annual series of events aimed at highlighting key national defense and homeland security issues.* (US Navy/Wikimedia Commons)

influence; it became prominent during the conservative era of Ronald Reagan's presidency. Reagan's policies were influenced by a foundation-published book, Mandate for Leadership (1981), that provided policy, budget, and administrative action recommendations and advocated limited government. The foundation continues to play a significant role in public policy and is considered one of the most influential research organizations in the United States.

Significance for Climate Change

The Heritage Foundation researches a broad range of policy issues, including issues related to the environment and global warming. It maintains a searchable online database of policy experts at www.policyexperts.org that includes several renowned climate change skeptics. The foundation's policy statements and legislative analysis related to climate change have included documents on greenhouse gas emission and other issues related to global warming. Most statements published by the foundation argue against mainstream views of the severity and extent of global warming and downplay its potential impacts.

Policy statements issued by the Heritage Foundation suggest that proposed legislation directed toward mitigating global warming may be prohibitively expensive. The foundation concluded that the proposed Climate Security Bill of 2007 would have resulted in substantial costs for little gain. The foundation maintains that global warming is neither unprecedented nor a cause for major concern. A statement published on its website reads, "Global warming is a concern, not a crisis. Both the seriousness and imminence of the threat are overstated." The foundation's viewpoint is that anthropogenic emissions contribute little to greenhouse gases and that climate variability has been present throughout the ages, with current temperatures within the range of natural variability rather than evidence of anthropogenic warming.

C. J. Walsh

Further Reading

Ball, Molly. "The Fall of the Heritage Foundation and the Death of Republican Ideas." Atlantic. Atlantic Monthly, 25 Sept. 2013. Web. 24 Mar. 2015.

Darwall, Rupert. The Age of Global Warming: A History. London: Quartet, 2013. Print.

Ioffe, Julia. "A 31-Year-Old Is Tearing Apart the Heritage Foundation." New Republic. New Republic, 24 Sept. 2013. Web. 24 Mar. 2015.

McKibben, Bill. The Global Warming Reader: A Century of Writing about Climate Change. New York: Penguin, 2012. Print.

Moore, Stephen. "Climate Change Self-Delusion." Heritage Foundation. Heritage Foundation, 18 Nov. 2014. Web. 24 Mar. 2015.

Heterotrophic respiration

Categories: Animals; plants and vegetation

Definition

Heterotrophic respiration is a set of metabolic processes through which organisms produce carbon dioxide (CO_2) and release energy from organic compounds that they have ingested or otherwise incorporated from outside themselves. It may be differentiated from autotrophic respiration, in which energy-bearing compounds are produced by the organism through processes such as photosynthesis. Heterotrophic organisms include animals, fungi, and many types of bacteria.

Significance for Climate Change

CO_2 is a greenhouse gas (GHG), meaning that its increased atmospheric concentration may trap more heat on Earth and raise global temperatures. The amount of carbon in the top meter of Earth's soil has been estimated to be twice that present as CO_2 in the planet's atmosphere. Consequently, increased decomposition of organic matter in the soil—a type of heterotrophic respiration—could make a substantial contribution to global atmospheric CO_2 and thus to global warming.

As a result of these relationships, factors that increase heterotrophic respiration could affect global climate. Certain types of land use are reported to affect heterotrophic respiration in soil. For example, deforestation increases CO_2 release from soil. In addition, CO_2 released into the atmosphere from other sources, such as fossil fuel, can increase global temperatures, which in turn increases the rate of soil heterotrophic respiration, releasing more CO_2. The effect of temperature on heterotrophic respiration may be more pronounced in temperate climates than in tropical ones, in which the effect of temperature may already be at or near maximal.

The soil organisms involved in heterotrophic respiration include macro fauna, such as earthworms, insects, and burrowing mammals; micro and meso fauna, such as protozoa and nematodes; and microscopic fauna, such as bacteria and fungi. The major soil heterotrophs are bacteria, in terms of numbers, and fungi, in terms of mass. Cellulose is a major molecule that is transformed into CO_2 by the process, and fungi, termites, and bacteria are the main types of organisms that produce enzymes that break down cellulose into simpler compounds.

Oluseyi Adewale Vanderpuye

Further Reading

Bardgett, Richard D. *The Biology of Soil: A Community and Ecosystem Approach.* New York: Oxford University Press, 2005. Covers the diversity of organisms that live in soil, including how their ecology and decomposition activities relate to climate change.

Luo, Yiqui, and Xuhui Zhou. *Soil Respiration and the Environment.* New York: Elsevier, 2006. Discusses the roles and significance of soil respiration and its involvement in the global carbon cycle; includes a description of the carbon substrates and heterotrophic organisms.

Schlesinger, William H. *Biogeochemistry: An Analysis of Global Change.* New York: Academic Press, 1997. Extensive analysis of the effects of human activities and those of other forms of life on biogeochemical cycles. This multidisciplinary book places heterotrophic respiration in a wide context along with other factors affecting global temperatures.

See also: Amazon deforestation; Animal husbandry practices; Carbon cycle; Carbon dioxide; Deforestation; Ecosystems; Nitrogen cycle; Soil erosion.

High global warming potential

Category: Meteorology and atmospheric sciences

Definition

High global warming potential (HGWP) is a term assigned to industrially produced gases that have extremely high global warming potentials (GWPs). There are three major groups of HGWP gases: perfluorocarbons (PFCs), hydrofluorocarbons (HFCs), and sulfur hexafluoride (SF_6). Emissions of a given mass of one of these gases contribute significantly more to the greenhouse effect than does an equivalent mass of carbon dioxide (CO_2) or other, more common gases.

Global Warming Potentials of Major HGWP Gases

Gas	GWP
HFC-152a	140
CF_4	6,500
C_2F_6	9,200
HFC-23	11,700
SF_6	23,900

Source: U.S. Environmental Protection Agency.

Significance for Climate Change

Measurement of GWP was developed as a tool to quantify and compare the efficiency of different gases to trap heat in the atmosphere. GWP depends on both the potency of the substance as a greenhouse gas (GHG) and its atmospheric lifetime and is measured relative to the equivalent mass of CO_2 and described for a set time period. HGWP gases operate in a manner similar to all other GHGs, by absorbing and emitting radiation within the thermal infrared range. However, HGWP gases are hundreds to thousands of times more potent than is CO_2 with respect to their atmospheric heat-trapping properties over a set time period. For example, the most potent GHG, SF_6, has an atmospheric lifetime of thirty-two hundred years and a GWP of 16,300 over twenty years, 22,800 over one hundred years, and 32,600 over five hundred years (relative to CO_2, which is defined to have a GWP of 1 over all time periods).

There are few natural sources of HGWP emissions. The majority of anthropogenic emissions result from the decision to use HGWP chemicals instead of ozone-depleting substances for equivalent functions. Preventing ozone depletion is a higher priority than preventing increases in the greenhouse effect, as outlined in authoritative publications such as the Kyoto Protocol and the Montreal Protocol. HGWP gases are also emitted during industrial processes, such as aluminum and magnesium production, and from commercial products such as automobile air conditioning.

In the early twenty-first century, nitrogen trifluoride (NF_3) began receiving increased attention due to its initial marketing as a green alternative in the technological manufacturing sector. Subsequent data indicated that it is seventeen thousand times more potent as a GHG than is CO_2, but by the time this was known, NF_3 was used to produce a wide range of household products, including computer chips, flat-screen televisions, and thin-film solar photovoltaic cells. HGWP emissions are relatively low in developed countries, accounting for less than 2 percent of total emissions in the United States in 2001, for example. They are so potent, however, that even small amounts can have significant effects, especially given that very few sinks exist for these gases. There is therefore a global movement to reduce HGWP emissions by improving industrial processes.

Rena Christina Tabata

See also: Carbon dioxide; Carbon dioxide equivalent; Carbon footprint; Global warming potential; Hydrofluorocarbons; Kyoto Protocol; Montreal Protocol; Ozone; Perfluorocarbons; Sulfur hexafluoride.

High Park Group

Category: Organizations and agencies
Date: Registered January 22, 2003
Web address: http://www.highpark group.com

Mission

The High Park Group (HPG) is a public affairs and policy consulting group in Toronto and Ottawa that represents many energy industry clients. The firm appears to support skepticism of anthropogenic climate change.

The group claims to work in a wide range of areas, but its primary areas are energy, environment, and ethics. Among its services are policy and strategic consulting, direct lobbying, media relations, and issues management. In 2009 the president was

Timothy M. Egan. One of the directors was Kathleen McGinnis. The director of regulatory affairs was Julio Legos. The Ottawa director until September, 2006, was Tom Harris. The High Park Group is registered as lobbying for several energy clients, including Areva Canada Inc., the world's largest nuclear power company; ARISE Technologies Corporation, a solar technology company; the Canadian Electricity Association; and the Canadian Gas Association. Harris was a lobbyist for the Canadian Electricity Association and Canadian Gas Association. McGinnis was registered to represent the Canadian Electricity Association to lobby government agencies regarding activities of the National Energy Board related to electric transmission. Egan was senior adviser to the Canadian Electricity Association and monitored U.S. policy as it related to the electricity industry of Canada.

In September, 2006, Harris became head of the Natural Resources Stewardship Project (NRSP), a group that describes itself as "promot[ing] responsible environmental stewardship." Two of the three directors of the board of the Natural Resources Stewardship Project are Egan and Legos of the High Park Group. NRSP promotes global warming skepticism. Tom Harris wrote in the June 7, 2006, *National Post* that "the hypothesis that human release of CO_2 is a major contributor to global warming is just that—an unproven hypothesis, against which evidence is increasingly mounting." Critics have claimed that the Natural Resources Stewardship Project allows the High Park Group to lobby against climate change regulations on behalf of its energy clients without the clients being identified.

Although not trained in climatology, Harris has bachelor's and master's degrees in mechanical engineering. Even before the formation of High Park Group, he made statements such as

> I think most investors would sensibly conclude that until the science is more mature, pressuring companies to do something, or not to do something, about GHG emissions is a costly gamble and that they should wait until the verdict is in before deciding what to do (if anything).

In a March 8-14, 2001, *European Voice* article co-authored with geologist Tim Patterson, Harris asked, "Is the UN guilty of exaggerating fears over climate change?" In November, 2002, Harris was one of the organizers of an event in Ottawa at which climate change skeptics were "to reveal the science and technology flaws of the Kyoto Accord."

In the June 8, 2006, *Vancouver Sun*, Harris wrote an editorial titled "Environmental Heresy." The subtitle read: "Failing to question the scientific assumptions underlying Kyoto isn't fair to citizens concerned about climate change." After his associations with the High Park Group and the Natural Resources Stewardship Project, he became executive director of the International Climate Science Coalition (ICSC), in March, 2008. This coalition describes itself as an association of scientists, economists, and energy and policy experts working to promote better understanding of climate change science and policy. Patterson is chairman.

Significance for Climate Change

Some skepticism is good when considering conclusions made from scientific data, especially when different scientists reach conclusions that do not agree. Skepticism can also be used to confuse people. Critics of HPG, NRSP, and ICSC claim that they lobby against regulation to protect their clients' profits. They emphasize that the registered lobbyist HPG directs the flow of skepticism from NRSP and ICSC, so that government decision makers will be so confused as to make no changes or the wrong change. If the wrong decision is made, then the world and its climate could be drastically changed.

C. Alton Hassell

Further Reading

Harris, Tom. "Scientists Call for 'Reality Check' on Climate Change." *The Winsor Star*, June 5, 2006. Attempts to refute theories of anthropogenic climate change. Typical of the skeptic viewpoint, which is usually expressed in newspapers or Internet sources and seldom in books.

McCaffrey, Paul, ed. *Global Climate Change.* Bronx, N.Y.: H. W. Wilson, 2006. Collection of essays by different authors on climate change. Nonskeptical treatment of facts and results. Bibliography, index.

Pearce, Fred. *With Speed and Violence: Why Scientists Fear Tipping Points in Climate Change.* Boston: Beacon Press, 2008. Study of climate change that includes a section addressing the concerns of skeptics. Bibliography, index.

See also: Canada; Canadian Meteorological and Oceanographic Society; Skeptics.

Hockey stick graph

Categories: Meteorology and atmospheric sciences; climatic events and epochs

The hockey stick graph, named for its shape, represents a thousand-year period of essentially stable average temperatures in the Northern Hemisphere, followed by an upward spike around 1900. The graph, which suggests the impact of industrial activity on global warming, has been cited to support arguments that global warming has anthropogenic causes, but its accuracy has been questioned; it has thus been both prominent and controversial.

Key concepts

Medieval Warm Period: a period of warmer-than-average temperatures between the tenth and fourteenth centuries

multiproxy reconstruction: a method of estimating prehistoric climate conditions using a combination of proxy indicators

paleoclimatology: the study of prehistoric climate conditions

proxy indicators: tree rings, fossils, and other artifacts that provide indirect evidence of past temperatures

Background

The "hockey stick" is a nickname given to a dramatic graph of historic average temperatures in the Northern Hemisphere. The graph became a prominent—and controversial—symbol in the debate over whether or not recent global average temperatures are historically abnormal and thus more likely to have been caused by human activities. The graph gets its nickname from its shape, which represents a flat, thousand-year period of stability (the stick) followed by a sudden upward spike (the blade). The nickname is generally attributed to Jerry Mahlman, head of the Geophysical Fluid Dynamics Laboratory, part of the National Oceanographic and Atmospheric Administration.

The hockey stick graph was first published in *Nature* magazine in 1998 by a research team of Michael E. Mann, Raymond S. Bradley, and Malcolm K. Hughes and was featured prominently in the 2001 Second Assessment Report of the United Nations Intergovernmental Panel on Climate Change (IPCC). Mann and Bradley were researchers with the University of Massachusetts, while Hughes was a specialist in dendrochronology (tree-ring dating), working at the University of Arizona in Tucson. The scientists engaged in a type of historical temperature reconstruction generally referred to as "multiproxy" reconstruction. That is, they combined a broad set of measured temperatures with estimates based on several proxy temperature indicators, such as tree-ring growth, fossilized leaf stomata, boreholes, lake sediments, and tree pollen.

The graph published in *Nature* was a radical departure from previous temperature reconstructions, which had depicted a period of significant warmth (equal to or greater than current temperatures) during the Medieval Warm Period, which lasted from the tenth to the fourteenth century. Controversy over the hockey stick graph arose when two Canadian researchers—Steven McIntyre, a policy analyst with a background in mathematics and mineral exploration, and Ross McKitrick, an economist at the University of Guelph—sought to examine the underlying data and programming used by Mann and his colleagues in creating the graph.

The Criticism

McIntyre and McKitrick examined the data and attempted to reproduce the hockey stick model using the same methodology Mann's team had employed. They then published an article in the European journal *Energy and Environment* asserting that the earlier team's data set contained significant errors, omissions, and duplicated data. They also claimed that the computer program used to analyze the data gave undue importance to a single proxy series of dubious value. In later analyses, published outside of the scientific literature, McIntyre and McKitrick asserted that the analytical program had a built-in bias that would produce a hockey-stick shape even if it was analyzing completely random data. A continuing bone of contention was the resistance that Mann's group demonstrated to sharing its data or providing sufficient information about its computer program to facilitate others' efforts to reproduce the hockey stick.

Mann's Response

The claims made by McIntyre and McKitrick have been sharply rejected by Mann, Bradley, Hughes, and other members of the academic paleoclimatology community. In addition to questioning McIntyre and McKitrick's ability to understand the development of multiproxy temperature reconstructions, Mann and other paleoclimatologists argue that what McIntyre and McKitrick found were only trivial data errors that did not alter the fundamental shape of the graph. Claims regarding bias in the analytical program have also been disputed and attributed to improper mathematical analysis techniques, as well as to failures properly to use the program.

Third-Party Referees

The controversy over the hockey stick graph led several high-level review panels to be convened. One such panel was created by the National Academy of Sciences' National Research Council at the request of U.S. representative Sherwood Bohlert, chair of the House Committee on Science. It was led by climate modeler Gerald North. Another committee, assembled at the request of two other Republican members of the House, was directed by statistician Edward Wegman, head of the National Academy of Sciences' Committee on Applied and Theoretical Statistics. Both of these committees upheld many of the findings of McIntyre and McKitrick, although the North panel was slightly less critical of the hockey stick model than was the Wegman panel.

Context

Despite the findings of the two review panels, the hockey stick controversy continued to rage with considerable rancor. It was conducted primarily via claims and counterclaims about various elements of the debate on blogs such as that of Steven McIntyre (www.climateaudit.org) and Real Climate, a collective blog operated by a group of climate researchers (www.realclimate.org). Areas of contention involved assumptions regarding the accuracy and global representativeness of individual proxy data sets.

The hockey stick controversy affected public opinion and the methodology of climate reconstruction, as well as the representation of such reconstructions in scientific publications. It became a rallying point for climate change skeptics, especially those doubting that greenhouse gas emissions and other anthropogenic factors play a role in global warming. These skeptics saw the controversy as evidence that climate researchers were intentionally exaggerating recent warming by erasing warmth from prior centuries. They produced a huge number of publications proclaiming the "breaking" of the hockey stick and, with it, the refutation of the idea that recent temperatures are historically abnormal.

Many believe that the controversy over the hockey stick also led to changes in the way that the IPCC chose to represent historical temperature reconstructions in its Fourth Assessment Report on the science of climate change. Rather than relying on a single historical climate reconstruction, the report presented the results of an ensemble of reconstructions. Some of these reconstructions included the Medieval Warm Period; others did not.

Kenneth P. Green

Further Reading

Intergovernmental Panel on Climate Change. *Climate Change, 2001—The Scientific Basis: Contribution of Working Group I to the Third Assessment Report of the Intergovernmental Panel on Climate Change.* Edited by J. T. Houghton et al. New York: Cambridge University Press, 2001. The first volume in the IPCC's Third Assessment Report on climate change, in which the hockey stick graph features prominently.

_____. *Climate Change, 2007—The Physical Science Basis: Contribution of Working Group I to the Fourth Assessment Report of the Intergovernmental Panel on Climate Change.* Edited by Susan Solomon et al.

New York: Cambridge University Press, 2007. Discusses paleotemperature reconstruction with reference to a greater range of reconstructions.

Kerr, Richard A. "Politicians Attack, but Evidence for Global Warming Does Not Melt." *Science* 313, no. 5786 (July 28, 2006): 421. This short article summarizes the events at the House Committee on Energy and Commerce testimony in which the Wegman report was discussed.

Mann, Michael E. *The hockey stick and the climate wars: Dispatches from the front lines.* Columbia University Press, 2013. Personal account by the originator of the "hockey stick" metaphor about the controversy it created.

McIntyre, Steven, and Ross McKitrick. "Corrections to the Mann et al. (1998) Proxy Data Base and Northern Hemisphere Average Temperature Series." *Energy and Environment* 14, no. 6 (November 1, 2003): 751-772. The first peer-reviewed publication in the debate over the hockey stick reconstruction.

National Academy of Sciences. *Surface Temperature Reconstructions for the Last Two Thousand Years.* Washington, D.C.: National Academies Press, 2006. This report by the committee headed by climate modeler Gerald North upheld many of the criticisms that McIntyre and McKitrick raised regarding the hockey stick graph, while still affirming the general correctness of the graph over the last nine hundred years.

Wegman, Edward, et al. *Ad Hoc Committee Report on the "Hockey Stick" Global Climate Reconstruction.* Washington, D.C.: House Committee on Energy and Commerce, 2006. In many ways the strongest criticism of the hockey stick graph, the Wegman report consists of two parts: an analysis of statistical problems in creating paleoclimate reconstructions and a social networking analysis showing that the modeling community is too tightly linked for its members to be able to review one another's work objectively.

See also: Bayesian method; Climate models and modeling; Climate reconstruction; Dating methods; Industrial Revolution and global warming; Medieval Warm Period (MWP); Paleoclimates and paleoclimate change; Parameterization; Peer review; Preindustrial society; Scientific credentials; Skeptics.

Holocene climate

Category: Climatic events and epochs

Understanding climatic fluctuations since the last ice age provides a context for evaluating the extent to which the present global warming trend is an anthropogenic phenomenon. Correlation of archaeological and historical records allows projections of the impact of global warming on human society.

Key concepts

positive feedback loops: self-accelerating processes, such as increased snow cover increasing planetary albedo and promoting additional cooling, leading to more snow cover

proxies: preserved, measurable parameters that correlate with climate and serve as evidence of past climatic conditions

thermohaline circulation: the rising and sinking of water caused by differences in water density due to differences in temperature and salinity

Background

Climatic changes since the end of the last ice age form the backdrop for much of human prehistory and are viewed by some as a driving force in the rise and fall of civilizations. The retreat of continental glaciers began in earnest 13 million years ago, with a gradual warming trend that reached its peak around 6,000 years ago during a period known as the Hypsithermal or Holocene climatic optimum. Proxy records, supplemented by historical data in more recent times, suggest six periods of abrupt cooling in the Holocene, 9,000-8,000, 6,000-5,000, 4,200-3,800, 3,500-2,500, 1,200-1,000, and 650-150 years before the present. Within warm periods and cold periods, there is considerable fluctuation on scales ranging from decades to centuries. Temperature variations as measured by a variety of proxies are more dramatic near the poles, while variations in rainfall associated with temperature-induced fluctuations in oceanic currents predominate in the tropics. Overall, climatic variability in the Holocene is considerably less than it was in the Pleistocene, and what fluctuations have occurred in the

Holocene have decreased over the course of the period.

Climatologists are continually modifying the prevailing picture of climate change in the Holocene as more high-resolution studies become available from areas other than Europe and eastern North America. Climatic shifts typically appear earlier in the Southern Hemisphere than in the Northern Hemisphere. Signs of the 8.2ka event, a period of dramatic European cooling due to disruption of currents in the Atlantic Ocean, are much less evident in western North America and are absent in New Zealand.

Measuring Holocene Climate

Systematic instrumental records of weather in parts of Europe and North America exist for the last 150 years. Agricultural records, historic narratives, and even legendary sources chronicle catastrophic events throughout human history.

Until recently, much of the available information about climate in prehistoric times came from archaeological investigations. The study of artifacts and settlement patterns reveals a great deal about the climate in which ancient people lived. For example, prolonged drought in the American Southwest, corresponding to the culturally benign Medieval Warm Period in Europe, is evident in shifting patterns of cultivation, declining population, skeletal deformities due to malnutrition, compressed tree rings in construction timbers, and eventual abandonment of cliff dwellings.

Vegetation is a good climatic indicator. Leaves, woody material, and particularly pollen occur in abundance in bogs, lake sediments, and areas of human settlement. Pollen analysis is a powerful tool, because pollen is extraordinarily decay-resistant. Many pollen grains can be identified to genus, and relative abundance provides a fairly complete picture of a region's flora. Wind-pollinated plants with narrow ecological niches are particularly useful. In Europe, the arctic-alpine herb *Dryas octopetala* indicates arctic-alpine conditions, spruce (*Picea*) indicates a cold, humid climate, and oak (*Quercus*) provides evidence of a drier, warm climate. Pollen of *Plantago*, a weed in grain fields, suggests cultivation. In marine sediments, relative abundance of planktonic types serves as a proxy for water temperature.

Various geological formations permit high-resolution analysis of local climate. Moraines and scouring document the advance and retreat of glaciers. Varves, which are layers of sediment in lakebeds, provide a record of stream flow into lakes. When precipitation is high, increased runoff and sediment load create thick varves and rapid deposition of alluvial fans at the mouths of rivers. Terraces along lake and ocean shores document rises and falls in water level. In some areas, the land may also be rising or subsiding relative to sea level.

Ice cores taken from glaciers in Greenland and Antarctica provide evidence of climate over the last 400,000 years, including rates of precipitation, amounts of atmospheric dust, and concentrations of carbon dioxide (CO_2) in trapped air. Analysis of the ratios of carbon and oxygen isotopes in carbonates and of oxygen isotopes in ice also provides clues to climate, since both biotic and abiotic processes use isotopes selectively. Isotope ratios can also be used as proxies for sunspot activity, a suspected factor in warming and cooling trends.

Climatic Change in Human Prehistory

The tenure of modern humans on Earth encompasses the last Pleistocene glaciation and the ten thousand years of the Holocene, during which the Earth's climate has fluctuated, with a temperature maximum roughly six thousand years ago. There are many studies correlating prehistoric cultural changes with climatic changes. For agricultural societies, the droughts associated with colder periods are more devastating than are lower temperatures themselves. The Holocene historical and geologic records contain no compelling evidence of rapid rises in temperature such as the Earth is currently experiencing: Global warming in geologic time appears to be a gradual process to which life adapts. Cooling, on the other hand, can be extremely rapid and catastrophic.

Gradual warming can also produce catastrophic results, when rising waters overwhelm a natural dam, unleashing a flood of biblical proportions. Indeed, a controversial theory postulates that the biblical flood was just such an event, and that the fertile agricultural land surrounding the Black Sea was suddenly inundated when rising waters in the Mediterranean breached the Bosporus roughly seven

Holocene Climate Time Line

Years Before Present	Climate Trend
9,000 to 8,000	Cooling
8,000 to 6,000	Warming
6,000 to 5,000	Cooling
5,000 to 4,200	Warming
4,200 to 3,800	Cooling
3,800 to 3,500	Warming
3,500 to 2,500	Cooling
2,500 to 1,200	Warming
1,200 to 1,000	Cooling
1,000 to 650	Warming
650 to 150	Cooling
150 to present	Warming

thousand years ago. Similar floods occurred in the lower reaches of the Tigris and Euphrates rivers and the northern Red Sea at about the same time.

In North America, rapid draining of Glacial Lake Agassiz through the St. Lawrence River caused local devastation, as well as disrupting oceanic currents. The Missoula floods in the Columbia River basin resulted from periodic breaching and reforming of an ice dam at the glacial margin. The widespread occurrence of devastating warming-induced floods in prehistoric times holds a lesson for the present. If there is a dam protecting a city from rising waters, whether it is natural or artificial, every incremental temperature rise increases the likelihood of disaster.

The causes of cooling episodes are various. They include changes in solar radiation levels, volcanic aerosol concentration, greenhouse gas (GHG) concentration, the hydrologic cycle, sea level, sea ice extent, and forest cover. The six defined periods of cooling in the Holocene most probably all derive from several such factors acting in concert.

The rapid cooling that occurred between nine thousand and eight thousand years ago took place during a decline in insolation and a high level of volcanic SO_2 production. A massive infusion of glacial meltwater into the North Atlantic disrupted thermohaline circulation. A weakening Afro-Asian monsoon contributed to tropical aridity. Atmospheric methane concentrations declined because of the drought; this depletion of GHGs created a feedback loop prolonging cooling conditions.

Proxy records based on isotope ratios suggest a decline in solar radiation during the four subsequent cooling periods. The cooling corresponds to lows in solar radiation that correspond to variation in the Earth's orbit. In the case of the 8.2ka event and the onset of the Little Ice Age, there is also evidence of increased volcanic activity. Massive volcanic eruptions cause global cooling by ejecting fine ash and sulfates into the atmosphere. However, in the absence of reinforcement from orbital forcing, this effect dissipates in about three years.

Cultural Effects of Holocene Climate Change

As several analysts have pointed out, climate has helped shape civilization, but not by being benign. Human technological and social advances appear in the archaeological and historical record as responses to climate change that rendered older ways of life maladaptive. The general pattern appears to involve a buildup of population and associated infrastructure during periods characterized by warm temperatures, adequate rainfall, and low variability, such as the Hypsithermal or the Medieval Warm Period. This is followed by a population crash due to famine, war, and pestilence when the climate changes, and a period of rapid technological and social innovation as the population adapts to the new conditions.

In the Middle East, the 8.2ka event, which occurred at a time when agriculture had not yet entirely supplanted the hunter-gatherer economy, spurred the transition to a permanently settled mode of life, more intensive methods of cultivation, and the rise of towns. The 6ka cooling period corresponds to the beginnings of civilization in the Middle East, China, and Southeast Asia. The need to coordinate a large population base over a wide area to manage irrigation during a prolonged drought has been postulated as a key factor in the rise of city-states in Mesopotamia.

During the Hypsithermal, large portions of the Sahara Desert were savanna dotted with seasonal lakes, supporting abundant game and hunters who left rock art and artifacts in areas that are completely barren today. The present aridity of the Sahara and much of the Middle East was well established in classical times (twenty-five hundred to fifteen hundred years ago) and has persisted through warm periods of increased rainfall, in part as a result of human activities such as overgrazing. Global warming could increase rainfall in regions such as the Sahel. Although this would ultimately be beneficial, immediate effects would include flash flooding, erosion, and the proliferation of invasive species.

The cultural pattern is less clear-cut in the humid tropics, but studies of the rise and fall of lowland Mayan civilization in Central America suggest that, contrary to expectation, the peak of this civilization corresponded to the cooling period that occurred twelve hundred to one thousand years ago. Its subsequent decline came during the Medieval Warm Period. In the tropics, cold conditions at the poles cause a weakening of monsoons. Seasonally dry conditions advance toward the equator. Climates with strongly marked wet and dry seasons are more favorable to intensive, highly productive agriculture than is a permanently wet tropical rain forest.

Modern Western civilization is the product of the Little Ice Age. (The term "Little Ice Age" is used differently by different writers. Many use it to refer to the climate cooling from about 1300 to 1850, while others use it for the latter half of that interval, when cooling was greatest, beginning around 1550 or 1600.) The abrupt drop in temperatures that occurred in 1315 produced first a famine, then war, and finally a pestilence that wiped out one-third of Europe's population. Without a famine-weakened population and the disruptions of war, the bubonic plague might well have remained localized, as it did in North Africa in the eighteenth century. Temperatures remained low during the recovery period. Thus, a return to the population levels and living standards of the thirteenth century required first technological innovation and later exploitation of warmer areas through colonialism. For reasons not clearly understood, the Little Ice Age did not produce increased drought in the tropics.

Context

Probably the most important global warming lesson to be learned from the Holocene record is that of the disruption of North Atlantic currents and resulting deep freeze in Europe eighty-two hundred years ago. Very rapid melting of the Greenland ice cap and release of freshwater into the Atlantic could well produce a similar effect. Short-lived episodes of warming and cooling are documented in the climate proxy record. On a local level, such episodes undoubtedly produced dramatic effects, but they left no lasting impression on the world's flora and fauna, nor on human culture as a whole. In today's overpopulated and environmentally degraded world, the consequences of a temporary drop in global temperatures due to either massive volcanism or a disruption of thermohaline circulation in any of the Earth's oceans present a much grimmer prospect than they did at any other time in history. Both scenarios have sufficiently high probability to attract the attention of Pentagon analysts.

Martha A. Sherwood

Further Reading

Alverson, Keith D., Raymond S. Bradley, and Thomas Pedersen, eds. *Paleoclimate, Global Change, and the Future.* Berlin: Springer Verlag, 2003. Collection of scholarly papers comparing past climate changes with present anthropogenic trends; discusses ancient civilizations.

Anderson, David G., Kirk Maasch, and Daniel H. Sandweiss, eds. *Climate change and cultural dynamics: a global perspective on mid-Holocene transitions.* Academic Press, 2011. Between 8,000 and 3,000 years ago, climate became cooler at high latitudes and drier at low latitudes, at the same time that human societies became more organized. This book explores possible links between climate and human culture.

Fagan, Brian. *The Long Summer: How Climate Changed Civilization.* New York: Basic Books, 2004. Argues that civilization has evolved in response to climatological challenges and obstacles that had to be overcome. Semipopular, with a focus on the Middle East and the rise of agriculture.

Mayewsky, Paul, et al. "Holocene Climate Variability." *Quaternary Research* 62 (2004): 243-255. Technical summary of high-resolution proxy records from many parts of the globe; includes some discussion of the correlation of these records with human history.

Saltzman, Barry. *Dynamical Paleoclimatology: Generalized Theory of Global Climate Change.* New York: Academic Press, 2002. Treats interactions between abiotic, biotic, and anthropogenic variables; discusses controversies about the magnitude of human impacts on climate.

Stipp, David. "The Pentagon's Weather Nightmare." *Fortune*, February 9, 2004. Posits a plausible chain of events leading from a climatic shift comparable to the 8.2ka event to World War III.

See also: Abrupt climate change; Civilization and the environment; Climate reconstruction; Dating methods; 8.2ka event; Little Ice Age (LIA); Medieval Warm Period (MWP); Pleistocene climate.

Hubbert's peak

Categories: Fossil fuels; energy

Geophysicist Hubbert's prediction that American oil production would peak in 1972 proved accurate and presaged the rise of energy independence and security as crucial national and international issues.

Key concepts

fossil fuels: fuels, including coal, oil, and natural gas, produced by chemical alteration of organic matter under pressure

greenhouse gases (GHGs): atmospheric trace gases that trap heat, preventing it from escaping into space

nonrenewable resource: a resource that, once consumed, cannot be renewed

reserves: the estimated amount of a nonrenewable resource remaining to be consumed

Background

In 1956, Shell Oil Company geophysicist M. King Hubbert predicted that U.S. oil production in the lower forty-eight states would peak in 1972; it peaked in 1970. Other scholars have extended Hubbert's work to predict that world oil production would peak sometime between 2005 and 2011. Although there are critics of Hubbert's approach, most geologists and geophysicists agree with the basic premises of his work. World demand would exceed production capacity about the time of peak production, driving up the cost of oil. Burning oil is a major producer of carbon dioxide (CO_2), a greenhouse gas (GHG). The impending oil shortage and resulting increases in the price of oil may help spur people to turn to alternative energy sources, many of which do not produce CO_2.

Hubbert's Predictions

Basing his analysis on proven oil reserves in the United States and on an analysis of production patterns in the anthracite coal industry, Hubbert produced a bell curve that showed the growth, peak, and eventual decline in American oil production. He described what has come to be called Hubbert's peak, the one-hundred-year period in which oil was the driving force of the economy. Hubbert tied his analysis to patterns of production for other fossil fuels, most notably anthracite coal in the eastern U.S. Because fossil fuels are a nonrenewable resource, they cannot be re-created once used. The model that Hubbert produced contained two scenarios, one predicting that U.S. oil production would peak in 1965, and a more optimistic one indicating 1972. Actual U.S. oil production peaked in 1970, giving additional weight to Hubbert's approach.

Hubbert's bell curve indicated that there would continue to be significant American oil production after the early 1970's, as has occurred. However, as easily accessible reserves are consumed, oil producers are forced to turn to increasingly expensive means to produce oil. This has meant drilling in unlikely places, often at great expense and to no avail; expensive recovery processes that try to obtain what oil remains in an underground reservoir; and increased emphasis on offshore drilling, in some cases through 3,000 meters of water and then 6,000 meters below the ocean floor.

Several scholars, such as Kenneth Deffeyes, have extended Hubbert's approach and predicted the peak for world oil production would occur some

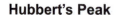

Hubbert's Peak

Reproduced below is a graph of M. King Hubbert's original 1956 prediction of global oil production over two and one half centuries.

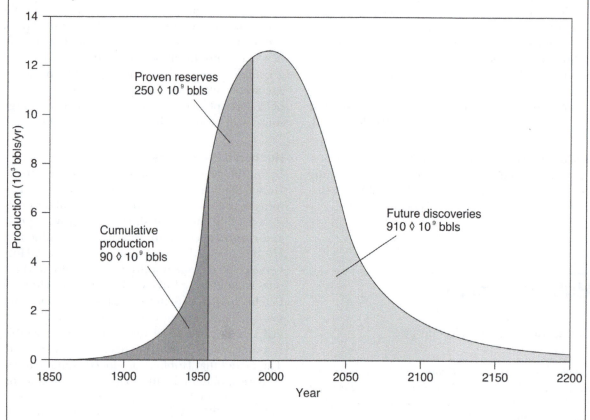

time during the first decade of the twenty-first century. Because the United States has come to rely on foreign oil, a decline in world oil production will have a significant impact on the American economy as well as climatic conditions. This second peak production point will lead to increasingly expensive means of trying to obtain oil.

A Question of Reserves

Hubbert's analysis was based on his knowledge of proven oil reserves taken from several sources. Oil is generally found in particular types of geologic formations between 2,300 and 4,600 meters beneath the surface of the Earth. These oil-bearing formations are found in both the land and beneath the ocean. In the past oil companies and government bodies have often overstated oil reserves for political and economic reasons. This situation may be changing, as Shell Oil sharply decreased its reserve predictions early in the twenty-first century. Although further exploration has led to higher estimates for reserves, by the early twenty-first century only the South China Sea region had not been fully explored, and most geologists do not expect any oil fields approaching the magnitude of the Saudi Arabian fields to be found there.

Economists often raise the principle of substitutability in dealing with oil reserve predictions.

They indicate that as it becomes more expensive to drill for oil in one area, oil companies will drill elsewhere distorting reserve predictions as oil still remains in the ground in the old fields. This was true for a time, but now oil companies are using new technologies to extract oil from old fields in places such as east Texas and Mexico. The criticism of Hubbert that he ignored substitutability may have been somewhat true at one point, but it is less true at present.

Oil is a finite resource, and as oil fields are depleted oil companies turn to more expensive means of production. For example, in 2008 some oil companies were paying more than $600,000 a day to lease drill ships for offshore drilling. In addition, some oil reserves are now found in politically unstable locations, such as off the east coast of Africa, a condition likely to drive up the costs of obtaining oil. Oil is also found in tar sands in Canada, but extraction is expensive and causes several environmental problems. Moreover, oil is used for more than energy. It is also the feedstock for the chemical industry for such products as plastics.

Context

Burning fossil fuels such as oil is a major source of GHGs such as CO_2. It is estimated that CO_2 generated by burning fossil fuels represents 57 percent of the GHGs emitted into the atmosphere. Although oil- and gasoline-burning comprise only parts of this figure (coal-burning is also a major source of CO_2), decreasing the burning of oil will help to reduce this source of CO_2 emissions. Oil consumption is so ingrained in industrial society in the use of oil for industrial and home energy and for transportation that simply turning to another source is not likely in the short run without the impetus of a steep price increase. With the exception of coal most other, cleaner energy sources have been more expensive than fossil fuels, so there has been little economic incentive to adopt them.

It is likely that the assumptions concerning world oil production derived from Hubbert's model are correct and the production of oil will become increasingly expensive in the next few years. The increasing cost of oil predicted by Hubbert's model may help to lower consumption of oil. Confronted by sharp increases in the prices of gasoline and fuel oil in 2008, more industries and consumers were searching for alternative fuels, fuels that often produced a smaller carbon footprint than oil.

John M. Theilmann

Further Reading

Adelman, M. A., and Michael C. Lynch. "Fixed View of Resource Limits Creates Undue Pressure." *Oil and Gas Journal*, April 7, 1997, pp. 56-60. Denies the validity of Hubbert's analysis. Written from an economist's perspective.

Deffeyes, Kenneth S. *Beyond Oil.* New York: Hill and Wang, 2005. Places oil production in the wider context of the energy industry in general and argues that the impact of a coming oil shortage will force a turn to other energy sources.

_____. *Hubbert's Peak.* Princeton, N.J.: Princeton University Press, 2001. Excellent explanation of Hubbert's model that extends it to examine world oil production.

Goodstein, David. *Out of Gas.* New York: W. W. Norton, 2004. Good description of energy and the coming depletion of oil that accepts the validity of Hubbert's model.

Hubbert, M. King. "The World's Evolving Energy System." *American Journal of Physics* 49 (1981): 1007-1029. Useful and comprehensible presentation of Hubbert's views concerning petroleum production and the implications of his model.

Hughes, Larry, and Jacinda Rudolph. "Future world oil production: growth, plateau, or peak?." *Current Opinion in Environmental Sustainability* 3, no. 4 (2011): 225-234. Hubbert's prediction of a global peak of oil production by 2000 has not come true, partly because of downturns in demand and the appearance of non-conventional petroleum sources. Nevertheless, oil is finite and non-renewable. This study seeks to explore possible future growth scenarios.

See also: American Association of Petroleum Geologists; Biofuels; Fossil fuel reserves; Fossil fuels and global warming; Fuels, alternative; Gasoline prices; Oil industry.

Human behavior change

Categories: Popular culture and society

Climate change will affect human health and welfare. Changing human behavior may mitigate the threats. Technologies exist to facilitate the development of proactive behavior and problem solving, and the reduction of human contributions to global warming.

Key concepts

adaptation: the process through which entities adjust in response to external circumstances

altruism: behavior intended to benefit others rather than the self

behavioral economics: the study of the combined economic and psychological principles affecting human behavior and decision-making

environmental psychology: the study of the interaction between behavior and the environment among individuals, groups, organizations, and systems

negative reinforcement: the process by which removal of stimuli increases a given behavior

positive reinforcement: presentation of a stimuli to increase a given behavior

punishment: deployment of stimuli to make a behavior stop or decrease

reinforcement: stimuli that increase behavior

systems: entities of individuals or groups organized to work together

systems theory: approaches to understanding interdependent behavior within and among groups or systems

Background

Humans regularly adapt their behavior to daily and seasonal weather changes with ease. With global climate change, however, the adaptation demands facing humans are expected to be extraordinary. There is scientific consensus that human activities will need to adapt substantially to meet the anticipated changes of the near future and beyond. Such changes will demand local and regional adaptations in daily individual and community functioning. Sustained behavior changes by individuals, organizations, and systems to reduce or cease activities affecting the climate deleteriously in a more global and persisting way also will be needed.

Behavioral economics and environmental psychology provide tools for making such change occur.

Preparing for Climate Change Through Behavior Change

Global climate change will demand different preparation activities from different people. This variation is partly a function of the varied effects of climate change experienced by different regions of the globe. Some regions may become hotter or cooler, drier or wetter, or may have more or less seasonal variation in weather than previously known. Necessary individual, organizational, and systems-level preparations vary by location. Relevant factors include latitude, longitude, elevation above sea level, and the time span considered.

People may adapt by adjusting where and how they live, making home improvements to meet new climate challenges, and developing personal climate disaster plans. Assessment and acquisition of skills needed in the event of a climate-related crisis also will be necessary. Community organizations may seek to protect whole communities and neighborhoods. Organizations such as businesses and governments may adapt by changing locations for headquarters, road systems, and other construction, as well as arrangements for food, medical, and other natural resource distribution. They also may prepare on the systems level by engaging with other governments to coordinate responses to challenges. This includes such things as dispersing aid and assistance and ensuring vital communications.

Lessening the Impact of Climate Change Through Behavior Change

Individuals, organizations, and governments must consider what they can do to lessen the continued development and ultimate impact of global climate change using available knowledge and technologies. For individuals, education about carbon footprints, water-use habits, use of natural space, and resource use related to climate change will be a first step. After understanding their personal behavior, people may need to alter their lifestyles, travel patterns, food choices, and consumption of other goods to lessen carbon footprints and other environmental impacts. Such proactive and altruistic behavior will lessen the overall burden on those expected to be most affected by climate changes and will perhaps lessen the overall damage that may occur.

Individuals also may consider changing how they influence other individuals and the relevant organizations and systems in which they interact. They may examine their work and governmental behavior and policies affecting climate. They may advocate for change to initiate and reinforce larger systems adjustments, as well as smaller change in local communities, such as within a family or among friends and neighbors. Climate change is not affected only by individuals, however; organizational and systems behavior matters too.

Governments and businesses are the largest global consumers of energy, having much greater carbon footprints and environmental impacts than individuals. Therefore, attempts to lessen the overall impact of climate change must incorporate organizational and systems-level adjustment. For families, this may mean assessing the overall impact of living far away from schools and workplaces. Businesses and government could similarly assess their employees' air travel and consider ways to minimize or replace such travel. At a systems level, efforts such as the Montreal Protocol, Kyoto Protocol, and Paris Climate Agreement may need to be updated and expanded, encouraging greater coordinated efforts across larger systems of human behavior.

Climate Change, Behavioral Economics, and Managing Human Behavior

Individuals, organizations, and systems do not enact behavior merely based on morals, principles, or theories. Incentives, such as positive and negative reinforcement, as well as costs, such as punishments, play a role. In effect, human behavior is subject to behavioral economics—a valuable tool for changing human behavior. Examination of key behaviors needing change among individuals, organizations, and systems, coupled with identification of key incentives and costs, can be used to encourage change. Rebates, taxes, credits, budgets, fines, barter, and allowances for different climate-related behaviors are examples of how behavioral economics may be applied for both businesses and individuals. Similarly, organizations such as governments can be subject to incentives,

such as basing the right to import or export goods on climate-related practices within a particular country. At the systems level, governments can develop agreements with one another that suit the goals of preparing for climate change and enacting behavior to lessen its impact. Motivating behavior change also may include assessment of entities to stage their readiness to change. Different stages of readiness may need different intervention strategies. For instance, those more ready to change may need action-based strategies. In contrast, those less ready may need strategies focused more on information, values clarification, for decision-making.

Context

Scientific advances allow humans to predict, shape, and respond to future change by adapting both individual and collective behavior. The ability of humans to communicate globally, cooperatively, and quickly has increased dramatically as a function of modern communications technologies. These technologies facilitate the communication of complex environmental and behavioral information, along with insights into the dynamics of human behavioral economics, market forces, and the connectedness of people across the globe. Such abilities may be combined with psychological technology to affect and shape human behavior at the individual, organizational, and systems levels, and thoughtful and sustained leadership can provide solutions to the pressing climate problems ahead.

Nancy A. Piotrowski

Further Reading

Brown, Lester B. *Plan B 4.0: Mobilizing to Save Civilization.* 4th ed. New York: W. W. Norton, 2009. A leading social scientist engagingly describes global warming and its associated problems in a sociological context and proposes socioecological strategies to address them.

Camerer, Colin F., George Lowenstein, and Matthew Rabin, eds. *Advances in Behavioral Economics.* Princeton, N.J.: Princeton University Press, 2003. Introduces behavioral economics; describes the integration of psychological and economic theories in order to understand and predict human behavior.

Cartwright, Edward. *Behavioral Economics*. 2d. ed. New York, N.Y. Routledge, 2014. Provides basic tenets and terms of behavioral economics and easy to understand classic examples.

Gifford, R., 2011. The dragons of inaction: Psychological barriers that limit climate change mitigation and adaptation. *American Psychologist, 66*(4), p.290. Although many people favor action on climate change, their own actions tend to be less than they could be. To quote one aphorism, "Planning for the future reaps future benefits, but procrastination pays off right now."

Gifford, Robert. *Environmental Psychology: Principles and Practice*. 5th ed. Colville, Wash.: Optimal Press, 2013. Presents the scope and principles of environmental psychology, how the environment shapes human behavior, and how human behavior can be managed to achieve goals such as sustainability.

See also: Anthropogenic climate change; Climate and the climate system; Economics of global climate change; Environmental economics.

Human migration

Category: Ethics, human rights, and social justice

Definition
Most literature uses the term "environmental refugees" or "climate refugees" to conceptualize the complex relationships between human migration and climate change. In the 1970's, Lester Brown of the Worldwatch Institute defined environmental refugees as people forced to leave their traditional habitats, temporarily or permanently, because of marked environmental disruptions.

Significance for Climate Change
Migration caused by environmental degradation is not a new phenomenon. People throughout history have migrated in order to seek new opportunities and resources for survival. What makes climate-change-related displacement different is the scale of the problem and the size of the population affected by them. Both the Intergovernmental Panel on Climate Change (IPCC) and the United Nations Environment Programme warn that human mass migration caused by climate change can turn into humanitarian crises and a global security threat, because the change of settlement may cause conflict over resources within nations or in host communities.

The Office of the United Nations High Commissioner for Refugees estimates that 24 million people around the world will be displaced because of floods, famines, and other environmental factors by 2050. The *Stein Review* and Friends of the Earth both predict that by that year 200 million people will become climate change migrants worldwide, including one million from small island states affected by shoreline erosion, coastal flooding, and severe drought resulting from rising sea levels, deforestation, dry-land degradation, and natural disasters. Christian Aid makes the most extreme prediction: One billion people will be forcibly displaced by 2050.

The impact of global warming is twofold: First, climate change increases the variability of extreme weather events associated with the changes in surface temperature and precipitation. This results in floods, droughts, and a high incidence of diseases that affect both human and animal health. Second, rising sea levels mean loss of farmland, accelerating shoreline erosion, and disruption of agricultural production. For example, in Egypt, the anticipated rise in sea level could cause 12 to 15 percent loss of arable land by 2050, and consequently 14 million people would be forced to disperse. In Bangladesh, a 1 meter increase in sea level would inundate half of Bangladeshi rice land, forcing the relocation of 40 million people.

Studies show that climate-change-related displacement has a negative impact on the well-being of migrants. The process may cause severe long-term stress and psychological effects. Prolonged uprooting may result in culture and identity loss. Displaced people become landless, homeless, and unemployed, and they are restricted from getting access to common property resources in new areas.

The weakening of community ties will reduce their social networks and further reinforce social marginalization.

The IPCC warns that climate-related migration may increase the risk of group-identity conflict and political instability within states. Reduced water availability, for example, may induce conflict between different water users, such as pastoralists and farmers. People may resort to violence to gain dominant control over limited natural resources. The grievance could increase recruitment opportunities for rebel movements and lead to civil war. Environmental scarcity could also bring about conflict across borders. Water scarcity can cause transboundary disputes.

Migration would also bring problems to host communities or nations, especially if they are not well prepared for the influx of migrants. The rising numbers of migrants will exert pressure on resources and social services. The migrants themselves will compete for jobs, and that competition may increase social tensions and intergroup clashes. In order to curb the influx of migrants, host societies may change their immigration laws and restrict asylum. International law does not require states to provide asylum to those displaced by environmental degradation.

The literature suggests that the impact of climate change on human migration is complex. The causes of migration are multiple, and many problems created by climate change build on existing development problems, such as socioeconomic vulnerability, political suppression, and institutional weakening. The consequences of climate-change-induced migration also depend on the rate of recovery, adaptive capability, preparedness for disasters, and effectiveness of conflict resolution mechanisms. Questions as to whether the deterioration of environmental conditions by climate change is sudden or gradual, whether the displacement is temporary or permanent, and whether migrants are victims or strategists who choose to migrate because of foreseeable opportunities in a new environment need to be carefully examined.

To address these problems, disaster preparedness must be improved; this includes building better adverse-weather advance-warning systems for vulnerable areas. Effective conflict resolution mechanisms can be developed to settle disputes. Governments can increase community resilience by reducing soil erosion and deforestation and enhancing awareness about sustainable use of resources. Better coordinated assistance is needed to reconstruct the livelihoods of the displaced. Land-based resettlement, rehousing, social inclusion, improved health care, and building community assets and services are a few strategies to reduce the impact of migration.

Sam Wong

Further Reading

Forced Migration Review 31 (2008). This special issue titled "Climate Change and Displacement" offers updated discussion of the relationships between climate change and human migration. Highlights the differences between voluntary and involuntary displacement, as well as their different impacts on climate change policies.

Friends of the Earth Australia. *A Citizen's Guide to Climate Refugees.* Melbourne, Vic.: Author, 2007. Offers a critical examination of the concept of climate refugees, as well as recommendations to address migration problems.

McLeman, Robert A. *Climate and human migration: Past experiences, future challenges.* Cambridge University Press, 2013. Human migrations in the wake of past environmental disasters has not always followed theoretical models. An attempt to learn from past events and apply the lessons to future migrations.

Vine, David. "The Other Migrants: Cause and Prevention in Involuntary Displacement and the Question of 'Environmental Refugees.'" In *International Migration and the Millennium Development Goals.* New York: United Nations Population Fund, 2005. Offers a critical examination of the concept of environmental refugees. Questions whether the term is critically conceptualized and whether it is relevant to climate change.

See also: Displaced persons and refugees; Intergovernmental Panel on Climate Change; United Nations Environment Programme.

Humidity

Category: Meteorology and atmospheric sciences

Definition

Humidity is the amount of water present in the atmosphere in the form of vapor. As a gas, water vapor contributes to the local atmospheric pressure in accordance with Dalton's law of partial pressures: In any mixture of gases, the partial pressure of any one component is equal to the total pressure of the mixture multiplied by the fraction of the gas present in the mixture. For example, molecular oxygen constitutes 20 percent of the atmosphere, so the partial atmospheric pressure of oxygen is 20 percent of Earth's total atmospheric pressure. The total pressure is about 1.03 kilograms per square centimeter, so the partial atmospheric pressure of oxygen is about .20 kilograms per square centimeter.

Water normally exists in liquid and solid as well as vaporous form. Its vapor pressure is the pressure at which pure water vapor coexists in equilibrium with either the liquid or the solid state. At equilibrium, the liquid would not evaporate, the solid would not sublimate, and the vapor would not condense. By contrast, if the local partial pressure of water vapor is greater than its vapor pressure, the vapor condenses; if the local partial pressure is less than the vapor pressure, then the liquid evaporates and the solid sublimates. Vapor pressure is not a constant but rather is a function of temperature.

If the local partial pressure of water is exactly equal to its vapor pressure, the air is said to be saturated. This state is defined as 100 percent humidity, and the corresponding temperature is water's dew point. If the vapor pressure of water is equal to the total local atmospheric pressure, the water will evaporate without limit, and the corresponding temperature is water's boiling point. Relative humidity is the ratio, expressed as a percentage, of the local partial pressure of water vapor to the vapor pressure associated with the local temperature.

Humidity can exceed 100 percent, a condition known as supersaturation. In supersaturation, water vapor's partial pressure exceeds the theoretical vapor pressure at that temperature. Condensation cannot take place, however, unless condensation nuclei are present. Water droplets exceeding a certain critical size act as such nuclei, absorbing water vapor and growing; water droplets below the critical size evaporate. If no droplets larger than the critical size exist and no other condensation nuclei are present, then the supersaturated vapor is stable. Fine, dry particles, such as dust or pollutants, also act as condensation nuclei in supersaturated air.

Evaporation is an endothermic process, or one that requires an input of energy in order to occur. The change of phase from liquid to gas takes place at constant temperature. The energy consumed by the process is stored in the water vapor in the form of latent heat of vaporization. When the vapor condenses, all of the latent heat is released, which means that condensation is exothermic. Water has a latent heat of vaporization of 2,256,000 joules per kilogram, an unusually high value for such a simple compound.

Significance for Climate Change

Water vapor is the most abundant greenhouse gas (GHG) in Earth's atmosphere, exceeding the amount of CO_2 by a factor of one thousand. It is transparent to visible radiation but opaque to infrared radiation of 5.25 to 7.5 micrometers in wavelength, which is a lower frequency range than that of visible light. Incoming solar radiation peaks in the visible spectrum. Energy re-emitted by the Earth peaks in the infrared portion of the spectrum. Thus, incoming energy is better able to penetrate water vapor than is outgoing energy.

In order for the Earth to maintain thermal equilibrium, radiating as much energy back to space as it receives from the Sun, the global average temperature must rise higher than it would if there were no humidity in the atmosphere. The balance of the amounts of radiation received from the Sun and emitted back into space is called the radiation budget.

No simple statements about the effect of total atmospheric water vapor on climate change are possible, because the atmosphere is a nonequilibrium system. Water vapor resides in the air for fairly short periods before precipitating out as rain, snow, or dew. As a result, the amount of water vapor in the atmosphere itself responds quickly to changes in climate. Water vapor in turn affects Earth's radiation budget and, through it, surface temperatures, closing the loop and generating feedback. Layers of air near the planet's surface are warm enough and close enough to the oceans to stay relatively saturated. Their effect on the radiation budget is small, because they are nearly as warm as the surface itself. Upper layers of the atmosphere are cooler and moistened only by the water vapor that convects or diffuses upward from the layers below. Small quantities of water in these upper layers can have a disproportionate effect on the radiation budget, trapping enough infrared energy to significantly warm the climate.

This feedback is complicated, however, by the presence of water in the air as suspended droplets in the form of clouds and fog. These droplets scatter visible radiation in all directions, preventing an appreciable fraction of the incident solar energy from reaching the ground and contributing to Earth's albedo (the fraction of incident solar energy reflected back into space). Ice and snow have the same effect. Increases in albedo have a cooling effect and act to moderate any global average temperature increases.

Billy R. Smith, Jr.

Further Reading

Colman, B. R., and T. D. Potter, eds. *Handbook of Weather, Climate, and Water.* Hoboken, N.J.: Wiley-Interscience, 2003. Concise and thorough treatment of the hydrologic cycle and its effect on climate. Illustrations, figures, tables, references, index.

Schneider, T., and A. H. Sobel, eds. *The Global Circulation of the Atmosphere.* Princeton, N.J.: Princeton University Press, 2007. Chapter 6, "Relative Humidity of the Atmosphere," may be difficult for the nonexpert to read in depth, but there is value in skimming it to understand the authors' positions on the subject. Figures, tables, and references.

Taylor, F. W. *Elementary Climate Physics.* New York: Oxford University Press, 2005. Readers without significant prior knowledge of climate science should consult this book for help in understanding difficult topics. Illustrations, figures, tables, bibliography, index.

See also: Clouds and cloud feedback; Dew point; Greenhouse effect; Noctilucent clouds; Polar stratospheric clouds; Rainfall patterns; Water vapor.

Hybrid automobiles

Categories: Economics, industries, and products; energy

Hybrid automobiles derive their energy from two or more sources, such as gasoline and electricity. They are low- or zero-emission vehicles and greatly reduce or eliminate the toxic carbon emissions that are by-products of gasoline-powered internal combustion engines.

Key concepts

aerodynamic: designed to minimize drag from the air

compressed air: air that is forced into a small space under high pressure

fossil fuels: fuels derived from decayed plants and animals under geologic pressure over millions of years

hydrogen automobiles: vehicles that use electricity generated from combining hydrogen and oxygen to produce water

lithium-ion batteries: long-lasting, lightweight batteries made from lithium

prototype: a trial model of a product a manufacturer is considering mass-producing

regenerative braking: process that uses the friction caused by applying a car's brakes to produce energy to help power the vehicle

zero-emission automobiles: motor vehicles that do not produce toxic pollutants

Background

Hybrid vehicles combine electric motors and internal combustion engines in such a way that their drive shafts can be powered by the gasoline engine, by the lithium-ion batteries of their electrical engines, or by a combination of the two. Although such vehicles existed as experimental prototypes as early as the late nineteenth century, they did not become commercially viable until the last quarter of the twentieth century, when fuel shortages, problems with air pollution, and concern over global warming converged to make fuel-efficient, nonpolluting vehicles attractive to the public.

The Japanese automobile company Toyota launched the first mass-produced hybrid, the Prius, in 1997. Although this vehicle was not expected to attract many buyers, its initial American reception after it was introduced to the U.S. market in 2000 was enthusiastic. Toyota increased production and moved toward the manufacture of an aerodynamic Prius the size of Toyota's Corolla that would travel more than 21 kilometers on a liter of gasoline. Soon, Honda was producing comparable hybrids.

People who wanted to buy a Prius often had to wait four to six months to have their orders filled. Prius sales in the United States doubled between 2003 and 2004, and they doubled again in 2005. By 2007, some 250,000 hybrids were sold in the United States.

Hybrid Technology

Hybrids generate much of their own power and are essentially low- or zero-emission vehicles. Although their internal combustion engines can provide power in ways comparable to those of more conventional vehicles, these vehicles derive much of their power from a lithium-ion battery pack that is continuously charged and recharged while the vehicle is operated by its gasoline-powered engine and by the friction caused in normal braking. This regenerative process permits hybrids to deliver higher mileage in stop-start, urban driving than in long-distance, highway driving.

The key to producing hybrid vehicles that will deliver over 40 kilometers to a liter of gasoline is the development of increasingly light, rechargeable lithium-ion batteries. Toyota and other manufacturers have already developed lithium-ion batteries that do not encroach upon the vehicle's interior space, as earlier versions of such batteries did.

Other Kinds of Hybrids

A number of automobile manufacturers have produced electric cars, "plug-in" vehicles that usually have ancillary gasoline-powered engines. Most of these vehicles must be plugged in for periods of from four to six hours for recharging. Most of them have a limited range, generally at best about 160 kilometers, and some are incapable of operating at interstate speeds. Despite these limitations, many such vehicles are in service with the United States Postal Service and other groups. They are practical if they are required to be in service for eight or ten hours a day, after which they can be plugged in to have their batteries recharged before the next work day.

Hydrogen vehicles have also been developed. These carry hydrogen in their fuel tanks, where it combines with oxygen taken in from the air and produces electricity to power the vehicle. The by-product of this technology is water vapor, which results in zero emissions.

The Tata Motor Company of India has developed automobiles that will retail for under five thousand dollars and are directed toward buyers in India and China. It is doubtful that the initial Tata vehicles could meet U.S. safety standards. One of the most exciting Tata hybrids is being produced for developing nations. These vehicles, equipped with auxiliary gasoline engines, operate on compressed air using a technology developed by Guy Nègre, a French engineer. They are extremely economical to operate.

Context

The rapid development of hybrid vehicles has been stimulated by a number of compelling forces, not the least of which is the problem of the worldwide pollution being caused by the burning of fossil fuels. The toxic residues that the burning of such fuels produces are poisoning the environments of every nation in the world.

Even if one ignores the threat posed by burning fossil fuels, society worldwide is exhausting the supplies of such fuels. In a world dependent for its transportation upon vehicles powered by various sources of energy, it is crucial that these sources must be both nonpolluting and renewable. It is unrealistic to assume that the contemporary world will drastically reduce its consumption of energy in the foreseeable future. Thus, the development of technologies that will make optimal use of renewable energy is perhaps society's greatest hope for the future.

R. Baird Shuman

Further Reading

Bethscheider-Kieser, Ulrich. *Green Designed Future Cars: Bio Fuel, Hybrid, Electrical, Hydrogen, Fuel Economy in All Sizes and Shapes.* Los Angeles: Fusion, 2008. Useful account of the various contemporary hybrid automobiles, with specifics about how each type works.

Boschert, Sherry. *Plug-In Hybrids: The Cars That Will Recharge America.* Gabriola Island, B.C.: New Society, 2006. Contends that electric automobiles that plug into conventional electrical sources are likely to be available commercially before hydrogen automobiles are.

Ehsani, Mehrdad, et al. *Modern Electric, Hybrid Electric, and Fuel Cell Vehicles: Fundamentals, Theory, and Design.* Boca Raton, Fla.: CYC Press, 2005. Comprehensive, explicit account of the evolution of hybrid motor vehicles, replete with illustrations.

Husain, Iqbal. *Electric and hybrid vehicles: design fundamentals.* CRC press, 2011. Creating a hybrid vehicle is not simply a matter of replacing a fossil fuel engine with an electric motor. There are a large number of design factors involved, including the propulsion system architecture, drive train, vehicle balance and controls.

Juettner, Bonnie. *Hybrid Cars.* Chicago: Norawood House Press, 2009. Directed at young adult readers, this overview of the development of hybrid automobiles, clear and thorough, is easily accessible to newcomers to the field.

Povey, Karen. *Our Environment: Hybrid Cars.* Farmington Hills, Mich.: KidHaven, 2006. Aimed at young adult readers, this slim volume discusses intelligently the role that hybrid cars can serve in protecting the environment and reducing global warming.

Walker, Niki. *Hydrogen: Running on Water.* New York: Crabtree, 2007. Explains how a technology that separates the two elements of water, hydrogen and oxygen, may eventually enable motor vehicles to run on water, whose exhaust would not be toxic.

See also: Automobile technology; Motor vehicles; Transportation.

Hydroelectricity

Category: Energy

Because hydroelectricity uses falling water rather than fossil fuels for its production, it does not contribute GHGs to the atmosphere. It is both a clean and a renewable energy source.

Key Concepts

fossil fuels: fuels, such as coal, gas, or oil, created during early geologic eras

generator: a mechanical device whose rotational movement around magnets produces an electrical current

head: the fall of the water at a site selected for production of hydroelectricity

reservoir: a gathering place for water held in reserve for later hydroelectric production

turbine: an enclosed vessel containing rotating parts turned by the passage of a fluid, such as water or air

Background

Hydroelectricity is a twentieth-century phenomenon. Although the use of moving water for the production of mechanical energy is ancient (water wheels go back to ancient times), the use of falling water to create electricity awaited knowledge about using electricity as a motive power. Although Michael Faraday invented a dynamo that produced an electric current in 1831, the concept of using an electric current to move energy from one place to another awaited the appearance of Thomas Alva Edison and the electric light in the 1870s. Critical for hydroelectricity were the understanding that electricity could be produced by other kinds of mechanical energy, the development of machines that could use mechanical energy to induce an electric current (generators), and knowledge of the transmission systems that made it possible to move electricity across significant distances.

Early History of Hydroelectricity

The invention of the turbine allowed water running through a containing vessel to produce mechanical energy that could, in turn, generate an electrical current. The turbine designed by James Francis was the earliest to appear, but modifications made by Lester Allan Pelton and later Viktor Kaplan adapted the turbine to use in situations involving falls of significant distance. Generally, hydraulic installations are classified into either low-head facilities, where the drop is on the order of 15 meters, and high-head facilities, with a drop of more than 60 meters. The height of the head determines whether a vertical- or horizontal-impulse turbine is used. The determining factor is the continuous volume of water passing through the turbine.

Initially, use was made of sites that both were close to manufacturing facilities and involved large volumes of water falling substantial distances. The most notable example was Niagara Falls. Sites with this combination of features are relatively rare, however. Most suitable sites tend to be located at the base of chains of mountains, where they can make use of the water generated by large amounts of melting snow. The drawback to such sites is that the amount of water they generate is seasonally variable. The very high capital costs of constructing such installations make it essential that they be sited where the flow of water is steady, allowing them to operate year-round. Thus, in the twentieth century, operators of hydroelectric facilities began to build holding ponds or reservoirs. These structures allowed them to collect water when its flow was great and store it for later use during drier seasons.

Sayano–Shushenskaya hydroelectric power station in Russia. (АлександрВв/Wikimedia Commons)

Major Facilities

The common preference, especially among governments, for economies of scale has led to the development of a number of very large hydroelectric facilities, of which the largest, in China, is the Three Gorges Dam in Hubei province. Governmental priorities also entailed building a large reservoir to ensure that there would always be sufficient water to operate the facility's generators. This construction in turn flooded some one hundred thousand of China's towns and villages, forced one million people to move, and submerged about 400 hundred square kilometers of farmland.

Other large facilities include the Itaipu, serving both Paraguay and Brazil; the Yacyreta, serving Argentina and Brazil; the Krasnoyarsk, the Bratsk, the Ust-Ilim, and the Volgograd in Russia; the Minami-aiki in Japan; and the Chief Joseph in the United States. The United States also has a number of facilities built during the 1920s and 1930s, notably the Hoover Dam, the Glen Canyon Dam, the Grand Coulee, and the Bonneville facility. Although many of these were built in sparsely inhabited areas, they have had an effect on the human populations in their area, though perhaps none so much as the Three Gorges Dam in China.

Environmental Effects of Hydroelectricity

Although the construction of large hydroelectric projects has had some benefit for the population served by the electricity they produce, the heavy use of hydroelectricity in some parts of the world has also had some negative effects. Among the advantages are the production of low-cost electricity without burning fossil fuels (although many hydroelectric plants have back-up fossil fuel generators so as to keep the generation steady); the creation of dams that can also be used for flood control, controlling water supply in urban areas with large populations; and the creation of artificial lakes that can provide many recreational opportunities.

The disadvantages include high construction costs, conflicting demands for flood control and hydroelectric generating capacity, the withdrawal of substantial amounts of land from agricultural use (as in China), degradation of fish habitat, and the elimination of trees on the flooded land. In several cases, aboriginal communities have lost the land that they traditionally used for their sustenance.

Context

The steadily growing need of human populations for electricity has created a continuing demand for large facilities that appear to place little burden on the environment. By producing electricity without burning fossil fuels and at low cost, hydroelectric plants are seductive, especially to the politicians who make public policy. Although the best sites have already been taken, there are still a number of lesser locations that may yet be claimed. For many countries in the developing world, hydroelectricity has a large attraction, especially if those countries lack local sources of fossil fuels.

Nancy M. Gordon

Further Reading

Hausman, William, Peter Hertner, and Mira Wilkins. Global Electrification: Multinational Enterprise and International Finance in the History of Light and Power, 1878–2007. New York: Cambridge University Press, 2008.

International Energy Agency/OECD. Projected Costs of Generating Electricity: Update, 1998. Paris: Author, 1998.

Johansson, Per-Olov, and Bengt Kriström. *The economics of evaluating water projects: hydroelectricity versus other uses.* Springer Science & Business Media, 2012. Not everyone sees hydroelectricity as a completely benign solution. Hydroelectric projects can displace populations, and conflict with fisheries, irrigation, and agriculture.

Manore, Jean. Cross-Currents: Hydroelectricity and the Engineering of Northern Ontario. Waterloo, Ont.: Wilfrid Laurier University Press, 1999.

Smil, Vaclav. Energy at the Crossroads. Cambridge, Mass.: MIT Press, 2003.

———. Transforming the Twentieth Century: Technical Innovations and Their Consequences. New York: Oxford University Press, 2006.

Hydrofluorocarbons

Category: Chemistry and geochemistry

Definition

Hydrofluorocarbons (HFCs) are a family of organic chemical compounds composed entirely of hydrogen, fluorine, and carbon. They are generally colorless, odorless, and chemically unreactive gases at room temperature. HFCs fall under the broader classification of haloalkanes. While HFCs do not harm the ozone layer, they contribute to global warming as greenhouse gases (GHGs) and are considered one of the major groups of high global warming potential (HGWP) gases.

Significance for Climate Change

Many of the HFCs were developed for use in industrial, commercial, and consumer products as alternatives to ozone-depleting substances such as chlorofluorocarbons (CFCs) and hydrochlorofluorocarbons (HCFCs). Common HFCs, in order of atmospheric abundance, include HFC-23 (fluoroform), HFC 134a (1,1,1,2-Tetrafluoroethane, tetrafluoroethane, R-134a, Genetron 134a, or Suva 134a), and HFC-152a (1,1-Difluoroethane, difluoroethane, or R-152a). HFC-23 is used in a wide range of industrial processes and is a by-product of Teflon(TM) production. HFC-134a is primarily used as a refrigerant for domestic refrigeration and automobile air conditioners. HFC-152a is commonly used in refrigeration, electronic cleaning products, and automobile applications as an alternative to HFC-134a.

The global warming potentials (GWPs) of HFCs range from 140 (for HFC-152a) to 11,700 (for HFC-23). These GWPs are significantly lower than those of the gases the HFCs are designed to replace. The atmospheric lifetimes of the same two HFCs are just over 1 year and 260 years, respectively. HFCs are one of two groups of haloalkanes targeted in the Kyoto Protocol. HFC emissions are projected to increase in the coming years, as industry continues to strive to decrease CFC and HCFC production.

HFCs are preferred over CFCs and HCFCs, because HFCs lack chlorine. When CFCs are emitted into the atmosphere, chlorine (Cl) atoms contained in those CFCs become disassociated through interaction with ultraviolet light. The resulting free Cl atoms decompose ozone into oxygen, and regenerated Cl atoms go on to degrade more ozone molecules. This reaction continues for the atmospheric lifetime of the Cl atom, which ranges from one to two years. On average, a single Cl atom destroys 100,000 ozone molecules. Thus, the lack of Cl in HFCs makes them a desirable alternative.

Some studies indicate that excessive exposure to HFCs may affect the brain and heart. However, this has only been established for concentrations higher than those found in the atmosphere.

Rena Christina Tabata

See also: Aerosols; Air conditioning and air-cooling technology; Chemical industry; Chlorofluorocarbons and related compounds; Kyoto Protocol; Ozone.

Hydrogen power

Category: Energy

Definition

Molecular hydrogen (H_2) is an ideal fuel to be used for transportation, since the energy content of hydrogen is three times greater than that of gasoline and four times greater than that of ethanol. Hydrogen power powered rockets launched by the National Aeronautics and Space Administration (NASA) for many years. Today, a growing number of automobile manufactures around the world are making hydrogen-powered vehicles. Because of depleting supplies and growing demand for oil, H_2 may become an alternative to gasoline.

The idea of hydrogen as the fuel of the future was expressed long ago by Jules Verne in his novel L'Île mystérieuse (1874–75; The Mysterious Island, 1875). However, compared to oil, H_2 is not abundant on Earth. Its atmospheric concentration is only 0.00001 percent, and there is even less of it in the oceans. Though many microorganisms produce H_2 during

Hydrogen-powered Mazda RX-8. (Taisyo/Wikimedia Commons)

fermentation, it is such a good source of energy that it is used almost immediately by other microbes. Thus, in order for humans to use hydrogen as fuel, it must be generated using other energy sources.

While molecular hydrogen is rare, the chemical element hydrogen is the most basic and plentiful element in the universe. It also forms a part of the most abundant chemical compound on Earth, water. Therefore, the challenge posed is to find a cost-effective and environmentally friendly way to generate H_2 from water or other chemical compounds. At present, H_2 is obtained mainly from natural gas (methane and propane) via steam reforming. Although this approach is practically attractive, it is not sustainable. Molecular hydrogen can be also produced by electrolysis. In this process, electric energy is employed to split water into H_2 and O_2. The requisite electricity can be obtained using clean, sustainable energy technologies such as wind and solar power. However, the process is not efficient, requiring significant expenditure of energy and purified water.

There are other technological and economic obstacles to hydrogen power. These obstacles include safety issues, as well as the lack of effective solutions for storage and distribution of H_2. Hydrogen has gained an unwarranted reputation as a highly dangerous substance among the general public. Like all fuels, H_2 may produce an explosion, but it has been used for years in industry and earned an excellent safety record when handled properly.

Hydrogen is the lightest chemical, so it has a much lower energy density by volume than do other fuels. As a gas, it requires three thousand times more space for storage than does gasoline. Thus, hydrogen storage, especially in cars, represents a challenge for scientists and engineers. For storage, H_2 is pressurized in cylinders or liquefied in cryotanks at −253° Celsius. Both processes require a significant expenditure of energy and generate large quantities of waste carbon dioxide (CO_2). In most contemporary hydrogen-powered vehicles, H_2 is stored as compressed gas. Because compressed gas cannot be delivered in the same fashion as liquid fuel, gasoline stations and pumps cannot simply be converted into hydrogen stations. Thus, the distribution system for the new fuel would have to be constructed from scratch, requiring considerable monetary investment.

Significance for Climate Change

Fossil fuels generate CO_2, contributing to the greenhouse effect. Switching from fossil fuels to H_2 would eliminate that source of greenhouse gas (GHG) emissions, provided that the new fuel could be produced without carbon-emitting technologies. Burning H_2 for an energy source produces only water as a by-product, and H_2 is also a renewable fuel, since it can be made from water again. Unfortunately, current methods of H_2 production from natural gas also generate CO_2. The ultimate goal is to generate H_2 without emitting GHGs into the atmosphere, perhaps by using wind or solar power.

One promising green method of H_2 production is a biological approach: A great number of microorganisms produce H_2 from inorganic materials, such as water, or from organic materials, such as sugar, in reactions catalyzed by the enzymes hydrogenase and nitrogenase. Hydrogen produced by microorganisms is called biohydrogen. The most attractive biohydrogen for industrial applications is that produced by photosynthetic microbes. These microorganisms, such as microscopic algae, cyanobacteria, and photosynthetic bacteria, use sunlight as an energy source and water to generate hydrogen. Hydrogen production by photosynthetic microbes holds the promise of generating a renewable hydrogen fuel using the plentiful resources of solar light and water.

It is possible to use hydrogen to fuel internal combustion engines. Doing so produces at least one GHG, nitrogen oxide, because the burning of hydrogen requires air, which is almost 80 percent nitrogen (N_2). To use hydrogen power in climate-friendly ways, it will be necessary to replace the internal combustion engine with fuel cells, which produces electricity to power vehicles without motors. Fuel cells are like batteries: They generate electricity via chemical reactions between H_2 and O_2. Fuel cells emit water and heat, not CO_2 or other GHGs. In addition, fuel cells are 2.5 to 3 times more efficient in converting H_2 energy than are internal combustion engines. Hydrogen fuel-cell cars could even provide power for homes and offices if necessary. As of 2015, Toyota was set for a limited commercial release of a hydrogen fuel cell vehicle.

For hydrogen power to become a reality, tremendous research and investment efforts are necessary. The fate of hydrogen power technology will also depend on consumers' willingness to spend money on climate-friendly technologies.

Sergei Arlenovich Markov

Further Reading

Gahleitner, Gerda. "Hydrogen from renewable electricity: An international review of power-to-gas pilot plants for stationary applications." *International Journal of Hydrogen Energy* 38, no. 5 (2013): 2039-2061. Electricity from variable output systems can be used to electrolyze water to create hydrogen for energy storage. The design of the system can materially affect the efficiency of energy storage.

Ogden, Joan. "High Hopes for Hydrogen." Scientific American Sept. 2006: 94–99. Print.

Service, Robert F. "The Hydrogen Backlash." Science 305.5686 (2004): 958–61. Print.

Zaborsky, Oskar R., ed. Biohydrogen. New York: Plenum, 1998. Print.

Hydrologic cycle

Category: Environmentalism, conservation, and ecosystems

Definition

The hydrologic cycle comprises the movement of water on Earth: It evaporates from the oceans, precipitates back to the surface, flows in the form of runoff to bodies of water, and infiltrates and is stored underground. A warming climate affects this cycle and therefore all life on the planet.

Steady warming increases the atmosphere's moisture-holding capacity, altering the hydrologic cycle and the characteristics of precipitation, most notably intensity. Changes in the global rate and distribution of precipitation may have a greater direct effect on human well-being and ecosystem dynamics than do changes in temperature. While temperature increases in a linear fashion, scientific research indicates that the hydrologic cycle may change exponentially, according to David Easterling of the U.S. National Climatic Data Center.

Atmospheric moisture increases rapidly as temperatures rise; over the United States and Europe, atmospheric moisture increased by 10 to 20 percent from 1980 to 2000. Studies of tree rings have provided a reconstruction of precipitation variability in the high mountains of northern Pakistan going back one thousand years that indicates a large-scale intensification of the hydrologic cycle coincided with the onset of industrialization and global warming. The unprecedented amplitude of this change strongly suggests a human role. This study suggests that an unprecedented intensification of the hydrologic cycle in western Central Asia occurred during the twentieth century.

Significance for Climate Change

Theory as well as an increasing number of daily weather reports strongly indicates that changes in precipitation patterns may vary widely across time and space. Such changes will be highly uneven, episodic, and sometimes damaging. Both droughts and deluges are likely to become more severe. They may even alternate in some regions. By 2007, the hydrologic cycle seemed to be changing more rapidly than temperatures. With sustained warming, usually wet places often seemed to be receiving more rain than before; dry places often were experiencing less rain and becoming subject to more persistent droughts. Some drought-stricken regions occasionally were doused with brief deluges that washed away cracking earth.

In many places, the daily weather increasingly was becoming a question of drought or deluge. Paradoxically, the intensification of the hydrologic cycle due to a warming climate may provoke both drought and deluge—sometimes alternately in the same location. The behavior of this cycle also figures into a lively debate regarding whether a generally warming climate may cause hurricanes to intensify, as well as the impact of climate change on the worldwide spread of deserts, which may contribute

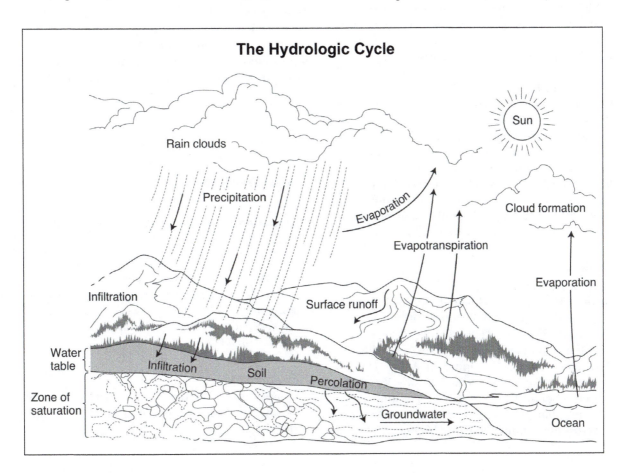

to the number of "environmental refugees," and on disaster relief generally.

By 2000, the hydrologic cycle seemed to be changing more quickly than temperatures. In 2005, for example, India's annual monsoon brought a 94-centimeter rainfall in 24 hours to Mumbai (Bombay). India is accustomed to a drought-and-deluge cycle because of its annual monsoon, but the rains have become more intense, as well as more variable. Excessive rain has deluged parts of Europe and Asia, swamping cities and villages and killing many people, while drought and heat have scorched other areas. Unprecedented precipitation events have occurred recently in the Arctic, where temperatures have risen rapidly. Winter rains with thunder and lightning have been experienced on Southern Baffin Island, the Alaskan North Slope, and southwestern Greenland. Climate-change skeptics argued that weather is always variable, but other observers noted that extremes seemed to be more frequent than before.

Violent weather has been hitting unusual areas with increasing frequency. Major tornadoes, for example, have been sighted in Michigan and Japan. At least nine people were killed, at least twenty injured, and others trapped under rubble when a tornado hit northern Japan on November 6, 2006, in a location where such storms had been all but unknown. The tornado demolished two prefabricated buildings at a construction site, about eight private houses, and a factory, according to an official at the fire department in Engaru, on the northeastern part of Japan's northern island of Hokkaido. Fire engines and ambulances were at the site, where crushed cars could be seen along with lumber and other scattered debris. Several houses were reduced to rubble.

Bruce E. Johansen

Further Reading

Barnett, Tim P., et al. "Human-Induced Changes in the Hydrology of the Western United States." *Science* 319 (February 22, 2008): 1080-1083. Discusses human influences on the hydrologic cycle.

Barringer, Felicity. "Precipitation Across U.S. Intensifies Over Fifty Years." *The New York Times*, December 5, 2007. A description of studies indicating that precipitation intensity is increasing across the continental United States.

Del Genio, Anthony D., Mao-Sung Yao, and Jeffrey Jonas. "Will Moist Convection Be Stronger in a Warmer Climate?" *Geophysical Research Letters* 34, no. 16 (August 17, 2007). Describes the theoretical basis of changes in the hydrologic cycle in a warmer climate.

Sivapalan, Murugesu, Hubert HG Savenije, and Günter Blöschl. "Socio–hydrology: A new science of people and water." *Hydrological Processes* 26, no. 8 (2012): 1270-1276. As environmentalists purchase water rights formerly used for irrigation, there is a need to consider humans as an integral part of the hydrologic cycle.

Trenberth, Kevin E., Aiguo Dai, Roy M. Rassmussen, and David B. Parsons. "The Changing Character of Precipitation." *Bulletin of the American Meteorological Society*, September, 2003, 1205-1217. Scientifically solid treatment of how warming conditions influence intensity of rain and snowfall.

See also: Aral Sea desiccation; Desertification; Deserts; Ekman transport and pumping; Freshwater; Groundwater; Hydrosphere; Ocean-atmosphere coupling; Saltwater intrusion; Sea surface temperatures; Water resources, global; Water vapor.

Hydrosphere

Categories: Water resources; oceanography; cryology and glaciology

The hydrosphere, the collective designation of all water on Earth, significantly affects Earth's greenhouse effect and global temperatures generally. Atmospheric water vapor is the major GHG, and the oceans are major storage sites for water, heat, and CO_2.

Key concepts

cryosphere: water on the Earth's surface in solid form, especially as snow, glaciers, ice caps, and ice sheets *evaporation:* the transformation of liquid water into water vapor

hydrologic cycle: water's movement among the spheres of the climate system

thermal expansion: the increase in water volume as its temperature increases and density decreases

water vapor: water in the gaseous state

Background

The range of Earth's surface temperatures and atmospheric pressures permits water to occur naturally as a solid, liquid, and vapor. Water is present in all Earth's climate subsystems. In broadest terms, the hydrosphere includes liquid water at the Earth's surface, water in the atmosphere in all its forms, and water in the cryosphere stored as ice. It extends several kilometers below Earth's surface and reaches 12 kilometers into the atmosphere.

Distribution of Water

The hydrosphere contains an estimated 1.36 billion cubic kilometers of water, and liquid water covers about 71 percent of the Earth's surface. Most of the world's water is contained in the oceans as salt water; only 2.5 percent is freshwater. Ice sheets and glaciers concentrated on about 10 percent of the Earth's land area, mainly in Antarctica and Greenland, constitute less than 2 percent of all water, but they account for 69 percent of freshwater. Groundwater and soil water constitute the subsurface water storage, which is important as a water supply for humans in arid regions and as a moisture source for plants. Groundwater and soil water together account for 0.5 percent of all water, but 30 percent of freshwater. Rivers, lakes, and streams contain 0.02 percent of all water and 0.3 percent of freshwater. Water vapor and cloud droplets in the atmosphere account for 0.0001 percent of all water and 0.04 percent of freshwater, but water vapor and cloud droplets contribute between 66 and 85 percent of the greenhouse effect, which traps heat in the atmosphere, preventing it from escaping into space.

Formation of the Hydrosphere

The hydrosphere's formation required a series of steps occurring over millions of years to reach the sequence of energy and moisture transfers currently linking the hydrosphere and Earth's climate. Water from outgassing by volcanic eruptions and possibly from incoming comets collected in the early atmosphere as water vapor along with other gases. Eventual cooling of the Earth's surface and the atmosphere caused the water vapor to condense into droplets and fall as rain to the Earth's surface. The rain collected in topographic depressions in the Earth's crust and formed rivers, lakes, and oceans. Some water drained vertically through the rocks and collected in large underground reservoirs, and some water was held by the soil that developed as rocks disintegrated.

The Earth's original hydrosphere probably was freshwater exclusively, with the hydrologic cycle as the centerpiece. Water cycling was sustained by evaporation mainly from the oceans that returned water vapor to the atmosphere to fall as rain. Dissolved gases fell with the rain from the early atmosphere, and the oceans absorbed large amounts of atmospheric carbon dioxide (CO_2). The transition to today's salt-dominated oceans spanned hundreds of millions of years and involved crustal differentiation, life processes, and a series of mineral equilibriums. Rain on the land dissolved minerals from rocks; rivers and streams carried the minerals to the oceans; chlorine leached from the oceanic crust accumulated in seawater; and dissolved carbonates were gradually removed. During periods of global cooling, snow accumulated on the continents, forming glaciers and ice sheets, and sea levels were lowered.

Effects of a Warmer Climate

Evidence of warmer temperatures exists in all the hydrosphere's components, but it is most apparent in the oceans. Oceans store more than 90 percent of the heat in the Earth's climate system, and they buffer the effects of climate change. Sea surface temperatures since 1860 show essentially the same trends as land temperatures, and ocean temperature changes contribute about one-half Earth's net temperature gain. Ocean circulation transfers energy from warm tropical regions toward the poles and brings colder water back to the tropics. Ocean circulation is accomplished by powerful, wind-driven surface currents coupled with deep-ocean currents responding to density differences produced by temperature and salinity variations. British oceanographers have reported that circulation in a section of the Atlantic Ocean has slowed 30 percent in the past five decades, which raises concern that the global transfer of energy by the oceans may be weakening.

Warmer oceans produce thermal expansion, which is a major contributor to rising sea levels of 15-20 centimeters over the last century. Observed twentieth century melting of glaciers and ice sheets contributed to sea-level rise, but the volume of melting accounts for about 20 percent of the observed increase. An observed 10 percent per decade decrease in Arctic sea ice probably contributes more to radiation balance changes and warming the air and water than to sea-level rise.

Oceans are the source for 85 percent of the water vapor transferred into the atmosphere, and warmer temperatures increase evaporation. Increased atmospheric water vapor supports greater precipitation, an intensification of the greenhouse effect, increased cloud cover, and increased global albedo. Changes in the timing, amount, and location of precipitation and runoff are occurring as global temperatures increase, and these changes alter water availability and quality and the duration and intensity of floods and drought.

CO_2 in the oceans increases as atmospheric CO_2 increases, but warm oceans are a less efficient absorber of CO_2. About half of the known CO_2 produced by fossil fuel consumption over the past fifty years has been absorbed by the oceans.

Context

Water is essential for humans and for natural ecosystems, but the relatively small proportion of Earth's total water represented by freshwater is unevenly distributed globally. Approximately one-third of the world's population lives in regions where the freshwater supply is less than the recommended per capita minimum, and 70 percent of all freshwater withdrawals are used for crop irrigation. The freshwater resource relies on precipitation delivered by the hydrologic cycle. Precipitation is sustained through the conversion of ocean water into freshwater by a complex system responding to global-scale climatic influences. Hydrological processes, related energy exchanges, and interactions of these exchanges with clouds and the radiation balance are highly variable at the regional scale where the global warming impact on the freshwater resource is greatest. Quantifying regional temperature and precipitation patterns and the effects of altered evaporation, transpiration, soil moisture storage, and runoff on the freshwater supply is a great challenge facing scientists as they strive to understand global warming at the regional scale.

Marlyn L. Shelton

Further Reading

Barnett, Tim P., David W. Peirce, and Reiner Schnur. "Detection of Anthropogenic Climate Change in the World's Oceans." *Science* 292, no. 5515 (April 13, 2001): 270-274. Observed ocean heat content and estimates from a global climate model are analyzed for evidence of global temperature change in deep ocean water. Both data sets indicate increases in heat content in the major ocean basins, but vertical temperature changes are markedly different.

Bigg, Grant R. *The Oceans and Climate.* New York: Cambridge University Press, 1996. Emphasizes the complex links between the atmosphere and oceans in driving climate and climate change, as well as interactions with other planetary spheres. By a highly regarded researcher in marine climate change. Illustrations, tables, glossary, bibliography, index.

Ernst, W. G., ed. *Earth Systems: Processes and Issues.* New York: Cambridge University Press, 2000. Integrated overview of Earth systems, from a process-oriented perspective, written by twenty-eight international scientists. The hydrosphere section contains four chapters focused on the hydrologic cycle and the oceans. Illustrations, tables, glossary, index.

Valero, Antonio, Andrés Agudelo, and Alicia Valero. "The crepuscular planet. A model for the exhausted atmosphere and hydrosphere." *Energy* 36, no. 6 (2011): 3745-3753. "Crepuscular" means "pertaining to twilight," and refers to an attempt to model the earth after all resources have been extracted.

See also: Evapotranspiration; Freshwater; Glaciers; Groundwater; Hydrologic cycle; Ice shelves; Ocean-atmosphere coupling; Water resources, global; Water resources, North American; Water vapor.

Ice cores

Category: Cryology and glaciology

Ice cores contain evidence of an orderly sequence of changes that occurred in the Earth's atmosphere, including changes in temperature and trace gases. They therefore provide a history of the Earth's climates.

Key concepts

Dome C: the Antarctic Plateau location of the Concordia Research Station

glacial ice: ice created by the compression of snow into glaciers

paleoclimatology: the study of past climates

Vostok: a Russian Antarctic research station built in 1957 during the First Geophysical Year

Background

An ice core is a cylinder of ice measuring 10 centimeters in diameter and obtained by drilling vertically into a glacier. Glacial ice is produced by the natural, gradual transformation of snow into ice. Snow is made of fragile ice crystals surrounded by air. In the process that transforms snow into ice, this air is trapped in tiny bubbles that are identical in composition to the atmosphere that produced the snowfall. As snow falls year after year and is compacted into ice, each layer can be identified visually or electronically. In this way, glaciers preserve a record of atmospheric changes that have occurred over time, and ice cores can be drilled to retrieve that record.

Variation in the Atmosphere's Composition

Because they contain a chronologically ordered record of atmospheric composition, ice cores can be used to reconstruct the history of both natural and anthropogenic pollution. The chemical composition of Earth's atmosphere varies over time. It reacts to volcanic eruptions, for instance, which add carbon monoxide, carbon dioxide (CO_2), and sulfur dioxide to its normal content of nitrogen, oxygen, argon, CO_2, and methane. Thus, changes in the chemical composition of the air trapped in ice cores make it possible to date past volcanic eruptions.

In the same way, some human activities can be identified and dated. Analysis of ice cores from the Alps and Greenland has identified the pollution resulting from the smelting of lead ore by the Romans more than two thousand years ago. More recently, traces of chemicals linked to nuclear explosions or to the Chernobyl nuclear reactor meltdown have been detected in glacial ice. Today, at a time when issues of climate change and pollution are uppermost in the minds of many, glaciologists have sought to retrieve the longest possible ice cores in order to produce a substantial historical record to contextualize current atmospheric trends.

The Longest Ice Cores

The first giant ice core was drilled in Vostok, the Russian base in Antarctica. The region of Antarctica in which Vostok is located is extremely cold and dry. The advantage of low precipitation in glaciology is that, because each year produces less ice, a large number of years is recorded relative to core length. The disadvantage of such low precipitation is that, because each annual layer of ice is extremely thin, it can be very difficult to produce an accurate count. When all analyses were complete, Vostok glaciologists concluded that the 3,623-meter core drilled in 1996 represented the last 420,000 years of the Earth's history.

The normal analysis of an ice core consists of analyzing the content of the ice for the presence of deuterium (an isotope of hydrogen) and the presence of oxygen 18 (O^{18}; an isotope of oxygen), because both are reliable indicators of the Earth's temperature at the time of the snowfall. The temperature records obtained thanks to the Vostok ice core compare well with cores obtained in the center of Greenland. Since the Greenland ice sheet is smaller and thinner (1,600 meters deep on average) than the Antarctic ice sheet (2,400 meters deep on average), the paleoclimatic record derivable from Greenland ice cores is shorter. However, that record is consistent with the Vostok ice core record. Greenland also receives more precipitation than does Vostok, resulting in thicker annual layers of ice; a layer as old as 100,000 years can still be as much as 1 centimeter thick. Greenland ice cores therefore are outstanding at establishing a chronology for the last glacial period for which the estimated error does not exceed 2 percent for the first forty thousand years.

In an effort to increase the length of the paleo-climatic record, scientists from Europe have drilled in two other regions of the East Antarctic Ice Sheet under the European Project for Ice Coring in Antarctic (EPICA). A site on Dome C located at 75°06¢ south latitude and longitude 123°21¢ east was selected. There, where the ice sheet is thicker while both the ice and the air remain extremely cold, a permanent research base, Concordia, was built by France and Italy (partly financed by the European Union). Dome C is ideal for the thickness of its ice and its low precipitation; it initially provided a 3,140-meter ice core, revealing 740,000 years of climatic history. Since then, even deeper ice has been retrieved, extending knowledge of climatic history to over 800,000 years.

Testing Theories of Climate Change

In order to analyze the ice core samples that have been retrieved, scientists crush the ice in a vacuum, releasing the air trapped within the sample without allowing it to mix with the modern atmosphere. They can then analyze these air samples, compute the proportion of greenhouse gases (GHGs) such as CO_2 and methane in the air, and compare their proportions to the temperature record derived from the deuterium and O^{18} analyses. The goal is to see if an increase of GHGs in ice always leads to an increase in the Earth's temperature or, conversely, if an increase in the Earth's temperature releases CO_2 and methane into the atmosphere.

Context

Scientists continue to search for deeper deposits of ice from which even longer historical sequences can be identified. The Dome C site holds the promise of ice cores that will extend the scientific record to more than one million years into the past. Meanwhile, research also continues on ways to recover more, and more precise, information from ice cores. Analysis of core samples containing volcanic dust taken at Dome C, for instance, has demonstrated that the magnetic polarity of the Earth reversed about 780,000 years ago. As ice cores get longer and scientists find ways to learn

more from them, the historical record will become more detailed over longer periods of time.

Denyse Lemaire and David Kasserman

Further Reading

Alley, Richard B. *The Two-Mile Time Machine: Ice Cores, Abrupt Climate Change, and Our Future.* Princeton, N.J.: Princeton University Press, 2000. Provides ice coring research information and discusses the work of geoscientists in relation to the cryosphere. The text is concise and clear, with superb artwork and photographs.

Masson-Delmotte, Valérie, D. Buiron, A. Ekaykin, M. Frezzotti, H. Gallée, Jean Jouzel, G. Krinner et al. "A comparison of the present and last interglacial periods in six Antarctic ice cores." *Climate of the Past* 7, no. 2 (2011): 397-423. Climate conditions appear to have been fairly uniform across much of Antarctica although two sites showed sudden changes possibly due to the growth of sea ice.

Trewby, Mary. *Antarctica: An Encyclopedia from Abbott Ice Shelf to Zooplankton.* Toronto: Firefly Books, 2002. General encyclopedia of Antarctica. Explains all the features found on the continent using a clear and simple language.

Turney, Chris. *Ice, Mud, and Blood: Lessons from Climates Past.* New York: Macmillan, 2008. Describes the discoveries derived from ice cores and how these discoveries led to a better understanding of paleoclimates.

See also: Climate reconstruction; Dating methods; Glaciers; Greenland ice cap; Oxygen isotopes; Paleoclimates and paleoclimate change.

Ice shelves

Category: Cryology and glaciology

The debate over global warming has prompted a renewed interest in the study of ice shelves, particularly after the collapse of the Larsen A and Larsen B shelves in 1995 and 2002, respectively. These collapses have been seen not only as products of global warming but also as part of a process by which the movement of Antarctic ice toward the ocean has been accelerated.

Key concepts

glacier: a mass of ice that flows downhill, usually within the confines of a former stream valley

ice sheets: masses of ice covering large areas of land

ice shelf: a platform of freshwater ice floating over the ocean

West Antarctic ice sheet: the smallest ice sheet in Antarctica, located west of the Transantarctic Mountains

Background

Ice shelves are the seaward extensions of continental ice sheets. They are found extending from 50 percent of the coast of Antarctica and account for 11 percent of the mass of ice of the Antarctic Ice Sheets. The two largest ice shelves—Ross (472,960 square kilometers) and Ronne-Filchner (422,420 square kilometers)—together represent 67 percent of the total area of Antarctic ice shelves. Ice shelves are typically thinner at their seaward edge; the largest shelves reach a depth of 1,300 meters near the grounding line (the location where the ice begins to float), but thin to 200 meters at the leading edge.

Formation of Ice Shelves

Ice shelves gain mass in three different ways. The primary source is the ice sheet that moves off the land surface onto the water, but accumulated snow on their upper surface and the freezing of seawater to the lower one also contribute. On the other hand, ice shelves lose mass in five ways: by calving of icebergs, by the melting of ice in contact with the sea, by snow blowing off the edge of ice shelves, by enhanced sublimation of ice under the very strong Antarctic winds, and by superficial melting. Since surface melting in Antarctica is negligible, and since wind ablation is limited in general, nearly all mass loss is the result of either iceberg calving at the seaward margin or bottom melting of ice in contact with the sea.

Ice Shelves of the West Antarctic Peninsula

With the widespread availability of video clips on the Internet, many people have observed the breaking of very large tabular icebergs off the Antarctic Peninsula. The calving of icebergs is a natural phenomenon that occurs regularly under normal climatic conditions. This calving should not be interpreted as a sure sign of global warming. In the extreme north of the Antarctic Peninsula, ice shelves are called Larsen A, B, C, and D, named in sequence from north to south. The Larsen A ice shelf began to retreat at a very rapid pace in January, 1995. The breakup of Larsen A was irrefutably a response to the rise in ocean temperature. South of Larsen A, Larsen B calved large tabular icebergs beginning in 1995, prior to a larger disintegration that occurred in the first three months of 2002. More recently, in 2008, the Wilkins Ice Shelf that had broken up in 1998 partially disintegrated. The culprit seemed to be the presence of warm ocean waters melting ice shelves from beneath and destabilizing the ice shelf at the grounding line.

Ice Shelves as Buttresses

Because ice shelves may be very thick, they are often anchored on the continental shelf, where they begin to float. The grounding of these ice shelves allows them to slow down the outward movement of ice sheets by acting as a buttress if they are thick enough. This effect is lost if the ice shelf disintegrates. In the case of the Larsen B Ice Shelf, its disintegration in 2002 led to an acceleration of the outward flow of ice from the glacier that feeds it.

Context

Although there is no doubt that ice shelves in the West Antarctic Peninsula have decreased in size over the last three decades, their complete melting would not raise sea level by more than a few millimeters. Their destruction could, however, contribute significantly to rising sea level if the glaciers feeding these ice shelves were to speed up once the ice shelf stops being grounded. A 2004 study by researchers from the National Snow and Ice Data Center and NASA described accelerations of such glaciers reaching four to six times their normal speed.

Interestingly, the ice shelves of the southern part of the West Antarctic ice sheet and those of the East Antarctic ice sheet do not seem to be affected by global climate change. Their temperature remains below freezing, probably because they are surrounded by abundant sea ice that acts as a buffer, protecting them from the advection of warmer water.

Denyse Lemaire and David Kasserman

Further Reading

Copeland, Sebastian. *The Global Warning*. San Rafael, Calif.: Earth Aware Editions, 2007. This abundantly illustrated book about Antarctica tends to focus on the potential calamitous effects of global climate change using a clear and simple language.

McGonigal, David. *Antarctica: Secrets of the Southern Continent*. Richmond Hill, Ont.: Firefly Books, 2008. Describes the discoveries of the International Polar Year, 2007-2008, in various fields, including geology, geography, and climatology. Discusses potential effects of global warming.

Paolo, Fernando S., Helen A. Fricker, and Laurie Padman. "Volume loss from Antarctic ice shelves is accelerating." *Science* 348, no. 6232 (2015): 327-331. Ice shelves slow the flow of Antarctic ice into the sea, but thinning rates have increased dramatically in the last two decades.

Trewby, Mary. *Antarctica: An Encyclopedia from Abbott Ice Shelf to Zooplankton*. Toronto: Firefly Books, 2002. General encyclopedia of Antarctica that explains all major features of the continent.

Turney, Chris. *Ice, Mud, and Blood: Lessons from Climates Past*. New York: Macmillan, 2008. Examines various discoveries derived from ice cores and how these discoveries led to a better understanding of paleoclimates.

See also: Antarctica: threats and responses; Glaciers; Greenland ice cap; Sea ice.

Ice-out studies

Categories: Arctic and Antarctic; cryology and glaciology

Definition

Across the northern United States, a popular tradition for many decades has been guessing the date of ice breakup ("ice-out") on local lakes. Not only is it a local news story and harbinger of spring, but ice-out has numerous practical ramifications as well. It marks the end of winter activities on the lake, such as ice fishing and travel across the ice, and the beginning of summer activities, such as open-water fishing.

In earlier days, ice-out was important for commercial activities such as log drives and steamship navigation. As a result, for many lakes ice-out data are available since the middle of the nineteenth century. Moosehead Lake, the largest lake in Maine, has continuous data since 1848; Sebago Lake in southern Maine has sporadic data since 1807. Areas outside New England also maintain ice-out statistics. Minnesota has at least four lakes with more than 100 years of data, and the longest record is 139 years. Lakes Mendota and Monona, which flank the city of Madison, Wisconsin, have records extending back more than 150 years.

Ice-out dates are quite variable from year to year and depend on air temperature, the flux of meltwater into the lake, and the discharge of the outlet stream. For example, the earliest and latest ice-out dates for Moosehead Lake differ by forty-five days. Nevertheless, scientists from the U.S. Geological Survey analyzed ice-out dates for New England lakes and found that the average date of ice-out was nine days earlier in 2000 than it was in 1850. Generally, ice-out data are tabulated in terms of day of the year (the Julian date) rather than the calendar date. For example, February 20 is Julian day 51. The data indicate generally steady ice-out dates until 1900, decreasing (earlier) dates until roughly 1950, a slight increase in ice-out dates until about 1970, and decreasing dates thereafter. Despite the wide variability from year to year, the overall pattern is consistent across numerous lakes.

Significance for Climate Change

There are many complexities in the use of ice-out data, largely due to variations in what defined "ice-out." Sometimes a marker placed on the ice has been used to signal ice breakup. In some places the marker was even a junked car (an environmentally dubious practice). In other places the criterion has been the ability to cross the lake by boat without being blocked by ice. Still other places have used visual estimates of ice cover. Such estimates can be quite subjective. It may take several days for ice to clear out, and ice may persist in restricted coves long after most of a lake is ice-free.

Unfortunately, it is not always clear what criterion was used for many early records. Also, ice-out data are extremely "noisy"—that is, the range in the data is much larger than the long-term change in average date. For example, for Maine's Moosehead Lake, the best-fit trend line for the ice-out data shows a nine-day decrease in ice-out day between 1849 and 2005, but the variation between earliest and latest ice-out is forty-five days. Even from year to year, the variations can be quite large: In 2001 ice-out occurred on day 124, dropping to day 110 in 2002 and increasing to day 127 in 2003.

Ice-out data for any individual lake are usually so variable that there is a significant possibility that the data, just by chance, happen to show a decrease over time. For example, if one flips a coin and keeps score of the difference between heads and tails, even though the flips are completely random there may be long runs where the difference increases or decreases steadily. However, when data from many lakes, especially those separated by large distances, show the same pattern, the statistical significance of the data becomes far greater.

Taking all these considerations into account, ice-out data for lakes across the northern United States show a broad trend of earlier ice breakup. Climate change skeptics have not paid much attention to ice-out data. Their most common approach has been to use anecdotal data—that is, point to unusually late ice-out on some particular lake, rather than examine long-term trends.

Steven I. Dutch

Further Reading

Hodgkins, Glenn A., and Ivan C. James. "Historical Ice-Out Dates for Twenty-nine Lakes in New England." U.S. Geological Survey Open-File Report 02-34. Denver, Colo.: U.S. Geological Survey, 2002. Listings of ice-out dates for all years with data and discussion of methods and interpretation. The data listings enable users to perform their own analyses of the data.

Hodgkins, Glenn A. "The importance of record length in estimating the magnitude of climatic changes: an example using 175 years of lake ice-out dates in New England." *Climatic change* 119, no. 3-4 (2013): 705-718. New England has the longest collection of ice-out data and recent changes have been faster than previously. Early ice-out is also correlated with decreased dissolved oxygen, so that climate change may affect the aquatic biology.

Hodgkins, Glenn A., Ivan C. James, and T. G. Huntington. "Historical Changes in Lake Ice-Out Dates as Indicators of Climate Change in New England, 1850-2000." U.S. Geological Survey Fact Sheet FS 2005-3002. Available online at http://pubs.usgs.gov/fs/2005/3002/. A summary of ice-out trends in New England lakes, with graphs of data trends but no data.

U.S. Geological Survey. "Lake Ice-Out Data for New England." 2007. Available online at http://me.water.usgs.gov/iceout.html. Contains links to other articles and an interactive map that links to data for each lake. The data are in simple columnar format, making it easy to copy and paste into a spreadsheet.

Wisconsin State Climatology Office. "Wisconsin Lake Ice Climatologies: Duration of Lake Ice." 2004. Available online at http://www.aos.wisc.edu/~sco/lakes/WI-lake_ice-1.html. Links to data for twenty-seven Wisconsin lakes. The links show duration of ice cover rather than specific dates of freezing and breakup. Lakes Mendota and Monona, with more than 150 years of data, show clearly decreasing trends, as do numerous other lakes.

See also: Arctic; Cryosphere; Deglaciation; Glaciations; Glaciers; Greenland ice cap; Ground ice; Hydrosphere; Ice cores; Last Glacial Maximum (LGM); Little Ice Age (LIA); Mass balance; Permafrost; Polar climate; Sea ice.

Inconvenient Truth, An

Category: Popular culture and society
Dates: Book published and film released 2006
Author: Al Gore (1948-   )
Director: Davis Guggenheim (1963-   )

An Inconvenient Truth emphasizes the need to stop global warming in order to prevent the destruction of Earth's natural resources and the accompanying loss of biodiversity. A centerpiece of Al Gore's work on climate change, it was the primary factor behind Gore's receipt of the 2007 Nobel Peace Prize.

Key concepts

carbon dioxide (CO₂): a greenhouse gas produced by burning fossil fuels

climate change: a lasting, significant change in the average weather of a given area

fossil fuels: hydrocarbon deposits, such as petroleum or coal, that have been derived from the remains of ancient plants and animals and are useful sources of energy

global warming: a significant increase in the mean temperature of the Earth's air and water over time

greenhouse effect: the ability of certain gases in the atmosphere to trap heat emitted from the surface of the Earth, resulting in the insulation and warming of the planet

greenhouse gases (GHGs): gases present in the Earth's atmosphere that reduce the loss of heat into space, thereby contributing to global warming through the greenhouse effect

Background

Throughout his time as vice president of the United States, and for many years before, Al Gore advocated for changes in environmental policies, especially as they related to the climate, most notably global warming. Over the years, he had prepared and worked to perfect a slide show dealing with the problem of global warming. This slide show, which he presented many times, became the basis for the movie *An Inconvenient Truth.*

The Film

The one-hundred-minute documentary film made its debut on May 24, 2006, in New York and Los Angeles. Directed by Davis Guggenheim, the movie features Gore. In the slide show within the movie, Gore presents scientific evidence of Earth's changing climate and looks at the economic and political impact of global warming on the planet. Gore goes on to discuss likely consequences of global warming

if society does not find a way to reduce the production of greenhouse gases (GHGs). Throughout the film, Gore makes an impassioned plea to society to take the steps necessary to halt global warming in order to stop a worldwide crisis. The film garnered the 2007 Academy Award for Best Documentary.

The Book

Titled *An Inconvenient Truth: The Planetary Emergency of Global Warming and What We Can Do About It,* Gore's book was published by Rodale Press in conjunction with the release of the movie. The 328-page book highlights most of the scientific evidence concerning GHGs and global warming, and it issues the same plea for major environmental changes to avoid global crisis. In 2007, Gore and Rodale Press in conjunction with Viking Press published a version of the book aimed at adolescents and young adults, *An Inconvenient Truth: The Crisis of Global Warming.* Gore believed that the younger generation would have to be involved in bringing about the technological and policy changes necessary to halt devastating climate change.

Origins of Gore's Interest in Climate Change

Gore's interest in global warming began when, as a student at Harvard University, he took a course dealing with climate change and the greenhouse effect. When he became a congressman, he introduced the first congressional hearings on global warming. When congressional response was slow, Gore responded by writing his first book dealing with environmental issues, *Earth in Balance* (1992), published the year he was elected vice president. As vice president, Gore encouraged passage of a tax designed to decrease the use of fossil fuels, which would thereby decrease carbon dioxide (CO₂) buildup and global warming. In 1997, he was also instrumental in the development of the Kyoto Protocol, an international treaty aimed at decreasing the emission of GHGs.

When he ran for president in 2000, Gore made saving the environment one of his platform issues, as he hoped to convince the Senate to ratify the Kyoto treaty. After losing the presidential election in 2000, Gore began to focus his attention fully on environmental concerns, revising his slide show and lecturing on global warming around the world. This slide show evolved into the basis for *An Inconvenient Truth*. For his work focused on the prevention of global warming, Al Gore was nominated for and received the Nobel Peace Prize in October of 2007.

Context

According to the Nobel Prize committee, Al Gore is probably the one man who has done the most to create an understanding of global warming and its potential planetary impact. By promoting this understanding and advocating for the changes that need to be made to avoid further climate change, Gore is leading the march against planetary destruction. This battle against global warming is one Gore believes must be won if society is to continue.

When people are exposed to *An Inconvenient Truth* (either the movie or book), they are made aware of the problems of climate change and are urged to focus on means that can be used to prevent the greenhouse effect and global warming. Since the first showing of the movie, Gore has run hundreds of training sessions, in which he has taught thousands of people to be ambassadors to the planet for *An Inconvenient Truth*. These ambassadors, known as climate project presenters, have traveled the planet, giving abridged versions of Gore's slide show and answering any questions that arise from the audience. Thanks to Gore and those that have been trained by him, many more people are aware of the problems of global warming and are taking steps to prevent it.

Robin Kamienny Montvilo

Further Reading

Al Gore, The Future: Six Drivers of Global Change, New York, Random House, 2013. Gore focuses on economic globalization, digital communications, shifting balances of power, a "flawed economic compass," biotechnology, neuroscience, and changes in the relationship between humans and the earth.

Gore, Al. *Earth in the Balance: Ecology and the Human Spirit*. Boston: Houghton Mifflin, 2000. In this book, the precursor to *An Inconvenient Truth*, Gore focuses attention on the environment and the danger of its imminent destruction unless humankind changes its behavior.

_____. *Our Purpose: The Nobel Peace Prize Lecture, 2007*. New York: Rodale, 2008. Gore's speech upon accepting the Nobel Prize, presenting his thoughts on the nature and obligations of humanity.

Maraniss, David, and Ellen Nakashima. *The Prince of Tennessee: Al Gore Meets His Fate*. New York: Simon & Schuster, 2001. Political biography of Gore, following his career and the place of environmentalism within it.

See also: Gore, Al; Greenhouse effect; Greenhouse gases and global warming.

India

Category: Nations and peoples

Key facts

Population: 1,148,000,000 (2008), 1,293,057, 000 (2016 estimate)

Area: 3,287,590 square kilometers

Gross domestic product (GDP): $3.267 trillion (purchasing power parity, 2008 estimate) $8.727 trillion (purchasing power parity, 2016 estimate)

Greenhouse gas (GHG) emissions in millions of metric tons of carbon dioxide equivalent (CO_2e): 1,000-1,300 in 2001, 3,014 in 2012

Kyoto Protocol status: Ratified August, 2002

History and Political Context

Great Britain's colonial occupation of India destroyed and replaced the region's traditional, cottage-based textile, metal, and crafts industries with plantations and infrastructure to export raw materials and mined ores through a few ports to the factories of the British Empire. By the early 1900's, India had a desperately poor subsistence economy that was dependent on the monsoons and experienced frequent famines when crops failed. The world wars necessitated steel plants to be located in India to build railways and armaments to supply the war effort, and a few entrepreneurs such as the

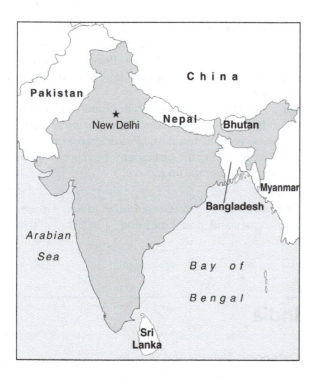

Tatas and Birlas won access to the capital needed to create heavy industry.

Upon independence in 1947, India boosted irrigation, agriculture, education, public transport, and heavy industry through a series of central five-year plans. Capital and technology came from the United Kingdom, United States, West Germany, and the Soviet Union. Public-sector plants produced steel, cement, fertilizers, chemicals, aluminum, titanium, railway equipment, heavy electricals and electronics, aircraft, ships, and refined oil. Democratic governments tried to invest scarce resources in education and to improve the desperately poor standard of living of the entire population, with the result that a socialist economy with top-tier tax rates as high as 90 percent and many restrictions on private enterprise evolved and stagnated. Free primary schooling was instituted in many states, and central institutes imparted essentially free, world-class scientific and technical education to students selected on merit.

The investments in education and broad-based opportunity paid off. By 1980, Indian agriculture had outpaced population growth and turned the famine nation into a net exporter of food. Even as the rupee plunged in the 1980s as a result of rising

energy costs and the inability of government enterprises to compete in the world market, remittances from skilled expatriates in the Middle East oil economies and the West boosted the foreign exchange reserves and the demand for good housing and consumer goods. In 1991, as the nation was facing a debt crisis, the Indian economy was opened to free enterprise and began to advance rapidly.

Globe-trotting Indian software engineers and entrepreneurs led competitive knowledge-based and global service industries and drove explosive growth in information technology, communications, consumer goods, manufacturing, and construction. Investment in space research spawned a thriving remote sensing, educational, communications, and weather satellite system leap-frogging the terrestrial communications grid to the villages. Today there is growing industry and export in chemicals, food processing, steel, transportation equipment, cement, mining, petroleum, machinery, and software. Automobile manufacturing is expanding for the domestic and export markets. Indian-owned steel manufacturing is among the leaders in the world. Modernization is creeping into the hinterland, with rising land values, education, food prices, and expectations.

Impact of Indian Policies on Climate Change

As of 2007, India had proven resources of 668 billion liters of oil and 1 trillion cubic meters of natural gas. National consumption was 98 billion liters of oil and 31 billion cubic meters of gas per year, mostly imports. Recoverable coal reserves were 92 billion metric tons, consumed at 434 million metric tons per year. Installed electric capacity in 2004 was 131 gigawatts, consumed at 588 billion kilowatt-hours per year. Total energy consumption was 16.3 quadrillion kilojoules. Coal accounted for 53 percent of energy consumption, oil accounted for 33 percent, natural gas 8 percent, and hydroelectricity 5 percent. Kerosene, a staple fuel for cooking and lighting of rural and lower-income homes, and diesel fuel for trucks were heavily subsidized. Over 56 percent of Indian homes and some 112,000 villages still lacked grid access to electric power.

India has a central ministry for renewable energy (RE). In 2008, the RE industry turnover was $500 million, with about $3 billion in investments. The government aimed to electrify eighteen thousand villages by 2012, with a 10 percent RE contribution. Microhydel plants of 2 to 25 megawatts exploit mountain streams and account for 2,500 megawatts, lighting many remote villages. By 2008, 3.94 million family-level biogas plants and over 2.15 million square meters of solar collectors had been installed, and 617,000 solar cookers were in use.

Some 3,365 villages and 830 hamlets had been electrified using renewable off-grid sources. Renewable power installation totaled over eleven gigawatts. India had become fourth in the world in installed wind power capacity, with 7.8 gigawatts. Large solar thermal plants operate in Rajasthan. Wind farms of 500 kilowatt turbines, limited by Indian roads, were being erected in Tamil Nadu, Gujarat, and Karnataka and along the Western Ghat mountain range. Indian Railways cultivates the jatropha plant along its right-of-way to generate biodiesel.

Indian Nuclear Energy

Since 1955, India has developed a series of nuclear power plants. In 2007, nuclear power supplied 15.8 billion kilowatt-hours (2.5 percent of the total) from 3.7 gigawatt-hours of capacity. In 2008, with approval from the International Atomic Energy Agency and the Nuclear Supplier Group cartel, India signed civilian nuclear cooperation agreements with the United States, France, and Russia to embark on 40 gigawatt-hours of civilian nuclear power by 2030. This would form the core of clean baseload generation for advanced industry and critical services. The ultimate aim was a three-stage cycle to use vast indigenous reserves of thorium, before uranium was depleted.

Clean Development Mechanism Projects and Greenhouse Gas Alleviation

The energy sector accounts for 61 percent of Indian greenhouse gas (GHG) emissions. As of 2008, India has 815 projects under the clean development mechanism, of which 536 deal with RE. Other projects focus on sequestering carbon dioxide (CO_2) by creating green spaces in urban areas and replanting forests. Pollution-control laws in major cities such as New Delhi have converted buses and taxis to compressed natural gas fuel. The

competition to build a small, cheap automobile has excited public interest, opening the market for hybrid and electric cars suited to regenerating power from the frequent braking needed on Indian roads. Mandatory rainwater collection for most new construction is making a huge impact on urban water supply and helping alleviate the flood-drought cycle and soil erosion. Tube wells must be supplemented with replenishment tubes.

India as a GHG Emitter

Home to more than 16 percent of the world's population, India accounts for only 4.9 percent of the world's GHG emissions. This places it in fifth place, behind the United States (22.2 percent), China and Taiwan (18.4 percent), the European Union (14.7 percent), and Russia (5.6 percent) and followed by Japan (4.6 percent) and Germany (3.1 percent). The average Indian's carbon footprint is 0.87 metric ton per year, compared to the average American's 18.1 metric tons, the average Australian's 25.5 metric tons, and the average Chinese's 2.5 metric tons. The Indian carbon footprint is growing rapidly and expected to reach the level of today's China by 2030. India is also expected to overtake China as the most populous nation at about the same time.

Low Carbon Footprint but High Pollution

Southern and central India are in the tropical climate zone, and even in the northern plains winter home heating is rarely needed. Residential air conditioning is generally impractical because of power shortages in summer. Houses built to resist summer heat and monsoon rains have efficient through-flow ventilation. Windows are kept open.

Most Indians live in small villages. The cities are densely populated, with few zoning restrictions. Many people can walk or ride bicycles or scooters to work. Commuters use public transportation heavily. Long-distance travel is mostly by rail or bus. The Indian staple diet is organically grown, mostly in small farms, and consists of grain, vegetables, fish, and meat products marketed and consumed locally, minimizing energy use in processing, packaging, and transportation. Few homes have large appliances. Much Indian clothing is made from cotton, suited for the hot, humid climate, rather than from synthetic petroleum products. For these reasons, the personal and discretionary parts of the Indian carbon footprint are minimal.

The primary power resource in India is coal, powering much of Indian industry with thermal plants. With heavy industry near cities, GHG emissions and urban air pollution are severe. Much of the population uses open, wood-burning stoves or other biomass burners for cooking. Composting and outdoor burning generate GHGs, fumes, and particulates. Fleets of two- and three-wheelers powered by two-stroke engines and ancient diesel trucks that predate vehicle emission laws generate noxious fumes in gridlocked traffic on narrow, monsoon-damaged roads.

Scenarios for GHG Emissions

Moderate economic growth is likely to double GHG emissions by 2030. Assuming that the population reaches 1.3 billion, the Indian CO_2 footprint will reach 2.4 metric tons. Scenarios assuming sustainable technologies and clean coal burning reduce the growth rate of emissions by roughly 50 percent, leveling it off at a much lower footprint.

Summary and Outlook

India has signed and ratified most of the United Nations treaties related to climate change, including the Kyoto Protocol, but understanding Indian progress requires a deeper perspective. This vast, densely populated nation has a common culture but a diverse population. The vibrant democracy holds regular, free elections and frequent, vociferous, and even violent protests. In every city, dozens of newspapers in English and Indian languages present diverse viewpoints, and hundreds if not thousands of political parties span the entire spectrum. The Internet has caught on rapidly. The transistor radio and newspaper are now supplemented by satellite television even in the villages, and Indians are connected to a worldwide diaspora through bicycle-delivered email, but also cell phones, text messaging, and voice-over-Internet.

In the 1980s, the Panchayati Raj law devolved many powers from the central government to elected village Panchayats (originally meaning an assembly of five or more wise people). Grassroots movements draw power partially from an ancient native respect for nature, partially from a fear of foreign enslavement, and partially from the class war ambitions of the communist parties. Skepticism about big corporations has evolved into critical examination of large-scale development and its effects on the environment. Intense and sophisticated opposition ostensibly based on concern for people displaced from fertile areas by dam reservoirs or large industrial plants has raised public awareness and interest in climate change and sustainable development issues all over India. This situation creates unique opportunities for game-changing developments.

Middle-class initiatives may revolutionize renewable power and sustainable development far beyond government targets, adapting advanced technology to Indian needs and preferences. Millions of Indian homes and businesses, forced by the abysmally unreliable power grid, already have grid-connected inverters, battery storage, voltage regulators, and auxiliary generators, and they are ready to incorporate renewable sources and perhaps a hydrogen economy. Highly efficient light emitting diodes are becoming popular for home lighting. Cell phones have rendered the landline telephone network obsolete, and beamed power may do the same to the power grid. Clean-burning liquefied petroleum gas delivered in cylinders has become popular in Indian urban homes, but rising fossil fuel prices may replace these with biomass natural gas. Family-level solar thermal and biomass system installations are reaching critical mass numbers for explosive growth. With these, India has the potential rapidly to become a leader in sustainable development.

Padma Komerath

Further Reading

Abdul Kalam, A. P. J., and Y. S. Rajan. *India, 2020: A Vision for the New Millennium.* New York: Penguin Books, 1998. Former president Dr. Abdul Kalam, an aerospace engineer, sets out his ambitious dream of India becoming a developed nation by 2020. A must-read for anyone who wants to understand Indian aspirations and plans.

Kamalapur, G. D., and R. Y. Udaykumar. "Rural electrification in India and feasibility of photovoltaic solar home systems." *International Journal of Electrical Power & Energy Systems* 33, no. 3 (2011): 594-599. Electrification in India faces challenges of many villages being far from the grid, in difficult terrain, and having low power demand. Solar power is too expensive for most villagers but can supply decentralize power.

Lala, R. M. *Beyond The Last Blue Mountain: A Life of J. R. D. Tata.* New York: Penguin Books, 1993. Through the biography of a leading industrialist, this book lays out the history of Indian technological development and heavy industry from the insider perspective of the largest industrial house in India.

Sharma, S., S. Bhattacharya, and A. Garg. "Greenhouse Gas Emissions from India: A Perspective." *Current Science* 90, no. 3 (February 10, 2006): 326-334. Deals with the methods and data used in estimating inventory of greenhouse gas emissions, and hence the results on growth rate of GHG emissions in India.

Shukla, P. R. "India's GFG Emission Scenarios: Aligning Development and Stabilization Paths." *Current Science* 90, no. 3 (February 10, 2006): 384 -396. Gives different scenarios of economic growth and incorporation of sustainable technologies. The values used are very conservative, with carbon footprints well beyond those reported by U.S. or Indian sources in 2008.

Solar or Coal? The Energy India Picks May Decide Earth's Fate. WIRED 2015 www.wired.com/2015/11/climate-change-in-india/ [6 July 2015]. India is on track to become the most populous nation on earth, its largest economy, and its largest greenhouse emitter. Like China, it relies on coal for much of its energy.

Srinivasan, M. R. *From Fission to Fusion: The Story of India's Atomic Energy Programme.* New York: Penguin Books, 2002. A view of the development of the Indian nuclear energy program, from an insider who started with the program. Gives a detailed view of how such projects get done in India.

Greenhouse gases; Kyoto mechanisms; Kyoto Protocol; Sustainable development.

Industrial ecology

Categories: Economics, industries, and products; environmentalism, conservation, and ecosystems

Definition

Industrial ecology is the study of the flows of materials and energy in industrial and consumer activities and their impact on the environment. It includes the effects of economic, political, and social factors bearing on those flows and their environmental consequences. The discipline is an adaptation of the concept of a biological ecosystem, which has been at the core of environmental studies for many decades, applying the same analysis to the human-created world as had been applied to the biological world. It is also an adaptation of the idea of materials flow, which has been applied to geologic and industrial evaluations, applying the concept to industrial production and to the consumption of industrial products.

The roots of industrial ecology go back to the "tragedy of the commons," defined by Garrett Hardin in 1968 in an article of the same name. Hardin pointed out that while humans are very sensitive to the costs of generating income from producing goods that they personally own, they are largely insensitive to the costs of production that derive from assets owned in common. That same self-centered approach has been adapted to industrial production, particularly in the framework of a market-based economy (though the unraveling of the Soviet system in the late 1980's and the 1990's arguably showed that it also applied in systems of collective ownership).

The result of this self-centered approach has been that environmental assets have not enjoyed the kind of value recognition that has gone into private ownership. The reevaluation of those assets' value has been driven by those who maintain that failure to maintain and protect the environment has caused environmental assets to lose a great deal of value. This loss of value has been documented by individuals such as Rachel Carson, whose paradigm-changing work on the effects of DDT pointed out that environmental damage can harm humans

Rachel Carson, whose groundbreaking work on the effects of pesticides on humans transformed understandings of industrial ecology. (Library of Congress)

as well. Industrialists responding to such studies applied the methodology of systems analysis to industrial and ecological processes, producing "materials flow assessments." The techniques involved came to be called "life cycle assessment," or LCA.

In the 1990's, scientists began to observe what looked like systemic changes to the atmosphere and began looking for causes and defining the consequences through the absorption of carbon into the atmosphere. As a result, people began to realize that the human-created world might have a very important part to play in climate. The conclusions of the Intergovernmental Panel on Climate Change, that atmospheric changes were to a substantial extent caused by human activities, put a focus on evaluating those activities, especially those carried out by industry. Industrial ecology offered a method of analysis that, by tracing the flow of materials through the industrial system, could provide detailed information on the anthropogenic influences on global climate.

Because of the multiplicity of industries that convert various materials into saleable products, performing LCAs for the manufactured world involves several kinds of disciplinary expertise. Raw materials, very often taken from the ground, are subjected to a number of alterations to convert them into a form that humans need, so geologists play a significant part in the analysis, as they have specialized knowledge that reveals how the raw materials can be altered. Very often, chemical processes are employed in this alteration, and for this to be fully understood requires sophisticated knowledge of those processes. Moreover, in most cases, industrial production involves the application of forces generated by energy, for which an understanding of the physics of energy use is essential. Finally, most manufacturing processes create waste by-products. Indeed, life in the industrialized world in general generates enormous amounts of waste, and its disposition, without harming the environment or generating environmental changes that affect the climate, is a major political and social concern requiring knowledge of ecology and biogeochemistry to achieve.

Significance for Climate Change

Since anthropogenic changes to the environment have been identified as major contributors to the changing climate, ways need to be found to minimize this impact. Full knowledge of the effects of the various parts of industrial production is essential to changing those effects. Industrial ecology, a specialized branch of environmental studies, has its own journal, the *Journal of Industrial Ecology*, edited at the Yale School of Environmental Studies. It draws on contributions from academics, members of the business community, and government officials. Through this journal and in other academic, private, and public forums, ideas and processes that can serve to modify industrial production and thereby reduce its negative impact on the climate are spread.

Nancy M. Gordon

Further Reading

Ayres, Robert U., and Leslie W. Ayres, eds. *A Handbook of Industrial Ecology.* Cheltenham, Gloucestershire, England: Edward Elgar, 2002. A collection of chapters by experts in many of the fields that contribute to industrial ecology, covering the broad range of disciplines that industrial ecology employs, including economics, national policy, and applications.

Graedel, Thomas, and B. R. Allenby. *Industrial Ecology.* 2d ed. Upper Saddle River, N.J.: Pearson Education, 2003. The standard text in industrial ecology.

Lifset, Reid. "Industrial Ecology and Forestry." *Journal of Forestry* 98, no. 10 (October, 2000). Special inset provides a very concise explanation by the editor of the *Journal of Industrial Ecology.*

Shi, Han, Jinping Tian, and Lujun Chen. "China's Quest for Eco–industrial Parks, Part I." *Journal of Industrial Ecology* 16, no. 1 (2012): p. 8-10. China is creating a nationwide system of industrial parks that combine ecological principles with traditional industry.

See also: Carson, Rachel; Civilization and the environment; Conservation and preservation; Ecological impact of global climate change; Ecosystems; Environmental economics; Environmental movement; Industrial emission controls; Industrial greenhouse emissions; Industrial Revolution and global warming; Sustainable development.

Industrial emission controls

Category: Pollution and waste

Governmental regulations and technological innovations have helped reduce the amount of environmental pollutants and GHGs emitted into the environment as by-products of industrial manufacturing.

Key concepts

capture: to secure and contain harmful particles and chemicals

combustion: the burning of fuels to produce energy

emissions: gases and particulates released by industrial and other processes

filtration: separation or removal of contaminants from emissions

industrial: related to large-scale manufacturing or energy production

neutralization: elimination of harmful characteristics or properties to render contaminants benign

scrubbers: liquids that remove pollutants from emissions

sequestration: isolation of hazardous emissions to prevent them from polluting the atmosphere

Background

Since the Industrial Revolution began in the mid-nineteenth century, factories using combustion to power machinery manufacturing products have released chemicals detrimental to the environment in their emissions. By the twentieth century, power plants generating energy through burning fuels such as coal and natural gas added to this pollution, and industrial greenhouse gases (GHGs) accelerated climate change. Motivated by economic, legislative, and environmental incentives, many industry operators sought ways to control industrial emissions. Engineers and scientists innovated and devised technology or methods to minimize, remove, convert, or store chemicals emitted during industrial combustion activities.

Industrial Emissions and Control Strategies

Turbines, boilers, generators, engines, and furnaces powered by burning fuels release GHGs produced during combustion. Emissions frequently associated with industries include nitrogen oxides, sulfur dioxide, carbon dioxide (CO_2), and methane. The U.S. Environmental Protection Agency identified petrochemical, ammonia, aluminum, steel, iron, and cement manufacturers as emitters of large amounts of GHGs.

Political and social demands to reduce emissions resulted in many industry leaders evaluating how to alter production methods and technology in order to satisfy laws limiting emissions while not experiencing profit losses. Intergovernmental Panel on Climate Change (IPCC) reports discussed how to control industrial emissions, recommending industry managers seek control strategies and technology appropriate for manufacturing processes and fuels their factories utilized.

Carbon Capture and Storage

Industries have successfully controlled emissions with carbon-capture-and-storage (CCS) methods by securing carbons released during combustion and then compressing and sequestering them in remote areas, usually underground, distant from the Earth's atmosphere. CCS is especially effective for minimizing CO_2 released in emissions from petroleum, iron, cement, and ammonia industrial processes and refineries.

Norwegian industries were early users of CCS because Norway's government began taxing carbon emissions in 1991. Norwegian engineers and scientists created CCS technology and procedures to store CO_2 in sandstone approximately 1,000 meters beneath the North Sea. Starting in 1996, the Norwegian industry StatoilHydro sequestered almost one million metric tons of carbon emissions annually.

Experts emphasized CCS technology is essential to achieve projected emission reductions by 2100. Researchers collaborated on CCS projects. Scientists experimented using chemicals to enhance CCS effectiveness and burning biomass to power equipment used to capture CO_2. In 2007, researchers demonstrated how algae capture carbon.

Scrubbers

Scrubber technology cleans exhaust and emissions from industrial sources by removing particulates from acidic gases. A typical scrubbing procedure results in chemicals in emissions being altered, sometimes undergoing reactions to transform into other compounds, or lessening their strength.

Scrubbing equipment designs incorporate a tank and recirculation system cycling liquid scrubbers into the presence of emissions. The basic particulate scrubbing process involves the swift movement, from 45 to 120 meters per second, of emissions inside a tank constructed from fiberglass or metals that will not corrode. In this vessel, a liquid, often water, serving as the scrubber impacts the fast moving emissions and transforms into small drops that absorb particles in emissions.

Engineers designed scrubbers to meet specific industrial needs. Scientists identified chemical solutions, including chlorine dioxide, hydrogen peroxide, sodium chlorate, and sulfuric acid, as effective as scrubbers to minimize sulfur oxides, nitrogen oxides, and heavy metals, such as mercury, in flue gas emissions.

Filtration

Industrial emissions can be controlled by filtering contaminants produced during combustion. Filtration technology consists of an insulated metal chamber, usually made from stainless steel or an alloy, and mesh filters, mostly constructed with copper, silicon, or aluminum. Tanks store water before and after filtration. Sprayers and pipes transport water during filtration.

Water and temperatures control industrial emissions during filtration. Inside the chamber, sprayers coat water that has been cooled to 2° Celsius in an adjacent refrigerator tank on one or more mesh filters near the top of the chamber prior to hot emissions rising beneath the filter in the chamber. The dripping water hits the emissions, cooling them, and capturing particulates or liquefying such gases as sulfur dioxide and CO_2 when they reach the filter. The water containing particulates and gases is expelled into a dump tank.

Neutralization and Absorption

Some industrial emissions are managed by neutralizing them. Researchers innovated methods to extract toxic chemicals prior to combustion. Engineers developed technology to impede nitrogen oxidization during combustion. In selective catalytic reduction (SCR), the reaction of ammonia with flue gases, aided by use of a catalyst such as tungsten oxide, breaks nitrogen oxides into nitrogen molecules and water. SCR effectively reduces emissions by 80 to 90 percent but is costly due to catalyst expenses.

Fluidized bed combustion (FBC) keeps nitrogen oxides from being produced because chamber temperatures are lowered to 750° to 950° Celsius by water tubes in the bed absorbing heat. FBC control methods used when burning coal achieve 80 to 90 percent reduction of sulfur oxides. Various flue gas desulfurization (FGD) methods utilize chemicals or minerals such as limestone that absorb emission contaminants, particularly sulfur dioxide.

Context

Images of smoke rising from industrial parks often are used to symbolize global warming. Endeavors to control industrial emissions exemplify international focus on enhancing and promoting the use of clean technology, particularly due to the expansion of industry because of economic incentives to produce more goods and energy to support expanding populations. Legislation such as the U.S. Clean Air Acts (1963-1990) outlined requirements for industries to control emissions. The Kyoto Protocol addressed industrial emissions control and suggested reductions. As global warming worsened into the twenty-first century, governments worldwide, such as the European Union, revised limits previously set for GHGs produced by industries.

Many industrial leaders recognized their environmental responsibilities and willingly limited emissions from factories and acquired updated equipment, trained operators, and enforced stricter procedures to minimize the impact of industrial emissions on climate change. Other industries, however, continued to release excessive GHGs because of apathy, ignorance, or inability to afford or attain access to emissions control technologies.

Elizabeth D. Schafer

Further Reading

Intergovernmental Panel on Climate Change. *Climate Change, 2007—Mitigation of Climate Change: Contribution of Working Group III to the Fourth Assessment Report of the Intergovernmental Panel on Climate Change*. Edited by Beth Metz et al. New York: Cambridge University Press, 2007. Chapter 7 of this IPCC report focuses on industry. Addresses industrial GHGs and the technology and techniques industries use to control emissions. Bibliography.

Klimont, Z., Steven J. Smith, and Janusz Cofala. "The last decade of global anthropogenic sulfur dioxide: 2000–2011 emissions." *Environmental Research Letters* 8, no. 1 (2013): 014003. North American and European emissions have declined, but Asian emissions and those of international shipping have increased. China's emissions have stabilized and shippin emissions are expected to decline as low-sulfur fuel is employed.

Maroto-Valer, M. M. Mercedes, John M. Andrésen, and Yinzhi Zhang. "Toward a Green Chemistry and Engineering Solution for the U.S. Energy Industry: Reducing Emissions and Converting Waste Streams into Value-Added Products." In *Advancing Sustainability Through Green Chemistry and Engineering*, edited by Rebecca L. Lankey and Paul T. Anastas. Washington, D.C.: American Chemical Society, 2002. Recommends using low-nitrogen-oxides burners to decrease emissions, in addition to disposing of any unburned carbonaceous ash.

Wald, Matthew L. "In a Test of Capturing Carbon Dioxide, Perhaps a Way to Temper Global Warming." *The New York Times*, March 15, 2007, p. C3. Describes American Electric Power's strategies for sequestering CO_2 using chilled ammonia to absorb it, a technique using less energy than do other carbon-capture methods.

Wilson, Elizabeth J., and David Gerard, eds. *Carbon Capture and Sequestration: Integrating Technology, Monitoring, and Regulation*. New York: John Wiley and Sons, 2007. Energy policy experts present technical descriptions of CCS processes, engineering concerns (including seepage), and legalities such as storage-site ownership.

See also: Carbon dioxide; Emission scenario; Emissions standards; Greenhouse gases and global warming; Motor vehicles; U.S. legislation.

Industrial greenhouse emissions

Categories: Pollution and waste; economics, industries, and products

Since the Industrial Revolution, increased amounts of anthropogenic industrial GHG emissions from factories and power plants have combined with emissions from other sources to raise the global temperature and change the climate. Modern technology has accelerated the global warming process with the production of new industrial GHGs.

Key concepts

albedo: the fraction of incident light reflected from a body such as Earth

anthropogenic: derived from human sources or activities

biodiversity: the variety of organisms at a particular geographic location

biosecurity: measures required or taken to protect organisms from risk or danger

fossil fuels: fuels formed by the chemical alteration of plant and animal matter under geologic pressure over long periods, including coal, oil, and natural gas

greenhouse gases (GHGs): gaseous constituents of the atmosphere, both natural and anthropogenic, that contribute to the greenhouse effect

industrial: related to large-scale manufacturing or energy production

Intergovernmental Panel on Climate Change (IPCC): a scientific intergovernmental body set up by the World Meteorological Organization (WMO) and by the United Nations Environment Programme (UNEP)

ozone layer: a region of the upper atmosphere with high concentration of ozone (a gaseous form of oxygen) that protects the Earth by absorbing solar ultraviolet radiation

Background

Industrial greenhouse gases (GHGs) are anthropogenic gaseous constituents of the atmosphere produced by human industrial processes that absorb and emit radiation at specific wavelengths within the spectrum of thermal infrared radiation, emitted by the Earth's surface, the atmosphere, and by clouds. This phenomenon causes the greenhouse effect, which contributes to global warming. The concentrations of carbon dioxide (CO_2) among such emissions are now more than one-third higher than they were before the Industrial Revolution. The Earth is artificially made even warmer by hydrofluorocarbons (HFCs) from industrial activities that trap extra heat from the sun, causing changes in weather patterns around the globe.

The Sources and Levels of Industrial Greenhouse Emissions

CO_2, the most abundant GHG, is emitted from products and by-products of manufacturing sites,

and from burning fossil fuels at large industrial facilities. Natural gas, coal, and oil are the three types of polluting power plants, and they produce 40 percent of CO_2 in the United States, coal being the biggest contributor. Brown coal produces more CO_2 than black coal. In the United States, industrial CO_2 emissions, resulting both directly from the combustion of fossil fuels and indirectly from the generation of electricity that is consumed by industry, accounted for 28 percent of CO_2 from fossil fuel combustion in 2006.

Industrial processes are among the six major sources of pollution in the United States and are responsible for over 60 percent of Canada's total emissions. In the United States, energy-related activities account for over three-quarters of anthropogenic GHG emissions, more than half of which comes from power plants and industrial processes. When emissions from electricity were distributed among various sectors, industry accounted for the largest share of U.S. GHG emissions (29 percent) in 2006. Overall, emission sources from industrial processes accounted for 4.5 percent of U.S. GHG emissions in 2006.

The six major industrial GHGs are CO_2, methane, nitrous oxide, hydrofluorocarbons, perfluorocarbons (PFCs), and sulfur hexafluoride (SF_6). The major industrial greenhouse gases are released from manufacturing and production processes related to iron ore pelletizing, lime, iron and steel, titanium, pulp and paper, aluminum and alumina, cement, petroleum refining, chemicals and fertilizers, electricity (produced with oil, coal, and gas), natural gas pipelines, potash, base metal smelters, silicon carbide, nitric acid, ammonia, urea, limestone and dolomite use (such as flux stone, flue gas desulfurization, and glass manufacturing), soda ash, ferroalloy, zinc, phosphoric acid, titanium dioxide production, lead, coal mining, wastewater treatment, stationary combustion, composting, manure management, semiconductors, magnesium, and adipic acid.

At present, electric power generation is the largest contributor to industrial GHG emissions. Industrial methane comes from petroleum systems and coal mining. Unintentional fugitive industrial methane emissions come from equipment leaks, natural gas distribution, and storage facilities.

Nitrous oxide is generated during the production of nitric acid and adipic acid. The manufacture of liquid crystal display (LCD) screens releases nitrogen trifluoride. Reports in 2008 indicated a rise in airborne levels of nitrogen trifluoride from flat-panel screen technology (manufacture of LCD screens).

Industrial activities also produce several classes of greenhouse halogenated substances that contain fluorine, chlorine, or bromine, such as potent greenhouse HFCs, PFCs, and SF_6 gases. Sulfur hexafluoride is the most potent GHG the IPCC has evaluated. Chlorofluorocarbons, or CFCs, are wholly human-made, new to the atmosphere, and widely used in aerosols, foam manufacture, air conditioning, and refrigeration. Other industrial sources include HCFC-22 production, semiconductor manufacturing, aluminum production, and magnesium production and processing.

Capacity of Industrial Emissions to Affect Climate Change

Industrial emissions form a significant percentage of anthropogenic sources of GHGs that are blanketing the Earth and increasing the average global air temperature near the Earth's surface. Industrial processes can chemically transform raw materials, which often release waste gases such as CO_2, CH_4, and N_2O, gases that can influence the atmospheric lifetimes of other gases. These emissions can affect atmospheric processes that alter the radiative balance of the Earth, such as cloud formation or albedo. The IPCC reported in 2007 that methane is twenty times as effective as CO_2 at trapping heat in the atmosphere, and its concentration in the atmosphere has increased by 148 percent over the last 250 years.

Nitrous oxide is over three hundred times more powerful than CO_2. Nitrogen trifluoride is thousands of times stronger at trapping atmospheric heat than CO_2. Human industrial activity has produced very potent GHGs such as hydrofluorocarbons that trap more heat from the Sun than CO_2 and other GHGs, making the Earth artificially warmer. As a result, extreme weather events will become more frequent, more widespread, and more intense in the times ahead. In addition to having high global warming potentials, SF_6 and PFCs have

Contribution of Anthropogenic GHG Emissions to Earth's Greenhouse Effect	
GHG	*Recent Enhanced Greenhouse Effect*
Carbon dioxide	69.6
Methane	22.9
Nitrous oxide	7.1
HGWP gases	0.4

Data from U.S. Environmental Protection Agency.

extremely long atmospheric lifetimes, resulting in their irreversible accumulation in the atmosphere once emitted.

How Industrial Greenhouse Emissions Affect Climate and Global Events

Industrial greenhouse emissions play a major role in the gaseous blanketing effects that create higher temperatures around the Earth. Chlorofluorocarbons emitted from factories destroy the protective ozone layer, causing more intense solar radiation to bombard the Earth. Increasing global temperature is expected to cause sea levels to rise, increase the intensity of extreme weather events, and create significant changes to the amount and pattern of precipitation, likely leading to an expanse of tropical areas, loss of biodiversity, and increased pace of desertification.

The effects of climate change have serious implications for the economy and environment of nations. Winters are becoming milder, and summers are becoming hotter. Snowpacks are shrinking, and unseasonably warm temperatures are leading to rapid spring melts, depleting the supply of summer water for agriculture and stream flows for wildlife. Storms and forest fires are becoming more severe while the risk of coastal flooding is increasing. These climatic changes will potentially affect native ecosystems, industries, infrastructure, health, biosecurity, and the economy.

Industrial smog emissions provoke responses of the atmosphere that cause health hazards, and affect agricultural as well as drinking water supplies. Other expected effects include modifications of trade routes, glacier retreat and disappearance, mass species extinctions, increases in the ranges of disease vectors, changes in quantity and quality of agricultural yields, and impacts on land use due to significant alterations in temperature and rainfall patterns. The health of livestock and other animals will be affected, as pests, pathogens, and diseases become modified or multiply. Land erosion and impoverishment will abound, with increase in soil infertility and proliferation of weeds. All these factors will have poverty impacts on people and jeopardize the global economy.

Direct and Indirect Industrial Emission Impacts on the Atmosphere and Environment

Direct radiative effects of industrial GHGs occur when the gases themselves absorb radiation and warm the Earth. Strong industrial greenhouse emissions such as CFCs directly contribute significantly to the depletion of the protective ozone layer around the Earth.

The IPCC reported in 2007 that from the preindustrial era (ending about 1750) to 2005, concentrations of CO_2, CH_4, and N_2O have increased globally by 36, 148, and 18 percent, respectively. Their emissions from industrial sources directly pollute groundwater after rainfall has washed the pollutants into the soil. Toxins from the pollutants directly affect the health, growth, and reproduction of plants and animals.

Indirect atmospheric modification occurs when chemical transformations of the substances emitted from industrial processes produce other GHGs. Industries contain by-product or fugitive emissions of GHGs from industrial processes not directly related to energy activities. Industrial gases can also indirectly influence the atmospheric lifetimes of other gases. There are several gases that do not have a direct global warming effect but indirectly affect terrestrial and solar radiation absorption by causing some GHGs to persist longer in the atmosphere, and influencing the formation or destruction of other GHGs, including tropospheric and stratospheric ozone depletion. These gases include carbon monoxide (CO), oxides of nitrogen (NOx), and non-CH_4 volatile organic compounds (NMVOCs).

Context

Today, Earth is hotter than it has been in two thousand years. The global temperature will rise higher, to destructive levels, than at any time in the past two million years if industrial emissions are not reduced. The IPCC predicts a further rise of 1.1° to 6.4 ° Celsius in the average global surface temperature. According to the *Science Daily* of September 17, 2008, the Earth will warm about 2.4° Celsius above preindustrial levels even under extremely conservative GHG emission conditions and under the assumption that efforts to clean up pollution continue to be successful. Although most studies focus on the period up to 2100, warming and sea-level rise are expected to continue for more than a thousand years even if GHG levels are stabilized. The delay in reaching equilibrium is due to the large heat capacity of the oceans.

As a result of the 1987 Montreal Protocol to curb CFCs, the IPCC reported in 2007 that the production of ozone-depleting substances (ODS) is being phased out, as CFCs and HCFCs are replaced by ODS substitutes such as HFCs and PFCs. Unfortunately, the substitutes, while being relatively harmless to the ozone layer, are equally potent GHGs, and at present their phase-out dates are not due for another 20 to 30 years.

Technological advancement and innovation are critical to achieving significant, long-term reductions in industrial GHG emissions. Investments in biofuels and alternative energy sources would promote clean electricity. Cement producers could use waste material from other industries in place of emission-intensive clinker. Some industries have installed new technology to cut emissions, and use filters that improve the quality of the air released into the atmosphere.

Samuel V. A. Kisseadoo

Further Reading

Clarke, A. G. *Industrial Air Pollution Monitoring: Gaseous and Particulate Emissions.* Environment Management Series 8. New York: Springer, 1997. Provides details of the types, effects, and emission rates of industrial pollutants.

Dolan, Stacey L., and Garvin A. Heath. "Life cycle greenhouse gas emissions of utility–scale wind power." *Journal of Industrial Ecology* 16, no. s1 (2012): S136-S154. Wind power still entails carbon dioxide emissions in manufacturing and construction. Estimates average about 10 grams of carbon dioxide equivalent per 1000 kilowatt hours.

Friedrich, R., and S. Reis, eds. *Emissions of Air Pollutants: Measurements, Calculations, and Uncertainties.* New York: Springer, 2004. Provides information intended to guide efforts to reduce and mitigate air pollution and to design experiments related to specific pollution sources.

Hardy, John T. *Climate Change: Causes, Effects, and Solutions.* New York: John Wiley, 2003. Emphasizes and explains potential global effects of anthropogenic climate change and provides scientific findings, case studies, and discussions of the broader issues of the ecological, economic, and human effects of such change.

Quigley, John T., Howard E. Hesketh, and Frank L. Cross, Jr. *Emission Control from Industrial Boilers.* Lancaster, Pa.: Technomic, 1995. Provides indepth treatment of industrial production related to the Clean Air Act, boiler systems, combustion fundamentals, scrubbers, filters, disposal of residuals, and emission monitoring.

Reay, Dave. *Climate Change Begins at Home: Life on the Two-Way Street of Global Warming.* New York: Palgrave Macmillan, 2006. Analyzes case studies, emissions trends, lifestyle comparisons, and calculations describing climate change as a great twenty-first century human threat.

See also: Aerosols; Carbon dioxide; Chlorofluorocarbons and related compounds; Emissions standards; Environmental law; Greenhouse effect; Greenhouse gases and global warming; Hydrofluorocarbons; Methane; Sulfur hexafluoride.

Industrial Revolution and global warming

Category: Economics, industries, and products

The Industrial Revolution involved a shift from human and animal power to reliance on coal for manufacturing and transportation, and it was marked by an escalating rate of population increase. The early industrial period saw global cooling rather than warming.

Key Concepts

carrying capacity: the number of people the environment can support

climate feedback loops: self-reinforcing and self-negating processes that accelerate or retard climatic trends

fossil fuels: combustible products of ancient photosynthesis, such as coal

planetary albedo: the reflectiveness of the Earth's surface, including cloud cover, to sunlight

Background

The Industrial Revolution began around the middle of the eighteenth century in England. Increasingly sophisticated power-driven machines augmented and replaced human and animal labor, greatly increasing individual worker productivity. This led to an overall increase in per capita resource consumption at a time when population was also beginning to increase at an exponential rate.

Some economic historians speak of three or four separate Industrial Revolutions. The first, roughly 1750–1850, centered on mechanization of the textile and metallurgical industries. Mechanization of transportation in the form of railroads and steamships dominated the second phase. The United States overtook Britain as the world's leading consumer of fossil fuels, and Japan became the first non-Western nation to embrace industrialization. The post–World War II era of globalization and rapid growth in information technology has been called a third Industrial Revolution. Finally, some economists call for a fourth Industrial Revolution that would drastically reduce depletion of nonrenewable resources.

Of the revolutions that convulsed Europe and North America in the last quarter of the eighteenth century, the Industrial Revolution arguably had the most profound effect on the mass of people who experienced it. Most of its environmental effects, however, were either local or indirect. Compared with the first half of the eighteenth century and the latter half of the nineteenth, temperatures were cooler. Industrial carbon and sulfur emissions, smoke, and deforestation probably contributed, but their importance relative to other factors is uncertain.

The Process of Industrialization

The roots of Europe's industrialization must be sought in population fluxes and changing agricultural practices of the preceding centuries. Plague caused Europe's population to crash in the mid-fourteenth century. By 1700, the population again reached carrying capacity, supporting urban growth and creating pressure to improve agricultural productivity. Rapid rural population growth also allowed colonial expansion, which in turn provided a source of raw materials for industry.

By 1719, when Isaac Watt patented the first steam engine, London and other English cities mainly used coal for domestic heating. Pithead steam engines used to pump water from coal mines greatly increased production capabilities. Not long afterward, a series of innovations in the textile industry set the stage for massive movement from cottage-based industries to factories relying on external power sources. Other key inventions included improvements in iron founding, mechanization of tool machining, gas lighting, efficient papermaking and printing, and Portland cement, as well as the birth of the chemical industry.

Energy Consumption

Prior to about 1790, the mechanized textile industry in Britain relied upon water power. After 1790, coal-fired steam engines powered most factories. The use of coal to produce gas for illumination and cooking began in the first decade of the nineteenth century. Steamships started to replace sail for river and coastal traffic by 1820, while the first passenger railroad opened in 1828.

In the United States, consumption of fossil fuels was negligible before the Civil War. The textile industry used water power, and heating and transportation relied on wood from seemingly limitless forests.

The carbon dioxide (CO_2) content of the Earth's atmosphere remained nearly constant from 1700 to 1850, which correlates with a lack of observable greenhouse effect in temperature measurements from this period. Explanations for this stability in the face of greater fossil fuel use include sequestration in oceanic carbonates and an increased level of photosynthesis. In any event, although the rate of increase in fossil fuel consumption was impressive, the absolute numbers are low. The 45 million metric tons of coal consumed in 1850 are dwarfed by nearly 5.4 billion metric tons of coal consumed worldwide in 2006.

Other Environmental Effects

Historically, industrialization stimulated population growth. In early eighteenth-century Britain, improved food production and better control of epidemic disease reduced infant mortality. Employment in cities and the opportunity to emigrate to colonies meant that these children married early and produced large families. While the population of the world as a whole roughly doubled between 1750 and 1860, that of Great Britain went from 7,500,000 to 23,130,000, and that of the United States went from 1,500,000 to 31,400,000. While the effects of immigration dominate US statistics, the bulk of it was from the British Isles. Industrialization played a significant part in creating the expanding resource base making such growth possible. Although per capita resource consumption among the working poor did not begin to rise significantly until after 1850, the rise in numbers meant it took more energy and raw materials to support the population.

Contemporary accounts of early industrial cities paint a vivid picture of belching smokestacks and sulfurous fumes. Pollution controls were nonexistent. This large volume of pollutants tended to offset any greenhouse effect due to CO_2. Pouring quantities of soot and sulfur dioxide into the atmosphere produces atmospheric cooling. Although sulfur dioxide (SO_2) is a greenhouse gas (GHG), it rapidly combines with water to form sulfuric acid, which reflects sunlight. Soot particles also reflect solar radiation. Both can serve as nuclei for cloud formation, increasing planetary albedo and producing additional cooling. Thus, a single pulse of sulfur and particulate matter, injected into the atmosphere, has the potential to set up a climatic feedback loop.

Particulate matter remains in the atmosphere less than a year, and sulfuric acid no more than three or four years, whereas CO_2 builds up over decades. Consequently, the dirtier and more polluting an industry, the less its energy consumption will contribute to global warming in the short term. This helps explain why there is no global warming effect due to industrialization in the early nineteenth century.

The period 1750–1850 also saw at least two massive volcanic eruptions that contributed to global cooling: Laki in Iceland in 1783–1784 and Mount Tambora in the East Indies, which precipitated 1816's "year without a summer." Neither eruption set up a feedback loop which would account for the generally depressed temperature levels recorded in Europe and confirmed by Greenland ice cores and North American tree ring data for the entire period 1780–1850.

Context

Comparison of industrialization rates, temperature profiles, and CO2 levels in the period 1750–1850 with more recent trends suggests an insight into why global temperatures have not responded in a linear fashion to increases in atmospheric CO_2 levels, but instead show a marked acceleration in the 1980s, when acid rain became an environmental issue of grave concern, and western countries began aggressively curbing sulfur emissions from coal-fired electrical plants. Once the natural SO_2 pulse from Mount Pinatubo cleared the atmosphere, the Earth experienced the full effect of GHGs unmitigated by pollution. The importance of SO_2 pollution in counteracting greenhouse warming has also been recognized by British scientists studying drought cycles in the Amazon basin.

Martha A. Sherwood

Further Reading

Alverson, Keith D., Raymond S. Bradley, and Thomas Pedersen, eds. Paleoclimate, Global Change, and the Future. Berlin: Springer Verlag, 2003.

Griffin, Emma. *Liberty's Dawn: A people's history of the Industrial Revolution.* Yale University Press, 2013. While not minimizing the problems of the Industrial Revolution, the autobiographies cited by Griffin paint a much more positive picture than often believed. "Yet even with a government who did nothing there is an uncomfortable truth we should confront: industrialisation had a remarkable power to put food on the table."

Singer, S. Fred, and Dennis T. Avery. Unstoppable Global Warming: Every Fifteen Hundred Years. Rev. ed. Blue Ridge Summit, Pa.: Rowman & Littlefield, 2008.

Stearns, Peter N. The Industrial Revolution in World History. Boulder, Colo.: Westview Press, 2007.

Information Council on the Environment

Category: Organizations and agencies
Date: Established 1991

Mission

The Information Council on the Environment (ICE) was a skeptic organization created to promote the idea that there is no scientific consensus on global warming. ICE was a public information group funded by the Edison Electrical Institute, the National Coal Association, and the Western Fuels Association. Its mission, supported by a $500,000 advertising budget, was to "reposition global warming as theory rather than fact," according to internal documents of Cambridge Reports, the Council's polling firm. The group employed a scientific advisory panel comprising geography professor Robert Balling, research physicist Sherwood B. Idso, and climatologist Patrick J. Michaels, as well as environmental scientist S. Fred Singer, to promote skepticism in television and newspaper interviews, opinion columns, and advertisements. The polling firm advised the council to emphasize that "some members of the media scare the public about global warming to increase their audience and their influence."

Significance for Climate Change

In addition to the messages carried by its spokespersons, ICE conducted polls about global climate change and greenhouse gas emissions, purporting to gauge public opinion about these issues. In fact, according to strategy documents, pollsters targeted "older, less-educated men from larger households who are not typically information seekers" and "young, low-income women" in areas whose electricity was generated by coal, in an attempt to skew the poll results in favor of less regulation of coal.

ICE also purchased magazine and newspaper ads in key sections of the country, including one that ran in Minnesota reading, "If the earth is getting warmer, why is Minneapolis getting colder?" Another ad showed a ship sailing off the edge of the Earth, with an open-mouthed dragon waiting to catch it. The caption read, "Some say the earth is warming. Some also said the earth was flat." The Arizona Public Service Company declined to join ICE, declaring that the ads over-simplified a complex issue.

ICE was disbanded after only a few months, after a packet of internal memos describing the organization's public relations strategies was leaked to the *Energy Daily,* a trade publication, and follow-up stories appeared in the *Arizona Daily Sun,* the *National Journal,* and *The New York Times.* After the press stories, Balling and Michaels broke with the council. In 1999, a former board member of ICE expressed regret that the campaign had not lasted longer, writing that ICE had resulted in a "dramatic turnaround in how people viewed the issue of global warming."

Cynthia A. Bily

See also: Media; Pseudoscience and junk science; Skeptics.

Institute for Trade, Standards, and Sustainable Development

Categories: Organizations and agencies; economics, industries, and products
Date: Established 2001
Web address: http://www.itssd.org

Mission

The Institute for Trade, Standards, and Sustainable Development (ITSSD) advocates sustainable development without absolute protection of natural resources and discounts anti-global warming measures as not scientifically or economically justified. Headquartered in Princeton, New Jersey, the ITSSD is a nonprofit organization dedicated to educating government, industry, and the public about science, technology and innovation policy, private property rights, and international trade. It is operated by its chief executive officer-president, vice president, and secretary and has a seventeen-member advisory board. The officers write white papers and articles, serve on a variety of discussion panels, administer an internship program for university students, and publish journals on economic freedom, intellectual property rights, trade barriers, the United Nations Convention on the Law of the Sea, "pathological communalism," and women's property rights.

The ITSSD's principal policy calls for positive sustainable development, which, it argues, has a general capacity to create well-being for present and future generations. This sustainable development is positive in that it eschews regulation "devoid of scientific and economic benchmarks" and "disguised trade barriers premised solely on cultural preferences" in favor of strongly protected property rights, free market (neoliberal) economics, decentralization, economic growth, and local, regional, or national (rather than supranational) institutions encouraging individual initiative. The ITSSD contends that regulations and standards must be developed "based on empirical science and economic cost-benefit analysis" under public scrutiny and free of the dominating influence of scientific fashion or sociopolitical ideology.

Significance for Climate Change

Informing its stance on climate change, ITSSD relies on the arguments of global warming critics, including (according to ITSSD) "established scientists." Most prominently cited is British businessman, politician, and inventor Christopher Monckton, who argues that global warming derives from natural cycles and is misrepresented by the scientific community. In the white paper "Europe's

Warnings on Climate Change Belie More Nuanced Concerns" (2007), ITSSD president Lawrence A. Kogan accuses leaders of sidestepping what he portrays as the ongoing scientific debate. This debate concerns the extent to which certain human activities can be shown to cause measurable global warming or merely to correlate with a barely observable rise in global temperatures that may or may not prove to be cyclical in nature. The failure of European leaders to discuss this issue in the ITSSD's eyes suggests a nuanced effort to base intergovernmental regulatory policy on popularly fanned fears about largely hypothetical, unpredictable or unknowable future natural and anthropogenic hazards that have not yet been shown to pose direct ascertainable risks to human health or the environment.

Roger Smith

See also: American Enterprise Institute; Cato Institute; Competitive Enterprise Institute; Cooler Heads Coalition; Skeptics.

Inter-Tropical Convergence Zone

Category: Meteorology and atmospheric sciences

Definition

The Inter-Tropical Convergence Zone (ITCZ) is a low-pressure belt located near and parallel to the Earth's equator, within a few degrees of latitude. Because it is a low-pressure belt and located in the vicinity of the equator, near-surface winds on both sides of this belt tend to converge into the belt. A band of clouds and accompanying precipitation is associated with the ITCZ, which can therefore be easily identified from satellite pictures.

The wind on the north side of the ITCZ comes from the northeast (blowing toward the southwest), and therefore, by meteorological convention, is called the "northeasterly." The obverse wind on the south side of the ITCZ is the "southeasterly." The northeasterly and southeasterly are also historically called "trade winds," because explorers, navigators, and merchant fleets once used these winds to sail to many tropical regions for trade and adventure. Thus, the ITCZ is the belt where the trade winds meet.

The ITCZ migrates slightly, oscillating across the equator with the change of seasons. It is primarily located north of the equator, although during winter in the Northern Hemisphere it sometimes moves to the south of the equator. The ITCZ can often be well organized, in the sense that the low pressure, clouds, and precipitation associated with the ITCZ form a solid band structure and make almost an entire circle around the globe. However, this well-organized structure can break down from time to time. When it breaks down, the ITCZ becomes separated into several blocks or a few segments.

Significance for Climate Change

The ITCZ is an important atmospheric phenomenon in tropical meteorology. In meteorology, a low-pressure center always draws air flows around it coming in, causes convection, and therefore generates a weather system. The ITCZ serves precisely this role. As the trade winds converge into the low-pressure belt, air starts to ascend to higher altitudes.

Because the ITCZ is typically formed over the tropical oceans, the air in the zone is very moist. The ascending of this moist air in the ITCZ typically leads to cloud formation as the air condenses at higher, colder altitudes. As the air continues to rise and condense, water will eventually precipitate back to the surface. In the ITCZ, this asending stream of tropical air forms the ascending branch of the Hadley cell. This air-circulation cell ascends in the tropics as a result of convection in the ITCZ, moves outward toward the poles, descends in the subtropical regions, and returns inward toward the equator at lower altitudes.

Hadley circulation thus typically transports warm, moist, tropical air to higher latitudes at upper levels. The air gradually cools and dries out as it travels poleward. In the subtropics, about 30° latitude north and south of the equator, the air begins

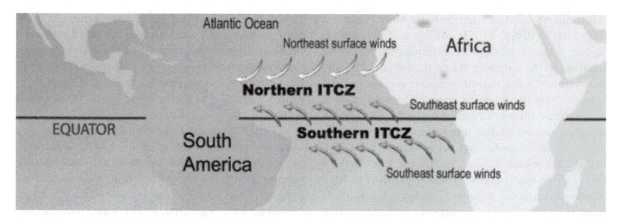

The northern Inter-Tropical Convergence Zone is created by the collision of the northern and southern trade winds, whereas the southern Inter-Tropical Convergence Zone results from two groups of southern winds squeezing together. (NASA-JPL)

to descend. This effect is believed to cause many of the large deserts in the subtropics, such as the Sahara Desert, because the descending air has lost most of its moisture to precipitation before it descends. Thus, the ITCZ plays an important role in global atmospheric circulation and global climate.

As a driving engine of global atmospheric circulation, the ITCZ plays a great enough role in Earth's climate that any change in the position and intensity of the ITCZ may have significant effects upon the climate. The ITCZ itself comprises the planet's deepest convective cloud systems with the most frequent thunderstorm activities and produces the greatest rainfall on the Earth. The Hadley circulation, driven in party by the ITCZ, has a profound impact on a global scale. For example, the intensification of the ITCZ may cause more evaporation of water vapor from tropical oceans and enhance the transport of water vapor to high latitudes.

Many scientists believe that the ITCZ may intensify and produce even more precipitation with the increase of global surface temperature. The position, movement, and intensification of the ITCZ are also influenced by various climate regimes, such as the El Niño-Southern Oscillation and La Niña events. Scientists have also shown that when the ITCZ breaks down, some of its segments can form into tropical disturbances, the initial stage of tropical cyclones, hurricanes, and typhoons. However, it is still unclear whether global warming can cause more frequent breakdown of the ITCZ, thereby producing more tropical storms, although scientists do believe that hurricane intensity will increase as the climate warms.

Chungu Lu

Further Reading

Ahrens, Donald C. *Essentials of Meteorology: An Invitation to the Atmosphere.* 4th ed. New York: Brooks/Cole, 2005. One of the most widely used introductory books on atmospheric science. The book covers a wide range of topics on weather and climate and places the ITCZ in the context of global weather patterns.

Henson, Robert. *The Rough Guide to Weather.* 2d ed. New York: Rough Guides, 2007. General, popular overview of meteorology that includes discussion of the ITCZ.

Levinson, D. H., et al. "State of the Climate in 2007." *Bulletin of the American Meteorological Society* 89, no. 7 (July, 2008): special supplement. Comprehensive overview of climatological issues and advances in the early twenty-first century. This annual assessment of global and regional climate status includes the status of the ITCZ.

Lu, J., G. A. Vecchi, and T. Reichler. "Expansion of the Hadley Cell Under Global Warming." *Geophysical Research Letters* 34 (2007). This paper examines the change in the Hadley circulation forced by the ITCZ in response to global warming, indicating that the Hadley cell experienced a northward expansion.

Lutgens, Frederick K., and Edward J. Tarbuck. *The Atmosphere.* 10th ed. Upper Saddle River, N.J.: Pearson Prentice Hall, 2007. Textbook for students of atmospheric sciences; includes discussion of the ITCZ.

Schneider, Tapio, Tobias Bischoff, and Gerald H. Haug. "Migrations and dynamics of the intertropical convergence zone." *Nature* 513, no. 7516 (2014): 45-53. The intertropical convergence zone is generally north of the equator because of poleward heat transfer in the Atlantic, but is position migrates toward whichever hemisphere is warmer.

See also: Atmospheric dynamics; Atmospheric structure and evolution; Average weather; Climate and the climate system; Climate variability; Climate zones; El Niño-Southern Oscillation; Global climate; Hadley circulation; La Niña; North Atlantic Oscillation (NAO); Rainfall patterns; Tropical climate; Tropical storms.

Intergenerational equity

Category: Ethics, human rights, and social justice

Definition

The principle of intergenerational equity is strongly embedded in the U.N. Framework Convention on Climate Change. Historically, it owes its recognition to the Brundtland Report, *Our Common Future* (1987), which popularized the principle of "development that meets the needs of the present without compromising the ability of future generations to meet their own needs." The Declaration of Rio on Environment and Development (1992) adds a rights-based perspective to the principle, suggesting that "the right to development must be fulfilled so as to equitably meet development and environmental needs of present and future generations."

Significance for Climate Change

The moral concerns over intergenerational equity have strong implications for climate-change-related policy making. The concept of intergenerational equity acknowledges the obligations of present generations in protecting the opportunities of future generations as well as the rights of future generations in enjoying the same level of opportunities and stocks of assets. This concept raises questions about how the interests of current generations are weighed against those of future generations and about the costs of inaction. Economists use discounting factors to address climate change by attempting to weigh the costs and benefits of alternative policies to different generations. The Stern Review, a report released by the British government in 2006, uses a similar approach to calculate optimal policy response and to highlight the consequences of inaction for future damage.

The concept of intergenerational equity is not without problems. It is hard to know the precise magnitude and distribution of the effects of climate change because of scientific uncertainty, as well as limited current knowledge about the future aspirations of particular societies. For instance, it is not clear how evenly the impact of climate change would be distributed over successive future generations. There is potential conflict between intragenerational and intergenerational needs. If current needs have not been met, this creates a dilemma about whether there is an obligation to meet future needs. Furthermore, countries have different capacities in meeting present and future needs. There are doubts about the reliability of discount rates. Research shows that unless the discount rate is very low, the benefits of climate change mitigation policies in future centuries are almost worthless in present value terms. There is also an implicit asymmetrical power in the concept. While current generations can influence how future generations live, future generations cannot exercise such an influence on presently living people.

Sam Wong

See also: Bioethics; Brundtland Commission and Report; United Nations Framework Convention on Climate Change.

Interglacials

Categories: Climatic events and epochs; cryology and glaciology

The question of when the current interglacial period will end and a new glacial period will begin is of paramount importance when attempting to predict the future of Earth's climate.

Key concepts

Holocene: the current interglacial, which began 11,700 years ago

isotopes: variants of an element that are chemically identical but have different atomic mass numbers and vary in radioactivity

marine isotope stage: half of a glacial cycle, as identified in the oxygen isotope data from ocean cores; advances are given even numbers, and retreats are given odd numbers

Milankovi6 cycle: cyclical variance in Earth's orbital parameters, including axial inclination, climatic precession, and orbital eccentricity

MIS 11: an interglacial that may be the best analogue for the Holocene; also called the Holsteinian or Termination V

MIS 5e: the most recent interglacial before the Holocene, also known as the Eemian, LIG (Last Inter-Glacial), or Termination II

O^{18}/O^{16} ratio: ratio between two oxygen isotopes that is altered by the advance and retreat of continental glaciers

Background

How does the present climate compare to the climate during other interglacials? Are current temperatures, sea levels, and carbon dioxide (CO_2) concentrations unprecedented, or should they be considered within the expected range? When, if ever, is the current interglacial going to end? These are among the important questions that climate science seeks to answer. Ice cores, loess deposits, pollen analysis, and other data and techniques provide information that comes together to form a detailed picture that will help answer them.

Biological Data

Over the last three million years, there have been forty to fifty glacial/interglacial cycles. Scientists have sought to characterize conditions during these periods using a variety of approaches. Plant remains, particularly pollen, have been analyzed to estimate temperature and humidity. The carbonate shells of planktonic marine organisms, preserved as fossils in the sediments beneath the sea, have been analyzed to infer sea surface temperatures. It is difficult to separate regional effects from global ones, and often data from different time periods is only available in different locations, so acquiring a global picture is not easy.

In general, interglacial climate is seen as being quite similar to the current climate. In fact, the term is often restricted to periods during which temperatures were at least as high as they have been during the Holocene, the name given to the current interglacial.

Ice Core Data

Cores have been drilled out of the ice in Greenland and Antarctica. Within the ice are bubbles of air that have been preserved since the ice was formed. The deepest core, from an area called Dome C in Antarctica, has samples of air from 800,000 years ago. These core samples provide detailed information on eight complete glacial cycles. Ratios of oxygen istopes O^{18} to O^{16} are used to infer how much of the Earth's water was tied up in glacial ice, since glaciers sequester O^{16} and thereby increase the proportion of O^{18} in the ocean. The concentration of deuterium (H^2) in a sample, moreover, can be used to infer the temperature of the air when the snow formed. Age is determined by combining depth, snow accumulation rates, ice flow rates, compaction rates, and so on, in a complex but reproducible way and calibrating the results by using radiometric dating techniques on volcanic dust incorporated in the ice. Dust is examined in detail, and often its place of origin can be determined. While no two cycles are identical, they all share a number of traits.

Each cycle has a saw-tooth shape. A glacial advance ends abruptly, with rapid melting of the ice and a rapid rise in air temperatures. For example, in MIS 9, the temperature rose 13° Celsius in eight thousand years. This warming can stop abruptly, with cooling starting almost immediately, or it can taper off, warming at a slower rate for perhaps a dozen millennia, before cooling begins. Of the eight previous terminations in the Dome C core, four started cooling immediately and four tapered off. Once cooling begins, it is far more gradual than is warming, continuing, with some reversals, for about 100,000 years. Ice is sequestered as cooling occurs, and CO_2 levels fall.

The air temperatures in Antarctica were higher than the average for the current millennium during each of the last four interglacials, but lower during those more than 500,000 years old. The Holocene (MIS 1) began 11,700 years ago. The three most recent interglacials before the Holocene (MIS 5, 7, and 9) had durations of about twelve thousand years or less, so simply by looking at the ice core data one might guess that the next glacial advance may be imminent. However, a theory proposed by Milutin Milankovi6 suggests things may not be this simple.

Periodic variations in some of the orbital parameters of the Earth are known to control the timing of glacial cycles. One past interglacial during which those variations were similar to today's was MIS 11, which lasted for twenty-eight thousand years and had warming taper off after an early rapid rise. However, MIS 19, which also had similar orbital variations, saw cooling right after its initial warming and was of shorter duration. Perhaps the Milankovi6 cycles are similar to lit matches thrown into the woods. When they are thrown will determine the timing of the fires produced, but the scale and intensity of the fires are dependent on how much fuel is available, how dry it is, the weather, relative humidity, and a host of other factors.

Context

The Earth is currently in an interglacial state and has been in that state for nearly twelve thousand years. Geology and oceanography have shown that over the last three million years the Earth has switched between glacial and interglacial states some forty to fifty times. With no anthropogenic influence, one would conclude that continental glaciers will advance again, the only question being when. In the presence of anthropogenic influences, however, it is not clear that the Earth will return to another glacial state.

Although there is some discussion over whether the orbital conditions for triggering a glacial advance have already occurred or are yet to occur, it is still prudent to examine the conditions on the planet today to evaluate the risk posed by another glacial advance. Of particular interest is why, during times of elevated CO_2 and elevated temperatures, former interglacials succumbed to the minor fluctuations of solar energy produced by those orbital conditions.

Otto H. Muller

Further Reading

Donald, R. A. P. P. *Ice ages and interglacials: measurements, interpretation, and models.* Springer Science & Business Media, 2012. An up to date compilation of ideas on the causes and functioning of ice ages. To establish a perspective on extreme climate change, it also includes a discussion of the snowball earth and Cretaceous hothouse.

Luthi, Dieter, et al. "High-Resolution Carbon Dioxide Concentration Record 650,000-800,000 Years Before Present." *Nature* 453, no. 7193 (2008): 379-82. Includes a plot of CO_2 concentrations and temperatures over eight glacial cycles. Charts, tables, bibliography.

Ruddiman, William F. *Earth's Climate Past and Future.* 2d ed. New York: W. H. Freeman, 2008. This elementary college textbook has a thorough discussion of how scientists have learned about the climate of interglacials. Illustrations, figures, tables, maps, bibliography, index.

_____. *Plows, Plagues, and Petroleum: How Humans Took Control of Climate.* Princeton, N.J.: Princeton University Press, 2005. Written for the lay public, this book provides the background and thinking behind the theory that humans have influenced the climate for the last nine thousand years, primarily through agriculture. Illustrations, figures, tables, maps, bibliography, index.

See also: Cryosphere; Deglaciation; Glaciations; Glaciers; Holocene climate; Ice cores; Last Glacial Maximum (LGM); Little Ice Age (LIA).

Intergovernmental Panel on Climate Change

Categories: Organizations and agencies; meteorology and atmospheric sciences
Date: Established 1988
Web address: http://www.ipcc.ch

Since its creation in 1988, the IPCC has proven to be one of the most credible sources of information regarding the state of scientific knowledge about climate change, its impacts, and possible policy responses. Its input has been instrumental in the global effort to reduce GHG emissions.

Key concepts

anthropogenic: caused by human activity

greenhouse gases (GHGs): gases that allow solar radiation to reach Earth's surface but trap heat, preventing it from escaping into space

International Council for Science: a global nongovernmental organization that fosters cooperation between national and international scientific unions and organizations in the advancement of science

Kyoto Protocol: an international treaty in which industrialized signatories made legally binding commitments to reduce their GHG emissions

United Nations Environment Programme (UNEP): a United Nations agency that promotes sensible environmental policies as the key to sustainable development in developing nations

United Nations Framework Convention on Climate Change (UNFCCC): an international agreement creating the overall context for future agreements designed to reduce GHG emissions

World Meteorological Organization (WMO): a United Nations agency charged with assessing the effects of atmospheric changes on weather, climate, and seas

Mission

The first call for an international effort to study the effects of anthropogenic climate change on the global community was issued by the World Meteorological Organization (WMO) at its 1979 World Climate Conference. The statement urged world governments to use scientifically generated knowledge to direct policy initiatives designed to slow the progression of global warming. In 1985, the Advisory Group on Greenhouse Gases (AGGG) was established by the International Council of Scientific Unions (ICSU; later the International Council for Science), the WMO, and the United Nations Environment Programme (UNEP). The group was to periodically evaluate scientific data relating to climate change, following a joint conference. These events created the impetus for the establishment of the Intergovernmental Panel on Climate Change (IPCC).

In 1987, the WMO and UNEP agreed that an organization should be created to coordinate an ongoing international effort to evaluate the results of scientific research on the climatic and socioeconomic effects of greenhouse gas (GHG) emissions. This organization would both evaluate research findings and suggest appropriate and effective policy responses based on those findings. The WMO established the IPCC in 1988 and gave it its mission. The panel was to develop a strategy to increase scientific information on global warming, use that information to assess possible policy initiatives for addressing climate change, evaluate policies already in place or proposed, and report its findings to governments and international organizations. UNEP and the United Nations General Assembly endorsed the IPCC.

The IPCC convenes annually in meetings that include hundreds of government officials and researchers from government agencies and nongovernmental organizations (NGOs) from countries that are members of the WMO and UNEP. At these meetings, the IPCC's objectives and activities are determined, the election of its chair is held, and members of its bureau, Task Force Bureau, secretariat, working groups, and Task Force on National Greenhouse Gas Inventories are selected. The thirty-member bureau, with representatives from all regions of the world, oversees the three working groups, and the Task Force Bureau directs the work of the task force. Working Group I (WG I) analyzes the scientific evidence regarding the causes of climate change, Working Group II (WG II) deals with the effects of climate change, and Working Group III (WG III) examines possible ways to reduce the negative effects. The task force is charged with developing better ways of measuring and reporting countries' GHG emissions. In addition, temporary special topic groups may be formed as necessary. The secretariat oversees and organizes all IPCC functions.

Activities of the IPCC

The main activities of the IPCC involve producing assessment reports (ARs) and methodology reports. The First Assessment Report (FAR) was requested by the U.N. General Assembly in 1989 to form the basis for the creation of the United Nations Framework Convention on Climate Change (UNFCCC). The FAR was completed in 1990. WG I concluded that anthropogenic GHG emissions, principally those of carbon dioxide (CO_2), would increase and would be responsible for global warming and sea-level rise during the twenty-first century. WG II asserted that this would have negative impacts on land and water ecosystems, coastal areas and cities, forestry and agriculture, and weather. WG III suggested short- and long-term policy responses. An IPCC supplementary report, prepared in 1992 to provide updated information for the newly created UNFCCC, supported the conclusions of the FAR.

Further IPCC ARs were published in 1995, 2001, and 2007. Each contained the three working group sections, with summaries for policy makers, as well as a synthesis report that summarizes the overall findings. These ARs reiterated and expanded the findings of the FAR and expressed greater confidence in the accuracy of the simulation models used to project future climate change, including the ability to better distinguish between natural and anthropogenic GHG emissions. The findings of these improved models indicated that human activity is the primary cause of past and future increases in global warming.

The Fourth Assessment Report, *Climate Change, 2007* (AR4), was the most strongly stated, asserting that mostly anthropogenic global warming will likely result in significant increases in global temperatures, ocean levels, hurricanes, droughts, heat waves, and other negative effects during the twenty-first century. It further stated that these effects will continue for hundreds of years under all simulated scenarios, even if GHG emissions could be limited to their current levels. All ARs undergo a rigorous review and revision process by the working groups, government officials, expert scientists, and the panel before receiving final approval.

In addition to the ARs produced by the working groups, the task force produced *Revised 1996 IPCC Guidelines for National Greenhouse Gas Inventories* to inform governments about the available methods for measuring GHG emissions. In 2006, an updated version was prepared that detailed improvements in software and methods for measuring emissions. The IPCC sponsors expert meetings and workshops to find more effective ways to do its work. It also creates special topic and function groups, such as the Task Group on Data and Scenario Support for Impacts and Climate Analysis, which provides the latest data from climate change studies that use different models for use by those who evaluate their impacts and develop response strategies.

Controversies Involving the IPCC

The IPCC's activities and processes have not been without their critics. Some have said that the ARs are inaccurate, alarmist, and politically driven. For example, Christopher W. Landsea, science and operations officer at the National Hurricane Center, resigned from participation in AR4 in protest when a lead author of both the 1995 AR and AR4, Kevin E. Trenberth, director of the Climate Analysis Section at the National Center for Atmospheric Research, asserted that increased hurricane activity was caused by global warming, which Landsea strongly disputed. Landsea claimed that AR4's conclusions were a product of political pressure and scientific community consensus, and that they stated the research questions as though they had empirical support, which, in his opinion, they did not. In 2005, the British House of Lords expressed concerns that the IPCC was exaggerating the future magnitude and impacts of global warming. It commissioned a report to perform a more objective cost/benefit analysis regarding the IPCC's suggested responses and the expense of possible damage. Ironically, the resulting Stern Review concluded that IPCC ARs and reports from other sources had underestimated the future risks of global warming.

In fact, another general criticism of the IPCC has been that it is too conservative in its analytical approach and downplays the need for immediate aggressive action. Critics cite other analyses that forecast higher future increases in GHG emissions, global temperatures, and sea-level rise than do the IPCC. Some claim that political pressure from the United States resulted in Robert T. Watson being replaced by Rajendra K. Pachauri as IPCC chair in 2002, because the latter was seen as being more desirable by conservative politicians and large oil companies. At the heart of these concerns are the IPCC review, revision, and acceptance processes, which are viewed by some as being unnecessarily cumbersome, leading to more conservative, bureaucratically acceptable findings.

Critics have objected to line-by-line reviews and revisions of the summary sections of the ARs by the panel, with some saying that they downplayed and distorted findings that had been validated through the peer-review process. The heads of major atmospheric research and meteorological organizations have responded by saying that this was a media campaign attacking rigorously reviewed data compilations and summarizations because of the biases of the critics. The IPCC has stood by its position that the revisions make the ARs more understandable, rather than being politically motivated, and that the strength of the AR processes lies in its objective, cautious, analytical approach, following the ideals of the scientific method.

Critics have also asserted that the submission deadlines for AR material are too far in advance of their publication. As a result, in order to allow for the lengthy review processes, the latest research findings are excluded. On the eve of the release of AR4, Pachauri responded by acknowledging that important research that projected more pronounced future global warming had been completed since the submission deadlines, indicating that aggressive policy initiatives should be initiated sooner than recommended in the report.

Context

The goal of the IPCC is to foster international cooperation between scientists working on climate modeling, those assessing climate impacts, and policy advisers in order to determine the causes and dangers of climate change and develop sensible response strategies. This has proven to be a challenge because of the disagreements and debates among scientists

and the frequent tension between political dynamics and research findings and conclusions. Nevertheless, the IPCC's work has had significant impacts.

Since 1988, multitudes of scientists have volunteered to participate in the panel's work without compensation, resulting in a dramatic increase in global warming research, and the IPCC has ensured that participants are from both developing and industrialized countries. Over 3,750 contributing and lead authors and expert reviewers from more than 130 countries contributed to AR4. The FAR brought about the creation of the UNFCCC and the Koyoto Protocol, which has been ratified by 183 nations.

Criticisms notwithstanding, a strong consensus of support for the IPCC has emerged within the global scientific establishment. A majority of scientific organizations from all parts of the world, including the United States' National Research Council, the Network of African Science Academies, the European Geosciences Union, and the Royal Meteorological Society have voiced strong support for the IPCC, calling it the foremost authority on the state of scientific knowledge relating to climate change. IPCC's ARs have become a primary source of information in climate change policy debates worldwide, and many conservative political factions that had denied the existence of anthropogenic global warming for decades have now conceded that it is real, in the face of the growing body of scientific evidence.

In 2007, the IPCC was awarded the Nobel Peace Prize for its efforts to compile, analyze, and disseminate scientific data about climate change as the basis for formulating strategies for mitigating against its negative global impacts. The co-winner of this award was former United States vice president Al Gore, who was also recognized for his work on global warming. The IPCC will undoubtedly continue to be an important force in the ongoing international efforts to understand and counteract the current and future effects of climate change. The Fifth Assessment Report (AR5) is scheduled for release in 2014.

Jack Carter

Further Reading

Brolin, Bert. *A History of the Science and Politics of Climate Change: The Role of the Intergovernmental*

Panel on Climate Change. New York: Cambridge University Press, 2008. The IPCC's first chairman discusses the mission and history of the organization and the tension between the science and politics of climate change that has undermined the implementation of decisive action. Notes, references, and index.

DiMento, Joseph F. C., and Pamela M. Doughman, eds. Climate Change: What It Means for Us, Our Children, and Our Grandchildren. Cambridge, Mass.: MIT Press, 2007. Collection of articles examining numerous interest- and activist groups' positions on climate change, emphasizing the role of the IPCC in these ongoing debates. Glossary and index.

Intergovernmental Panel on Climate Change. Climate Change 2014–Impacts, Adaptation and Vulnerability: Regional Aspects. Cambridge University Press, 2014. Annual summary of global climate change indicators and mitigation actions.

Intergovernmental Panel on Climate Change. Climate Change, 2007—Impacts, Adaptation, and Vulnerability: Contribution of Working Group II to the Fourth Assessment Report of the Intergovernmental Panel on Climate Change. Edited by Martin Parry et al. New York: Cambridge University Press, 2007. Details the research on climate change that has led to the IPCC's calls for policy initiatives to combat global warming. Bibliographies, appendixes, index, and a CD-ROM with a summary for policy makers, technical summary, and regional and subject databases of references.

Skodvin, Tora. Structure and Agent in the Scientific Diplomacy of Climate Change: An Empirical Case Study of Science-Policy Interaction in the Intergovernmental Panel on Climate Change. Norwell, Mass.: Kluwer Academic, 2000. Comprehensive analysis of the IPCC's organization, assessment processes, and attempts to resolve conflicts between policy and science. References, appendix, and index.

See also: Greenhouse gases and global warming; International Council for Science; Kyoto Protocol; Skeptics; United Nations Environment Programme; United Nations Framework Convention on Climate Change; World Meteorological Organization.

Intergovernmental Panel on Forests

Category: Organizations and agencies
Date: 1995-1997; succeeded by the Intergovernmental Forum on Forests, 1997-2000
Web address: http://www.un.org/esa/forests/ipf_iff.html

Mission

As a U.N.-sponsored body on which all United Nations members were entitled to representation, the Intergovernmental Panel on Forests (IPF) strove to reach a consensus among all its members that could have provided the basis for a world forest convention. Discussions of the role played by forests in controlling climate change, specifically global warming, began in earnest at the United Nations Conference on Environment and Development (UNCED), which took place in Rio de Janeiro in June of 1992. This meeting highlighted the existence of two opposing groups in discussions of the role of forests in global warming. The object of U.N. climate discussions had already been defined as "sustainable development" by the Brundtland Commission of 1987, but the meeting in Rio brought the two opposing camps face to face, as developed nations advocated for a convention that would constrain deforestation while developing countries, notably Brazil and Malaysia, opposed the idea of a world forest convention. Their opposition rested on the fact that much of the deforestation then taking place was occurring in the developing world.

Following the meeting in Rio de Janeiro, a United Nations body was formed called the Commission on Sustainable Development (CSD), which authorized the establishment of the IPF to conduct research that could be used to provide a factual basis for a convention on forests. Such a convention could devise rules that would constrain further deforestation and thereby slow the addition of carbon dioxide to the atmosphere. The CSD began with fifty-three member states, but other U.N. members were free to join.

Meetings of the IPF, which continued to function as a subsidiary of the CSD, received input from a variety of other bodies, especially the Food and Agriculture Organization (FAO), which hoped to take the lead in defining the framework issues of a forest convention. Other bodies, including nongovernmental organizations (NGOs) representing various indigenous tribes, also sought to take part in the discussions. The result was a divisive set of meetings that was dominated by the fundamental division between the developed countries and the developing countries. Since the rules governing the CSD called for a consensus to be reached before a convention could be created, little if any progress was made.

In 1997, the CSD replaced the IPF with another body, the Intergovernmental Forum on Forests (IFF). Like its predecessor, the IFF was an "open-ended" body, which meant that any U.N. member could take part in its discussions. It met four times, in October, 1997, in New York; from August to September, 1998, in Geneva, Switzerland; in May of 1999 in Geneva; and from January to February of 2000 in New York. Discussions by the participants ranged over many issues, including the causes of deforestation, economic issues related to forests, and particularly how to value the services that forests provide. These services include not just possible climate stabilization but also biodiversity (especially in tropical forests) and preservation of indigenous cultures.

A consensus that could lead to a forest convention with terms that might be legally enforceable proved as elusive to the IFF as it had to the IPF. Meanwhile, deforestation of tropical forests continued, while at the same time the United States joined the developing nations in opposing any convention or binding agreement on the grounds that participation had to be "voluntary." Canada led the developed nations, including the European Union, in advocating for a binding convention. In 2001, the IFF was effectively replaced by the U.N. Forum on Forests, a subsidiary of the Economic and Social Council, where discussion of forest issues continued.

Significance for Climate Change

The IPF and the IFF served to publicize the environmental impact of deforestation of tropical forests,

and even though they did not develop a consensus that could lead to a convention on forests, they did bring to light a great deal of factual information about forests and their role in climate change. Forests cover about 30 percent of the world's land mass, but the percentage is declining. Between 15 and 30 percent of the carbon contributed to the atmosphere globally results from the reduction of the world's forests. Tropical forests contain almost as much carbon as do the temperate and boreal forests combined. Stopping deforestation could do much to stabilize Earth's climate.

Nancy M. Gordon

Further Reading

Humphreys, David. *Logjam: Deforestation and the Crisis of Global Governance.* London: Earthscan, 2006. A comprehensive account of the various international efforts to control deforestation.

McDermott, Constance L. "REDDuced: From sustainability to legality to units of carbon—The search for common interests in international forest governance." *Environmental Science & Policy* 35 (2014): 12-19. The quest for metrics creates a reductionist climate where variables get simplified and more definable, but also lose crucial information along the way.

Ramakrishna, Kiloparti, and George M. Woodwell, eds. *World Forests for the Future: Their Use and Conservation.* New Haven, Conn.: Yale University Press, 1993. Various articles present much of the detail about forests and their role in climate change.

Ruddell, Steven, et al. "The Role for Sustainably Managed Forests in Climate Change Mitigation." *Journal of Forestry* 105, no. 6 (September, 2007). Surveys the state of the field.

See also: Amazon deforestation; Deforestation; Forestry and forest management; Forests; United Nations Conference on Environment and Development; United Nations Environment Programme; United Nations Framework Convention on Climate Change.

International agreements and cooperation

Category: Laws, treaties, and protocols

Definition

International law connects nation states in a framework of agreements that implement recognized principles, norms, rules, and procedures. Subjects of public international law include sovereign nation states, international organizations, and movements of national liberation. Sources of international law are customs, accepted standards of human behavior, and treaties. The design of international agreements is voluntary, but once concluded they are binding instruments to achieve collective benefits that states would not achieve unilaterally. To either open a new set of options that offer beneficial opportunities or to restrain options that are recognized as harmful, state governments abide by the terms of an international agreement, which in either case is a recognition of mutual interdependence. An ultimate goal of international law is to facilitate win-win solutions through cooperation.

The importance of international agreements is well demonstrated by the prisoner's dilemma in game theory. Here two actors are in a better situation if they both cooperate rather than do not cooperate. If either side individually gives up cooperation for individual benefits, the joint outcome is worse for both. To prevent actors from non-cooperation to maintain mutual benefits requires agreement based on communication, verification, and enforcement. To discourage violations of an agreement, they need to be detected and the expected benefits from violation to be denied. It is often efficient to transfer the tasks of verification and enforcement to institutions that are not bound by individual capabilities and interests.

Significance for Climate Change

International agreements are essential in climate policy to address the tragedy of the commons regarding the atmosphere. They impose emission constraints for each member state and create opportunities for technology cooperation, capital flows, and trading markets.

The law of treaties defines the rules for legally binding agreements between states, which is codified in the 1969 Vienna Convention on the Law of Treaties. A treaty is a written agreement between two or more sovereign states in which the parties involved agree to abide by certain specified procedures and standards of conduct, which include the *signature* by an authorized State representative; the *ratification* process by parliamentarian bodies to

GHG and CO$_2$ Emissions, by Nation

Nation	Percent of Global CO$_2$ Emissions, 1850-2000	Percent of Global GHG Emissions, 2004
United States	30	19
EU-25	27	13
Russia	8	5
China	7	17
Germany	7	3
United Kingdom	6	2
Japan	4	4
France	3	1
India	2	5
Canada	2	2
Italy	2	2
Ukraine	2	1
Poland	2	<1
Australia	1	1
South Africa	1	1
Brazil	<1	3
Mexico	<1	2
Indonesia	<1	2
South Korea	<1	1
Spain	<1	1

Source: Pew Center on Global Climate Change.

enter a treaty into binding national legislation, independent of the political leadership; and the *entry into force* upon fulfillment of specified conditions, such as the number of ratifying states. During this process, the status of a member state changes from a negotiating state to a state signatory, ratifying state, and ultimately state party after entry into force. A state that has signed a treaty is bound to it and is obliged to refrain from acts that would defeat the object and purpose of a treaty even if it has not yet ratified it. A state can change agreements before ratification and announce that it is withdrawing its signature, after which it is no longer bound.

After ratification, a state is obligated to announce to the world in advance that it plans to withdraw from a treaty, following the advanced notice required. Usually treaties hold for a limited number of parties (bilateral, trilateral, multilateral) and for a given period (for example, the Kyoto Protocol, with obligations for industrialized countries by 2012). Rules of treaties are binding, regardless of the name (for example, protocol, accord, covenant, convention, memorandum of understanding, or exchange of letters or notes). A challenge for international agreements is to design mechanisms that strike a balance between the required commitments and the degree of support for an agreement.

The threat of climate change cannot be resolved or prevented by a single nation, but requires an unprecedented degree of international cooperation to foster emission reductions and technological change in the energy sector. Instruments of international law define the rules of emissions, climate impacts, and policies to diminish the risks and enhance cooperation, providing mutual assurance that policies are pursued in an integrated, coordinated, and effective way. An adaptive regulatory framework would address the collective action problem in long-term climate policy, designing efficient legal mechanisms to improve the effectiveness of institutions, to reduce conflict and codify cooperation and compliance, including the international transfer of investments and technologies to shift the composition and learning rates of the energy system toward emission reductions. Coalitions are emerging where individual action is insufficient or inefficient—for example, in achieving a critical number of votes, a critical mass of investment to realize projects, or a threshold of emissions.

The legal framework of climate policy has been defined by the 1992 U.N. Framework Convention on Climate Change (UNFCCC). With the signing of the Kyoto Protocol in 1997 and its entry into force in 2005, the international community has established a first set of instruments through a combination of cooperative governance and market-based incentives. A lack of agreement on the underlying causes, expected risks, and required actions related to long-term climate change, as well as the expected costs and partial interests, has impeded progress during the past decade. To overcome the hurdles for post-Kyoto agreements, an evaluation and negotiation process across all levels is needed, involving citizens, firms, institutions, and states in a multi-stakeholder environment, which became visible during the Bali climate summit in December 2008. Moving beyond Kyoto is a challenge for the legal and policy process that is supposed to implement the longer-term objectives and manage the potentially severe implications in case of failure. It is also a challenge for the scientific community, which becomes involved in value judgments that require innovative integrated approaches.

Jurgen Scheffran

Further Reading

Chandy, Laurence, and Homi Kharas. "Why can't we all just get along? The practical limits to international development cooperation." *Journal of International Development* 23, no. 5 (2011): 739-751. The incentives for donors and recipients in aid programs frequently differ, complicating agreement.

Feaver, D., and N. Durrant. "A Regulatory Analysis of International Climate Change Regulation." *Law and Policy* 30, no. 4 (October, 2008): 394-422. Examines the regulatory architecture and coherence of global climate change regulations.

Scheffran, J. "Preventing Dangerous Climate Change." In *Global Warming and Climate Change*, edited by V. I. Grover. Enfield, N.H.: Science Publishers, 2008. Analyzes the implications of Article 2 of the UNFCCC.

Stein, J. von. "The International Law and Politics of Climate Change: Ratification of the United

Nations Framework Convention and the Kyoto Protocol." *Journal of Conflict Resolution* 52 (2008): 243. Discusses the challenge to design mechanisms that deter defection without deterring participation; emphasizes international social networks and domestic nongovernmental organizations.

Verheyen, R. *Climate Change Damage and International Law: Prevention Duties and State Responsibility.* Boston: Martinus Nijhoff, 2005. Comprehensive assessment of the legal duties of states regarding damage from anthropogenic climate change. Analyzes the legal duties of states to prevent climate change damage and international liability to breaches of these duties. Advocates an internationally negotiated solution to the climate issue.

Yamin, F. *The International Climate Change Regime: A Guide to Rules, Institutions, and Procedures.* New York: Cambridge University Press, 2004. Provides a guide to the rapid developments in international law in climate change and explains the regime participants and commitments, adaptation, finance and capacity building, reporting and compliance, institutions and negotiation processes, and scientific and technical input.

See also: Antarctic Treaty; Basel Convention; Berlin Mandate; Convention on Biological Diversity; Convention on International Trade in Endangered Species; Convention on Long-Range Transboundary Air Pollution; Intergovernmental Panel on Climate Change; Kyoto Protocol; Montreal Protocol; United Nations Convention to Combat Desertification; United Nations Framework Convention on Climate Change; U.S. and European politics; World Trade Organization.

International Atomic Energy Agency

Categories: Organizations and agencies; energy
Date: Established July 29, 1957
Web address: http://www.iaea.org

Mission

The International Atomic Energy Agency (IAEA) promotes the peace, health, and prosperity of the people of the world through the use of atomic (nuclear) energy. The IAEA is independent of the United Nations but reports to the United Nations General Assembly, and if violations involving nuclear materials are detected, it reports to the U.N. Security Council. The agency, headquartered in Vienna, Austria, has over twenty-two hundred trained personnel and their support staff. Member nations pay a fee each year as part of their membership but there is also a voluntary fund, the Technical Cooperation Fund, that is used especially in developing countries. The mission of the IAEA is based on three work areas: safety and security, science and technology, and safeguards and verification.

In the safety and security area, the work is focused on helping countries develop standards and procedures for nuclear safety and security. That involves training sessions for untrained personnel, meetings for experts to share information, and planning sessions and training for emergencies. IAEA's budget is geared to the sharing of information with anyone needing nuclear information.

The science and technology area works toward the use of peaceful applications of nuclear science and technology to solve problems in health, poverty, pollution, and other areas of life. The work can be described in three components: technical cooperation, research and development, and energy and electricity. The technical cooperation includes expert advice, personnel, materials, and equipment to support projects that benefit people in developing countries socially or economically. The research and development includes not only the joint research done with IAEA, but also the support of research by different groups all over the world. Any area in which nuclear technology can be valuable is researched but issues in health, environment, poverty, and food are the main focus points. The energy and electricity area of work includes dealing with the energy needs of a country. If the needs can be met with nuclear power, IAEA provides support in equipment, personnel, training, and materials. Ideas that are innovative are especially encouraged in this work. In 2004 IAEA developed the Program of Action for Cancer

Therapy (PACT) to aid developing countries in cancer detection and treatment using radioisotope and other nuclear techniques.

Safeguards and verification is the area of IAEA's mission that is most important for world peace. For over fifty years, IAEA has monitored the use of nuclear materials all over the world. The monitoring is not only to check that countries are not using nuclear materials to build weapons, but also that the nuclear materials used in a peaceful manner are used, stored, and secured in a manner that causes no chance of radiation exposure to humans. The United Nations Security Council has requested that IAEA keep a watch on Iran and its use of nuclear materials. The IAEA and its director were awarded the Nobel Peace Prize in 2005.

Significance for Climate Change

The use of nuclear energy to generate electricity does not generate greenhouse gases to cause global warming. It does not generate smog-producing materials. The use of nuclear energy should not cause any change in the climate unless a disaster spreads radiation into the atmosphere. After the Chernobyl disaster in 1986, the IAEA increased its focus on nuclear safety.

C. Alton Hassell

Further Reading

Fischer, David. *History of the International Atomic Energy Agency: The First Forty Years.* Vienna, Austria: International Atomic Energy Agency, 1997. This history of the IAEA covers the birth and growth of the agency and the history of nuclear energy during the same period of time. Illustrations, bibliography, index.

Florini, Ann, and Benjamin K. Sovacool. "Bridging the gaps in global energy governance." *Global governance: a review of multilateralism and international organizations* 17, no. 1 (2011): 57-74. Thinking about energy has mostly been dominated by economic, geopolitical and security considerations. Numerous other issues like environmental and human rights issues, have been much less examined.

U.S. Congress. House. Committee on International Relations. Subcommittee on International Operations and Human Rights. *The Role of the International Atomic Energy Agency in Safeguarding Against Acts of Terrorism.* Washington, D.C.: Government Printing Office, 2001. In addition to discussing the role of the IAEA, this book contains reports on the safety measures necessary to prevent radioactive pollution in the use of nuclear energy.

U.S. Congress. Senate. Committee on Foreign Relations. *Safeguarding the Atom: Nuclear Energy and Nonproliferation Challenges.* Washington, D.C.: Government Printing Office, 2008. Reports the governmental hearings on the IAEA, international security, international cooperation in nuclear nonproliferation, and U.S. policies on nuclear arms control and nuclear nonproliferation. Illustrations.

_____. *The January 27 UNMOVIC and IAEA Reports to the U.N. Security Council on Inspections in Iraq.* Washington, D.C.: Government Printing Office, 2003. Contains the reports to the Senate Foreign Relations Committee of the U.N. Monitoring, Verification, and Inspection Commission and the IAEA on the on-site inspection of Iraq. Bibliography.

See also: International agreements and cooperation; Nuclear energy; United Nations Framework Convention on Climate Change.

International Council for Science

Category: Organizations and agencies
Date: Established 1931
Web address: http://www.icsu.org/

Mission

The International Council for Science (ICS) was formerly known as the International Council of Scientific Unions (ICSU). It is a nongovernmental organization of research councils or scientific academies, which are national, multidisciplinary bodies, and scientific unions, which are international, disciplinary organizations. These two complementary types of groups provide a wide range of scientific expertise. The mission of ICS includes research, information access, information exchange, and scientific freedom. The research area includes coordinating research by interdisciplinary groups in topics important to mankind's well-being. Climate change is one of those topics.

The ICS develops Internet Web sites to disseminate data and technology and publishes newsletters, books, journals, and proceedings. It supports over six hundred meetings of discipline-specific scientists, as well as multidisciplinary, topic-driven meetings. Support for science education and for exchange of ideas is also part of the group's mission, as is initiation of regional networks of scientists. Scientific freedom is supported by ICS, which defines it as the freedom to research scientific topics without gender, racial, economic, or geographical limitations. ICS cooperates closely with many other national, international, governmental, and nongovernmental groups, especially the agencies of the United Nations.

The funding for ICS comes from member contributions, grants, and contracts. The grants and contracts come from foundations, agencies, and other bodies that financially support scientific research and information sharing. Each member is part of the General Assembly, which elects an executive board and a slate of officers. The officers are members of the executive board and are responsible for the day-to-day operation of ICS. The officers are supported in the day-to-day operations by the Secretariat, the officers, advisors, and staff of ICS located in Paris, France. To expedite scientific research and education in developing countries, ICS is creating four regional offices for Africa, Latin America and the Caribbean, Asia, and the Arab region.

Significance for Climate Change

With part of its mission to research areas that affect the well-being of the world's population, climate change is one of the major areas of study by the ICS. One of the major committees of ICS is the Scientific Committee on Problems of the Environment (SCOPE). SCOPE not only researches different factors concerning the environment but also publishes reports on the different factors. Other reports, such as *Invasive Alien Species, Sustaining Biodiversity and Ecosystem Services in Soils and Sediments,* and *Interactions of the Major Biogeochemical Cycles,* all deal with the environment; many deal with the climate and man's interaction with the climate, both how humans change the climate and how humans adapt to climate change.

In 2008, ICS initiated a major research program on natural disasters, including those caused by weather and climate. The study is designed to help countries plan for disasters, and to determine changes in lifestyle that will reduce loss of life and lessen economic impact of disasters. The study will also research how man may cause or increase the extent of disasters. Weather and climate disasters include events such as droughts, floods, hurricanes, tornados, cyclones, mudslides, and pollution. Another major study initiated in 2008 is to research the human impact on Earth's life-support systems. This study is to fill in the gaps in the knowledge about the global ecosystem, including the climate.

In each regional office, there are programs about climate or the ecosystem. In the Africa region, one of the four science plans is on global environmental change. The plan includes both climate change and adaptation to climate change. The plan is to develop projects in six areas: degradation of land and biodiversity and how it affects humans, the effect of climate change on rainfall, food system resilience, water resources, atmospheric change, and the effect of the ocean on Africa. The Latin America and the Caribbean regional office held a conference on sustainable energy and another on the world's geosphere/biosphere. The Asia regional office held a meeting to form plans for the coastal cities to adapt to climate change. Other meetings were for sustainable energy and changes in the ecosystem. Training sessions were held for individuals to learn to study environmental problems.

C. Alton Hassell

Further Reading

Committee on Data for Science and Technology. *Data Science Journal.* Paris, France: Author, 2002. Provides continuing coverage of the state of data science. Produced by the Committee on Data for Science and Technology (CODATA), a part of the International Council for Science.

Ernster, Lars. *ICSU: The First Sixty Years.* Paris, France: International Council of Scientific Unions, 1991. Details the establishment and growth of the International Council for Science.

Hak, Tomas, Bedrich Moldan, and Arthur Lyon Dahl. *Sustainability Indicators: A Scientific Assessment.* Scientific Committee on Problems of the Environment 67. Washington, D.C.: Island Press, 2007. Focuses on the evaluation of sustainable development, environmental monitoring, and environmental indicators. Illustrations, maps, bibliography, index.

Kates, Robert W. "What kind of a science is sustainability science?." *Proceedings of the National Academy of Sciences* 108, no. 49 (2011): 19449-19450. A meta-analysis of articles on sustainability shows that authors frequently come from countries outside the dominant science centers, and outlets include social science, biology and engineering journals. Sustainability science is highly interdisciplinary.

Krupnik, Igor, Michael A. Lang, and Scott E. Miller, eds. *Smithsonian at the Poles: Contributions to International Polar Year Science.* Washington, D.C.: Smithsonian Institution Scholarly Press, 2009. Contains the record of the proceedings of a symposium organized as part of the International Polar Year (2007-2008) and sponsored by the International Council for Science and the World Meteorological Organization. Illustrations, maps, bibliography, index.

See also: International agreements and cooperation; Science and Public Policy Institute; Scientific Alliance; Union of Concerned Scientists.

International Geosphere-Biosphere Programme

Category: Organizations and agencies
Date: Established 1987
Web address: http://www.igbp.kva.se

Mission

The International Geosphere-Biosphere Programme (IGBP) is an international research program devoted to the study of global environmental change, with a focus on the interactions between the solid Earth and its living organisms. The IGBP's purpose is to work toward improving the sustainability of Earth's biosphere. A central secretariat, hosted by the Royal Swedish Academy of Sciences in Stockholm, Sweden, coordinates IGBP's scientific program. The organization has seventy-six member countries and international project offices in North America, Europe, and Australia.

The IGBP studies the interactions occurring among Earth's natural biological, chemical, and physical processes and the effects human activities have on these processes. Sponsored by the International Council for Science, the IGBP collaborates with other programs to gain and disseminate knowledge regarding global environmental change and to make recommendations regarding how best to respond to it. The IGBP, which emphasizes networking and integration, seeks to enhance scientific understanding by encouraging scientists to transcend disciplinary, institutional, and political boundaries in their research.

The International Council for Science established the IGBP in 1987. The international scientific community had recognized that the research efforts of a single country, region, or scientific discipline would not yield sufficient understanding of global environmental change; there was a clear need for international collaborative research. Key findings from the IGBP's initial studies (IGBP-I, 1990-1999) provided the foundations for a second phase of research (IGBP-II, 2004-2013). Phase II, like Phase I, concerns the Earth systems of land, ocean, and atmosphere and the interfaces among them.

Significance for Climate Change

As of early 2009, IGBP research comprises nine projects: four focused on the major Earth system components of land, ocean, and atmosphere; three on the interfaces between those components (land-ocean, land-atmosphere, and ocean-atmosphere); and two on system-wide integration (Earth system modeling and paleoenvironmental research). The Global Land Project

(GLP), cosponsored by the International Human Dimensions Programme on Global Environmental Change (IHDP), looks at how humans transform terrestrial ecosystems and landscapes. The Global Ocean Ecosystem Dynamics (GLOBEC) project studies the effects of global change on the abundance, diversity, and productivity of marine populations. Integrated Marine Biogeochemistry and Ecosystem Research (IMBER) studies how marine ecosystems and biogeochemical cycles react to global change over time periods ranging from years to decades.

Both marine projects are cosponsored by the Scientific Committee on Oceanic Research (SCOR). International Global Atmospheric Chemistry (IGAC), a project cosponsored by the Commission on Atmospheric Chemistry and Global Pollution (CACGP), focuses on the atmospheric chemistry issues facing society. The project seeks to gain an understanding of the role of atmospheric chemistry in the Earth system while determining how changing regional emissions and depositions, long-range transport, and chemical transformations affect air quality.

Land-Ocean Interactions in the Coastal Zone (LOICZ) studies the coastal zone—an interface where land, atmosphere, and ocean interact—as a key player in the functioning of the Earth system. The project is cosponsored by IHDP. The Integrated Land Ecosystem-Atmosphere Processes Study (iLEAPS) concerns the physical, chemical, and biological processes that transport and transform energy and matter at the land-atmosphere interface. These processes are tightly coupled and highly responsive to climate change. The Surface Ocean-Lower Atmosphere Study (SOLAS) looks at how the key biogeochemical-physical interactions and feedbacks between the ocean and atmosphere affect and are affected by climate and environmental change. Cosponsors are CACGP, SCOR, and the World Climate Research Programme (WCRP).

System-wide integration projects include Analysis, Integration, and Modelling of the Earth System (AIMES) and Past Global Changes (PAGES). AIMES focuses on the use of models and observations in reaching a better and more quantitative understanding of the role human action plays in biogeochemical cycles. PAGES supports study of the Earth's past environment as a means for making sound predictions regarding the future.

In addition to its project cosponsors, IGBP collaborates with a number of other international science organizations. IGBP participates in global assessments such as the Intergovernmental Panel on Climate Change (IPCC) and the Millennium Ecosystem Assessment (MEA), and it is part of the Earth System Science Partnership (ESSP).

In July, 2001, during the Global Change Open Science Conference in Amsterdam, the Netherlands, IGBP joined IHDP, WCRP, and the international biodiversity program DIVERSITAS (all ESSP members) to issue the Amsterdam Declaration on Global Change. The statement acknowledged the increasing effect of human activity on food, water, clean air, and the environment and likened some anthropogenic changes to great natural forces in terms of their extent and impact. Warning of the possibility of abrupt, irreversible, and inhospitable environmental changes in response to human actions, the declaration called for a new, multidisciplinary, multinational, and multicultural system of global environmental science to respond to the complex challenges of global change with good and ethical stewardship of the Earth. IGBP was one of the convening organizations of the Second Symposium on the Ocean in a High-CO_2 World, held in Monaco in October, 2008.

Karen N. Kähler

Further Reading

Kabat, Pavel, et al., eds. *Vegetation, Water, Humans, and the Climate: A New Perspective on an Interactive System.* Berlin: Springer-Verlag, 2004. Part of IGBP's Global Change Series, this volume looks at how the land surface is related to weather and the climate—near-ground, regional, and global. Figures, references, index.

Steffen, Will, et al. *Global Change and the Earth System: A Planet Under Pressure.* Berlin: Springer-Verlag, 2005. Considers the dynamics of the Earth system before it experienced significant human influence, as well as the changes humans have wrought since then. Part of IGBP's Global Change Series. Figures, references, appendix, index.

Steffen, Will, Wendy Broadgate, Lisa Deutsch, Owen Gaffney, and Cornelia Ludwig. "The trajectory of the Anthropocene: the great acceleration." *The Anthropocene Review* 2, no. 1 (2015): 81-98. Since 1950 a broad range of social, economic and environmental indicators have shown rapid acceleration. Many of the most beneficial changes have been largely confined to the developed world, however.

Steffen, Will, et al., eds. *Challenges of a Changing Earth.* Berlin: Springer-Verlag, 2002. Collection of papers presented at the Global Change Open Science Conference in Amsterdam in July, 2001. Includes the text of the Amsterdam Declaration on Global Change. Figures, references, index.

See also: Atmosphere; Climate and the climate system; Climatology; Earth structure and development; International agreements and cooperation; Ocean-atmosphere coupling.

International Human Dimensions Programme on Global Environmental Change

Category: Organizations and agencies
Date: Established 1996
Web address: http://www.ihdp.unu.edu

Mission

The International Human Dimensions Programme (IHDP) is an umbrella organization that funds and coordinates studies relating to the human impact on all biological, physical, and chemical systems and how those impacts relate to patterns of human social organization. An international, interdisciplinary science program specializing in a social science perspective on global change, it is a joint program of the International Council for Science (ICS), UNESCO's International Social Science Council (ISSC), and the United Nations University (UNU).

All IHDP projects address three major questions:
• How do human lifestyle choices and patterns of consumption contribute to global environmental change?
• How are humans affected by changes in the natural environment?
• How can humans reduce their negative effects on natural ecosystems necessary to support continued human existence?

Significance for Climate Change

IHDP brings together scientific researchers as well as social scientists and political and economic policy makers in order to consider all aspects of human contributions to pressing environmental problems, probable consequences of continued anthropogenic stresses on a variety of ecosystems, and possible response to mitigate and even eliminate some anthropogenic environmental problems. All IHDP programs share a mission to help translate scientific research on environmental change into practical advice for policy makers. These programs and research projects may be local, regional, or global in scale.

IHDP administers six core scientific research projects. Earth System Governance (ESG) studies anthropogenic changes, local to global, in all physical and biological systems on the planet. The overarching goal of all research projects under ESG is to construct appropriate sustainable development responses that can actually be implemented. Some finds of sponsored research projects are published in the journal *Climate and Development*, as well as in the IHDP publication *Institutions and Environmental Change.*

Global Environmental Change and Human Security (GECHS) studies the impact of environmental changes on vulnerable human populations primarily located in the developing world. Flooding, drought, soil erosion, conflicts over access to natural resources, and human population issues are all research topics sponsored by GECHS, which publishes its research reports in the journal *AVISO*. GECHS encourages cooperation and communication between climate researchers and political leaders in the developing world.

The Global Land Project (GLP) sponsors projects on local and regional levels to study how human interact with both land and water-based systems. GLP projects also give direction on how humans might interact with those systems in more sustainable ways. The GLP projects also consider the negative impact global economic changes produce on local and regional environmental systems, primarily in the developing world.

Industrial transformation (IT) research initiatives look for ways to reduce the negative impact technology developments will continue to have on the natural environment, specifically in the developing world. IT-sponsored projects target energy production, distribution and usage, food production and consumption patterns, and urban development.

Land-Ocean Interactions in the Coastal Zone (LOICZ) is more narrowly focused than the other IHDP core projects. LOICZ research projects focus on coastal communities that are among the first to feel negative effects of environmental changes in the form of soil erosion, depletion of fishing stocks, increased water salinity, overdevelopment, decreased freshwater supplies, and increased vulnerability to catastrophic weather events.

Urbanization and Global Environmental Change (UGEC) projects study urban planning and lifestyle choices as well as patterns of consumption of energy, food, and goods and services. The research results are then used to suggest planning policies that would allow urban areas to become more environmentally sustainable and reduce the impact of environmental change on the urban poor.

In addition to the six core scientific research projects and their subprojects, IHDP shares joint responsibility with Earth System Science Partnership (ESSP) to promote research into climate change, agricultural responses and food security. IHDP also funds smaller pilot projects that allow younger researchers opportunities to secure funding and collaborate on an international scale. The goal of many of the pilot projects is to seek ways to reduce carbon production. IHDP also participates in a variety of research networks. The four primary research networks include the Mountain Research Initiative (MRI), Population and Environment Research Network (PERN), System for Analysis Research and Training (START), and Young Human Dimensions Researchers (YHDR) including students associated with research teams. The development of research networks allows researchers in developed countries to share funding, access to research information, and expertise with colleagues in the developing world. IHDP specifically funds research projects on issues of concern to societies in both the developed and developing world.

In its second decade, 2006-2016, IHDP looked forward to increased funding for pilot projects in order to increase the scale of the projects until some could become additional core scientific projects. As environmental concerns become more urgent and the capacity and economic feasibility of sustainable technologies more widely available, IHDP anticipates playing a wider, more public role in drafting policy positions related to environmental change and its impact on the human community.

Victoria Erhart

Further Reading

Houghton, John. *Global Warming: The Complete Briefing.* 4th ed. New York: Cambridge University Press, 2009. A readily accessible, nontechnical introduction to the scientific research supporting theories of global warming and the impacts of climate change.

Kasperson, Roger E., and Mimi Berberian. *Integrating science and policy: Vulnerability and resilience in global environmental change.* Routledge, 2011. There is a persistent gap between scientific understanding of problems and practice in addressing them. This book surveys a number of problem areas.

Weart, Spencer. *The Discovery of Global Warming.* Rev. and expanded ed. Cambridge, Mass.: Harvard University Press, 2008. Covers global warming from both a scientific as well as a political perspective. Includes a discussion on the history of global warming theory.

Young, Oran R. *The Institutional Dimensions of Environmental Change: Fit, Interplay, and Scale.* Cambridge, Mass.: MIT Press, 2002. Study conducted under the auspices of the IHDP that addresses both the contributions of human institutions to environmental changes and the ability of those institutions to respond to those changes.

See also: Climatology; Intergovernmental Panel on Climate Change; International agreements and cooperation; International Council for Science; International Geosphere-Biosphere Programme; International Institute for Applied Systems Analysis; Scientific credentials.

International Institute for Applied Systems Analysis

Category: Organizations and agencies
Date: Established 1972
Web address: http://www.iiasa.ac.at

Mission

The International Institute for Applied Systems Analysis (IIASA) is an interdisciplinary organization founded in 1972 with the aim of promoting East-West cooperation during the Cold War. It now devotes its efforts to complex systems problems of international or global scope. Through its various programs, IIASA conducts research into major economic, demographic, social, and environmental issues, including global climate change. The scope of IIASA research projects is huge, beyond the capabilities of any one country or national research institution to manage. IIASA research projects collect massive amounts of data on a variety of factors, such as the connection between climate change and air pollution, and then build computer simulations and statistical models based on various "what if" future scenarios. Since the days of its founding in the midst of the Cold War between the United States and the Soviet Union, IIASA has maintained a position of strict political neutrality in its research projects. At the same time, IIASA tries to build scientific consensus in order to present credible research findings to technical advisors for political policy makers.

IIASA is financially supported by contributions from fifteen member countries in both the developed and the developing world. In the past, IIASA conducted research projects of topics requested by member countries. Increasingly, however, IIASA is shifting its expertise in quantitative modeling and computational technologies to focus on problems of more urgency for the developing world. This allows IIASA to assist in the construction and implementation of policies relevant to global change. Most research projects focus on one or more problems and interconnections between climate policy, energy production and consumption, water management, agricultural policy, sustainable development, pollution studies, and the interplay between environmental and economic policies. IIASA devises models to reduce future uncertainties about the consequences of current policy options. IIASA divides its core research projects into three very broad categories: energy and technology, population and society, and environment and natural resources.

Significance for Climate Change

Since the 1990's, IIASA has helped develop models of global energy production, consumption, and pollutant emissions. As developing economies become increasingly industrialized, IIASA provides information on the economic and environmental consequences of these changes, particularly how rising air pollution will affect the ability of boreal forests to function as carbon sinks to mitigate negative impacts of climate change. IIASA projects also provide models to forecast the environmental impact of changes in the global economy, particularly as previously centrally planned economies move in the direction of free-market economies.

IIASA projects under this heading study and analyze the very complex interdependence among population groups, economic development, and environmental impact. Computer simulations indicate the probable social, political, economic, and environmental consequences of probable population projections. Water management, long-term food security, biodiversity, and other environmental impacts related to population change are forecasted and analyzed. Research projects in this category also study age disparity among various national populations and what economic scenarios will be necessary to fund the social security, pension obligations, and health care costs of those populations.

IIASA research projects in this category study all aspects of the human impact on the natural environment, including climate change. Since its foundation, IIASA has been involved in forestry research and the impact of land use and land cover change research. IIASA continues its long-range studies on the global management of boreal forest areas and its impact on global carbon emissions. Recently IIASA projects have focused on cost-effective methods to reduce air pollution, particularly in the developing world. One of the larger projects is the Transboundary Air Pollution Project with India and China. Increasingly, IIASA addresses regional environmental problems in order to provide a scientific basis upon which countries can negotiate specific reductions in carbon dioxide, sulfur dioxide, and other greenhouse gas emissions that have a negative impact on the global climate.

All IIASA projects study global change, whether that change is demographic, political, economic, or environmental. Through increasingly sophisticated and comprehensive computer simulations, IIASA provides scientifically credible information on global greenhouse gas emissions, energy supply and demand, population increases and migration patterns, land use change, and transboundary air and water pollution. It then forecasts probable consequences of both current policies and various policy options to mitigate the harmful effects of anthropogenic impacts on all aspects of the natural environment.

Victoria Erhart

Further Reading

Battarbee, Richard. *Natural Climate Variability and Global Warming: A Holocene Perspective.* New York: Wiley-Blackwell, 2008. Studies the Earth's naturally variable climate system and global warming on a variety of timescales, from the near future to a millennium away. Carefully lays out the paradigms and problems in Holocene climate research.

Palsson, Gisli, Bronislaw Szerszynski, Sverker Sörlin, John Marks, Bernard Avril, Carole Crumley, Heide Hackmann et al. "Reconceptualizing the 'Anthropos' in the Anthropocene: Integrating the social sciences and humanities in global environmental change research." *Environmental Science & Policy* 28 (2013): 3-13. Although "Anthropocene" comes from the root for "human," discussions of environmental problems typically focus on the non-human scientific and technical aspects. The authors call for inclusion off social science and humanities perspectives as well.

Schrattenholzer, Leo, et al. *Achieving a Sustainable Global Energy System: Identifying Possibilities Using Long-Term Energy Scenarios.* Northampton, Mass.: Edward Elgar, 2004. This study of responsible long-term energy policy was produced in association with the IIASA and exemplifies the group's contributions to such research.

Yoshimoto, Atsushi, ed. *Global Concerns for Forest Research Utilization: Sustainable Use and Management.* New York: Springer, 1998. Collects papers presented at a 1998 conference on global forest issues and climate change in Japan. Covers various modeling efforts to measure the impact of climate change and global warming on the global forest sector on a variety of timescales. Includes case studies of regional forest resources, as well as considerations of global forest resources as a whole.

See also: Climate models and modeling; Climate prediction and projection; Geographic Information Systems in Climatology; International agreements and cooperation; Microwave sounding units; Nonlinear processes.

International Policy Network

Category: Organizations and agencies
Date: Established 2001
Web address: http://www.policynetwork.net

Mission

A nongovernmental, nonpartisan organization that is charity-based in the United Kingdom and a nonprofit in the United States, the International Policy Network (IPN) highlights the dangers of linking environmental regulations to trade rules. The IPN claims to achieve its vision by "promoting the role of market institutions in certain key international policy debates: sustainable development, health, and globalization and trade." In the area of sustainable development, the group takes the stand that, too often, sustainable development involves policies that may, in fact, perpetuate environmental problems and poverty. Although acknowledging a relationship between health, wealth, and a clean environment, IPN believes that sustainable development policies focus too much on sustainability at the expense of development.

To achieve its mission, IPN pursues a number of activities: It supports individuals and groups who profess similar beliefs with Web sites, advice, and small grants. It sponsors workshops, seminars, and conferences and cooperates with groups to produce books, monographs, and articles that express opinions on issues for the media. At international conferences, it helps coordinate the participation of experts in various fields that relate to the interests of IPN and promotes its partners among the news media.

In the area of health, IPN addresses issues such as access to medicine, barriers that reduce the spread of infectious diseases, and the impact of regulation on modern technologies related to nutrition and health. The group feels that the policies of governments and other agencies are often enthusiastic but not practical, thus wasting funds and labor hours on ill-conceived initiatives that ultimately cost lives. In the area of trade, IPN espouses the benefits of freedom to trade; trade is considered fundamental to eliminating poverty and gaining economic freedom. The organization also, however, sees as dangerous the linking of environmental and labor regulations with trade rules. IPN bases its vision of sustainable development on achieving both environmental and human well-being through eliminating poverty and promoting progress.

Significance for Climate Change

IPN released a report in 2004 saying "climate change is 'a myth'" and that warnings of environmental disaster due to climate change are "fatally flawed." The activities of the organization are focused on resisting attempts to stave off global warming at the expense of economic interests.

Victoria Price

See also: American Enterprise Institute; Catastrophist-cornucopian debate; Cato Institute; Competitive Enterprise Institute; Heartland Institute; Heritage Foundation; Institute for Trade, Standards, and Sustainable Development; Nongovernmental organizations (NGOs) and climate change; Skeptics.

International Union for Conservation of Nature

Categories: Organizations and agencies; environmentalism, conservation, and ecosystems
Date: Established October, 1948, as International Union for the Protection of Nature; name changed to International Union for Conservation of Nature and Natural Resources in 1956
Web address: http://www.iucn.org

Mission

The International Union for Conservation of Nature (IUCN), formerly the World Conservation Union, is a hybrid organization of states, state agencies, and nongovernmental organizations that facilitates and encourages the conservation and equitable use of nature and natural resources. It was founded in 1948 and has gone by several names in its history (recently reverting from the World Conservation Union to its former title, the International Union for Conservation of Nature). Its Secretariat is headquartered in Gland, Switzerland, but it has a staff of one thousand experts in sixty nations which coordinates several theme-based programs that focus on issues such as forests, gender, and business and biodiversity. Under the IUCN are six Commissions, the most prominent being the Species Survival Commission, which updates the Red List of threatened species; the Commission on Environmental Law, which has facilitated negotiations on several conservation treaties; and the World Commission on Protected Areas. The IUCN is supported by over eleven thousand volunteer scientists in over 160 nations.

Significance for Climate Change

The IUCN position on climate change is that nations should reduce greenhouse gas emissions by 50 percent below 1990 levels by 2050; use ecosystem-based management to mitigate and adapt to warming; and prioritize efforts on behalf of vulnerable peoples and ecosystems. Its research and communications emphasize linkages between conservation, energy use, globalization, and climate change. Internally, much of this work is coordinated by its Climate Change Initiative.

In its role as an expert adviser, the IUCN has advised members to the UN Framework Convention on Climate Change (UNFCCC), the Convention on Biological Diversity, and other agreements on climate change mitigation, adaptation, and impacts, *inter alia*. It has alerted parties to the UNFCCC that if global average temperature were to rise 2° Celsius or more above preindustrial levels, massive extinctions and profound ecosystem changes would result. As a complement to its Red Lists, the IUCN has calculated the number of species that are currently of favorable conservation status that would be endangered by global warming (for example, 51 percent of currently unthreatened corals could be endangered by climate change).

The IUCN is one of the most prominent international organizations drawing attention to the linkages between warming, conservation, and human well-being. It assesses and promotes natural resource management practices that would aid mitigation and adaptation to global warming. For example, it advocates inclusion of REDD (Reduced Emissions from Deforestation and Degradation) in forthcoming climate agreements.

In conjunction with other organizations, the IUCN has worked to mainstream issues that have been sidelined by the dominant discussion, such as the role of indigenous peoples in climate policy and the relationship between gender and natural resource use.

Adam B. Smith

See also: Conservation and preservation; Environmental law; Environmental movement; International agreements and cooperation.

International waters

Category: Laws, treaties, and protocols

International waters are less regulated than national waters, allowing environmentally damaging activities to occur. In addition, as global warming continues, changes in shorelines and the melting of the Arctic ice cap may create international complications.

Key concepts

exclusive economic zone: a zone extending about 320 kilometers from a nation's shore in which all economic rights are granted to that nation

freedom of the seas: the principle that outside of water adjoining nations' shorelines, all nations have the right of free passage and use of the ocean's resources

territorial waters: areas within bodies of water that are within a nation's borders and subject to all laws and regulations of that nation

United Nations Convention on the Law of the Sea: an international agreement outlining the rights of nations regarding territorial waters, economic zones, and rights of passage

Background

Nearly three-fourths of the world's surface is covered by water. Each terrestrial or oceanic ecosystem has developed depending upon the availability of a certain type and quantity of water. Major bodies of water and the atmosphere work together to form a circular system. The atmosphere affects glaciers, ice caps, rivers, lakes, seas, and oceans, while simultaneously these bodies of water affect the atmosphere. Changes in one directly result in changes in the other. Because international waters lie outside national boundaries, caring for them is often a low priority.

Pre-Twentieth Century History

With the Portuguese and Spanish explorations of the fifteenth century, Europeans began to have aspirations for global dominance. New claims over the ocean were put forward, first by Portugal and Spain, which were quickly followed by the British and

Dutch. In the early seventeenth century, to reduce conflict, Dutch jurist Hugo Grotius put forward the freedom of the seas theory, that all oceans and adjoining seas could be used by everyone for any purpose. By the beginning of the eighteenth century, this principle was accepted by all European nations, with the provision that each nation controlled the first 4.8 kilometers of water adjacent to its shoreline. In certain strategic straits, all vessels were allowed to travel. This was the norm for the next two hundred years. The assumption during this period was that the oceans were so vast that nothing people could do would cause any harm to them.

Changes in the Twentieth Century

By the dawn of the twentieth century, it was becoming clear that not all the ocean's resources were inexhaustible. Not wanting to lose resources, in 1945 the United States claimed the entire continental shelf and all the water above it. Because of this, other countries extended their territorial claims to 19.3 kilometers off their coasts. The United Nations organized the Conference on the Sea in 1958. Two more conferences were held, leading to the negotiation of the United Nations Convention on the Law of the Sea. This agreement recognized new economic realities, as well as assigning responsibility for preserving the ocean's resources to all nations. A 19-kilometer territorial limit from the shoreline was formally recognized, as well as the 320-kilometer exclusive economic zone. The country controlling a given economic zone was formally charged with protecting the oceanic environment within it. Free navigation by all countries was allowed in this zone, in international waters, and through strategic straits. It was formally agreed that in international waters, all nations would share both the right to use environmental resources and the responsibility to protect them.

Effects of Climatic Changes on International Waters

The assumption that human actions can have little effect upon oceans has been discarded. The direct effect of Earth's oceans upon the climate has become well known, as the global effects of weather patterns such as the El Niño-Southern Oscillation

and La Niña have become understood. In the Northern Hemisphere, most of the Atlantic Ocean and much of the western Pacific Ocean increased in temperature by at least 1° Celsius between 1990 and 2004. In some locations, water temperatures have increased by more than 2° Celsius. One of the most visible maritime biological changes is the bleaching of coral and the slow destruction of coral reefs caused by changes in both temperatures and currents.

The currents are a necessary part of the ocean's ecosystem. They help equalize the salinity and other aspects of seawater. Some parts of the ocean are becoming fresher, and some are becoming more acidic. Aquatic life is often more sensitive to subtle environmental changes than are land-based creatures, and changes in their habitats can be devastating both for sea creatures and for the humans that depend on them.

The melting of the polar ice caps also poses potential problems for island nations and for some continental nations that have large amounts of territory close to sea level. Many of the islands in the Pacific Ocean are not very high above sea level. With the projected sea-level rise during the twenty-first century, these islands face the prospect of losing much of their land. This loss will place great pressure on their inhabitants to take drastic steps to survive. It may become necessary for other nations to accept immigrants from these areas.

Context

The oceans constitute an extremely complex system in themselves, and that system is merely a component in the larger and more complex global climate system. The full effect that climate change will have upon international waters remains unclear. Because historically much less study has been done in the world's oceans than on the land, scientists do not have the breadth of knowledge needed fully to understand physical and biological oceanic systems. Thus, it cannot be said with certainty when climatic changes will have catastrophic effects in maritime ecosystems. However, it is certain that those effects will not respect national boundaries. So much of the oceans lie outside of territorial waters that efforts to protect them must take place largely in international waters.

Climate change is already causing dramatic changes in places such as the Arctic Ocean. This not only affects the plants and animals in the region, but it is also stirring up new conflicts among the nations that border this ocean. Once again, the desire for economic supremacy is pushing nations to enforce their rights in the territorial and international waters of the Arctic. For example, in 2008, Norway detained Russian fishing trawlers that they claimed were in their territorial waters, while Russia claimed the area was open to all nations. Thus, climate change is causing not only physical changes but political changes as well. As changes in other international waters become more pronounced, similar conflicts might arise over resources in those areas that countries believe are vital to their well-being. The effect of climate change on international waters can directly touch people in all parts of the world.

Donald A. Watt

Further Reading

Gore, Al. *An Inconvenient Truth: The Crisis of Global Warming.* Emmaus, Pa.: Rodale, 2006. This popular book contains chapters on the effects of global warming on the polar ice caps and how they will affect the oceans.

Grotius, Hugo. *The Free Sea.* Reprint. Indianapolis: Liberty Fund, 2004. Along with the classic work on the subject, this volume includes a contemporary critique and Grotius' response.

Henn, Cathryn. "Trouble with Treasure: Historic Shipwrecks Discovered in International Waters, The." *University of Miami International and Comparative Law Review* 19 (2011): 141-196. Until recently, salvage of shipwrecks took place almost entirely within waters governed by some nation state. New technology allows detection and salvage of wrecks in deep waters outside any national boundaries. There is no existing legal framework for governing such salvage.

Strati, Anastasia, et al., eds. *Unresolved Issues and New Challenges to the Law of the Sea: Time Before and Time After.* Leiden, the Netherlands: Martinus Nijhoff, 2006. Focusing on areas of potential weakness in the treaty, this book examines how the document might apply to such issues as migrating fish populations.

United Nations. *United Nations Convention on Law of the Sea.* Hauppauge, N.Y.: Nova Science, 2008. The United Nations document that defines the international law of the sea.

See also: Arctic seafloor claims; El Niño-Southern Oscillation; La Niña; Marginal seas; Maritime climate; Ocean acidification; Ocean-atmosphere coupling; Ocean disposal; Ocean dynamics; Ocean life; Reefs; Sea-level change; United Nations Convention on the Law of the Sea.

Invasive exotic species

Categories: Animals; plants and vegetation

With climate change come changes in species distributions. Invasive exotic species in particular can have significant effects upon other species and the structure and function of entire ecosystems.

Key concepts

exotic species: a species found in an area where it is not native

habitat: an area normally inhabited by a particular species

invasive species: a species whose population grows out of control and negatively affects other species

niche: the role of a species within its ecosystem

preadaptation: the ability of a species to move into a new ecosystem or to assume a new niche as a result of adaptations it already possesses

Background

An exotic species is a creature found in an area to which it is not native. Species that succeed after arriving in new areas, whose population growth is not limited by natural controls, and that negatively affect other species in their new ecosystems are considered invasive exotics. Such species often evince preadaptations for surviving differences in climate, so global climate change increases the threat they pose of rapidly invading new ecosystems.

Many exotic plants can only exist in new areas with special care and conditions, such as soil fortified with nutrients, a special watering regime, artificial lighting, or protection from cold. Many landscape plants, agricultural plants, domesticated livestock, and pets are exotic plants or animals that benefit humans when brought to new areas. Invasive exotic species spread uncontrollably and become problems for native species and humans. Often, their invasive potential is not apparent until long after they have become established in a new niche.

Exotics and Global Warming

Global warming is one aspect of the more complex phenomenon of global climate change, and the bigger picture of climate change must be kept in mind relative to invasive exotics. Among other attributes, global climate change can include changes in temperature ranges; timing and amount of precipitation; wind patterns; frequency and intensity of storms; and frequency, intensity, and timing of lightning-started fires. Such changes can dramatically influence living creatures and favor invasive exotics.

Invasive exotic species are often adapted to be effective travelers and colonizers. Many plants can travel on animals, using hooked spines or sticky hairs that allow them or their seeds to adhere to clothes or fur. Lizards and insects may readily enter small openings in packing crates or suitcases and are inadvertently taken to new places. Other plants and animals can be dispersed over great

The grass Phragmites, shown here, probably always lived in North America but a recently arrived subspecies from Europe has become a nuisance since it out-competes native aquatic plants, including the native subspecies. On the other hand, it furnishes very poor habitat for other organisms. The only effective way to eradicate it is to burn it over several consecutive seasons, since the roots are deep and tough enough to survive a single burn. (© Steven I. Dutch)

distances by wind. Seeds often survive in the digestive tracts of animals that consume their fruit and are thus dispersed to distant locales. Wood pallets and goods, necklaces made with exotic seeds, and handicrafts made from palm leaves can contain insects and other plant pests or diseases that emerge after arrival at a new place.

Most creatures on the move do not survive the trip. Where they land is unpredictable, and most die, unable to cope with their new environment. They may be unable to reproduce for lack of a mate or appropriate breeding conditions. Those that survive are often adaptable and reproduce rapidly.

Biodiversity and Interspecies Relationships

Biodiversity is a result of adaptations to local conditions that hone the survival potential of each species. An ecosystem includes a community of living organisms interacting with their physical environment. The species fill unique niches, together fitting into the ecosystem like a hand fits into a glove. Each finger represents a species, and each has a defined space and role within the ecosystem. In a stable ecosystem, species work together smoothly. When a new species is introduced, it is like trying to fit a sixth finger into a five-fingered glove. The exotic species may displace a native species and take its place within the glove, or it may constrict the population of native species so that it carves out a new space within the glove. It may alter the glove in such a way as to form a suitable place in which to exist. These habitat alterations may affect the survival of native species and the ability of the ecosystem to function appropriately. Adaptation to specific conditions is beneficial only so long as those conditions prevail. Changes in the physical environment or in the species composition of an ecosystem can cause minor to major disruptions.

Global climate change reshuffles the deck of biodiversity, and new winners and losers are likely to emerge. Some species hold better cards than others: They are preadapted to specific changes in their ecosystems. Exotic species are often among those; they have already overcome numerous obstacles to establish themselves in their new habitat. Native species, especially specialists with specific

adaptations to an existing environment, are often unable to adapt to changed conditions.

Why Exotics Become Invasive

When exotic plants and animals move or are moved from their native habitat to a new area, they often leave competitors, predators, and diseases behind, allowing their populations to grow unimpeded. If a species produces many offspring and has an easy means of dispersing, its population can grow rapidly, displacing, disrupting, or eliminating other species. In such circumstances, it has become invasive. Native species can become invasive also, but only if changing conditions or human impacts eliminate their natural controls. Invasive exotic species are found everywhere in the world.

Prognosis

The melting pots of Florida, California, and Hawaii are experiencing severe problems from exotic invasives. They have subtropical climates that favor introduction of exotic plants and animals. Although often for local use, many such plants are also farmed for markets in more northern areas, where they might only survive indoors as pets, house plants, or outdoor perennial plants. Through the pet trade, many species of tropical fish, lizards, and parrots are breeding in Florida.

Interactions among exotics often exacerbate their effects upon an ecosystem, as introduced plants provide important food sources for introduced animals. In Florida, consumption of ornamental plants by exotic green iguanas (*Iguana iguana*), of plants and small animals by black spiny-tailed iguanas (*Ctenosaura similis*), and of animals as large as small dogs and cats by two-meter-long Nile monitors (*Varanus niloticus*) are regional problems that could spread northward with global warming. So, too, could the Burmese python (*Python molurus bivittatus*)—at up to six meters long and weighing ninety kilograms, one of the largest snakes in the world. It has been estimated that this snake could survive in the wild as far north as Washington, D.C. It presently breeds in the Everglades ecosystem, where more than two hundred have been captured. Sources for this population include pets lost or released by owners and some whose escapes were facilitated by hurricane-related damage.

Exotic plants once treasured for their blossoms or other attributes have invaded areas throughout North America and have the potential to spread farther north as the climate warms. Purple loosestrife (*Lythrum salicaria*), a native of Eurasia and widespread as an ornamental, replaces native plants and reduces habitat quality for wildlife as it forms dense stands in wetlands across North America. The Australian paperbark tree (*Melaleuca quinquefolia*) and Brazilian pepper (*Schinus terebinthefolius*), originally introduced as ornamentals, have formed dense monocultures in what had been habitats to a diversity of native plants and animals in south Florida. They are poised to spread northward as the climate warms. Kudzu (*Pueraria lobata*) vines, introduced from Asia to Gulf coast states in the 1800's to control erosion, often blanket ground, shrubs, and trees and are already expanding northward, following mild winters.

Major invasions of marine environments are also occurring around the world, as marine life is transported in the ballast water of ships. These invasions will alter natural ecosystems even further, as global climate change alters the temperature and salinity of the oceans, creating new opportunities for invaders.

Context

As ecosystems are altered as a result of global climate change and enhanced invasions of exotic plants and animals, dramatic shifts in the populations of native species will result, causing the extinction of many species that are unable to compete favorably with their new neighbors. Shifts in ranges will vary from species to species, depending on their dispersal abilities, breeding success, and a diversity of habitat needs. A warmer climate might favor extinction of cold-adapted species and northward shifts for southern species, but if favored food plants do not shift at the same time or if a new range includes a predator or disease not previously encountered, the move could mean extinction.

Jerome A. Jackson

Further Reading

Cox, George W. *Alien Species and Evolution: The Evolutionary Ecology of Exotic Plants, Animals, Microbes, and Interacting Native Species*. Washington, D.C.: Island Press, 2004. Considers the long-term ecosystem impacts of exotic, invasive species.

Occhipinti-Ambrogi, A. "Global Change and Marine Communities: Alien Species and Climate Change." *Marine Pollution Bulletin* 55 (2007): 342-352. Reviews linkages between exotic invasives in marine environments and climate change.

Randall, Roderick Peter. *A global compendium of weeds*. No. Ed. 2. Department of Agriculture and Food Western Australia, 2012. A compendium of 28,000 species. North American readers will recognize many as common and benign, but they have colonize places where they are not wanted.

Willis, K. J., and H. J. B. Birks. "What Is Natural? The Need for a Long-Term Perspective in Biodiversity Conservation." *Science* 314 (November 24, 2006): 1261-1265. Focuses on recognition and understanding of long-term versus short-term changes within ecosystems; addresses the need to disentangle impacts of long-term climate change from short-term impacts of invasive species and wildfires in order better to address conservation challenges.

See also: Convention on International Trade in Endangered Species; Ecosystems; Endangered and threatened species; Extinctions and mass extinctions.

Iran

Category: Nations and peoples

Key facts

Population: 66,429,284 (July, 2009, estimate) 81,824,270 (July 2015 estimate)

Area: 1,628,554 square kilometers

Gross domestic product (GDP): $842 billion (purchasing power parity, 2008 estimate) $1.371 trillion (purchasing power parity, 2015 estimate)

Greenhouse gas (GHG) emissions in millions of metric tons of carbon dioxide equivalent (CO2e): 288 in 1990; 480 in 2000, 715 in 2012.

Kyoto Protocol status: Ratified 2005

Historical and Geopolitical Context

In 1979, millions of Iranians succeeded in ousting Mohammad Reza Shah Pahlavi (1941–1979), the Shah of Iran and son of Reza Khan, who came to power in 1921 via a coup. While the Shah was popular in the West, some believe he tried to modernize Iran too rapidly and did not adequately adapt his political institutions to the economic and social changes that ensued. Inspired by hopes for democracy, economic prosperity for all classes, gender equality, and a leadership that would not allow Iranian culture to be swallowed up by the West, many Iranian women joined the revolt that stunned the world. The Iranian Revolution was led by Ayatollah R. Khomeini, who had been living in exile in Paris, France.

Today, the Islamic Republic of Iran is a unitary Islamic republic with a sole legislative body, the Islamic Consultative Assembly. Since 1989, the spiritual leader of Iran or Rahbar has been Ayatollah Sayyed Ali Khamenei. The capital of Iran is Tehran, and Farsi (Persian) is the official language. In 2000, crude and refined petroleum composed 85.4

percent of the nation's exports, which totaled more than $28 billion. Trading partners included Japan, China, Italy, and Taiwan. Petroleum and natural gas accounted for 57 percent of $181 trillion in revenues. Since 2005, Mahmoud Ahmadinejad has served as president of the Republic of Iran. While relations with the United States have been shaky, newly elected U.S. President Barack H. Obama emphasized the importance of communicating with Iran, a country the George W. Bush administration had called "dangerous."

In 2006, the International Atomic Energy Agency (IAEA) demanded that Iran halt uranium enrichment for its nuclear program, which Iran claimed was to be used for peaceful purposes. While President Obama eschewed Iran's pursuit of nuclear weapons, he said "it is important for us to be willing to talk to Iran, to express very clearly where our differences are, but where there are potential avenues for progress." In 2015 Iran signed an agreement to limit its nuclear programs, and some sanctions were lifted, but low global oil prices and concerns by foreign investors meant that economic improvements have been slow to materialize.

Impact of Iranian Policies on Climate Change

Iran ratified the Kyoto Protocol in 2005. The decision was endorsed by the Guardian Council, which discussed Iran's plan to adhere to the U.N. Stockholm Declaration regarding sustainable development; the Stockholm Declaration put forth principles that might guide the world's nations to preserve and enhance the human environment. As such, the Office of Climate Change in Iran's Environmental Protection Agency released a report stating that since the Kyoto Protocol came into force in February, 2005, new activities designed to decrease greenhouse gas (GHG) emissions were occurring under the U.N. flexible mechanisms program of the protocol. In addition, it was thought that there was a need to immediately sign the protocol, because an upcoming summit in 2005 was to focus on technology transfer and financial aid to be given by developed, Annex I nations to developing, non-Annex I nations, and only parties to the Kyoto Protocol could participate.

Iran as a GHG Emitter

Iran is ranked eighteenth of the highest GHG emitters. Developing nations, all of whose emissions rose during the first decade of the Kyoto Protocol, include China (now the top GHG emitter), India, Brazil, Mexico, South Korea, Indonesia, South Africa, and Iran. These eight nations accounted for 30.1 percent of global GHG emissions in 2000; the share of GHG emissions was 40 percent for the developing, non-Annex I countries. The developing countries have been increasing their shares; in 1990, the relative share was 32 percent. For the year 2000, twelve of the top twenty countries were Annex I countries, including seven of the top ten emitters. In 2000, the Annex I countries accounted for about 60 percent of the top-twenty GHG emissions. The number-one emitters of each group were the top two emitters overall: the leading developing, non-Annex I country, China, and the leading developed, Annex I country, the United States, now the second highest of the top-twenty GHG emitters. Together, China and the United States account for over one-third of total global GHG emissions.

While the Organization of Petroleum Exporting Countries (OPEC) Gulf States have the highest GHG emissions, data from 2000 show that—of the top twenty emitters—those with highest per capita emissions were the Annex I countries. Australia, the United States, and Canada ranked fifth, seventh, and ninth, respectively. Their per capita emissions (7.0, 6.6, and 6.1 metric tons per person) were approximately double the emissions of the highest-ranked developing country in the top twenty (South Korea, at 3.0 metric tons), and they were six times those of China (1.1 metric tons). Of note, the population density of Iran is sixteen persons per square kilometer in a country that is relatively small.

Summary and Outlook

It is indeed a paradox that OPEC countries concerned with manipulating the production of petroleum products so as to affect global financial markets have expressed concern with green energy. In January, 2009, Iran inaugurated its first solar energy power plant, adding a modest 250 kilowatts of solar energy to the country's energy grid. The power plant uses parabolic mirrored troughs to gather sunlight where it is used to produce steam and generate electricity. This solar thermal plant is part of 4,075 small-scale solar thermal installations throughout Iran, which comprise 3,781 solar water heaters for residential use, and 294 public baths that are heated with solar thermal energy. Putting an economic value on renewable resources makes Iran, with its abundance of sunlight, rich in "solar energy potential." Iran took its first step toward the large-scale realization of that potential with the inauguration of its first solar energy plant, which was constructed with domestic materials and labor in Shiraz.

Cynthia F. Racer

Further Reading

Al-Jazeera. Iran nuclear deal one year on: Expectations vs reality. http://www.aljazeera.com/programmes/countingthecost/2016/07/iran-nuclear-deal-year-expectations-reality-160716140251975.html?src=ilaw [16 July 2016]

Ardehali, M. M. "Rural Energy Development in Iran: Non-renewable and Renewable Resources." Renewable Energy 31, no. 5 (April, 2006): 655–662. Identifies problems and difficulties encountered in the socioeconomic infrastructure as related to rural energy development; presents the nonrenewable and renewable energy resources.

Axworthy, M. A History of Iran: Empire of the Mind. New York: Basic Books, 2008. Comprehensive history from Zoroaster and the Greeks to the Pahlavi dynasty.

Gary Samor, Belfer Center for Science and International Affairs. "The Iran Nuclear Deal: A definitive Guide" (2015). The proposed agreement renders it unlikely Iran will be able to produce plutonium weapons, and makes it harder to construct covert facilities, but cannot entirely rule them out.

Sagar, A. D. "Wealth, Responsibility, and Equity: Exploring an Allocation Framework for Global GHG Emissions." Climatic Change 45, nos. 3/4 (June, 2000): 511–527. Explores the framework derived from the climate convention regarding OPEC nations.

See also: Kyoto Protocol; Organization of Petroleum Exporting Countries; Saudi Arabia; United Nations Framework Convention on Climate Change.

Islands

Category: Geology and geography

Global warming causes thermal expansion and ice melt, increasing seawater volume and provoking more frequent and more severe storms. As a result, small, low-lying islands are threatened with ecological degradation and submergence.

Key concepts

brackish water: mixture of freshwater and seawater

calving: separation of a large portion of ice from a glacier or ice shelf, creating an iceberg

coral atoll: an island or islet composed of a lagoon encircled wholly or partly by a coral reef

methane: a gas whose greenhouse effect is considerably stronger than that of carbon dioxide

nissology: the study of islands

thermal expansion: a heat-induced increase in the volume of a liquid or gas

Background

Global warming threatens the survival of low-lying islands, deltas, and beaches. Over two thousand scientists, in conjunction with the United Nations, have estimated that by the year 2100 sea levels will rise by between 9 and 88 centimeters. Rising sea levels threaten human, animal, and plant life, as arable land and potable water are compromised. Refugee migrations and species extinctions result.

Conditions

For millennia before the nineteenth century, Earth's sea levels remained relatively stable. However, as the pace of industrialization and urbanization accelerated around the globe, sea levels began to rise by milliliters per year. Tide gauges and satellite monitoring indicate that this rise has increased in recent decades. Two factors contribute to increases in sea volume. One is thermal expansion, whereby warming waters increase in volume. The other is the melting of ice sheets and glaciers at the North and South Poles and in the mountains of the Andes, Alps, and Himalayas.

Retreating ice has been noted in all these regions. Ice melt in Greenland begins earlier each summer, leaving ever smaller areas to refreeze by winter. The runoff water from the melt penetrates glaciers, loosening their attachment to rock surfaces. Large chunks then break away, in a process known as "calving," launching masses of ice several square kilometers in area that float away, eventually melting in warmer waters. The snow and ice of ski fields in Switzerland have disappeared, as the snow line rises in altitude. Less snow and ice on the surface of the Earth reduces the deflection of sunlight, or albedo, thereby causing more warming in a postive feedback loop. Furthermore, thawing allows the escape of methane gases that have been locked in the frozen Earth, further strengthening the greenhouse effect.

Consequences

Islands are bodies of land surrounded by water, ranging in size from continents to tiny atolls. They can be found on all continents, oceans, and seas. One type of island rises from an underwater, oceanic volcano, as the tip emerges above the water surface to form the island. The other type is an elevation of land on a continental shelf that rises above surrounding waters. Hawaii is an example of the former; the British Isles are instances of the latter. The smaller and lower in height an island is, the more vulnerable it is to rising seas. In addition to threats from rising sea levels, small islands are endangered by tropical storms, including hurricanes and cyclones. If the frequency and ferocity of these storms increase, the risk to islands does as well.

The consequences of this vulnerability are devastating in several respects. In addition to being submerged, small land bodies are being eaten away at their edges by erosion. Moreover, as salt water penetrates an island, salinity enters the water table below the island's surface. The more salinity that freshwater absorbs, the less potable it is for drinking. Moreover, brackish water stunts or kills crops, reducing the food supply. Not only is the sustenance of humans threatened but that of wildlife is as well.

Some islands are being abandoned by their inhabitants as the sea consumes the land. The two thousand inhabitants of the Carteret Islands of Papua New Guinea are refugees from their devastated habitat. Lohachara Island in the Bay of Bengal disappeared under water in 2006; its residents fled to the mainland. Tuvalu, with a population under ten thousand, is a country of nine narrow coral atolls in the South Pacific. New Zealand has agreed to accept thirty-five of its inhabitants per year as environmental refugees.

Prevention

Small island countries produce negligible amounts of greenhouse gases (GHGs). However, they suffer most directly and critically the consequences of global warming that are due to the major emitters. Small island countries act cooperatively through a number of organizations. The Global Islands Network (GIN) is an information clearinghouse and resources cooperative for islands in all regions of the world. Complementing it is the Small Islands Development Network (SIDSnet), specializing in communications and information technology to support island maintenance and development. Members of the Alliance of Small Island States (AOSIS) coordinate their efforts through their respective U.N. diplomatic missions. The International Small Islands Studies Association (ISISA), which has held ten world conferences, is a professional organization that supports research about small islands. The International Scientific Council for Island Development (INSULA) supports the economic, technical, ecological, social, and cultural programs of the world's islands and publishes the journal *INSULA: The International Journal of Island Affairs*, which specializes in nissology. Another periodical is the *Island Studies Journal.*

Context

Small, low-lying islands are both idyllic and fragile fragments of the Earth's surface. Somewhat like a canary in a coal mine, their extinction is an early warning sign of ecological danger. Before they submerge, erosion and saltwater intrusion render them uninhabitable. Larger, low-lying surfaces, such as delta regions and beaches, are also threatened. As more islands lose the means to sustain their populations, the number of environmental refugees in the world increases, putting further pressure on the planet's remaining resources.

The inhabitants of small islands have contributed the least to global warming, yet they suffer the worst of its initial consequences. They therefore unavoidably raise issues of social and environmental justice. Global warming admits of several feedback loops. The more the conditions for warming accumulate, the more such conditions are strengthened. For any inhabitant of a low-lying island or mainland surface, the question of the survival of one's physical environment is a daily concern.

Edward A. Riedinger

Further Reading

Adamson, Joni, Mei Mei Evans, and Rachel Stein, eds. *The Environmental Justice Reader: Politics, Poetics, and Pedagogy.* Tucson: University of Arizona Press, 2002. Collection of texts that considers the legal and diplomatic issues and consequences of ecological damages.

Diamond, Jared M. *Collapse: How Societies Choose to Fail or Succeed.* New York: Viking, 2005. Noted author examines ecological and environmental causes of collapse of various societies, including several island civilizations, particularly that of Easter Island.

Gerdes, Louise I. *Endangered Oceans: Opposing Viewpoints.* San Diego, Calif.: Greenhaven Press, 2004. Collection of articles examining various ocean-related issues, including whether sea levels are rising.

Lisagor, Kimberly, and Heather Hansen. *Disappearing Destinations: Thirty-seven Places in Peril and What Can Be Done to Help Save Them.* New York: Vintage Departures, 2008. Places the threat of disappearing islands within the context of other endangered ecosystems such as retreating glaciers, thawing permafrost and peatland, and diminishing boreal forest.

Rudiak-Gould, Peter. "Promiscuous corroboration and climate change translation: A case study from the Marshall Islands." *Global Environmental Change* 22, no. 1 (2012): 46-54. The closest word for "climate" in the language of some islands has a much broader meaning and includes the entire cosmos as well as the social environment. Overly literal interpretation of local terms for climate can reinforce simplistic narratives of climate change while missing the actual local understanding.

Tompkins, Emma L., et al. *Surviving Climate Change in Small Islands: A Guidebook.* Norwich, Norfolk, England: Tyndale Centre for Climate Change Research, 2005. Describes climate change adaptation scenarios for small islands and offers plans and implementation strategies.

See also: Alliance of Small Island States; Barrier islands; Coastal impacts of global climate change; Easter Island; Human migration; Maldives; Saltwater intrusion; Sea-level change; Tuvalu.

Isostasy

Category: Geology and geography

The theory of isostasy allows geologists and cryologists to understand the growth or retreat of ice sheets, based upon the movement of the land in response to its increasing or decreasing glacial load.

Key concepts

continental crust: the uppermost layer of the Earth, granitic in composition, with a density of about 2,700 kilograms per cubic meter

flexure: the way in which an elastic plate bends

forebulge: an elevated area just beyond the area depressed by an ice sheet

glacial isostatic adjustments: vertical motions of the crust to restore it to preglacial elevations

mantle: the layer of the Earth beneath the crust, peridotite in composition, with a density of about 3,400 kilograms per cubic meter

viscosity: a measure of how easily a material flows

Background

Earth's continental crust floats on a geological layer called the mantle. A continental ice sheet weighs a great deal, so an advancing ice sheet will cause the crust to subside beneath it. When the ice melts, the land it had depressed will rebound upward. These vertical motions of hundreds of meters control the growth and retreat of the glaciers, as well as the directions in which meltwater drainage flows. Models of mantle deformation based on data from the last glaciation can be applied to areas beneath contemporary ice sheets to better understand their dynamics.

Theory of Isostasy

During the survey of Mount Everest, investigators found that their plumb bob was not attracted by the mass of the Himalayan mountain range as much as they had expected. They concluded that the mountain range was supported by some sort of mass deficiency beneath it. This led to the theory of isostasy, which holds that Earth's mantle is denser than its crust, which floats upon it; it is thicker or less dense beneath mountain ranges. This theory has been found to be generally true, except for some unusual circumstances such as the trenches near subduction zones. These exceptions have been explained by treating the surface of the Earth as an elastic plate that flexes and spreads out gravitational loads.

The continental ice sheets that developed in Canada and Northern Europe placed enormous loads on the crust of the Earth beneath them. Just as a boat rides lower in the water when fully loaded than when empty, the crust of the Earth sank

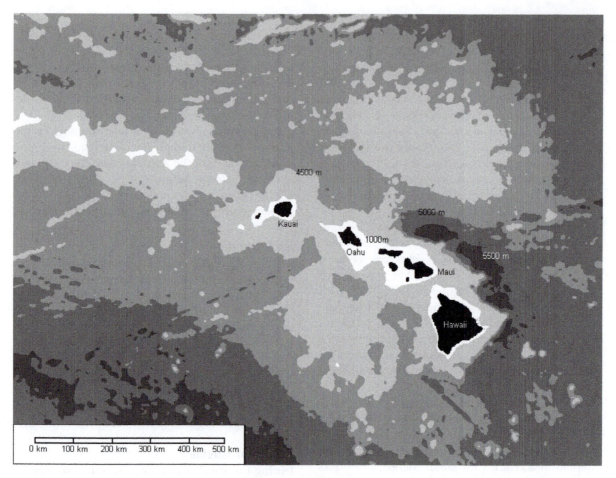

The deep trough northeast of Hawaii and the depressions south of the island chain are due to the immense weight of the volcanoes depressing the crust. Further away the sea floor is actually higher than the regional average because the crust flexes upward, a feature known as a forebulge. As erosion stripped away older volcanoes, the crust rebounded and sediment filled the trough, so the trough is not visible beyond Oahu. (© Steven I. Dutch)

further into the mantle to accommodate this additional load. The flexure produced by this motion on the elastic plate caused parts of it just beyond the loads to rise up as a forebulge. When the ice melted, the load disappeared, areas that had been depressed began to rise up, and the forebulge started to subside. These changes in elevation are often referred to as glacial isostatic adjustments. Although the ice disappeared thousands of years ago, the crust continues to move vertically today, because the mantle is incredibly viscous. Glacial rebound analysis suggests it has a viscosity between $10^{20.5}$ and 10^{22} Pascal seconds. (For comparison, the viscosity of water is 10^{-3} Pascal seconds, and the vicosity of molasses is 10^4 Pascal seconds.)

Application of the Theory

Waves coming ashore often cut terraces, erode cliffs, and produce beaches at elevations close to sea or lake level. If the land rises relative to these levels, a series of terraces may result. Organic remains can be used to date these terraces, and geologists have discovered that terraces of the same age at different locations are not at the same elevation. Those closest to where the glacial load was greatest are higher, as they have experienced greater isostatic rebound. Furthermore, terraces formed long ago will show a greater slope, presently, than those formed more recently. By studying such patterns of strandlines, a history of isostatic rebound can be reconstructed. For example, areas around Hudson Bay, in Canada, have already rebounded over 300 meters, continue to rise at rates on the order of 1 centimeter per year, and probably have another 100 meters of uplift to go before isostatic balance is obtained.

The history of isostatic rebound in a given area can be used to make estimates of mantle viscosity and of how this viscosity changes with depth. Computer models can help fill in the gaps where data are absent. Where there is a significant lack of data, suites of scenarios may be obtained, all of which fit the model equally well, and all of which are internally consistent. As additional data become available, some scenarios will be eliminated, and the preferred scenario can shift rather dramatically. For example, preferred scenarios in the mid-1990's envisioned 4 kilometers of ice in Greenland but only 2.5 kilometers of ice in central Canada during the last glacial maximum. A few new data points led to revisions in 2004 that reversed this distribution and placed parts of Greenland in the area of the glacial forebulge.

These scenarios are used by climate scientists to try to determine how rapidly the ice sheets on Greenland and Antarctica might be shrinking. The elevation of the top of the ice sheets and the gravitational attraction at points above the ice sheets can be determined from satellites. However, to calculate how much the volume of ice is changing, the elevation change of the bottom of the ice sheet must also be considered, and isostatic adjustment models are used to estimate this change.

Isostatic adjustments may also have influenced the growth and retreat of the continental glaciers. The accumulation of ice occurs much more rapidly than can be accommodated by isostatic subsidence. Just as the crust of the Earth is still rebounding thousands of years after the load was removed, it subsided over a period of thousands of years during and after the addition of the load. These changes in elevation, frequently called "bedrock lag," affect local climatic conditions and thus the growth or retreat of the continental ice sheets.

Another important consequence of isostatic adjustment concerns drainage of the large lakes at the southern margin of the continental glacier in North America. The route and ultimate destination of the drainage were likely important factors influencing global temperatures near the end of the last glaciation. These lakes consisted of freshwater, most at 4° Celsius, which would have floated on the more saline water in the North Atlantic. Such a freshwater cap would be unlikely to sink initially, and it would take a long time for the winds to mix it sufficiently to permit sinking. The cap may have slowed or turned off the thermohaline circulation, also called the global conveyor belt or the meridional overturning circulation (MOC). The North American lakes drained into the Atlantic Ocean through the Mississippi River Valley, the Hudson River Valley, and the St. Lawrence River Valley, as well as draining northwest to the Arctic Ocean.

Context

Isostasy provides a theory to explain the vertical movements of Earth's crust as the load of continental ice sheets is added or removed from large areas. It permits us to estimate volume changes of contemporary ice sheets by providing estimates of the crustal movements taking place beneath them. By constraining drainage histories of the glacial lakes that formed near the end of the last glaciation, it makes it possible to determine where and when cold, freshwater pulses may have flooded out over the oceans, with their potential radically to modify Earth's climate.

Otto H. Muller

Further Reading

Fowler, C. M. R. *The Solid Earth: An Introduction to Global Geophysics*. 2d ed. New York: Cambridge University Press, 2005. A somewhat technical text that thoroughly covers the flexure of elastic plates. Illustrations, figures, tables, map of the isostatic uplift of Fenno-Scandia, bibliography, index.

Molnar, Peter, Philip C. England, and Craig H. Jones. "Mantle dynamics, isostasy, and the support of high terrain." *Journal of Geophysical Research: Solid Earth* 120, no. 3 (2015): 1932-1957. Analysis of gravity data show that isostasy is sufficient to explain the elevation of high terrain, and that forces within the mantle account for only a small amount of elevation.

Peltier, W. R. "Chapter 4: Global Glacial Isostatic Adjustment and Modern Instrumental Records of Relative Sea Level History." In *Sea Level Rise: History and Consequences*, edited by B. C. Douglas, M. S. Kearney, and S. P. Leatherman. San Diego, Calif.: Academic Press, 2001. Explains how isostatic computations are done. Charts, figures.

Ruddiman, William F. *Earth's Climate Past and Future*. 2d ed. New York: W. H. Freeman and Company, 2008. Ruddiman's treatment of bedrock lag, which he calls "Delayed Bedrock Response Beneath Ice Sheets," explains this phenomenon with many excellent figures, making it easy to understand. Illustrations, figures, tables, maps, bibliography, index.

See also: Coastline changes; Deglaciation; Earth history; Earth structure and development; Freshwater; Glaciations; Glaciers; Ice shelves; Meridional overturning circulation (MOC); Plate tectonics; Thermocline; Thermohaline circulation.

Italy

Category: Nations and peoples

Key facts

Population: 58,126,212 (July, 2009, estimate); 60,674,003 (2015 estimate)

Area: 301,230 square kilometers

Gross domestic product (GDP): $1.82 trillion (purchasing power parity, 2008 estimate) $2.17 trillion (purchasing power parity, 2015 estimate)

Greenhouse gas (GHG) emissions in millions of metric tons of carbon dioxide equivalent (CO2e): 516.9 in 1990; 582.5 in 2004; 567.9 in 2006; 496.3 in 2010, 465.2 in 2012

Kyoto Protocol status: Ratified 2002

Historical and Political Context

Italy has played a major role in the religious, philosophical, and artistic development of the world. For centuries, Italy served as the central point from which empire, religion, thought, and artistic creation emanated. The Roman Empire spread its political system and its culture throughout Europe and the Mediterranean. After its collapse, Rome became the center of Catholicism, and it was in Italy that the Renaissance began in the fourteenth and fifteenth centuries, eventually spreading across Europe and into the British Isles. It revived the thought and art of Antiquity with Humanism and rediscovered the human creative ability.

Politically, Italy has had many periods of political unrest. During the thirteenth century there were fierce battles between the imperial and religious powers. The eighteenth century saw regional factions competing for political power. And again in the twentieth century, Italy's government experienced serious instability. Under the republic established in 1946, some fifty different political groups have been in power. In contrast to its internal political volatility, Italy is a consistent supporter of international organizations. Italy joined the United Nations in 1955 and is a member of five other international organizations, including the North Atlantic Treaty Organization, the organization of industrialized nations known as the G-8, and the World Trade Organization. Italy was a founding member of the European Union (formerly the European Community). Italy's main trading partners are in the European Union; the country trades mostly with Germany, France, and the United Kingdom. It also trades with the United States and ranked twelfth among the United States, trading partners in 2008.

Although Italy has the world's eighth-largest economy, its economy stagnated in the 2000s. In order to stimulate the economy, the government spent large sums, pushing the public debt level to 135 percent of GDP by 2014. Italy's debt was second only to that of Greece, but in contrast to

economy in the world. While Italy possesses few natural resources other than natural gas, which is primarily in the Po Valley and in the Adriatic Sea, it imports a large quantity of raw materials for processing. In addition approximately 80 percent of its energy resources are also imported. Its major areas of manufacturing include precision machinery, motor vehicles, chemicals, and clothing. Italy also has a significant oil refining industry. This activity is one of the main sources of increased CO_2 emissions, amounting to a gain of 2.4 million metric tons. The other major source of increased greenhouse gas (GHG) emissions is road transport with a substantial increase in the consumption of diesel fuel. The increase in the use of fossil fuels to produce electricity has also greatly increased the GHG emissions. Production of the energy supply and energy use accounted for 60 percent of the GHG emissions in 2006. The transport sector was the source of another 23.5 percent.

Italy as a GHG Emitter
According to data collected by the European Environment Agency, Italy ranked third among the members of the European Union in GHG emissions. In 1990, the base year for the Kyoto Protocol, Italy had GHG emissions totaling 516.9 million metric tons of carbon dioxide equivalents. Its emissions target under the Kyoto Protocol is 483 million metric tons. From 1990 to 1995, emissions remained more or less the same; the following years were then a period of substantial increase. From 2002 to 2006, Italy's average GHG emissions were 571.4 million metric tons, which was an average increase of 10.6 percent for the period. The amount of GHGs emitted by Italy in 2006 showed a small decrease to 567.9 million metric tons but was still an increase of 9.9 percent over its 1990 emissions and well in excess of its Kyoto Protocol target. Its 2012 level of 465.2 met its Kyoto target. Italy was responsible for 11 percent of the European Union's GHG emissions.

Summary and Outlook
Among the measures and policies in place in Italy are the use of abatement technologies in the

Greece, Italy's debt is mostly owned internally. Italy also has a serious economic division between north and south, with the industrialized north having high levels of income and low unemployment, while the mostly agrarian south is poorer and has higher unemployment.

Impact of Italian Policies on Climate Change
Although Italy underwent some industrialization during the nineteenth century, it remained primarily an agricultural economy until after World War II. Agriculture is still important to the country. Italy produces a wide variety of products, including grapes, olives, rice, sugar beets, and soybeans. It is the largest producer of wine in the world. The agriculture sector employs a large portion of Italy's workforce. Approximately 1.4 million people are employed in farming. Tourism continues to be important in the Italian economy. The northern section of Italy where manufacturing is concentrated has one of the highest per capita incomes in the world, while the south remains poor by world standards. After the war, Italy became more and more involved in manufacturing and is now the sixth largest market

production of adipic acid and the replacement of fossil fuels with biomass for heating in the household sector. Italy plans to initiate additional measures, Kyoto mechanisms, and carbon sink activities to reduce emission further. However, the proposed measures have not quite met the –6.5 percent target of the Kyoto Protocol.

Italy has been slow to regulate electrical consumption, and austerity is not a popular idea among Italian citizens, but the European Union is proposing a program for changing the source of energy production to that of renewable resources. Other significant reductions in GHGs emitted by Italian industries should result from the restrictions that will be placed upon them under the European Union Emission Trading Scheme. Reduction of GHG emissions and control of global warming is important to Italy, which is already seeing the effects of global warming on its land area. The melting of Alpine glaciers is causing problems in regard to the Swiss/Italian border, which will probably need to be re-delineated.

Shawncey Webb

Further Reading

Archer, David. *Global Warming: Understanding the Forecast.* Malden, Mass.: Blackwell, 2007. Targeted to undergraduates. Excellent discussion of global warming and its effects, combined with a concise scientific explanation of the phenomenon. Also looks at economic effects of global warming. Chapters on GHGs and fossil fuels.

Charnovitz, Steve, and Gary Clyde Hufbauer. *Global Warming and the World Trading System.* Washington, D.C.: Peterson Institute for International Economics, 2009. Discusses the reduction of GHGs and its relation to trade and trade organizations, especially the World Trade Organization. Suggests methods to reduce GHGs without harming carbon-intensive industries, both domestic and international. Appendix on biofuel.

Clapp, Jennifer, and Peter Dauvergne. *Paths to a Green World: The Political Economy of the Global Environment.* Cambridge, Mass.: MIT Press, 2005. Good discussion of environmental issues and global warming in respect to trade and the world economy.

Economist, The. The Italian Job, 30 January 2016. http://www.economist.com/news/finance-and-economics/21689630-reviving-italys-economy-will-require-sacrifices-not-just-italians-also [9 July 2016] Summarizes the economic problems plaguing Italy, which is one of the largest economies in the European Union but suffering from severe debt and dysfunctional regulations.

European Environment Agency. SOER 2015 The European Environment: Italy, 2015. http://www.eea.europa.eu/soer-2015/countries/italy (3 July 2016). Summary of environmental statistics and issues for Italy, with links to key sources (many in Italian).

Locke, Richard M. *Remaking the Italian Economy.* Ithaca, N.Y.: Cornell University Press, 1997. Good for understanding changes in the Italian economy that have increased Italy's GHG emissions. Pays special attention to the automobile and textile industries.

Shogren, Jason F. *The Benefits and Costs of the Kyoto Protocol.* Washington, D.C.: AEI Press, 1999. Excellent for understanding the Kyoto Protocol.

See also: Annex B of the Kyoto Protocol; Europe and the European Union; Kyoto Protocol; United Nations Framework Convention on Climate Change; Venice; World Trade Organization.

Japan

Category: Nations and Peoples

Key facts

Population: 127,078,679 (July, 2009, estimate); 127,110,047 (2015 census)

Area: 377,873 square kilometers

Gross domestic product (GDP): $4.348 trillion (purchasing power parity, 2008 estimate); $4.901 trillion (purchasing power parity, 2015 estimate)

Greenhouse gas (GHG) emissions in millions of metric tons of carbon dioxide equivalent (CO_2e): 1,154 in 1990; 1,223 in 2000; 1,216 in 2006; 1,345 in 2012.

Kyoto Protocol status: Ratified 2002

Historical and Political Context

A strong Asian island nation fiercely protective of its national identity and independence, Japan always sought to be at least on equal footing with the known leading nations of the ages. Thus, when industrialization with its attendant side effect of massive emission of carbon dioxide (CO_2) into the atmosphere gave Western nations a decisive military and economic edge by the middle of the nineteenth century, Japan chose to rapidly catch up and industrialize during the Meiji Restoration announced in 1868.

On the other hand, when concern for global warming led the leading nations of the world to seek for means to reduce the negative impact of their industry and consumers and traffic on the Earth's atmosphere by the end of the twentieth century, Japan took a leading role in researching and committing to less environmentally damaging alternatives. It certainly spoke to the Japanese self-understanding as a leading innovative nation that the decisive protocol to the United Nations Framework Convention on Climate Change was adopted in Kyoto, an ancient capital of Japan, on December 11, 1997. Known since as the Kyoto Protocol, the agreement represents a strong international commitment to combat global warming.

In a pattern still typical by the early twenty-first century, once Japan decided to industrialize, the government worked very closely with private enterprise and could draw on a well-educated, hardworking, and frugal workforce. This public-private partnership contributed to Japan's fast industrial rise, albeit at the price of environmental damage.

Japan's first encounter with industrial pollution occurred in 1878. Runoff water contaminated with inorganic copper compounds from the privately owned Ashio Mine north of Tokyo caused copper poisoning among farmers and villagers living downstream. Tanaka Shozo, the local member of the lower house of Japan's new parliament, publicly raised the issue. As a result the company paid compensation, and the government erected runoff prevention, yet some problems remained. Well into the middle of the twentieth century, most Japanese pollution cases followed this pattern of privately caused pollution, public reaction, and increasingly vigorous financial, governmental, and legal redress.

An extremely negative side effect of Japan's drive to industrialization despite a surprising lack of natural resources such as coal, oil, iron, and nonferrous metals was its embarkation on imperialism and colonialism. Copying the West, Japan created a colonial empire that at its peak in 1932 encompassed Korea, Taiwan, and the puppet state of Manchukuo, carved out of Northeast China. Conflict with the United States over Japan's war in China after 1937 led to an American oil embargo of Japan in 1941 that persuaded Japanese militarists to attack the United States at Pearl Harbor on December 7, 1941, drawing the Americans into World War II. After surrendering unconditionally on August 15, 1945, Japan lost its colonies. To become an economic leader again was a major challenge.

In 2011, Japan suffered one of the worst national disasters in its history. A magnitude 9.0 earthquake, the fourth largest worldwide since modern record keeping began, struck northern Japan. The resulting tsunami reached heights of 6 meters in places with run-up heights over 30 meters. Total casualties numbered almost 16,000 deaths, almost all drowned by the tsunami. But the worst environmental effect was the meltdown of reactors at the Fukushima nuclear plant. The plant survived the earthquake but the tsunami flooded the plant's backup generators, and the plant lost cooling

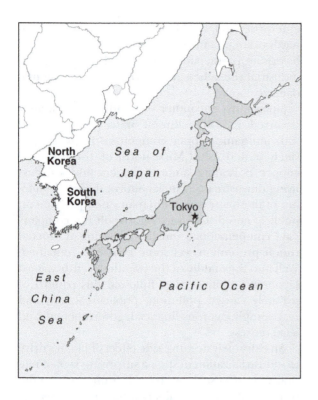

capacity. The radiation leakage was second only to the Chernobyl (Ukraine) disaster of 1986, but affected fewer people because most of the radiation blew out to sea. This event is variously known as the Sendai earthquake after the nearest city, or Tohoku earthquake after the prefecture.

Impact of Japanese Policies on Climate Change

Japan's post-World War II political decision to promote Japanese economic recovery included a broad national consensus to rebuild and expand Japan's industry. This national aim was aided by the United States, who wanted a strong Japan as a free world counterpart to communist aggression in Europe and Asia during the Cold War. With virtually no natural fuel resources such as oil and coal, Japan had to import these from abroad, as well as almost all raw materials for industrial production of steel and industrial manufacturing. In this situation, Japanese policy promoted manufacturing of consumer goods, chemicals, and machines primarily for export to pay for its import of fuel and raw materials.

The strong pro-industry policies of Japan led to an initial postwar neglect of environmental protection and the outbreak of three major pollution incidents in 1956, 1961, and 1965. There was strong public reaction to these incidents. Two involved mercury poisoning, and the 1961 case resulted in poisonous smog created by unregulated burning of petroleum and crude oil waste by the Showa Yokkaichi Oil Corporation. As a result of public pressure and successful lawsuits based on irrevocable scientific evidence for the anthropogenic cause of the pollution incidents, Japan's public policy shifted toward more effective environmental protection.

In 1970, Japan passed six environmental laws and put much more teeth into eight existing ones. In 1971, the Environmental Agency was established and given a broad mandate to protect Japan's environment from air, soil, water, and other pollution. Exhaust by cars was also addressed aggressively and industrial polluters were held accountable by passage of the 1973 Pollution Health Damage Compensation Law. Ironically, the drastic oil price increases of the first oil crisis of 1973 forced Japanese companies, utilities, and consumers to find immediate means of conserving fuel to survive economically in the face of Japan's near total dependency on fuel imports. Its industry became more fuel efficient with lower emissions.

The first white paper of Japan's Environmental Agency in 1984 strengthened environmental awareness among Japanese corporations and citizens, as did further policy papers, campaigns, and surveys. The 1993 Basic Environment Law of Japan explicitly restricted industrial emissions considered harmful for the climate, as well as calling for improved energy conservation, recycling, and pollution control programs.

Indicative of Japan's serious commitment to environmental protection, including addressing and combating climate change, in 2001, the Environmental Agency was upgraded to the Ministry of the Environment. In 2008, Tetsuo Saito became its Minister. He was chosen because of his scientific background, holding a Ph.D. in physics, and his knowledgeable commitment to clean energy technology.

Japan as a GHG Emitter

According to Japan's emission statistics reported to the United Nations in 2008 for the years from 1990 to 2006, with 1.22 billion metric tons of greenhouse gases (GHGs) emitted in 2006, Japan was the world's fifth largest emitter of CO_2 and its equivalents. By comparison, the top two emitters were the United States with 6.37 billion metric tons in 2006, a figure closely matched or perhaps even surpassed by 8 percent by the People's Republic of China, according to a 2007 estimate by the Netherlands Environmental Assessment Agency. When Japan's emissions were compared to the size and output of its national economy, the Japanese economy proved remarkably emissions-effective. This was particularly in contrast to the number three and four GHG emitters, India and Russia. In 2008, Japan had the world's second largest gross domestic product, $4.84 trillion, just behind the United States, and remained in fifth place among GHG-emitting countries.

Under the complex United Nations measuring system for GHGs, nations can earn some credit or debit to their overall emissions when they offset emissions through land use, land-use change, or forestry (abbreviated as LULUCF). This was an incentive for nations to promote reforestation and protect forests, as was done in Japan vigorously since the 1980s. As a result, Japan's overall 2006 emission was lowered by 82.6 million metric tons of GHGs. In addition, only anthropogenic emissions are considered, excluding, for example, CO_2 from volcanic eruptions.

In Japan, as elsewhere, CO_2 was the major GHG emitted, accounting for 1.156 billion metric tons in 2006 (without LULUCF credits). Under United Nations standards, other GHG emissions are added to those of CO_2 with a weighted formula. For example, each ton of methane is multiplied by 21, each ton of nitrous oxide by 310, and fluorocarbons each have specific values. In 2006, Japan emitted a CO_2 equivalent of 21.8 million metric tons of methane, a 23.6-million-metric-ton equivalent of nitrous oxide and just 15.4 million metric tons of CO_2 equivalents of the ozone-layer-damaging fluorocarbons and sulfur hexafluoride (all excluding LULUCF credits).

As an industrialized, Annex I country that signed and ratified the Kyoto Protocol, which became effective on February 16, 2005, Japan has committed to reduce its emissions to 6 percent below their 1990 level no later than 2012. This commitment is slightly higher than the 5.2 percent average for industrialized nations. However, Japan's 2006 data show an increase of 5.3 percent from the 1.154 billion metric tons of GHGs emitted in 1990, increasing to 1,345 by 2012. When the figures are used that account for possible LULUCF credits, Japan even increased its emissions by 5.8 percent as reforestation and other measures had reached their intended limits in the 1990s and had leveled off.

Japanese politicians and experts have pointed out that Japan had relatively low emission levels in 1990 already. Ever since the 1973 oil crisis, Japan had invested great resources in cleaner factories and power plants. Japan reduced emissions considerably earlier than other countries. Despite this relative disadvantage, Japan ratified the Kyoto Protocol in 2002 and remained committed to its targets.

Summary and Outlook

As a leading industrial nation creating the world's second largest gross domestic product, behind the United States, and running the third largest economy, behind the United States and the People's Republic of China, if measured by purchasing power parity, in 2008, Japan would continue to emit a huge share of the world's GHGs. Japan has been particularly motivated to earn the money it needs to import its food and fuel, both of which it is lacking, through export of high-value goods that require considerable energy consumption and thus, GHG emission in their manufacture. At the same time, Japanese politicians and the populace were seriously committed to combating global warming and willing to spend considerable resources to achieve this aim.

Far from the days of permitting rampant industrial pollution just after World War II, when economic recovery was on the top of the agenda, Japan has become a leader in environmental protection. By 2002, the second environmental performance review of Japan by the Organization for Economic Cooperation and Development praised Japan for its highly effective implementation of positive environmental policies and its strict, well-enforced, and seriously monitored regulations. In 2006, Japan's Ministry of the Environment made prevention of global warming, ozone layer protection, and conservation of the air, water, and soil environments issues of top concern.

Japan has become strongly committed to an international approach to Earth-spanning challenges such as global warming. Japan is a founding member of the Asia-Pacific Partnership on Clean Development and Climate founded on January 12, 2006, in Sydney, Australia. In addition, Japan continues to stress environmental protection in the countries of its trade partners.

Japan has been actively involved on the cutting edge of meteorological research. A major scientific project concerns the Earth Simulator, a supercomputer used for developing and running global climate models since 2002 at the Earth Simulator Center in Yokohama. The Earth Simulator is projected to be replaced by an even more powerful model after 2009, indicating Japan's continuous commitment to climate research.

R. C. Lutz

Further Reading

Bianchi, Adriana, et al., eds. *Local Approaches to Environmental Compliance: Japanese Case Studies and Lessons for Developing Countries.* Washington, D.C.: International Bank for Reconstruction and Development, 2005. Illustrated description and analysis of five successful cases of arresting pollution in Japan since the 1960s. Chapter 2, "Successful Air Pollution Control in Japan: History and Implications" by Ryo Fujikura, is of special relevance to combating causes of global warming.

Cruz, Wilfrido, et al., eds. *Protecting the Global Environment: Initiatives by Japanese Business.* Washington, D.C.: International Bank for Reconstruction and Development, 2002. Three case studies covering the steel industry, power plants, and forestry show how Japanese companies were successful in improving environmental protection while simultaneously increasing productivity.

Flath, David. *The Japanese Economy.* 2d ed. New York: Oxford University Press, 2005. Chapter 11 addresses Japanese environmental policy in the context of economic issues.

Ichikawa, Atsunobu, ed. *Global Warming—The Research Challenges: A Report of Japan's Global Warming Initiative.* New York: Springer Verlag, 2005. Presents the results of Japan's top scientific research. Covers climate monitoring, modeling, and projections; impact and risk assessment; and options for response policies. Japanese climate policy is presented and analyzed.

Kameyama, Yasuko. "Will Global Warming Affect Sino-Japanese Relations?" In *Japan and China,* edited by Hanns Gunther Hilpert. New York: Palgrave Macmillan, 2002. Analyzes the costs and benefits of regional environmental cooperation in northern Asia; considers the potential for Japanese-Chinese cooperation to reduce GHG emissions through successful bilateral treaties and agreements.

Oshitani, Shizuka. *Global Warming Policy in Japan and Britain: Interactions Between Institutions and Issue Characteristics.* Manchester, England: Manchester University Press, 2006. Comparative analysis of the relationship between government and industry in Japan and the United Kingdom. Contrasts national approaches to official decision making and implementation of new policies; explores how public institutions pursue climate policy in both nations.

Norio, Okada, Tao Ye, Yoshio Kajitani, Peijun Shi, and Hirokazu Tatano. "The 2011 eastern Japan great earthquake disaster: Overview and comments."*International Journal of Disaster Risk Science* 2, no. 1 (2011): 34–42. The great 2011 earthquake, tsunami, and consequent nuclear disaster exceeded even the capabilities of an advanced and seismically prepared nation like Japan. Previous preparedness measures mitigated the effects of this event but revealed a need for additional improvements.

Perrin, Noel. *Giving up the gun: Japan's reversion to the sword, 1543–1879.* David R. Godine Publisher, 1979. Although focused on the specific issue of Japan's abandonment of firearms, Perrin ranges broadly across Japanese culture and history and makes it clear that Japan has always been one of the most technologically advanced nations on Earth.

See also: Carbon dioxide; Carbon dioxide equivalent; Greenhouse effect; Greenhouse gases and global warming; International agreements and cooperation; Kyoto Protocol; United Nations Framework Convention on Climate Change.

Jet stream

Category: Meteorology and atmospheric sciences

Definition

A jet stream is a band of high-velocity atmospheric current that encircles the Earth. This band of strong winds is typically found in the upper troposphere and lower stratosphere. In both the Northern and the Southern Hemisphere, there are two distinctive jet streams: one is located just outside the tropical latitudes, in the subtropics; the other is located at the boundary of the midlatitudes and the polar region. In these latitudes, the jet-stream winds are westerlies, blowing from west to east. The two jet streams are named, respectively, "the subtropical jet" and "the polar jet." These jets shift locations seasonally.

The rotation of the Earth around its own axis causes the air that surrounds the Earth to move as a result of the drag exerted by the Earth's solid surface. This movement of air is called "wind." Wind can blow in any direction, even though the Earth rotates from west to east. Two of the major forces in the atmosphere that determine wind direction are the pressure gradient force and the Coriolis effect. Because of differential solar heating between the tropics and a polar region, there is a strong tendency of the atmosphere to move outward from the warmer equator to the colder poles, distributing heat. This tendency would cause winds to blow in the north-south directions in the Northern and Southern Hemispheres at different altitudes.

Owing to the Earth's rotation, these two major north-south circulations break up, forming several smaller cells of circulation. These smaller cells include the Hadley circulation, the Ferrel circulation,

and the polar circulation, listing them from the equator to the poles in each hemisphere. At upper atmospheric levels, when air travels to the north in the Northern Hemisphere, the Coriolis effect turns this airstream to its right, turning a southerly wind into a westerly wind. Furthermore, because the Coriolis effect becomes stronger as air travels further north and the effect of Earth's surface drag is smaller at upper atmospheric levels, these westerly winds become very strong. As a result, jet streams form outside of the tropics, toward high latitudes, at upper atmospheric levels.

Significance for Climate Change

Because it is a high-velocity wind current, a jet stream can be an important energy source for extratropical cyclones and storms. The position of a jet stream serves as an important indicator for the location of surface front and storm development. Both the subtropical and the polar jet are located at the boundaries of two different air masses. The boundaries of different air masses are the locations of surface fronts, at which extratropical cyclones develop. Therefore, a jet stream is an important component of a synoptic weather system.

Inside a jet stream, there is often a core region where winds become even stronger than they are in other parts of the jet. This core region is called the "jet streak." Because winds increase toward the jet streak, large horizontal and vertical wind shears exist at the boundaries of a jet stream. The sheared winds can cause a flow current to become curved and unstable. Therefore, in the vicinity of a jet stream, atmospheric waves and turbulence are often generated. Upper air turbulence may threaten aircraft.

Several studies have indicated that global warming tends to cause jet streams to move poleward. The implication of this migration is that extratropical storms or "storm tracks" will extend poleward as well. Furthermore, the combined effects of global warming and the poleward movement of jet streams leave a larger spatial area in which tropical storms may develop. Therefore, the area of coastal land that is vulnerable to hurricanes may increase.

Some scientists have pointed out that, as a result of a continuous increase in Earth's average surface temperature and of the temperature throughout

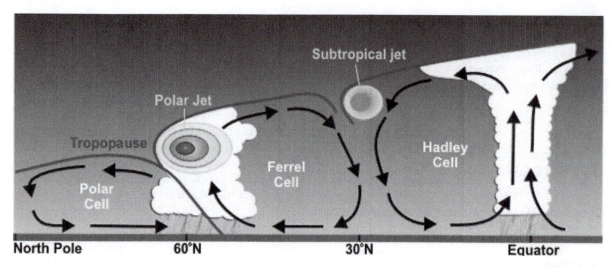

This cross-section of the polar and subtropical jet streams illustrates their relationship to Earth's major atmospheric circulation cells. (NOAA)

the troposphere, the temperature gradient from the equator to the poles may be decreased. This possibility seems to suggest that the jet stream's intensity may decrease as the climate warms. The weakening of jet streams may in turn decrease the activity of extratropical cyclones. However, other scientists have argued that global warming may not lead to a warming of the entire troposphere and a corresponding decrease in jet-stream intensity. Moreover, because global warming would also tend to increase atmospheric humidity, latent heat release may supply additional energy for extratropical cyclones despite a decreased global temperature gradient.

Chungu Lu

Further Reading

Ahrens, C. Donald. *Essentials of Meteorology: An Invitation to the Atmosphere.* 5th ed. Belmont, Calif.: Thomson Brooks/Cole, 2008. Widely used introductory textbook on atmospheric science; covers a wide range of topics on weather and climate.

Archer, Cristina L., and Ken Caldeira. "Historical Trends in the Jet Streams." *Geophysical Research Letters* 35, no. 24 (2008). Reports on an investigation of changes in the location of the jet stream in response to global warming.

Barnes, E.A. and Screen, J.A., 2015. The impact of Arctic warming on the midlatitude jet-stream: Can it? Has it? Will it?. *Wiley Interdisciplinary Reviews: Climate Change,* 6(3), pp.277-286. Research on the effects of a warming Arctic on the jet stream has been contradictory. Modeling suggests that a warmer Arctic is likely to push the jet stream equatorward, but the effects of Arctic warming are difficult to separate from other climate variables in practice.

Intergovernmental Panel on Climate Change. *Climate Change, 2007—Synthesis Report: Contribution of Working Groups I, II, and III to the Fourth Assessment Report of the Intergovernmental Panel on Climate Change.* Edited by the Core Writing Team, Rajendra K. Pachauri, and Andy Reisinger. Geneva, Switzerland: Author, 2008. Comprehensive overview of global climate change published by a network of the world's leading climate change scientists under the auspices of the World Meteorological Organization and the United Nations Environment Programme.

Lutgens, Frederick K., and Edward J. Tarbuck. *The Atmosphere.* 10th ed. Upper Saddle River, N.J.: Pearson Prentice Hall, 2007. Introductory textbook that covers a wide range of atmospheric sciences.

Yin, Jeffrey H. "A Consistent Poleward Shift of the Storm Tracks in Simulation of Twenty-first Century Climate." *Geophysical Research Letters* 32, (2005). Investigates poleward shifting of storm tracks in relation to the jet stream under global warming conditions.

See also: Adiabatic processes; Atmospheric dynamics; Gulf Stream; Hadley circulation; Walker circulation.

Journalism and journalistic ethics

Categories: Popular culture and society; ethics, human rights, and social justice

The changing nature of journalism, including the competition between objective coverage and advocacy, has played a key role in the public perception of global warming.

Key concepts

code of ethics: an unofficial standard for proper journalistic behavior

media bias: presentation of news or information that favors one side of a particular issue

Society of Environmental Journalists: a group of media professionals who focus wholly or in part on a wide range of environmental stories and issues

Society of Professional Journalists: a voluntary association of journalists who try to further the profession

Background

American journalism is regarded as the Fourth Estate, or a final check and balance on government. Media professionals claim to have evolved a tradition of neutrality in news stories, but both the reality and the ideal of neutrality have been called into question. Journalists themselves are unsure of the role they should play in global warming debates. At the same time, the nature of journalism has blurred with the rise of the Internet and the focus on immediacy in news coverage.

Traditional Media Approaches to Reporting

Over time, American journalists have evolved an unofficial code of ethics that calls for unbiased or neutral reporting. This concept is integral to ordinary reporters being seen as chroniclers rather than advocates. The ideal has gained widespread but not universal acceptance.

Journalists use several means to enforce neutrality in stories. The traditional approach would include editors removing any comments that are not objective from a given story. Journalists also seek the help of professional organizations to support unbiased reporting. The Society of Professional Journalists says reporters must distinguish between advocacy and news in their reporting. The Society of American Business Editors and Writers tells its members to simply avoid any action that might "appear" to harm objectivity. The Society of Environmental Journalists has been criticized for taking sides in the global warming debate.

Many newspapers also have their own codes of behavior for journalists. *The New York Times* urges staff to remain free of any bias, as do many other organizations. Some media outlets even employ ombudsmen to serve as reader representatives. Media outlets also have detailed correction policies.

There are flaws in all of these approaches. Editors do not catch every instance of bias. Organizations are voluntary and do not police a code. News outlets also use codes as guidelines, but public examples of enforcement are rare. Ombudsmen generally lack any real authority in a newsroom.

Every one of these strategies is at issue in the global warming debate. Scientists and other experts complain that biases are present in news stories. The other methods of controlling bias, from policies to ombudsmen, are unable to keep up with a sophisticated and scientific debate involving professionals who often know far more than reporters or editors.

Other Strategies

The traditional approach is not the only approach. The term "news media" itself has become fluid. It encompasses alternative weeklies, freelance journalists, and opinion media that take an advocacy role. Even the mainstream media have a strong strain of advocacy. Many media outlets have long employed journalists who advocate for positions—from environmentalism to a higher minimum wage. There is a journalistic tradition that reporters should "afflict the comfortable and comfort the afflicted." That plays out in the climate change debate, as journalists try to show the potential harm of global warming—from species loss to climate disaster.

Several journalists have spoken out and complained about the neutrality requirements. Columbia Broadcasting System (CBS) reporter Scott Pelley famously compared global warming skeptics to Holocaust deniers. Other reporters and editors have taken an activist stand on the issue, claiming that too many lives are at risk because of global warming. *Time* magazine has repeatedly put global warming on its cover, warning that something must be done to stop it.

All major news outlets have found navigating the issue to be difficult. Fox News, whose motto is Fair and Balanced, was criticized by supporters of anthropogenic global warming theory for including too many skeptics. When Fox News ran a special that included their side, skeptics complained and Fox News then produced a second documentary highlighting flaws in the alleged global warming consensus.

Context

The issue of media neutrality is central to how ordinary Americans see global warming. Do they view it as an imminent threat or as debated science? Journalists have been much criticized by readers, as well as by advocates on the Left and Right.

The issue of media bias about global warming has become almost as big a story as climate change itself. Scientist James E. Hansen complained that the George W. Bush administration censored his work, but he received a great deal of media attention as a result. Skeptical scientists have complained about being either ignored or undermined in stories. *Time* magazine modified the famous photo of the Iwo Jima flag raising to replace it with a tree in a push for global warming action. That decision resulted in thousands of reader complaints as well as criticism from photojournalists. Because global warming is such a controversial issue, it highlights the many flaws in journalistic neutrality. That, in turn, undermines the idea that reporters are objective in other areas of news coverage.

Dan Gainor

Further Reading

Bowen, Mark. *Censoring Science: Inside the Political Attack on Dr. James Hansen and the Truth of Global Warming.* New York: Dutton Books, 2008. Explains how the Bush administration censored scientists who predicted global warming, effectively limiting their media access as well. Also lists facts that demonstrate the warming trend, primarily that recent years have reportedly been the warmest years on record.

Gelbspan, Ross. *Boiling Point: How Politicians, Big Oil and Coal, Journalists, and Activists Are Fueling the Climate Crisis—And What We Can Do to Avert Disaster.* New York: Basic Books, 2004. Describes climate change as a major crisis and refers to denial of it as "a crime against humanity." Gelbspan blames the media for delaying necessary action and calls for a global project to repair the planet. References, index.

Hanitzsch, Thomas, Folker Hanusch, Claudia Mellado, Maria Anikina, Rosa Berganza, Incilay Cangoz, Mihai Coman et al. "Mapping journalism cultures across nations: A comparative study of 18 countries." *Journalism Studies* 12, no. 3 (2011): 273-293. A cross cultural survey shows that basic journalistic ethical standards like impartiality and factual accuracy are universal but the degree to which journalists should be advocates differs among cultures.

Lomborg, Bjørn. *Cool It: The Skeptical Environmentalist's Guide to Global Warming.* New York: Knopf, 2007. Director of the Copenhagen Consensus Centre, Lomborg discusses exaggerations of global warming in the media and how scare-mongering tactics and misleading information are the media's primary tools. Rather than discount global warming theory, Lomborg suggests readers look at the costs of the suggested measures to stop it.

Michaels, Patrick J. *Meltdown: The Predictable Distortion of Global Warming by Scientists, Politicians, and the Media.* Washington, D.C.: Cato Institute, 2004. Covers the numerous and extreme exaggerations of climate change in the media. Argues that although there is warming occurring, it does not require panic from the public, and denies that humans can do much to affect the Earth's temperature.

See also: Carson, Rachel; *Collapse*; Gore, Al; *Inconvenient Truth, An*; *The Limits to Growth (book)* Thoreau, Henry David.

Kilimanjaro

Category: Geology and geography

During the political debates over global warming, Kilimanjaro's vanishing ice cap became an iconic image of the negative consequences of anthropogenic climate change, but climatologists and glaciologists have discovered that decreased snowfall, enhanced exposure to solar radiation, ice-sheet locations and shapes, and other factors are more important than climate change in explaining the disappearing ice.

Key concepts

ice cap: semipermanent glacial crown of ice atop a mountain or other geologic formation

sublimation: the conversion of a solid directly into a gas, bypassing its liquid state

volcano: a usually mountainous rift in Earth's crust caused by magma erupting through fissures onto the planet's surface

Background

Kilimanjaro—which, at 5,895 meters, is Africa's highest mountain—is of volcanic origin, based in its location alongside a rift zone created by the Earth's spreading crust. Its trio of volcanic peaks gradually became inactive, first Shira, then Mawenzi, and finally Kibo, the highest. Ice and snow gathered on Kilimanjaro's upper slopes during the late Pleistocene epoch, about eleven thousand years ago. Ice cores taken from the mountain's glaciers have revealed a history of expansions and contractions of the ice cap, with the contractions occurring during periods of drought eighty-three hundred, fifty-two hundred, and four thousand years ago. Periods of expansion resulted from climatic conditions that were warmer and wetter than those of today. Recently, Kilimanjaro's ice sheets have been confined to Kibo.

One-half million years ago, early human inhabitants of the Rift Valley first saw this majestic mountain, but its names derive from a much later period. African tribes called it the "white" or "shining" mountain, because of its ice cap. Its present name most likely comes from the Swahili, "Kilima Njaro," meaning "Mountain of Greatness" because of its massive height and bulk. The first European to see the mountain was a German missionary who published an account of his African explorations in 1849. This not only sparked interest in searches for the source of the Nile but also in the scientific study of Kilimanjaro and its ice sheets.

Vanishing Ice Sheets

In the late nineteenth century, explorers created maps and drawings of the mountain and its environs. From these, scholars derived an estimate that, in 1880, ice covered about 20 square kilometers of Kilimanjaro's principal peak. In 1889, two Europeans were the first to reach Kilimanjaro's summit, and they brought back information about the extent and depth of the ice sheets. In the twentieth century, photographs from the mountain's base and, later, from airplanes, provided benchmarks by which the shrinkage of the ice cap could be measured. In 1912, a precise map was constructed based on photogrammetric evidence. At that time, the ice sheets had diminished to 12.1 square kilometers, though they still existed on all sides and descended to about 4,400 meters. From 1912 to the early twenty-first century, Kilimanjaro's icefields were periodically surveyed, and the data gathered indicated that the time of greatest contraction had been from about 1880 to 1950.

Ernest Hemingway, an American writer, made Kilimanjaro's ice cap world-famous through his popular short story, "The Snows of Kilimanjaro," (1936) which was made into a successful Hollywood movie in 1952. By the early 1950's, the mountain's icefields had diminished to 6.7 square kilometers. Sufficient data had been collected to determine that the rarely observed trickles of meltwater were unable to account for the retreating glaciers. Some scientists began to attribute glacial contraction to the sublimation of summit ice into water vapor under the influence of the tropical sun in dry air at below-freezing temperatures. However, during the second half of the twentieth century, as evidence multiplied that midlatitude mountain glaciers were shrinking because of global warming, some scientists extended this explanation to such equatorial mountain glaciers as Kilimanjaro's.

Aerial photographs of Kibo's ice sheets had been supplemented by satellite pictures, and this new information showed that, although the ice cap's contraction had slowed since 1953, 75 to 80 percent of its area had vanished during the past century, along with deep reductions in the volume of its ice and snow. Some scientists predicted that the ice cap would be completely gone by 2015 or 2020. Since the "Snows of Kilimanjaro" were famous as well as photogenic, it was natural for advocates of greenhouse gas (GHG) reduction to use dramatic images of the mountain's disappearing ice cap to bolster their cause. Greenpeace advocates held a news conference from Kilimanjaro's summit, and, most significant, former vice president Al Gore argued in a very successful documentary film and companion book, both entitled *An Inconvenient Truth* (2006), that the vanishing ice cap was evidence of global warming. The movie's Academy Award and the book's best-seller status helped popularize the images of Kilimanjaro's vanishing ice, which Gore associated with carbon-dioxide-induced global warming.

Significance for Climate Change

Although many scientists agreed that temperate-zone mountain glaciers were retreating because of global warming, an increasing number of climatologists and glaciologists believed that the specific case of Kilimanjaro's ice loss was caused by other factors. Data collected from balloons, satellites, and an automatic weather station on one of Kibo's icefields revealed that reduced precipitation of ice and snow in dessicated air, along with extensive exposure to solar radiation in below-freezing temperatures, led to the accelerated sublimation of ice and snow. Fluctuating weather systems in the Indian Ocean, which influenced humidity and cloud cover over Kilimanjaro, may also have played a role.

Context

Most scientists familiar with the data held that global warming had little or no effect on the ice decline. These scientists chided global-warming enthusiasts for their misuse of Kilimanjaro's vanishing ice cap to support their views on climate change, but they also criticized global-warming

deniers who overgeneralized the Kilimanjaro case to include midlatitude glaciers. Kilimanjaro can therefore serve as a cautionary tale of how politics and overheated rhetoric often lead passionate advocates to distort scientific data and images in unacceptable ways.

Robert J. Paradowski

Further Reading

Bowen, Mark. *Thin Ice: Unlocking the Secrets of Climate in the World's Highest Mountains.* New York: Macmillan, 2005. This book has been called one of the best yet published on climate change. Part 6 is dedicated to Kilimanjaro. Notes, references, index.

Cullen, N. J., P. Sirguey, T. Mölg, G. Kaser, M. Winkler, and S. J. Fitzsimons. "A century of ice retreat on Kilimanjaro: the mapping reloaded." *The Cryosphere* 7, no. 2 (2013): 419-431. Detailed reanalysis of early ground based and aerial observations, plus modern satellite observations, paints a dramatic picture of ice loss on Kilimanjaro.

Gore, Al. *An Inconvenient Truth: The Planetary Emergency of Global Warming and What We Can Do About It.* Emmaus, Pa.: Rodale Books, 2006. Written as a complement to a very successful film. The great popularity of the film and book helped create the image of Kilimanjaro as an "icon" of global warming advocacy.

Michaels, Patrick. *Meltdown: The Predictable Distortion of Global Warming by Scientists, Politicians, and the Media.* Washington, D.C.: Cato Institute, 2004. Although Michaels, a climatologist, accepts the reality of global warming, he argues that alarmists have exaggerated its future effects, and he uses the debate over Kilimanjaro's shrinking ice cap to show how science can be distorted for political ends.

Mote, Philip W., and Georg Kaser. "The Shrinking Glaciers of Kilimanjaro: Can Global Warming Be Blamed?" *American Scientist* 95 (July/August, 2007): 318-325. This article by a climatologist and glaciologist musters much scientific evidence to show that factors other than global warming are responsible for the retreat of Kibo's ice sheets. Illustrated with photographs, graphs, and a map. Bibliography.

See also: Greenland ice cap; Hockey stick graph; *Inconvenient Truth, An*; Mount Pinatubo; Sahara Desert; Volcanoes.

Kyoto lands

Category: Laws, treaties, and protocols

Kyoto lands are lands within the territory of Kyoto Protocol party nations on which vegetation is planted that acts to sequester carbon. These lands enable protocol parties to earn carbon offset credits to balance their continuing GHG emissions.

Key concepts

biomass: the organic substance of life-forms, especially plants

carbon fixation: incorporation of carbon atoms into organic compounds within biomass, making it unavailable for CO_2 production

carbon offset credits: credits earned by funding projects that remove GHGs from the atmosphere and that can be used to replace reductions in GHG emissions

carbon sequestration: isolation of carbon from the atmosphere, either by converting it into a form other than CO_2 or by capturing CO_2 and storing it outside of the atmosphere

greenhouse gases (GHGs): trace atmospheric gases that trap heat, preventing it from escaping into space

Background

Forests and other types of vegetation that store or sequester carbons and are increasing in size are considered to be carbon sinks, or reservoirs. In the main, natural carbon sinks include trees, plants, and organisms that use photosynthesis to remove carbon dioxide (CO_2) from the atmosphere by incorporating, or fixing, carbon into biomass as carbohydrates $(CHOH)_n$ and releasing oxygen (O_2) into the air. Carbon sinks not only rid the atmosphere of greenhouse gases (GHGs) but also, by producing additional CO_2-consuming

biomass, can scavenge even more CO_2 from the atmosphere.

The Kyoto Protocol, a United Nations treaty, aims to reduce anthropogenic GHG emissions. It allows member countries to use carbon sinks as a form of carbon offset, gaining credit for their sinks against their continued GHG emissions. These so-called Kyoto lands may then be used to comply with emission-reduction requirements of the treaty.

According to the Intergovernmental Panel on Climate Change (IPCC), if GHG emissions continue to rise at their current rate, global warming will accelerate to levels that will increase both floods and droughts, raise sea levels, and obliterate thousands of plant and animal species. The use of carbon sinks to mitigate the effects of global warming may be useful to countries with large areas of forest or other vegetation for compliance with the Kyoto Protocol. Specific, legally binding quotas for reduction of GHG emissions have been established for the developed nations; developing countries are not compelled to restrict their GHG emissions, which may come, in large part, from land use such as cultivation of lands and destruction of forests. For developed nations, land use would have little effect in meeting Kyoto quotas, since most of their land has already been cultivated. Developing countries may benefit from offsetting their GHG emissions with carbon sinks via land use or green projects, and can help developed countries meet their quotas by trading carbon credits. Usually, an industrialized country will purchase carbon credits from a developing country to meet its quota.

Implementation

In implementing the Kyoto Protocol, parties to the treaty will most likely base their policy decisions on "definitions, accounting procedures . . . for carbon stocks, and changes in carbon stocks." Per the protocol, each party "must devise its own method of verifying its carbon emission reductions (CER) to account for carbon sequestration." In the clean development mechanism (CDM), only afforestation, "the establishment of forest on land that has been 'unforested' for a long period, decades to centuries," and reforestation, the conversion of 'nonforested' lands to forest, are allowable methods of producing CERs for the first period of the protocol, 2008 to 2012. Forest conservation or avoidance of deforestation, "the conversion of forest to nonforest," is ineligible.

The IPCC in its revised *Guidelines for National GHG Inventories* (1997) and the Kyoto Protocol itself call for a strict accounting of the carbon changes encompassing all carbon sinks within a given timeframe; they require that inclusion be limited to "land areas subject to direct human-induced activities." GHG emissions and removals via land use, land-use change, and forestry (LULCF) may be designed to help parties to Annex B of the protocol meet their obligations. In this case, the *Guidelines* are to serve as the basis for "transparent and verifiable" reportage of changes in forestry and related GHG emissions.

Significance for Climate Change

Changes in land use can appreciably change the GHG emission levels, with deforestation increasing them and afforestation reducing them. Anthropogenic emissions of CO_2 grew by 2.5 percent to record levels in 2006, as international efforts to fight global warming failed to curb the main gas blamed for rising temperatures. Burning fossil fuels and changing land use together produced 8.8 billion metric tons of CO_2 in 2006. The inclusion of land use as part of the calculations for GHG emissions tends to focus on certain developing nations, notably Indonesia and Brazil, which are significantly raising their GHG emission rankings. Between 1950 and 2000, when land use is taken into account these rankings rose from eighteenth to fifth and from twenty-seventh to fourth, respectively. Indonesia and Brazil have both cleared large parcels of forest, converting carbon sinks into timber and agricultural land and thereby emitting GHGs.

Cynthia F. Racer

Further Reading

Bloomfield, J., M. Ratchford, and S. Brown. "Land-Use Change and Forestry in the Kyoto Protocol." *Mitigation and Adaptation Strategies for Global Change* 5, no. 1 (March, 2000): 1381-2386. Rehearses the strategies employed to satisfy compliance with the Kyoto Protocol, emphasizing land use and forests.

Carswell, Fiona E., Norman WH Mason, Jacob McC Overton, Robbie Price, Lawrence E. Burrows, and Robert B. Allen. "Restricting new forests to conservation lands severely constrains carbon and biodiversity gains in New Zealand." *Biological Conservation* 181 (2015): 206-218. Conservation lands generally have low biodiversity, whereas non-conservation lands have little or no indigenous vegetation. Afforesting non-conservation lands results in much more rapid sequestration of carbon and improvements in biodiversity.

Kindermann, G., et al. "Global Cost Estimates of Reducing Carbon Emissions Through Avoided Deforestation." *PNAS* 105 (2008): 10302-10307. Explores the financial benefits accrued by using Kyoto lands to mitigate global warming through maintenance of forests.

Schulze, E.-D., et al. "Making Deforestation Pay Under the Kyoto Protocol?" *Science* 299 (March, 2003): 1669. Argues that changing regulations governing the "reforestation time limit" may compromise basic principles of the protocol, including preservation of pristine forests.

See also: Amazon deforestation; Annex B of the Kyoto Protocol; Deforestation; Forestry and forest management; Forests; Kyoto mechanisms; Kyoto Protocol; Reservoirs, pools, and stocks; Sinks; United Nations Framework Convention on Climate Change.

Kyoto mechanisms

Category: Laws, treaties, and protocols

The Kyoto Protocol established three mechanisms for GHG reduction: emissions trading, joint implementation, and the clean development mechanism. The aim of these mechanisms is to provide additional, cost-effective opportunities for industrialized countries to reduce their emissions of GHGs in order to mitigate climate change.

Key concepts

Annex I parties: industrialized nations listed in Annex I of the UNFCCC

assigned amount units: greenhouse gas emission allowances of Annex I parties

certified emissions reductions (CERs): credits for contributing to reduced emissions in developing nations that Annex I nations can substitute for domestic reductions

emission reduction units (ERUs): measurement of the amount of GHG emissions reduced by a joint implementation project

removal units (RMUs): emission reduction units generated by land use, land-use change, and forestry projects

Background

Article 3 of the Kyoto Protocol requires industrialized countries, identified in Annex I of the U.N. Framework Convention on Climate Change, to take on quantified emission limitation and reduction commitments (QELRCs). These QELRCs are set out in Annex B to the protocol and require parties listed in Annex I of the convention to reduce their emissions of six specified greenhouse gases (GHGs). Based on individual targets, Annex I parties are assigned the amount of carbon dioxide (CO_2) equivalent emission units, called assigned amount units (AAUs), they are allowed to emit during the protocol's first commitment period (2008-2012). These parties are obliged to meet their targets primarily through domestic measures. However, recognizing the need for flexibility, the protocol established three market-based mechanisms—emissions trading, joint implementation, and the clean development mechanism (CDM)—that are meant to provide additional, cost-effective ways for developed countries to meet their commitments.

Implementation and Participation Requirements

The mechanisms are under the authority and guidance of the Conference of the Parties to the Convention serving as the meeting of the Parties to the Protocol (COP/MOP or CMP). The relevant articles in the protocol and decisions of the COP/MOP contain the rules and implementation modalities for the mechanisms. Decision 2/CMP.1 contains some general requirements and provides that use of the mechanisms must be supplemental to domestic actions taken by Annex I parties, with these domestic actions constituting a significant element in Annex I parties' efforts to meet their targets. The decision further provides that Annex I parties' eligibility to participate in the mechanisms depends on their compliance with their methodological and reporting obligations under Articles 5 and 7 of the protocol. Participation in the Kyoto mechanisms may involve private or public entities, which must be authorized by a party to the Kyoto Protocol to participate in any of the mechanisms.

Joint Implementation

Under joint implementation, Article 6 of the protocol allows Annex I parties to earn or buy emission reduction units (ERUs) generated from emission reduction or removal projects implemented in other Annex I parties. These ERUs, each representing one metric ton of CO_2 equivalent emission reduction, can be used by Annex I parties to contribute to meeting their targets under the protocol. A joint implementation project could, for example, involve installing solar panels in homes in a developed country to replace power generation from fossil fuels. The use of these solar panels will result in reduced GHG emissions, and the investor can use the reductions achieved, issued in the form of ERUs, toward its own reduction target. The joint implementation mechanism is overseen by a supervisory committee.

The protocol sets out some of the eligibility criteria for participating in joint implementation. Article 6.1 provides that joint implementation projects must have the approval of the parties involved and must result in a reduction in emissions by sources, or removal by sinks, additional to what would have occurred in the absence of the project. The detailed participation requirements are set out in the Annex to Decision 9/CMP.1.

Clean Development Mechanism

The CDM was established by Article 12 of the protocol. Under the CDM, Annex I parties can finance or invest in emission reduction or removal projects in developing countries. These projects generate certified emissions reductions (CERs), which Annex I parties can add to their assigned amounts and use toward meeting their targets under the protocol. Again, each CER represents one metric ton of CO_2 equivalent emission reduction. A CDM project could involve the same activities as a joint implementation. The only differences would be that the project would be implemented in a developing country, and the emission reductions would be issued as CERs. The CDM is overseen by an executive board.

Article 12 sets out some of the participation requirements for the CDM. It provides for the certification of emission reductions resulting from project activities by operational entities, on the basis of voluntary participation of the parties involved; real, measurable, and long-term benefits related to the mitigation of climate change; and emission reductions additional to those that would have occurred in the absence of the project activity. The Annex to Decision 3/CMP.1 further provides that participants must be parties to the protocol and must designate a national authority. These requirements are additional to the general participation requirements described above.

Emissions Trading

Article 17 of the protocol provides that parties with QELRCs may participate in emissions trading in order to fulfill their Article 3 commitments. Emissions trading involves Annex I parties trading in their AAUs. Here, Annex I parties who have not used up all their AAUs are allowed to sell them to other Annex I parties. These other Annex I parties are generally parties who either have exceeded or believe they will exceed their assigned amounts. Under the emissions trading scheme, Annex I parties can buy or sell any of the emission reduction units, including AAUs, CERs, ERUs, and removal units (RMUs). RMUs are ERUs generated under land use, land-use change, and forestry projects. This trade in emission reduction units or carbon credits has led to the creation of what is called the carbon market, where carbon emissions are traded like other commodities.

The rules governing emissions trading are contained in the Annex of Decision 11/CMP.1. In addition to the general participation requirements, Annex I parties are required to maintain a commitment period reserve of at least 90 percent of their assigned amounts. They are not allowed to make any transfer that would result in their holdings falling below this reserve.

Context

The United Nations recognized that differing circumstances exist in different countries. Some countries may be relatively energy efficient, making it more difficult for them to reduce their GHG emissions sufficiently to meet their targets. In some countries, the potential cost of reducing their emissions may be very high. The need for flexibility was recognized, although the protocol provides that the use of the flexible mechanisms must be supplemental to domestic actions taken by Annex I parties. The Kyoto mechanisms were established to provide this flexibility, to enhance Annex I parties' ability to achieve their targets, and to provide them with cost-effective means of achieving these targets. In addition, other advantages of the mechanisms include contribution to the sustainable development of recipient countries through technology transfer and enhanced investment flows.

Tomi Akanle

Further Reading

Freestone, David, and Charlotte Streck. *Legal Aspects of Implementing the Kyoto Protocol Mechanisms: Making Kyoto Work.* New York: Oxford University Press, 2005. Provides a broad overview of the Kyoto mechanisms, including the history of negotiations and provisions, and outstanding issues needing to be addressed.

Grubb, Michael, Christiaan Vrolijk, and Duncan Brack. *The Kyoto Protocol: A Guide and Assessment.* London: Earthscan, 1999. Part 2 of this book analyzes the commitments taken on by Annex I parties and the mechanisms established to assist them in achieving the commitments, assessing the pros and cons of the mechanisms.

Hoffmann, Matthew J. *Climate governance at the crossroads: experimenting with a global response after Kyoto.* Oxford University Press, 2011. Although the Kyoto accords have not resulted in large scale multilateral changes, there have been a large number of local "experiments" at addressing climate change.

Yamin, Farhana, ed. *Climate Change and Carbon Markets: A Handbook of Emissions Reductions Mechanisms.* London: Earthscan, 2005. Provides a comprehensive description of the carbon market, including the rules governing the three Kyoto mechanisms and related rules, such as those of the European Union Emission Trade Scheme.

See also: Annex B of the Kyoto Protocol; Carbon dioxide equivalent; Certified emissions reduction; Clean development mechanism; Kyoto lands; Kyoto Protocol; United Nations Framework Convention on Climate Change.

Kyoto Protocol

Category: Laws, treaties, and protocols
Date: Entered into force February 16, 2005

The Kyoto Protocol establishes clear targets for reducing GHG emissions, along with accounting mechanisms that encourage nations to undertake projects that reduce emissions. It balances national economic development plans with emissions reductions and sets up free market mechanisms to achieve the targets set by intergovernmental negotiations.

Key concepts

Annex I nation: an industrialized nation listed under Annex I of the UNFCCC and committed to achieve a certain level of GHG emissions reduction within a specified period

certified emissions reduction (CER): a unit of GHG reduction, equal to 1 metric ton of carbon dioxide emissions, certified by the UNFCCC and tradeable on the carbon market

clean development mechanism (CDM): a system accounted by the UNFCCC, whereby an Annex I country sponsors a greenhouse gas reduction project in a developing country to earn CERs

joint implementation: a system in which Annex I countries can earn CERs by working on CDM projects in an economically transitional nation, generally one of the former Soviet or Balkan nations

United Nations Framework Convention on Climate Change (UNFCCC): an agreement between nations to participate in reducing anthropogenic effects on the climate

Participating nations: *1998:* Antigua and Barbuda, El Salvador, Fiji, Maldives, Tuvalu; *1999:* Bahamas, Bolivia, Cyprus, Georgia, Guatemala, Jamaica, Mongolia, Nicaragua, Niue, Palau, Panama, Paraguay, Trinidad and Tobago, Turkmenistan, Uzbekistan; *2000:* Azerbaijan, Barbados, Ecuador, Equatorial Guinea, Guinea, Honduras, Kiribati, Lesotho, Mexico, Samoa; *2001:* Argentina, Bangladesh, Burundi, Colombia, Cook Islands, Czech Republic, Gambia, Malawi, Malta, Mauritius, Nauru, Romania, Senegal, Uruguay, Vanuatu; *2002:* Austria, Belgium, Benin, Bhutan, Brazil, Bulgaria, Cambodia, Cameroon, Canada, Chile, China, Costa Rica, Cuba, Denmark, Djibouti, Dominican Republic, Estonia, European Community, Finland, France, Germany, Greece, Grenada, Hungary, Iceland, India, Ireland, Italy, Japan, Latvia, Liberia, Luxembourg, Malaysia, Mali, Morocco, Netherlands, New Zealand, Norway, Papua New Guinea, Peru, Poland, Portugal, Republic of Korea, Seychelles, Slovakia, Slovenia, South Africa, Spain, Sri Lanka, Sweden, Thailand, Uganda, United Kingdom, United Republic of Tanzania, Vietnam; *2003:* Armenia, Belize, Botswana, Ghana, Guyana, Jordan, Kyrgyzstan, Lao Democratic People's Republic, Lithuania, Madagascar, Marshall Islands, Myanmar, Namibia, Philippines, Republic of Moldova, Saint Lucia, Solomon Islands, Switzerland, Tunisia; *2004:* Indonesia, Israel, Liechtenstein, the former Yugoslav Republic of Macedonia, Niger, Nigeria, Russian Federation, Rwanda, Saint Vincent and the Grenadines, Sudan, Togo, Ukraine, Yemen; *2005:* Albania, Algeria, Belarus, Burkina Faso, Democratic People's Republic of Korea, Democratic Republic of Congo, Dominica, Egypt, Eritrea, Ethiopia, Guinea-Bissau, Haiti, Iran, Kenya, Kuwait, Mauritania, Mozambique, Nepal, Oman, Pakistan, Qatar, Saudi Arabia, United

Arab Emirates, Venezuela; *2006:* Bahrain, Cape Verde, Gabon, Lebanon, Libyan Arab Jamahiriya, Monaco, Sierra Leone, Singapore, Suriname, Swaziland, Syrian Arab Republic, Zambia; *2007:* Angola, Australia, Bosnia and Herzegovina, Congo, Côte D'Ivoire, Croatia, Montenegro, Serbia; *2008:* Central African Republic, Comoros, Saint Kitts and Nevis, Sao Tomé and Principe, Timor-Leste, Tonga; *2009:* Kazakhstan, Tajikistan, Turkey

Background

By the early 1990's, anthropogenic emissions had been identified as a significant contributor to the concentrations of greenhouse gases (GHGs) in the atmosphere and to climate changes. Such GHG emissions are generated by industrial activities, especially the burning of fossil fuels for energy, as well as agricultural processes in both developing and industrial nations. Thus, the United Nations sponsored the framing of the U.N. Framework Convention on Climate Change in 1992 and the Kyoto Protocol to the UNFCCC in 1997 to address and mitigate the generation of GHGs by the signatories to those treaties.

On a per capita basis, the industrialized nations are by far the more intense of the world's emitters. The per capita carbon footprint, or the number of metric tons of carbon dioxide (CO_2) or other equivalent GHG emissions per year per person, is a useful metric for comparing national emissions. The per capita carbon footprint of the United States, for example, is around 18 metric tons, while Australia's is 25.5 metric tons. The world average is around 2.8 metric tons, China's is 2.4 metric tons and India's is 0.87 metric ton.

When the world's nations began negotiating a response to the perceived threat represented by anthropogenic GHG emissions, it was soon decided that the brunt of the effort would have to be borne by industrialized nations. This decision was not without controversy, but most parties to the negotiations believed that developing nations could not incur the costs of cutting emissions. It was argued that the industrialized nations, which were largely responsible for the rise in GHG concentrations, could not expect developing nations

to curtail their attempts to improve their own people's standards of living by industrializing, nor could industrialized nations disavow their own primary responsibility for rising GHG levels and therefore their obligation to make amends. As a result of this decision, the United Nations Framework Convention on Climate Change (UNFCCC) divided its parties into Annex I (industrialized) nations and non-Annex I (developing) nations.

Annex I of the UNFCCC not only lists the industrialized parties to the convention but also commits them to reducing their GHG emissions by a specified date. It was left to the successor document, the Kyoto Protocol, to determine the details for each nation. That document spelled out the obligations of each party nation, as well as the specific accounting mechanisms that would be used both to measure a nation's progress toward reducing its emissions and to assign credits in lieu of actual reductions for certain complementary activities.

To meet the overall target of a 5.2 percent reduction in global GHG emissions from 1990 to 2012, the mandated reductions for individual nations are as large as an 8 percent reduction from the nation's 1990 emissions level. Some nations, including Australia, negotiated allowances for a rapid rate of growth and were allowed to increase their emissions by as much as 10 percent over their 1990 levels. They could also sell any unused portion of that quota as credits to other nations unable to meet their own reduction quotas. Countries set up national registries validated by the UNFCCC to monitor their emissions.

Each signatory to the Kyoto Protocol undertook to adopt policy measures to enhance energy efficiency; protect and enhance sinks and reservoirs of GHGs; and promote forest cover, sustainable forms of agriculture, and research on renewable energy and sequestration technologies. The signatories agreed progressively to remove policies and other incentives that support or encourage GHG emissions; to reduce emissions in the transport sector; and to reduce methane emissions through recovery and use in waste management. Emissions were also to be reduced in the production, transport, and distribution of energy.

Emissions Trading

Article 6 of the Kyoto Protocol, reproduced below, establishes the basic framework for parties to the treaty to trade pollution credits with one another, thereby employing market principles to drive international emission reductions.

1. For the purpose of meeting its commitments under Article 3, any Party included in Annex I may transfer to, or acquire from, any other such Party emission reduction units resulting from projects aimed at reducing anthropogenic emissions by sources or enhancing anthropogenic removals by sinks of greenhouse gases in any sector of the economy, provided that:

(a) Any such project has the approval of the Parties involved;

b) Any such project provides a reduction in emissions by sources, or an enhancement of removals by sinks, that is additional to any that would otherwise occur;

(c) It does not acquire any emission reduction units if it is not in compliance with its obligations under Articles 5 and 7; and

(d) The acquisition of emission reduction units shall be supplemental to domestic actions for the purposes of meeting commitments under Article 3.

2. The Conference of the Parties serving as the meeting of the Parties to this Protocol may, at its first session or as soon as practicable thereafter, further elaborate guidelines for the implementation of this Article, including for verification and reporting.

3. A Party included in Annex I may authorize legal entities to participate, under its responsibility, in actions leading to the generation, transfer or acquisition under this Article of emission reduction units.

4. If a question of implementation by a Party included in Annex I of the requirements referred to in this Article is identified in accordance with the relevant provisions of Article 8, transfers and acquisitions of emission reduction units may continue to be made after the question has been identified, provided that any such units may not be used by a Party to meet its commitments under Article 3 until any issue of compliance is resolved.

Primary GHGs

The specific gases covered by the Kyoto Protocol are listed in Annex A of the protocol. They are carbon dioxide (CO_2), methane (CH_4), nitrous oxide (N_2O), hydrofluorocarbons (HFCs), perfluorocarbons (PFCs), and sulfur hexafluoride (SF_6). The contribution of each gas to the greenhouse effect is converted to that of an equivalent amount of CO_2 for easy comparison. This standardized measure of a GHG's climatic effects is known as its global warming potential. Methane, for example, has a global warming potential of 20, meaning it is twenty times as potent per unit mass as CO_2 over a specific time horizon.

GHG emissions arise from fuel combustion, energy industries, manufacturing industries, construction, and transportation. High-value targets for reduction include fugitive emissions from stored fuels, such as oil and natural gas; industrial processes in the metal and chemical industry, such as solvent and other product use; and direct production of halocarbons and sulfur hexafluoride. In agriculture, enteric fermentation, manure management, rice cultivation, prescribed burning of savannas, field burning of agricultural residues, waste incineration, and wastewater handling are all targeted activities.

Parties in Annex I

The parties included in Annex I of the UNFCCC are Australia, Austria, Belgium, Canada, Denmark, the European Community, Finland, France, Germany, Greece, Iceland, Ireland, Italy, Japan, Liechtenstein, Luxembourg, Monaco, the Netherlands, New Zealand, Norway, Portugal, Spain, Sweden, Switzerland, the United Kingdom, and the United States. Countries listed in Annex I as undergoing transitions to market economies are Bulgaria, the Czech Republic, Estonia, Hungary, Latvia, Lithuania, Poland, Romania, the Russian

Federation, Slovakia, Slovenia, and Ukraine. Annex I nations have agreed to limit or reduce GHG emissions from aviation and marine bunker fuels and to ensure that their aggregate anthropogenic emissions of the GHGs listed in Annex A of the protocol do not exceed their assigned amounts. Each party included in Annex I was to have made demonstrable progress by 2005 in achieving its commitments under the protocol.

Credits Versus Taxes

The Kyoto Protocol did not institute taxes on carbon emissions. Instead, it created a system of credits and trading as the favored mechanism to reduce GHG emissions. By turning emissions into tradeable market commodities, the protocol enabled businesses to better manage their emission reduction strategies. The carbon market sets pricing, and market growth is driven by investors. Using flexible mechanisms, the Kyoto Protocol ensures that investments contribute to genuine, sustainable carbon reduction schemes.

Certified Emissions Reductions and the Carbon Market

One certified emissions reduction (CER) unit, also known as one carbon credit, is equivalent to 1 metric ton of equivalent reduction in CO_2 emissions. CERs can be sold privately or in the international market. Each international transfer is validated by the UNFCCC. The European Commission also validates each transfer of ownership within the European Union. Climate exchanges provide a spot market in allowances, as well as futures and options markets, to discover the market price and maintain liquidity.

Trading in CERs is done through at least four exchanges: the Chicago Climate Exchange, the European Climate Exchange, Nord Pool, and PowerNext. The carbon market has exceeded 60 billion euros and is projected to grow beyond 1 trillion euros within a decade. As the need to meet national quotas intensifies, the value of CERs may rise, encouraging more investment in generating new CERs. Speculation and derivatives trading in the value of CERs contributes to rising market activity.

Clean Development Mechanism, Joint Implementation, and International Emissions Trading

Annex I countries earn CERs for financing projects in less developed countries, including projects started ahead of the Kyoto trading period. Under the clean development mechanism (CDM), a developed country can sponsor a GHG reduction project in a developing country, where the cost of GHG reduction activities is usually much lower than it is in the developed country itself. The CDM is administered through an executive board, based in Bonn, Germany.

Joint implementation refers to CDM projects implemented by Annex I countries with the transitional economies of Eastern Europe and the Balkans. Carbon projects can be created by a national government or by an operator within the country. Under international emissions trading, countries can trade in the international carbon credit market to cover their shortfall in allowances.

Progress Since the Kyoto Protocol

A 4 percent reduction in GHG emissions had been achieved overall by 2004. Large reductions came from former Soviet states that were transitioning to market economies. As of May, 2008, 182 countries had ratified the Kyoto Protocol, representing 61.6 percent of global emissions of GHGs. The United States and Kazakhstan had signed but not yet ratified the treaty. Afghanistan, Andorra, Brunei, Chad, Iraq, the Palestinian Authority, the Sahrawi Arab Democratic Republic, San Marino, Somalia, Taiwan, Tajikistan, Vatican City, and Zimbabwe had not expressed any position.

U.S. Approach to Kyoto

In 1997, the U.S. Senate voted 95-0 against ratifying the Kyoto Protocol unless several conditions were met. The vote reflected lawmakers' concerns that American manufacturing would be severely affected by the costs of compliance, at a time when strong economic growth was driving emissions up. There was, moreover, widespread skepticism about the claims regarding anthropogenic global warming. The government took the position that since China and India, whose overall emissions are large despite very low per capita emission levels, were not subject to reduction requirements, the United States should not participate in the agreement.

On March 29, 2001, President George W. Bush withdrew the United States from the Kyoto Protocol. By 2004, the country had showed a 15.8 percent rise in emissions, against the target of a 7 percent reduction. Early studies projected a 4.2 percent drop in U.S. gross domestic product (GDP) if Kyoto requirements were met. Subsequent studies have projected far lower effects.

The U.S. approach at the beginning of the twenty-first century focused on using federal tax credits to encourage reduction in GHG emissions, accompanied by a phase-out of the most harmful products such as chlorofluorocarbons. Several states adopted so-called Green Tags, or renewable energy certificates. A renewable energy certificate is earned for each 1,000 kilowatt-hours of electrical energy produced from renewable sources. Some large renewable energy projects, such as wind farms on Native American reservations, have been funded partially through the sale of long-term rights to Green Tags. Green Tags can be used by corporations to reduce their net emissions profile.

A sharp rise in fossil energy prices beginning in 2004 created new market realities in the United States. Nuclear energy, which despite its other problems is a clean source from the GHG perspective, seemed poised for a comeback, along with an increased push toward renewable energy. Coupled with the drive toward efficiency in using costlier fossil fuels, the United States was moving along the path needed to bring GHG emissions down.

Context

The ambitious aspiration of the Kyoto Protocol is to hold anthropogenic emissions of GHGs at the levels of 1990, as the world economy grows. The technological and policy improvements required to achieve this goal imply dramatic changes in the way the world operates. Fundamentally, the emission of waste heat and hot gases implies thermodynamic inefficiency. The Kyoto Protocol provides the urgent motivation, means for accurate accounting, and substantial funding to improve efficiencies, which in turn reduces the need for energy.

Energy efficiency, as a cost-saving rather than a cost-inducing measure, is also more likely to gain favor among industrialized nations and businesses doubtful of their ability to compete globally while reducing GHG emissions. As the nations of the world begin to negotiate the successor treaty to the Kyoto Protocol, a new sense of urgency combined with a new belief that green technologies may become an economic growth sector may make the next treaty negotiated under the UNFCCC more successful than the Kyoto Protocol has been.

Narayanan M. Komerath

Further Reading

Douma, W. Th., L. Massai, and M. Montini, eds. *The Kyoto Protocol and Beyond: Legal and Policy Challenges of Climate Change.* West Nyack, N.Y.: Cambridge University Press, 2007. Includes general discussion of the implementation of the protocol, as well as several detailed studies of the experiences of specific signatory nations.

Fletcher, S. R. *Global Climate Change: The Kyoto Protocol.* Washington, D.C.: Congressional Research Service, 2005. Provides an overview of and updates to the status of the Kyoto Protocol and specifically the U.S. role in its negotiation. Includes an interesting summary of negotiations subsequent to the signing of the protocol.

Holte, S. J., et al. "Impacts of the Kyoto Protocol on U.S. Energy Markets and Economic Activity." Washington, D.C.: U.S. Energy Information Administration, 1998. This 247-page report prepared by a team of scientists and policy makers analyzes six cases of varying amounts of emission reduction. Summarizes data from various sources and aspects of the economy.

Newell, P. *The Kyoto Protocol and Beyond: The World After 2012.* New York: Human Development Report Office, 2008. Summarizes the many implications of climate change and the Kyoto Protocol for various nations from the viewpoint of the World Bank as a lender to development projects.

Peters, Glen P., Gregg Marland, Corinne Le Quéré, Thomas Boden, Josep G. Canadell, and Michael R. Raupach. "Rapid growth in CO_2 emissions after the 2008-2009 global financial crisis." *Nature Climate Change* 2, no. 1 (2012): 2-4. The 2008-2009 global financial crisis resulted in a sharp but transient decline in carbon emissions. Carbon dioxide emissions rapidly recovered to pre-crisis levels as the economy improved.

See also: Annex B of the Kyoto Protocol; Certified emissions reduction; Clean development mechanism; Emissions standards; Greenhouse gases and global warming; International agreements and cooperation; Kyoto lands; Kyoto mechanisms; United Nations Framework Convention on Climate Change.

La Niña

Category: Meteorology and atmospheric sciences

Definition

Two large weather anomalies alternate south of an equatorial band across the middle and eastern portions of the Pacific Ocean. One is El Niño, characterized by warmer-than-average surface waters; the other is La Niña, marked by cooler surface temperatures (the two are collectively designated as the El Niño-Southern Oscillation, or ENSO). La Niña is defined as Pacific surface temperatures of 0.5° Celsius or more below average for a period of at least five months. The cooling occurs as stronger-than-usual eastward trade winds blow across the Pacific, churning cold water from the ocean depths to the surface.

The Spanish names of these weather events originate from the appearance of El Niño off the coast of Peru at Christmastime; niño (boy child) refers to the birth of Jesus. The opposite pattern is termed niña (girl child), although at one time it was called El Viejo (the old man). ENSO does not occur on a scheduled basis but does happen regularly with vehement, widespread consequences.

Significance for Climate Change

The effects of global warming on ENSO, the causes of which are only incompletely known, remain the subject of scientific debate, particularly because so great a number of variables affect ENSO. To obtain more data and understand long-term weather patterns, a monitoring system in the Pacific collects data from buoys, satellites, and computer models. Hotter or colder surface Pacific waters alter overhead trade winds, thereby shifting the North American jet stream. Altering the track and strength of the jet stream produces exceptional rain or drought. The cold, heavy air from La Niña pushes the jet stream to the upper part of the United States.

On average, a La Niña event may last from nine to twelve months, appearing at the end of one year and extending into the next. Consequently, in the southeastern portions of the United States, winter temperatures are warmer and dryer than normal. In northwestern regions, they are cooler and wetter. Hurricanes in the Atlantic and tornadoes in the United States tend to increase in number and force during La Niña events. In South America, dryer-than-usual conditions prevail in southern regions.

El Niño events occur more frequently than do La Niña events, in a ratio of approximately two to one. However, La Niña events last longer. One such event lasted, with a brief interlude, from mid-1998 to early 2001. Another occurred during the latter half of 2007 through the first half of 2008, provoking epic rainfall in Australia and record snowfalls in parts of China.

Edward A. Riedinger

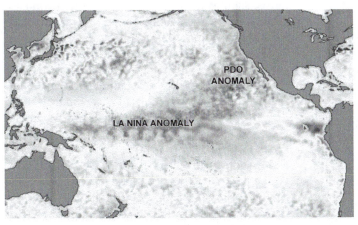

Sea Surface Temperature Anomaly (°C)

-5   0   5

A cool-water anomaly known as La Niña occupied the tropical Pacific Ocean throughout 2007 and early 2008. The cool water anomaly in the center of the image shows the lingering effect of the year-old La Niña. (NASA image by Jesse Allen)

Further Reading

Bell, Gerry. "Impacts of El Niño and La Niña on the Hurricane Season." Climate.gov. NOAA, 30 May 2014. Web. 24 Mar. 2015.

D'Aleo, J. S., and P. G. Grube. El Niño and La Niña. Westport: ORYX, 2002. Print.

Di Liberto, Tom. "ENSO + Climate Change = Headache." Climate.gov. NOAA, 11 Sept. 2014. Web. 24 Mar. 2015.

Glantz, M. H. Currents of Change: Impacts of El Niño and La Niña on Climate and Society. 2nd ed. New York: Cambridge UP, 2001. Print.

"La Niña." National Geographic Education. Natl. Geographic Soc., 1996–2015. Web. 24 Mar. 2015.

Lamb, Hubert Horace
English climatologist

Born: September 22, 1913; Bedford, England
Died: June 28, 1997; Holt, England

Lamb was a pioneer in examining climate change from a historical perspective.

Life

Hubert Horace Lamb spent much of his life as a working climatologist engaged in long-range weather forecasting and examining climate change with the United Kingdom Meteorological Office. In this capacity, he spent time in Antarctica, North Africa, and Malta, as well as becoming a member of the World Health Organization Working Group on Climate Fluctuations. In 1971, he became the founding director of the Climatic Research Unit in the School of Environmental Sciences at the University of East Anglia. As the director of the Climatic Research unit, Lamb did a great deal to encourage historical research into climate change as well as research into recent climatic developments. Among other awards, he received the Royal Meteorological Society's Darton Prize in 1964.

Climate Work

A prolific writer, Lamb authored or coauthored 145 publications in a career beginning in 1939 and lasting through the second edition of his *Climate,* *History, and the Modern World* (1982), which appeared in 1995. Although he wrote six books, *Climate, History, and the Modern World* is Lamb's most far-ranging and influential work. This book has served as a model of the interdisciplinary perspective necessary for historical climate research. Even though some of his conclusions may be disputed, his work along with E. LeRoy Ladurie's *Times of Feast, Times of Famine* introduced scholars and nonscholars alike to the importance that climate has played in human events.

Almost all of Lamb's written work has dealt with some aspect of climate change. Some of his publications are technical works written for scientists, while others are written for broader audiences. Gradually, Lamb turned to an examination of historical climate as a means to understanding climate change.

In 1982, the initial edition of *Climate, History, and the Modern World* laid out a research program for the historical study of climate that influenced many later works. In this book, Lamb first provided scientific background on how climate works and how it fluctuates. He then provided a useful discussion of the sources used by historical climatologists, such as meteorological records, chronicles, grain prices, archaeological material, pollen analysis, radiocarbon dating, tree ring data, and ocean bed deposits. By emphasizing the use of multiple sources, Lamb made it clear that reliance on one type of source could be misleading and that achieving a comprehensive record of past climate is often difficult. In carrying out his analysis of past climate change, Lamb often extrapolated from his sources, such as written records or tree ring data, to achieve a more complete picture.

The bulk of *Climate, History, and the Modern World* is devoted to an examination of climate from the time of early civilization to 1950 and beyond. Two highly influential chapters posited a medieval warm era that lasted until the end of the thirteenth century and a period encompassing the sixteenth and seventeenth centuries that he labeled the Little Ice Age. These chapters have spawned numerous other works dealing with topics such as the Medieval Warm Period or the Little Ice Age that have used Lamb's analysis as starting points for further work. Subsequent analysis has modified some of Lamb's conclusions, but his work remains the starting point for research.

As a practicing climatologist, not a historian, Lamb provided more of a discussion of the changing climate and how humans have affected climate than of the impact of climate on human affairs. In contrast to Le Roy Ladurie, Lamb argued that climatic changes have had an impact on humankind, a view that has come to be generally accepted. The final section of the book dealt with the human influences on climate change, forecasting future developments in the light of the past, and a chapter dealing with the lessons to be learned from the impact of humans on the climate and what can be done to lessen this impact. The second edition included a final chapter on more recent developments.

Lamb may be criticized for a narrowness of focus because he dealt primarily with Western Europe in his work. Although documentation is better for this region, there have been some interesting works that deal with other areas of the globe, such as China and North America. In addition he tended to paint with a broad brush, emphasizing climate change in general rather than dealing with climate change in a particular region that might even run counter to the overall trend that Lamb was describing.

Lamb deserves credit for helping to open up the historical study of climate change and human impact on climate and climatic impact on human endeavor. Although some of his interpretations have been superseded, he helped to pioneer a field and has helped enhance knowledge of the factors that have contributed to climate change.

John M. Theilmann

Further Reading

Brown, Neville. *History and Climate Change.* London: Routledge, 2001. Updates, extends, and critiques some of Lamb's work.

Fagan, Brian. *The Great Warming.* New York: Bloomsbury Press, 2008. Extends Lamb's analysis of the medieval warm epoch along with providing some criticisms of his methodology.

Lamb, H. H. *Climate, History, and the Modern World.* 2d ed. London: Routledge: 1995. Lamb's most influential work, one that continues to generate debate among scholars of the historical climate.

Martin-Nielsen, Janet. "Ways of knowing climate: Hubert H. Lamb and climate research in the UK." *Wiley Interdisciplinary Reviews: Climate Change* 6, no. 5 (2015): 465-477. The dominance of numerical modeling overshadows other approaches to understanding climate, and the historical approach of Lamb is used as a competing approach.

_____. *Weather, Climate, and Human Affairs.* London: Routledge, 1988. Useful collection of several of Lamb's articles dealing with the history of climate and its impact on the human population.

See also: Climate reconstruction; Dating methods; Deglaciation; 8.2ka event; Little Ice Age (LIA); Medieval Warm Period (MWP); Pleistocene climate; Younger Dryas.

Land use and reclamation

Categories: Economics, industries, and products; environmentalism, conservation, and ecosystems

Land use and reclamation affect and are affected by climate change in complex ways. The creation of new land from sea- or riverbeds or the restoration of an area to its previous natural state has the potential to shape the complex interaction between land cover and the atmosphere, an interaction that affects climate change as much as do GHG emissions.

Key concepts

anthropogenic: due to human sources or activities

greenhouse gases (GHGs): trace atmospheric gases that trap heat, preventing it from escaping into space

land reclamation: modification of unsuitable land to allow for some form of human use or for a return to its original natural state

land rehabilitation: restoration of land that has suffered environmental degradation

Background

The different types of land surface affect differently the interaction between the Earth and the atmosphere. Land cover and land-use patterns are affected by the distribution and density of human population, by its activities, and by natural causes. Land-use changes in urban and peri-urban areas, including land reclamation, often drive environmental change, reducing or increasing greenhouse gas (GHG) emissions. In turn, land use and land cover are affected by climate change through erosion, desertification, drought, heavy precipitation, ocean warming, sea-level rise, salinization, heat waves, increases in wind intensity, decline of glaciers and snow cover, and drying of wetlands, among other processes.

The role of land-use change in overall climate change is not well understood. However, if most of the observed increase in temperatures in recent decades can be attributed to GHG emissions, then urban and industrial land uses can be seen as co-responsible for global warming.

Land Creation, Land Rehabilitation, and Soil Remediation

Land reclamation as a vehicle of land-use change can be one of two different processes. One consists in the creation of new land where there was once water, either in the sea or in a riverbed. Many cities on all continents have used this process to create new residential or business areas, harbors, airports, and other infrastructure. Land reclamation

This recreation area near Black River Falls, Wisconsin was the site of an open pit iron mine that closed in 1983 due to low iron and steel prices. The waste piles were landscaped and vegetated and the former open pit is now a lake. (© Steven I. Dutch)

also includes the construction of artificial islands for tourism-related activities or the creation or restoration of beaches in areas affected by beach erosion. Another form of land reclamation is the draining of seasonally submerged lands to convert them to agricultural uses and, in some places, also to control the agent of malaria.

The other form of land reclamation, also called land rehabilitation, consists in restoring an area after it has suffered physical degradation. Such degradation may be natural (for example, erosion) or anthropogenic (for example, pollution, soil contamination, or mining operations). Land can be rehabilitated to allow a new use, such as housing or commercial development, or it may be restored to its natural state in the interests of conservation. In the latter case, knowledge of the land's original soil, flora, and fauna characteristics will facilitate their restoration. Soil remediation, a soft form of land rehabilitation, is the removal of industrial contaminants in soil and water to reduce the risks for human health and for the ecosystem.

Land-Use and Climate Change

GHG emissions and other anthropogenic and natural factors have the capacity to affect climate change. Urban and peri-urban areas have probably the greatest influence on climate change, being responsible for most GHG emissions. This influence is expected to increase, according to United Nations projections of urban population growth and urban sprawl. Outside urban areas, changes in land cover—including deforestation, reforestation, agriculture, and irrigation—affect temperatures, precipitation, and atmospheric circulation. Replacing a rain forest with agricultural lands reduces evaporation and consequently raises temperatures, while irrigation increases evaporation and, eventually, precipitation. Similar changes can be associated with reforestation of snow areas, with planting trees as carbon sinks, and with the creation of new areas for growing crops for fuel.

A changing climate can in turn affect land cover and land use. Climate change affects land cover by creating new risks associated with temperature rise, flooding of low areas, dike collapses, sea-level rise, storms and other natural disasters, erosion of coastal areas, salinization, droughts, the quantity and quality of groundwater resources, water scarcity, air quality, land subsidence, and deterioration of soils. These events can lead to habitat destruction, invasion by alien species, ecosystem fragmentation, and species loss, with profound and irreversible impacts on biodiversity. Climate change also directly affects many other economic and social activities that in turn affect land-use patterns. In some parts of the world, for example in small island states, the forecasted climate change and sea-level rise will make land one of the rarest and most vulnerable resources. In this context, cities will probably be the land-use type most affected by climate change, especially cities located on low coastal zones.

Land-Use Planning in Adaptation and Mitigation Strategies

The interaction between climate changes and land-use calls for innovative adaptation and mitigation measures. Land-use planning can have an important role in climate change adaptation strategies at the local level, and can contribute to wider mitigation strategies as well. These measures include, for example, the coordination of public and private stakeholders involved in land-use and land reclamation processes, tax benefits, innovative financing mechanisms, and other incentives for the use of renewable energy.

Building codes can also be changed to incorporate norms for GHG emissions in public buildings, green roofs, energy efficiency within household dwellings, and energy conservation to prevent the production of carbon dioxide. Land-use planning strategies, including land zoning, must discourage automobile use, provide adequate incentives for mass transit, promote mixed land use and infill development in sites previously used by other urban functions (including land rehabilitated or areas subjected to soil remediation), and promote higher densities and compact urban layouts, stimulating walkability.

Context

The interaction between landuse and land reclamation, on one side, and climate change, on the other, is exceptionally complex and is still insufficiently understood. However, the evidence available suggests that changes on each side of this relation affect the other, and, consequently, failure to consider these interactions and to act accordingly may have long-lasting negative consequences for the environment and for the economy and society. For that reason, measurable, verifiable, and reportable actions on adaptation and mitigation, in the field of land-use planning, must be taken sooner rather than later, in all countries, to reduce the environmental, economic, and social effects that would otherwise result from inaction. Special support for adaptation measures, including technology transfer and innovative finance, should be provided by the international community to the poorer and more vulnerable countries.

Carlos Nunes Silva

Further Reading

Intergovernmental Panel on Climate Change. *Climate Change, 2007—Synthesis Report: Contribution of Working Groups I, II, and III to the Fourth Assessment Report of the Intergovernmental Panel on Climate Change.* Edited by the Core Writing Team, Rajendra K. Pachauri, and Andy Reisinger. Geneva, Switzerland: Author, 2008. Suggests that climate change is already occurring, that it affects mainly the poorer and the more vulnerable, and that it is still possible to avoid cataclysms in the long term.
_____. *Emissions Scenarios.* Geneva, Switzerland: Author, 2000. Provides insights about the inter-linkages between environmental quality and development choices. Demonstrates the impacts of different social, economic, and technological development on emission trends.
_____. *Land Use, Land-Use Change, and Forestry: A Special Report of the IPCC.* Edited by Robert T. Watson et al. New York: Cambridge University Press, 2000. Examines the scientific and technical state of understanding of carbon sequestration strategies related to land use, land-use change, and forestry activities.
Organization for Economic Cooperation and Development. *Literature Review on Climate Change Impacts on Urban City Centres: Initial Findings.* Paris: Author, 2007. Offers a review of the literature on climate change's impacts on cities, evaluates the methods used, and offers relevant policy guidance.
_____. *OECD Environmental Outlook to 2030.* Paris: Author, 2008. Based on projections of economic and environmental trends to 2030; presents simulations of policy actions to address key challenges.
Verburg, Peter H., Kathleen Neumann, and Linda Nol. "Challenges in using land use and land cover data for global change studies." *Global Change Biology* 17, no. 2 (2011): 974-989. Uncertainties introduced by using convenient but inappropriate data sets can equal the uncertainties in the actual processes being studied. Proposed remedies include better documentation and data analysis, and data collection aimed at areas where data is especially uncertain.

See also: Agriculture and agricultural land; Kyoto lands; Soil erosion; Urban heat island.

Last Glacial Maximum (LGM)

Categories: Cryology and glaciology; Climatic events and epochs

Definition

The Last Glacial Maximum, also called the Wisconsin Glacial Stage, commenced about 35,000 years ago and ended about 12,900 years ago. At that time, the earth entered the current interglacial stage of warming. The Last Glacial Maximum reached its peak of cold about 20,000 years ago, when the continental glaciers reached their maximum extent of southern advance. In the Northern Hemisphere during the Last Glacial Maximum, continental glaciers flowed out over the land surface from three main centers, northern North America, Greenland, and Scandinavia. Much of the Northern Hemisphere was covered with ice,

Earth at the last glacial maximum of the last ice age. (Ittiz/Wikimedia Commons)

including the North Atlantic Ocean. Similar effects occurred in the Southern Hemisphere.

During the maximum, global sea levels dropped by as much as 120 meters below present levels. This drop had profound effects on the locations of shorelines, the gradients of rivers, and the hydrologic cycle on Earth. Sea level was much lower than present as a result of the amount of water that was locked up in the world's ice sheets covering the land.

Continental ice sheets of the Last Glacial Maximum consisted of individual large lobes, which expanded and contracted at different rates. Action of the individual lobes resulted in deposition of vast layers and mounds of rock, sand, and fine sediment, as well as erosive scour of the land. The Great Lakes of the United States and Canada formed as a result of such scour during the Last Glacial Maximum. Glacial ice covering the northern latitudes reconfigured the land's surface and

largely removed preexisting drainage patterns, which were reestablished after the ice melted away.

Significance for Climate Change

The presence of so much ice on the land had a huge effect on global climate and local weather. As a result of the uptake of water by glaciers, many arid areas of the earth became much more arid. An example of this occurred in the Sahara region of Africa, where deserts greatly expanded during the era. Rain forests, which were plentiful before the Wisconsin Stage, shrank considerably and were fragmented into small areas. One effect of this fragmentation of the rain forests was isolation of animal groups such as gorillas into separate geographic areas, where the various isolated populations diverged.

Reduced evaporation from cooler seas generally made the Last Glacial Maximum much drier, but there were areas where this was not the case. For example, in parts of North America, heavy rains from moist winds flowing over continental glaciers caused much heavier precipitation to occur in areas not covered by ice. For this reason, vast continental lakes formed in areas where lakes are not found today, such as the Great Basin of the United States. The largest such lake, Lake Bonneville, was the forebear of the Great Salt Lake.

The onset of the Last Glacial Maximum may have been triggered by the development of the Isthmus of Panama. Prior to the uplift of rocks and volcanic activity that formed the isthmus, waters of the Atlantic and Pacific Oceans freely mixed, and the salinity of these waters was therefore kept nearly equal. After the isthmus formed, dry winds from Africa evaporated Atlantic Ocean water and raised the Atlantic's salinity slightly. Mixing could no longer mitigate this effect, so Atlantic Ocean water that moved toward Earth's northern pole, which had helped warm that area in past, no longer reached the pole. Instead, the higher density resulting from the water's greater salinity caused that water to sink north of Iceland. The resulting cooling of the northern pole area is thought to have been enough to trigger the initial ice buildup that started the glacial development of the Last Glacial Maximum. Once the process of ice buildup began, increased reflectivity of the ice (known as albedo) sent increasing amounts of solar energy back into space, and the cycle of global cooling began to accelerate.

The history of the Last Glacial Maximum shows what happens when the earth descends into a glacial episode of ice accumulation and cooling. All global systems, both living and nonliving, are affected by this climate change. In many ways, the contemporary earth is the result of many vast changes that occurred during the Last Glacial Maximum, as well as the effects of the current interglacial episode of warming.

Sea-level rise since the end of the Last Glacial Maximum has caused nearly all of the world's coastlines to move and reestablish themselves. Most of the high-latitude rivers and streams have established new channels and drainage patterns since the continental ice sheets melted away. Temperature and rainfall patterns are now mostly different from those during the last glacial episode. Therefore, modern ecosystems have entirely shifted in most instances since then. In many instances, populations of plants and animals have been dramatically affected by this climatic shift. As a result, mass extinctions have occurred in some groups of organisms, and changes have occurred in others. For example, the ice age fauna of North America, including the saber-toothed cats, mammoths and mastodon, giant ground sloths, and small native horses—all of which were once plentiful— are now entirely gone. The physical and biotic changes related to the onset, duration, and end of the Last Glacial Maximum serve as reminders that the world and its living systems can be profoundly changed by global climatic shifts such as the ones that occurred as recently as 12,900 years ago.

David T. King, Jr.

Further Reading

Clark, Peter U., and Lev Tarasov. "Closing the sea level budget at the Last Glacial Maximum." PNAS 111.45 (2014): 15861–62. PDF file.

Gillespie, A. R., S. C. Porter, and B. F. Atwater. The Quaternary Period in the United States. Developments in Quaternary Science 1. Amsterdam: Elsevier, 2004.

Lambeck, Kurt, Hélène Rouby, Anthony Purcell, Yiying Sun, and Malcolm Sambridge. "Sea level and global ice volumes from the Last Glacial Maximum to the Holocene." *Proceedings of the National Academy of Sciences* 111, no. 43 (2014): 15296-15303. There were complex and sometimes abrupt changes in ice volume and sea level during deglaciation, but once sea level stabilized, it remained very stable until it began to rise in the last couple of centuries.

Madsen, D. B., ed. Entering America: Northeast Asia and Beringia Before the Last Glacial Maximum. Salt Lake City: University of Utah Press, 2004.

Rutter, Nat, et al. Glaciations in North and South America from the Miocene to the Last Glacial Maximum: Comparisons, Linkages and Uncertainties. Dordrecht: Springer, 2012. Print.

Stanley, Steven. Earth Systems History. 3d ed. New York: W. H. Freeman, 2009.

Latent heat flux

Category: Chemistry and geochemistry

Definition

Water can exist as a liquid, solid, or vapor. The latent heat flux (expressed in joules per kilogram) is the rate at which energy is released or absorbed when water changes from one to another of these three states. It is latent by contrast with the sensible (or dry) heat flux caused by the movement of air. The energy involved in the various phase changes of water are known as the latent heat of vaporization (evaporation and condensation of gas and liquid), latent heat of fusion (melting of ice) and latent heat of sublimation (transition of ice directly to vapor). Evaporation of water, melting of ice, and sublimation of ice are heat-absorbing processes. The heat energy used is held in a latent or dormant (hidden) state until it is released back into the atmosphere when freezing, condensation, or the sublimation of vapor to ice occurs.

Significance for Climate Change

A rise in the concentration of greenhouse gases (GHGs) in the atmosphere makes available more energy at the Earth's surface. This additional energy can be used either to heat the atmosphere by way of the sensible heat flux (increased warming) or evaporate water at the Earth's surface via the latent heat flux. If water is present, the latent heat flux will always be larger than the sensible heat flux, meaning that energy otherwise available to heat the atmosphere is used in evaporation. Given that over 70 percent of the Earth's surface is water, most of the additional energy at the surface due to GHG radiative forcing is used in an enhanced latent heat flux. The resulting warming of the atmosphere would be less in this case than if all the additional available energy was accounted for by the sensible heat flux alone.

The heating of air by the sensible heat flux at the Earth's surface causes the air to become buoyant and, as a result, to rise and mix with the cooler air above. Moisture is entrained in this rising air. As a result, the mean annual global climate impact of increased energy available due an enhanced greenhouse effect will manifest itself both as an increased sensible heat flux (warming) and increased evapotranspiration via the latent heat flux. The latter is dormant heat energy that can be transported great distances by wind and over time without loss. The impact on climate may occur in regions far from its source. This dormant heat energy does not radiate back to space until condensation occurs and returns the heat to the air.

C R de Freitas

See also: Aerosols; Alkalinity; Atmospheric chemistry; Dew point; Evapotranspiration; Greenhouse gases and global warming; Hydrologic cycle; Water vapor.

Lavoisier Group

Category: Organizations and agencies
Date: Established 2000
Web address: http://www.lavoisier.com.au

Mission

The Lavoisier Group is an independent, nonprofit organization that aims to promote debate within Australia on the science of global warming and climate change and the perceived need to reduce fossil fuel consumption through national and international regulation. The group's activities are motivated by the belief that some of the science on which national and international climate policy is based is not beyond reproach. In light of this belief, the organization encourages dialogue within Australia on the scientific merits of climate policies and agreements such as the Kyoto Protocol, as well as their economic consequences. The organization also seeks:

> to explore the consequences which any international treaty relating to global decarbonisation targets, and the methods of policing such treaties, would have on Australian sovereignty and independence, and for the World Trade Organization rules which protect Australia from the use of trade sanctions as an instrument of extraterritorial power.

The group is run by volunteers comprising mainly active and retired earth and atmospheric scientists, engineers, and other professionals.

The Lavoisier Group is named after Antoine-Laurent Lavoisier, a French scientist, economist, and public servant. Although he is best known for his discovery of the role oxygen plays in combustion, Lavoisier was politically liberal and persuaded of the need for social reform in France during the years leading up to the French Revolution. At age fifty, during the Reign of Terror, he was found guilty of *incivisme* (avoiding civic responsibility) by the French revolutionists and put to death by guillotine. Eighteen months after his death, Lavoisier was exonerated by the French government when it was formally declared he had been falsely convicted.

Significance for Climate Change

The Lavoisier Group highlights noteworthy developments in climatology, publishing information on its Web site and inviting speakers on the subject. The topics most frequently covered in the organization's site and seminars include greenhouse gas theory, solar and planetary influences on Earth's climate, predictions and projections of future climate, the history of climate change, the economics of energy and technology, energy security, the role of the Intergovernmental Panel on Climate Change in promoting global warming hysteria, and the response of Australia's governments to the decarbonization campaigns of the environmentalist movement and the media.

C R de Freitas

See also: Australia; Engineers Australia; Skeptics.

Levees

Category: Economics, industries, and products

Definition

Levees are embankments or engineered structures near bodies of water; they are designed to prevent flooding of the land behind them. Permanent levees are used along rivers, such as the Mississippi and Sacramento Rivers, and in coastal areas, such as New Orleans, Louisiana. Once constructed, levees are protected from erosion by planting them with vegetation such as grass or willows on the river side. Concrete abutments may also be used on the river side of a levee to protect the structure from strong currents. A breech occurs when a section of levee is washed away, letting water onto the adjacent land. On large floodplains, levees are built in a series, stepping back from the river, to provide an extra measure of protection against a breech.

Levee systems along coasts and rivers are built and maintained by a variety of state and federal agencies, including the U.S. Army Corps of Engineers (USACE), state and tribal levee boards, and

private groups. For example, in Missouri, a group of private landowners who own a majority of the land in a wetland or another region where flooding occurs may create their own levee district, obligating themselves to pay taxes to finance the building and maintenance of levees.

The National Levee Safety Act (NLSA) of 2007 includes several provisions related to the oversight of levee safety and maintenance—including of nonfederal (state and private) levees. One provision of the NLSA authorizes the USACE to develop a plan to create a national levee safety program and to inventory and inspect all federal levees. Nonfederal levees can be added to the USACE inspection and inventory program at the request of local levee boards, but the maintenance and safety of those levees remains the responsibility of the local board. The NLSA also created a National Committee on Levee Safety (NCLS), which consists of representatives from USACE; the Federal Emergency Management Administration (FEMA); and state, regional, and tribal levee boards and is charged with creating a national levee safety program.

Significance for Climate Change

The U.S. levee system is a large, complex patchwork of many different structures that have been built by a variety of organizations over many years. Some of the oldest levees in the United States were built more than 150 years ago. Most levees are earthen structures, built with a tapered profile, at a width-to-height ratio of seven to one. Thus, a maximum height must be determined by the time construction begins, based on analysis of historical data regarding maximum local flood heights and storm surges and the recurrence interval of these events.

Results of climate models, notably those used by the Intergovernmental Panel on Climate Change (IPCC), predict that future increased global temperatures will lead to higher sea levels, increased precipitation rates, and a greater potential for large hurricanes. Also, a report released by Environment Texas in 2007 concluded that weather data from 1948 to 2006 demonstrated that extreme precipitation events had increased by 24 percent. Together, such weather events have the potential to lead to higher flood levels and storm surges. For

Levees along the Galena River, Galena, Illinois. Although the river might flood, a greater danger is backflooding from the Mississippi River only a couple of kilometers away. (© Steven I. Dutch)

example, during the Upper Mississippi Valley flood of June, 2008, peak flows were approximately 1.8 meters above those previously recorded. The storm surge from Hurricane Katrina on August 29, 2005, was 7.6 meters. A report by University of California, Berkeley, scientists concluded that flooding in New Orleans was a result of the levees not being built high enough; however, this conclusion has been disputed by the USACE and independent investigators.

Following Hurricane Katrina, the United States Congress passed the Water Resources Development Act of 2007, which authorized creation of the National Levee Safety Program. The law mandated the formation of the National Committee on Levee Safety. The direct effects of global warming on levee safety have not been considered in initial plans for levee safety standards. A further concern for levee safety in coastal Louisiana, and in other cities located on deltas, is the combined effects of ground subsidence and projected sea-level rise. One potential solution is to replace levees with other flood control technologies, such as the movable water gates employed in Japan, England, and the Netherlands.

Anna M. Cruse

Further Reading

Cech, Thomas V. *Principles of Water Resources: History, Development, Management, and Policy.* Hoboken, N.J.: John Wiley & Sons, 2005. While levees are not given a separate chapter of their own, this book provides a comprehensive overview of water resources in the United States, including flood-control structures. Index, illustrations.

Fischetti, Mark. "Protecting New Orleans." *Scientific American* 294, no. 2 (February, 2006): 64-71.

The levees in New Orleans were too small to prevent flooding from the Hurricane Katrina storm surge. This article discusses levee systems used in the Netherlands and Great Britain, plans for new systems, and the potential impacts for wetland restoration. Figures, maps.

Intergovernmental Panel on Climate Change. *Climate Change, 2007—The Physical Science Basis: Contribution of Working Group I to the Fourth Assessment Report of the Intergovernmental Panel on Climate Change.* Edited by Susan Solomon et al. New York: Cambridge University Press, 2007. Comprehensive treatment of the causes of climate change, written for a wide audience. Figures, illustrations, glossary, index, references.

Ludy, Jessica, and G. Matt Kondolf. "Flood risk perception in lands "protected" by 100-year levees." *Natural hazards* 61, no. 2 (2012): 829-842. Lands protected by levees capable of withstanding a 100 year flood are considered not to lie in a floodplain, creating the paradox that some areas in California are below sea level but not considered to be in a flood plain. Surveys of residents show widespread unawareness of flood risk.

O'Neill, Karen M. *Rivers by Design: State Power and the Origins of U.S. Flood Control.* Durham, N.C.: Duke University Press, 2006. Comprehensive historical treatment of the creation of the flood control system—including levees—in the United States. Bibliography, index.

See also: Flood barriers, movable; Floods and flooding; Louisiana coast; New Orleans; Storm surges.

Level of scientific understanding

Category: Science and technology

Definition

The Intergovernmental Panel on Climate Change (IPCC) is a scientific body that represents a hybrid of science and government. Its members strive for maximum reliability in their measurements, but they are keenly aware of the virtual impossibility of reaching that goal. Earth's climate is simply too vast a system to allow for carefully controlled experiments. Rather than ignore the problem, IPCC workers have designed a compromise system that specifies the degree to which workers in the field are confident of the validity of reported data. A level of scientific understanding (LOSU) designation is applied to assessments of each of the influences on the Earth's temperature. These radiative forcing factors include the greenhouse gases (GHGs), aerosols, mineral dust, and contrails.

Each of the radiative forcing factors is given an assessment of A to C depending on the evidence supporting it and an assessment of 1 to 3 for the degree of consensus among climate scientists. LOSU is assigned on the basis of these two evaluations. The terms used are high, medium, medium-low, low, and very low. This scale and its terminology were deliberately chosen to avoid confusion with the measures of confidence associated with statistics. IPCC reports clearly state that such statistical precision is not possible in the present state of climate science. The LOSU designation is designed to avoid the unwarranted impressions of understanding.

Significance for Climate Change

Central to all scientific discussions is the reliability of the measurements that are used to support one's position. In the case of climate change or global warming, there are a number of extremely complex issues seldom encountered in scientific studies. These include the size of the system being studied; the constant changes undergone by that system; the large number of scientists conducting partial studies, each of whom has a particular objective; the lack of power to demand that specific procedures be followed; and so on. By contrast, scientists are accustomed to designing experiments with particular objectives and carrying them out under carefully controlled conditions. Added to these difficulties is the obvious necessity of collecting enormous amounts of data on a worldwide scale and attempting to draw from them meaningful leads for further investigation.

These problems pale in comparison to the realization that these scientific studies are conducted in a highly charged political arena. Imagine a range of scientists with little training and personal sympathies or antipathies toward policy makers presenting tentative conclusions to officials with little scientific background, and perhaps even a jaundiced view of civilians who never have to stand for reelection. Even assuming both groups earnestly desire to benefit humanity, it is clear that differences of viewpoint will often lead to discord over the appropriate strategies and tactics.

Those who disagree with a general position on global warming argue that science must not involve consensus and that the LOSU is therefore fatally flawed. This conclusion is based on a red herring and a profound lack of knowledge of the history of science. The most important reason for publication of the outcome of scientific experimentation is to allow other scientists to fulfill their obligation to question any new proposal. This is how new ideas are tested through peer review. The normal outcome of this process is debate over the validity and utility of that idea, and it inevitably leads to a mixture of acceptance, reservation, and rejection. It was long after John Dalton had proposed the atomic hypothesis that the majority of chemists accepted the existence of atoms. Even after an idea becomes textbook science and is accepted by nearly everyone, at least as a working hypothesis, it may still be modified or rejected when the consensus is broken. A consensus existed that the inert elements did not form compounds, until it was shown that they do. Science depends on consensus at every stage of its evolution.

With these technical and political problems in clear view, the IPCC has sought to bring common sense to bear on its mission. Nowhere is this effort clearer than in the LOSU approach. By considering both the strength of the evidence offered by scientists and the degree of acceptance it obtains from the scientific community, the group offers the public and the policy makers its best estimate of the real-world situation. It has been said that the only simple systems are those that have not yet been studied. It appears clear to informed people, scientists, and policy makers that the problems of climate change must be resolved using the best ideas and technology available. The level of scientific understanding represents a useful approach to evaluating and disseminating the current state of this aspect of climate science.

K. Thomas Finley

Further Reading

Curry, Judith A., and Peter J. Webster. "Climate science and the uncertainty monster." *Bulletin of the American Meteorological Society* 92, no. 12 (2011): 1667-1682. Presents a classification of types of uncertainty with numerous pithy and often humorous quotes from many sources.

Forster, P., et al. "Changes in Atmospheric Constituents and in Radiative Forcing." In *Climate Change, 2007: The Physical Basis,* edited by S. D. Solomon et al. New York: Cambridge University Press, 2008. A detailed analysis of the method of assigning the LOSU along with a complete set of current data. Technical in parts, but the main concepts are clear.

Haley, James, ed. *Global Warming: Opposing Viewpoints.* San Diego, Calif.: Greenhaven Press, 2002. An unusual approach, but perfect for this topic. Many questions are raised from both the positive and negative perspective. Experts in each area make their case for the particular viewpoint. Excellent bibliography.

Silver, Jerry. *Global Warming and Climate Change Demystified.* New York: McGraw-Hill, 2008. Comprehensible text filled with data that are all explained with a minimum of technical vocabulary. Good review quizzes for each chapter and a final exam with answers. A glossary, historical outline, and discussion of lingering doubts completes a worthwhile book.

See also: Falsifiability rule; Gaia hypothesis; Intergovernmental Panel on Climate Change; Peer review; Scientific credentials; Scientific proof.

Liberalism

Category: Popular culture and society

Liberalism is a political philosophy that, in contemporary usage, supports government action to redress social problems. Policies commonly associated with liberalism include social safety nets, public health care, civil rights legislation, regulations of potentially harmful activities, and support of individual rights over traditional cultural values.

Key Concepts

Left and Right: Alternate terms used as synonyms for liberal and conservative, respectively.

Progressive: Describes a period of liberal social reform in the United States in the late nineteenth and early twentieth centuries, and also used to describe liberals who advocate strong government action to restrain economic abuses and alleviate social problems.

Socialist: A liberal political movement that advocates strong government control of the economy and in many cases, ownership of sectors of the economy. Many Western European governments can be described as democratic socialist governments.

Marxist: Political philosophy modeled on the theories of Karl Marx. The government of the former Soviet Union and its allies regarded itself as Marxist, as do those of Cuba, China, and North Korea.

Background

Conservatism and Liberalism are commonly used to describe the spectrum of political opinion. As used in the United States, "liberal" generally signifies a greater acceptance of government regulation and intervention in social affairs and "conservative" signifies a desire for less government intervention and taxation, often coupled with calls for defense of traditional values and an emphasis on law and order. In the United States, the Democratic Party is the more liberal major party and the Republican Party is more conservative.

Liberalism is a political philosophy founded during the Enlightenment in Europe. Prominent founding liberal philosophers included John Locke, Adam Smith, Jean-Jacques Rousseau, and

John Stuart Mill. Liberalism is basically optimistic about the ability of humans to use reason to achieve social progress, as opposed to relying on traditional values and institutions for guidance Liberalism accepts the authority of the nation state on the grounds that the nation state is a necessary social institution to protect individual rights and liberty. Governments have a legitimate role in defending property rights and one's freedom to enjoy property according to one's will. Liberalism holds an optimistic view of human nature and enlightened self-interest. Individual rights are central to the liberal outlook. Liberalism stood in opposition to conservatism, which emphasized the importance of traditional institutions and mores in maintaining civil society, and often supported the use of government authority to preserve them. Although liberalism originally placed strong emphasis on property rights, at a time when individual rights were generally not widely recognized, modern liberals are more inclined than conservatives to allow government regulation of property rights, for example, to prevent discrimination, control pollution or prevent economic inequality.

At the same time, in France, the terms "right" and "left" came into use to describe political positions, since supporters of the Revolution in the National Assembly sat to the left of the president, and supporters of the monarchy sat to the right. The term "left" in politics now generally describes liberalism, and "right" describes conservatism.

Liberalism in United States History

The ideas of John Locke and his contemporaries strongly influenced the founders of the United States. Thomas Jefferson began the Declaration of Independence with an appeal to reason, by saying that the Colonies owed their fellow citizens and the world an explanation for their actions. The Declaration then went on to list ways the British government had failed to live up to its responsibilities to protect liberty.

In the decades leading to the Civil War, the great divisive issue was slavery, mostly supported by conservatives and opposed by liberals. However, the positions of the two major parties were opposite to what they are today. In the 1860 election, the Republicans supported Federal spending on a transcontinental railroad and opposed slavery (both liberal positions) while the Democrats defended slavery and opposed Federal spending on domestic improvements (both conservative positions).

Following the Civil War, American politics were a confused mosaic of liberal and conservative positions. In the North, the Republicans tended to be supported by industrial and financial interests, but in the South, they were supported by newly freed blacks. Democrats in the North were favored by the urban working class and immigrants, but in the South by whites intent on maintaining white supremacy. Thus, Democrats tended to be more liberal in the North and conservative in the South, and vice versa for Republicans. In the late nineteenth century, causes most supported by liberals were the growth of labor unions, the breakup of monopolies and the passage of laws guaranteeing rights to workers. This era in American politics is often called the "progressive era."

In the late nineteenth century, European reformers began pressing for reforms to correct economic inequality, as well as eliminating artificial social barriers to individual freedom. Many advocated common ownership of industry and resources. Writers like Karl Marx and Friedrich Engels took a more extreme stance and predicted the eventual overthrow of existing society and the emergence of a state governed by the working class. These varieties of liberal thought never gained much support in the United States and, in fact, caused many conservatives to suspect all liberal ideas as potentially dangerous to the stability of American society.

Following World War I, liberal President Woodrow Wilson advocated the formation of a League of Nations where international disputes could be peacefully resolved, but he was disappointed when Congress refused to allow the United States to become a member. During the Great Depression, Democrat Franklin D. Roosevelt launched into an unprecedented campaign of social engineering which eventually resulted in the Social Security program, minimum wage laws, and massive spending on infrastructure. His successor, Harry Truman, continued Roosevelt's policies and also took steps to desegregate the armed forces. The next several decades saw a steady stream of liberal policies enacted: Supreme Court bans of school segregation and official prayer, legalization of abortion, and Congressional enactment of civil rights and environmental protection laws.

Nevertheless, to some liberals, those successes failed to address what they saw as the unresolved core problems of American society: ingrained racism, repressive laws and social customs, and economic inequality. Protests on university campuses and urban riots began in the mid-1960s, and the escalation of the unpopular war in Vietnam, which confronted students with the threat of conscription into the armed forces, caused an eruption of civil strife and ideological division rarely seen in American history.

Implications for Climate Change

With comparatively few exceptions, liberals tend to accept the scientific evidence for climate change and support, or at least accept, efforts to mitigate climate change. There are some liberals who suspect the climate change debate is merely another guise for granting privileges to corporations. Others complain that efforts to combat climate change are merely superficial technological fixes, when what is really needed, in their view, are wholesale changes in lifestyles and values. Democrat Al Gore, who served as vice president from 1993 to 2001 and ran unsuccessfully for president in 2000, wrote the highly influential book *An Inconvenient Truth* and won the Nobel Peace Prize in 2007 for his efforts at organizing efforts to combat climate change.

Despite widespread liberal support for action on climate change, since the mid-1990s, liberals have not had sufficient power in Congress to make significant changes. They either failed to hold a majority, or if they did have a majority, it was too slender to overcome opposition. President Barack Obama (2008–2016), one of the most liberal Presidents in United States history, faced one of the most hostile and conservative Congresses. As a consequence, most of Obama's efforts to deal with climate change employed executive orders and regulatory actions.

Justin Ervin

Further Reading

Kernell, Samuel, Gary C. Jacobson, Thad Kousser, and Lynn Vavreck. *The logic of American politics.* Cq Press, 2015. A survey of how American politics operates, including the mechanisms established in law, and also informal processes such as the shaping of public opinion and the rise of political factions.

Klein, Danial. The Origin of Liberalism. The Atlantic, February 13, 2104. http://www.theatlantic.com/politics/archive/2014/02/the-origin-of-liberalism/283780/ A search of various uses of "liberal" in political contexts pinpoints the appearance of the word as a political term about 1770, and it originally meant something quite different from today's usage.

Lakoff, George. *Moral politics: How liberals and conservatives think.* University of Chicago Press, 2010. The author, a prolific writer on cognitive science and its applications, contends that contrasting views of the family explain differences in liberal and conservative thought.

Lakoff, George. *The political mind: why you can't understand 21st-century politics with an 18th-century brain.* Penguin, 2008. Cognitive research explains why attempts to communicate between political opposites fail. Words are connected to subconscious views of the world and mean quite different things emotionally to different people. Much of the book explores applications of these findings to contemporary political debates.

Locke, John. The Second Treatise on Civil Government. New York: Prometheus Books, 1986. A central exposition of liberal political theory. It has had an enormous impact on the political and economic culture of the United States.

See also: Conservatism, Libertarianism, Obama Administration

Libertarianism

Category: Popular culture and society

Libertarianism is a political philosophy that champions individual rights and is suspicious of government restrictions on private property. Many libertarians are skeptical of the scientific basis of belief in anthropogenic climate change. Other libertarians endorse the scientific claims but favor voluntary private-sector efforts at mitigation and adaptation.

Key concepts

geoengineering: the intentional production of large-scale changes to the Earth in order to promote human welfare

laissez-faire: an antiregulation policy that allows markets to operate free of government interference

private property: an institution through which legal rights to most or all tangible economic possessions are assigned to individual owners

revenue neutrality: an approach to fiscal policy in which new taxes (such as carbon taxes) are tied with dollar-for-dollar reductions in other taxes, in order to keep the total tax burden constant

tragedy of the commons: a term describing the perverse collective outcome of self-interested behavior in the absence of property rights

Background

Libertarianism is a political philosophy rooted in individual liberty. It stands in contrast to other political philosophies that place the interests of the nation, state, or other collective group above the interests of the individuals composing the group. There is no list of beliefs shared by all libertarians, though they generally oppose the Drug War and military draft, while favoring large tax and spending cuts. The libertarian position is often described as conservative on economic issues and liberal on social issues, but its proponents believe that libertarianism represents a consistent defense of individual freedoms against government encroachment.

Strains of Libertarianism

Historically, many different political philosophies have embraced the term *libertarian*. Some libertarians do not have strict rules limiting state power but generally favor reducing the size of government in most areas. Economist Milton Friedman (1912-2006) falls into this camp. Another group of libertarians are principled minarchists, who believe that the only legitimate role for the government is to provide judicial, police, and military services for its citizens. The novelist Ayn Rand (1905-1982) was a famous minarchist libertarian.

Other self-described libertarians are more radical anarcho-capitalists, who call for the abolition of the modern state and the privatization of all legitimate government services, including defense from foreign invasion. Economists Murray Rothbard and David Friedman are anarcho-capitalist theorists. Finally, there are libertarians who reject the system of capitalism as it exists today and believe that private property rights can lead to unjust power relations among the capitalists and the common workers. Pierre Proudhon (1809-1865) and Emma Goldman (1869-1940) are examples. However, in modern American political debate, libertarianism is usually associated with support for private property and laissez-faire capitalism.

Libertarianism and the Environment

In mainstream political debate, the libertarian position is often associated with the "pro-economy" forces and opposed to the "pro-environment" groups. For example, if there is a conflict between the livelihood of loggers and the natural habitat of the spotted owl, most libertarians would typically support the former. Despite this tendency, many libertarians embrace defense of the environment but believe that market solutions are more effective than government regulations. Such free market environmentalists argue that air and water pollution are not examples of market failure but rather government failure. Free market environmentalists argue that under the common law, a factory dumping chemicals into a river could be prosecuted by the homeowners living downstream. In the late nineteenth and early twentieth centuries,

> ### Major Libertarian Think Tanks and Climatological Organizations
>
> - Cato Institute
> - Fraser Institute
> - Heartland Institute
> - International Policy Network

however, British and American courts often threw out lawsuits against big businesses, in order to promote industry.

Libertarians explain environmental abuses using the economics concept of the tragedy of the commons, defined in the essay "The Tragedy of the Commons" (1968) by Garrett Hardin. Historically, when pastureland was treated as common property, herders would routinely allow their animals to overgraze the land, because any individual's restraint would simply allow another herder's animals to eat more grass. Only the introduction of private property boundaries (and barbed wire in the United States) eliminated harmful overgrazing. Likewise, libertarians often recommend the introduction of well-defined property rights as solutions to overfishing, water shortages, air pollution, and other environmental problems typically blamed on capitalism.

Libertarianism and Climate Change

In the political debate concerning anthropogenic climate change, most libertarians are skeptical of government solutions, which often involve new government powers and hundreds of billions of dollars. Many libertarians have embraced criticisms of the scientific evidence for anthropogenic climate change, especially as the consensus view is promulgated by a worldwide coalition of governments, the Intergovernmental Panel on Climate Change (IPCC). This group of libertarians does not believe continued carbon emissions pose any serious threat and therefore opposes new taxes or regulations because they would impose economic damages with no corresponding benefit.

A smaller but growing group of libertarians endorses the scientific evidence of anthropogenic global warming but supports voluntary, private-sector measures to address the problem. Many in this group rely on the public choice school of economic thought, which demonstrates that democratic systems often lead to unintended consequences. Even if climate change is a serious threat, these libertarians believe, politicians cannot be trusted to implement effective solutions. Instead, they argue that unfettered economic growth will allow mankind to adapt to rising sea levels, higher temperatures, and other possible changes. They also believe that if the situation requires it, the market can provide more extreme remedies such as geo-engineering solutions (mirrors in space, aerosols released into the atmosphere, and so on).

Finally, there are some libertarians who believe the nature of global climate change requires a political remedy. Such libertarians usually favor a revenue-neutral approach, in which the revenues raised from a new carbon tax, or from auctioning carbon permits, are used to reduce preexisting taxes.

Context

There has always been a strong emphasis on individual liberty and suspicion of bureaucratic government in American politics. Since Franklin Roosevelt's New Deal in the 1930's, Republican politicians have generally presented themselves as advocates of fewer taxes and regulations on business. Democrats, in contrast, have generally focused on protecting the rights of minority groups and others with little political power against abuses from employers, the police, and other powerful groups. In modern political debate, libertarians have combined both elements of this suspicion of government, emphasizing the freedom of the individual in both the marketplace and public arena. Because most of the proposed remedies for anthropogenic climate change involve more money and power for the federal government, many libertarians have been very skeptical of the growing calls for government-directed mitigation efforts.

Robert P. Murphy

Further Reading

Adler, Jonathan H., ed. *Ecology, Liberty, and Property: A Free Market Environmental Reader.* Washington, D.C.: Competitive Enterprise Institute, 2000. Adler is a leader in this field and has assembled a collection of essays dealing with topics such as species preservation and conventional pollution, as well as climate change. The book emphasizes the reasons for government failure at these important tasks.

Anderson, Terry L., and Donald R. Leal, eds. *Free Market Environmentalism.* New York: Palgrave, 2001. Similar to Adler's collection, this book covers several different applications, including the assignment of property rights to the world's oceans. Tables, figures, index.

Brennan, Jason. *Libertarianism: What everyone needs to know.* Oxford University Press, 2012. A series of roughly 100 essays on libertarianism in a question and answer format.

Boaz, David. *Libertarianism: A Primer.* New York: Free Press, 1997. Boaz, executive vice president of the Cato Institute, presents a very readable introduction to mainstream libertarian ideas. Further reading, index.

Dolan, Edwin. "Science, Public Policy, and Global Warming: Rethinking the Market-Liberal Position." *Cato Journal* 26, no. 3 (Fall, 2006): 445-468. In this provocative article, economist Dolan chides libertarians for parroting conservative Republican views in the global warming debate, and urges them to apply libertarian principles when formulating policy recommendations.

Rothbard, Murray. "Law, Property Rights, and Air Pollution." *Cato Journal* 2, no. 1 (Spring, 1982): 55-99. In this classic article, economist Rothbard summarizes the libertarian approach to legal remedies for violations of property rights, then applies the framework to the case of air pollution.

See also: Conservatism; Liberalism.

Lichens

Category: Plants and vegetation

Definition

Lichens are very common organisms found in widely diverse habitats, including rock surfaces, trees, and human-made structures. Lichens use structures on which they grow as hosts to support their growth. They are composites, composed of two different organisms, an alga and a fungus, resulting in a symbiotic organism that has a morphology very different from the two original organisms. In lichen symbiosis, the alga provides energy through photosynthesis and the fungus provides protection and support. Lichens come in three types: crustose, a crusty form that grows tightly on rocks or trees; foliose, which resembles foliage; and fruticose, which has the appearance of "fingers." Lichens provide food for animals, such as reindeer living in arctic regions, and habitat for invertebrates. Lichens absorb nutrients from air and rain and are an important part of nutrient recycling. They grow very slowly, with rates less than 5 millimeters per year. Some species even have growth rates of less than 0.5 millimeter per year. Lichens often live for long periods of time, with some species in the arctic estimated to be over five thousand years old.

Significance for Climate Change

Because of their sensitivity to environmental factors, including temperature changes, lichens can serve as indicator species (or canaries in the coal mine) to predict environmental changes. Historically, lichens have been important indicators of pollution and climate changes. Many long-term climate change studies have used lichen growth to estimate changes in environmental temperature over time. Mapping the distribution of climate-sensitive species provides an indication of current climatic conditions, whereas monitoring over time reveals past climate change effects. Lichen growth studies have been an important part of the debate over greenhouse effects by providing data that support climate change and global warming. Because of global warming, arctic-alpine species have been diminishing, while more tropical species have flourished.

Lichens grow at very slow rates, increasing in diameter as they grow. The size of individual lichen patches on rocks can be used to estimate age, with measuring the diameter of the largest lichen on a rock surface a method to determine the time period during which that rock was exposed. The study of lichen growth to determine the age of, or to "date," surfaces is called "lichenometry." The slow growth and longevity of crustose lichens have made them especially useful in lichenometry. Measurement of lichens has been used to document effects of global warming, such as glacial deposits, former extent of persistent snow cover, and avalanche activity. Most climate change studies using lichens have been conducted using a group of

Lichens on rock in Ontario, Canada. Colonies of lichens expand slowly at predictable rates, making them useful for estimating ages of exposed rock surfaces. (© Steven I. Dutch)

crustose lichens of the genus *Rhizocarpon*, which is very abundant in many Arctic environments.

C. J. Walsh

See also: Canaries in the coal mine; Carbon 4 plants; Carbon 3 plants; Forests; Mangroves; Photosynthesis.

Liming

Category: Economics, industries, and products

Definition

Liming is the addition of calcium and magnesium to soil to neutralize acidity and increase the activity of soil bacteria. The amount of lime needed to reach the desired acidity depends on the pH and the buffering capacity of the soil. The pH is a measurement of acidity or alkalinity. Buffering capacity refers to the soil's ability to resist change to its pH. Additionally, liming improves nutrient efficiency by encouraging fertilizer uptake. However, there are important limits to liming, as over-supply may cause harm to plant life. The most common liming materials come from grinding natural limestone. Limestone is composed mostly of calcium carbonate ($CaCO_3$) and by most state laws must contain 6 percent magnesium for purposes of liming. In addition to buffering the soil, the main way to reverse acidification in freshwater is through liming the water body or its surrounding drainage basin.

Significance for Climate Change

Carbon dioxide (CO_2) does not absorb the Sun's energy, but it does absorb the heat energy released from the Earth. Some of the released heat energy returns to Earth and some goes out into space. This describes the greenhouse effect of CO_2; that is, CO_2 lets light energy in, but not all of the heat energy can get out. Soil liming can act as either a source or a sink for CO_2. A heat sink absorbs and dissipates the heat energy.

Some studies indicate that rather than all of the carbon in lime becoming CO_2, the lime may sequester up to half of its carbon content, thus becoming an effective carbon sink. If this is so, liming might even be part of the strategy to moderate climate change. Streams draining agricultural watersheds generally show a net CO_2 uptake. However, as nitrate (a common fertilizer ingredient) concentrations increase, lime may switch from a net CO_2 sink to a CO_2 source.

Peat soil contains large amounts of carbon and therefore is a significant emission source of greenhouse gases—especially of methane. When farming acidic peat lands, adjustment of the acidity through liming is essential. However, when adding lime to these soils often rich in carbon and methane, CO_2 production increases. In addition, because of the enhanced decomposition of organic matter through liming, soluble organic carbon increases and may factor into climate change.

Finally, it is important to note that the amount of CO_2 emissions from liming is not completely clear. This is because the amount of carbonate lime applied to soils is unclear, as is the net amount of carbon from liming that is released as CO_2.

Richard S. Spira

See also: Agriculture and agricultural land; Composting; Land use and reclamation; Nitrogen fertilization; Ocean acidification; Pesticides and pest management; pH.

The Limits to Growth (book)

Category: Popular culture and society
Date: Published 1972; updated 2004
Authors: Donella H. Meadows (1941–2001), Dennis L. Meadows (1942–), and Jørgen Randers (1945–)

In 1972, a team at the Massachusetts Institute of Technology led by Donella Meadows developed a computer model called World3 that consisted of a complex set of interrelated differential equations that modeled the global future in terms of human population growth, the global economy, and feedbacks to the environment. The book describing this model, The Limits to Growth *(1972), was prominent among many publications at the time that raised a clarion call regarding biological productivity, resource exploitation and environmental degradation, increasing population growth, and economic activity.*

The five fundamental variables modeled in World3 were human population, industrialization, food production, pollution, and resource depletion. It focused on analysis of how increasing rates of resource use (tied to a growing global population) affect the depletion of known reserves of resources and therefore other aspects of the global economy. The computer simulation and the book describing it were not intended as concrete predictions of the Earth's future, but as a framework for examining the potential trends of the entire ecological system based on the key factors of population and economic growth and limited natural resources. Notably, among the speculated possible outcomes was the total collapse of the ecological system, a conclusion that lent support to critics of growth-focused policies.

The model was criticized by many neoclassical economists and others for myriad reasons, including its aspatial formulation, failure to account for technological innovation and discovery of new resources, and exponential growth assumptions. However, continued research has also provided support for the validity of some of the study's conclusions. The authors of The Limits to Growth *released several updates taking into account new data, including 2004's* The Limits to Growth: The 30-Year Update. *The work remains a staple of environmental, economic, and population studies.*

Background

Other environment-conscious books were published in the decades leading up to the release of *The Limits of Growth*, including Rachel Carson's *Silent Spring* (1962), Aldo Leopold's *A Sand County Almanac* (1949), and Paul Ehrlich's *Population Bomb* (1968). Indeed, debates about the sustainable limits of human society harkened all the way back to Thomas Robert Malthus' seminal *An Essay on the Principle of Population* (1798). Societal awareness of the potentially negative effects of humans on the environment rose in the late 1960s and early 1970s because of high profile environmental issues

such as the Cuyahoga River fires of 1969, oil spills in Santa Barbara (1969), and significant air quality problems in major cities such as Los Angeles. Many contemporary films also presented dystopic visions of the future based on problems associated with limits to growth, including *Soylent Green* (1973), *Silent Running* (1972), and *Logan's Run* (1976).

This heightened awareness of environmental degradation related to limits on sustainable growth contributed to the first Earth Day, which took place in 1970. Soon after, Republican president Richard Nixon signed a great deal of environmental legislation, including the Clean Air Act of 1970 and the Clean Water Act (1972), as well as issuing an executive order establishing the Environmental Protection Agency (1970). The environmental movement laid the foundation for many subsequent considerations of limits to growth. Civilized and uncivilized discussions of the idea of a human population "carrying capacity" increased in frequency and vitriol. Many would argue that Gro Harlem Brundtland's subsequent coinage of the phrase "sustainable development" (meeting the needs of the present generation without sacrificing the needs of future generations) is intimately related to ideas of limits to growth. Nonetheless, discussions of limits to growth and related ideas of carrying capacity remain controversial today. Perhaps not surprisingly, the specter of present and looming climate change resulting from human activity rekindles these old and ongoing debates.

Significance for Climate Change

Discussions of limits to growth are often centered around the limits represented by the interrelated resources of food, water, and energy. The causes and consequences of climate change are also intimately related to those resources. Most policy proposals regarding climate change involve adaptation to the impacts of climate change or reduction or elimination of its causes. Primary among these causes is the anthropogenic emission of greenhouse gases (GHGs), particularly carbon dioxide (CO_2). A chief concern is the use of fossil fuels such as coal and oil that contribute to this pollution; such fuels also present a potential limit to growth due to their finite supply. Increasing use of biofuels as a substitute for oil, however, demonstrates the problematic and interconnected nature of limits to growth.

When corn is used to produce ethanol as a substitute for gasoline, food supplies are affected, and the combustion of ethanol still produces CO_2 emissions. When densely populated coastal communities resort to the desalination of ocean water to obtain freshwater, they often burn fossil fuels to generate the energy needed to do so. Global warming also has the potential significantly to alter climate patterns in ways that will negatively affect both food and water supplies. The concepts of carbon footprints and carbon neutrality have gained currency in discussions of climate change. The idea of a carbon footprint borrows from a broader and older idea, than of an ecological footprint, which is itself essentially an analysis of limits.

Typical ecological-footprint analyses compare human demands on Earth's ecosystems and natural resources and contrast them with Earth's ability to provide for those demands through time. Many scholars and researchers are convinced that the present demands of the human population on Earth's resources exceed Earth's ability to provide them and that the human population is therefore in a state of ecological deficit. Increasingly, scholars are conducting various kinds of economic-ecological accounting studies that raise once again some very contentious and controversial questions about the limits to growth. The prospects of climate change, global warming, and their consequences have reignited the interest in and perhaps the urgency of resolving these important questions regarding the limits to growth.

Paul C. Sutton

Further Reading

Cohen, Joel E. How Many People Can the Earth Support? New York: Norton, 1995. Print.

Ehrlich, Paul. The Population Bomb. Rev ed. New York: Ballantine, 1978. Print.

Hardin, G. "The Tragedy of the Commons." Science 162.3859 (1968): 1243–1248. Print.

McKibben, Bill. Deep Economy: The Wealth of Communities and the Durable Future. New York: Times, 2007. Print.

Meadows, Donella, Jørgen Randers, and Dennis Meadows. Limits to Growth: The Thirty-Year Update. White River Junction: Chelsea Green, 2004. Print.

Lithosphere

Category: Geology and geography

Definition

Besides compositional classification, Earth is separated into layers based on mechanical properties. The topmost layer is called the lithosphere, composed of tectonic plates that float on top of another layer known as the asthenosphere. The term lithosphere is derived from the Greek words lithos, meaning rock, and sfaira, or sphere. The rigid, brittle lithosphere extends about 70 kilometers and is made up of Earth's crust and the upper part of the mantle underneath. It is broken into a mosaic of rigid plates that move parallel across Earth's surface relative to each other.

The lithosphere rests on a relatively ductile, partially molten layer known as the asthenosphere, which derives its name from the Greek word asthenes, meaning "without strength." The asthenosphere extends to a depth of about 400 kilometers in the mantle, over which the lithospheric plates slide along. Slow convection currents within the mantle, generated by radioactive decay of minerals, are the fundamental heat energy source that causes the lateral movements of the plates on top of the asthenosphere. According to the plate tectonic theory, there are approximately twenty lithospheric plates, each composed of a layer of continental crust or oceanic crust. These plates are separated by three types of plate boundaries. At divergent boundaries, tensional forces dominate the interaction between the lithospheric plates, and they move apart and new crust is created. At convergent boundaries, compression of lithospheric plate material dominates, and the

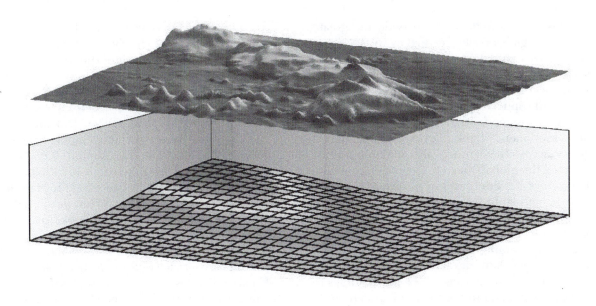

Schematic sketch of the lithosphere-asthenosphere boundary beneath the Hawaiian islands illustrating the progressive thinning of the lithosphere with island age caused by thermal impact of the mantle plume. (Iwoelbern/Wikimedia Commons)

plates move toward each other where crust is either destroyed by subduction or uplifted to form mountain chains. Lateral movements due to shearing forces between two lithospheric plates create transform fault boundaries. Earthquakes and volcanic activities are mostly the result of lithospheric plate movement and are concentrated at the plate boundaries.

Significance for Climate Change

Volcanic eruptions have severe effects on global climate. The greenhouse effect, icehouse effect, and ozone depletion by far have gained the most attention in climate research and planning. In addition to lava and pyroclastic materials (fragments of hot and molten rocks), volcanoes emit a variety of gases such as water vapor, carbon dioxide (CO_2), carbon monoxide (CO), chlorine, fluorine, and sulfur dioxide (SO_2). Both CO_2 and CO are greenhouse gases (GHGs) that contribute to global warming by creating a shield over the Earth that prevents heat from escaping into the atmosphere.

In contrast, SO_2 gas causes short-term cooling resulting from what is known as the icehouse effect. In the lower atmosphere, SO_2 gas is converted to sulfuric acid (H_2SO_4), which condenses to form a thick layer of sulfate aerosol. The suspended aerosols increase Earth's albedo by reflecting the Sun's rays back to space and cause cooling of the Earth's surface. An anomalous increase in SO_2 layers in the atmosphere and decrease in average temperature correlates significantly with several volcanic eruptions. The 1991 eruptions of Mount Pinatubo in the Philippines were responsible for about a 0.5 °Celsius decrease in global temperature and an unusually cold summer in 1992 in the intermediate latitude of the Northern Hemisphere.

Although volcanic activity increases the global temperature by adding CO_2 to the atmosphere, a much greater amount of CO_2 is added to the atmosphere by anthropogenic activities each year. Research by Terrence M. Gerlach indicates that anthropogenic CO_2 emissions are about 150 times greater than volcanic CO_2 emissions. A small amount of global warming caused by the GHGs from volcanic eruption can considerably supersede the greater amount of global cooling caused by volcano-generated aerosol particles in the atmosphere. Without such cooling effect, global warming due to GHGs would have been more pronounced.

The lithosphere's plates move at a rate of about 3 centimeters per year. The distribution and relative movement of the oceanic and continental plates across the latitude also have profoundly affected the global climate. The major contributing factors are differences in surface albedo, land area at high latitudes, the transfer of latent heat, restrictions on ocean currents, and the thermal inertia of continents and oceans. According to the present configuration of oceans and continents, the low latitudes have a greater influence on surface albedo because the lower latitudes receive a greater amount of solar radiation than the higher latitudes.

Whereas the continents in higher latitudes receive lesser solar radiation and accumulate snow that consecutively increases the albedo and decreases the Earth's surface temperature, the latent heat of evaporation influences the surface temperature at lower latitudes, where there is greater oceanic surface. The evaporation of water from the oceanic surface, a dominant mode of heat transfer, results in greater heat loss in lower latitudes. Oceanic circulation is a primary mechanism by which the heat due to solar radiation is spread from equatorial to polar latitudes. The continents in between work as barriers that restrict the oceanic heat transport toward the poles, and can influence the area and thickness of polar snow cover. The thermal inertias of continents and oceans are different. The continents respond quickly if there is a change in solar input, whereas oceans have high heat capacity and act slowly. In addition to the present positions of the continents and oceans, the higher elevations due to mountain orogeny also control the global climate.

Arpita Nandi

Further Reading

Artemieva, Irina. *The lithosphere: An interdisciplinary approach*. Cambridge University Press, 2011. The outermost layer of the earth can be defined by a number of different criteria. This book describes the seismic, chemical, mechanical and seismic properties of the lithosphere.

Artemieva, I. M. Lithosphere : An Interdisciplinary Approach. Cambridge: Cambridge UP, 2011. eBook Collection (EBSCOhost). Web. 20 Mar. 2015.

Gerlach, Terrence M., et al. "Carbon Dioxide Emission Rate of Kilauea Volcano: Implications for Primary Magma and the Summit Reservoir." Journal of Geophysical Research 107 (2002).

Pasquale, Vincenzo, Paolo Chiozzi, and Massimo Verdoya. Geothermics : Heat Flow in the Lithosphere. Cham: Springer, 2014. eBook Collection (EBSCOhost). Web. 20 Mar. 2015.

Schuiling, Roelof D. "Thermal Effects of Massive CO2 Emissions Associated with Subduction Volcanism." Comptes Rendus Geosciences 336 (2004).

Zielinski, Gregory A. "Use of Paleo-Records in Determining Variability Within the Volcanism-Climate System." Quaternary Science Reviews 19 (2000).

Little Ice Age (LIA)

Categories: Climatic events and epochs; Cryology and glaciology

A brief cold period may have occurred during the seventeenth and eighteenth centuries. Some researchers think that the lower temperatures resulted from a reduced solar energy output related to the Maunder Minimum in sunspot activity. The possibility that climate variations may be induced by solar activity complicates the conventional theory that human activity is causing global warming.

Key Concepts

luminosity of Sun: the total energy output of the Sun every second, measured in watts

Maunder Minimum: a period from about 1645 to about 1715 when very few sunspots were observed

sunspot cycle: also known as the solar activity cycle, an eleven-year cycle in the number of sunspots and amount of other solar magnetic activity *sunspot minimum/maximum:* the time when there is the minimum/maximum number of sunspots during the eleven-year sunspot cycle

sunspots: dark spots on the surface of the Sun caused by solar magnetic activity

Background

Indirect and anecdotal evidence indicates that it was colder than normal during the seventeenth century in Europe, and possibly worldwide as well. Without accurate weather records for this time period, the exact dates of the Little Ice Age are not known. The coldest period was the seventeenth century, but date estimates range from as early as about 1350 or 1400 to as late as 1850. Indeed, some writers use the term "Little Ice Age" to refer only to that coldest period, while others designate the entire period from 1350 to 1850 with the term. The coldest portion of the Little Ice Age occurred during the time of the Maunder Minimum, a period of virtually no sunspot activity. If the Sun's luminosity is slightly lower during periods of reduced sunspot activity, the Little Ice Age may have been caused by a temporary decrease in the Sun's luminosity.

Little Ice Age

The Little Ice Age was not an ice age. During the various ice ages, which occurred many millennia ago, glaciers invaded temperate midlatitude regions. The evidence for these ice ages is geologic, because they occurred before recorded history. The Little Ice Age occurred only a few centuries ago. It was colder than normal, but not nearly as cold as the ice ages. Historians know about the Little Ice Age from various anecdotal documents and indirect proxies. There are, however, no accurate weather records for this period, because the instrumentation needed to produce such records had for the most part not yet been invented. Therefore, scientific knowledge of the extent and severity of the Little Ice Age is not precise.

The exact dates of the start and end of the Little Ice Age are somewhat controversial. Authors generally agree that the Little Ice Age encompassed the seventeenth century, but there is some disagreement as to both how long afterward it lasted and how much sooner it started. This disagreement can be understood by examining reconstructed temperatures for the past millennium. The seventeenth century, which was about 0.5° Celsius cooler than the 1961–1990 average temperature, was both the longest-lasting cool period and the coldest period of the past one thousand years. Thus, nearly all researchers agree that this century was part of the Little Ice Age.

The entire period from about 1300 to the late nineteenth century was cooler than normal for the millennium. However, most of this time was not as cold as the seventeenth century, and there were some relatively warm periods during these centuries. The first half of the nineteenth century was colder than normal, but the eighteenth century was about as warm as the average for the millennium. Thus, the Little Ice Age may have lasted as late as 1850, but it may have ended in the early 1700's.

The period from 1000 to shortly after 1200, the Medieval Warm Period, was nearly as warm as the end of the twentieth century. Average temperatures then dropped fairly quickly during the late thirteenth century, making it possible to date the beginning of the Little Ice Age as early as about 1350 to 1400. The middle of the fourteenth and fifteenth centuries were nearly as cold as the seventeenth century, but there were warmer periods about 1400 and during the first portion of the sixteenth century. Thus, the Little Ice Age may not have started until the seventeenth century.

The uncertainty of dating the beginning and end of the Little Ice Age results from the facts that the cool climate from approximately 1350 to 1850 was interspersed with relatively warm periods and that the longest-lasting and coolest period was the seventeenth century. The exact time period of the Little Ice Age is therefore fairly loosely defined.

Evidence for the Little Ice Age

Thermometers were not invented until the end of the sixteenth century, and widespread, systematic use of accurate thermometers did not occur until much later. There are therefore no accurate weather records to verify the Little Ice Age. Climate researchers must use other lines of evidence, including both various proxies and anecdotal evidence.

The most common proxy studies for climate involve tree rings. The thickness and density of the rings varies with various climatic conditions, including temperature and rainfall. In polar regions, studies of various properties of ice cores provide climate information. The properties include the rate at which ice accumulates, layers that have melted, and isotope ratios. Growth thicknesses and other properties of corals can also provide climate information.

For all of these proxy studies, climate researchers statistically analyze the relationship between the proxy and climate conditions during recent periods for which accurate weather records exist. The researchers then extrapolate the climate conditions back to the dates before accurate weather records. The further back researchers extrapolate, the less accurate the proxy is in reconstructing climate conditions. Hence, climate estimates for the first half of the millennium are less accurate than are more recent estimates. Proxy studies are further complicated by the fact that multiple variables can affect the proxy. For example, tree rings are affected by both temperature and rainfall conditions.

In addition to proxy studies, there is anecdotal evidence for the Little Ice Age. Examples of this type of evidence include such things as diary entries and paintings. Diary entries might include reports of unusual freezings of various bodies of water, extreme snowfalls, and so forth. Paintings in eras when artists strove for realism can also depict frozen landscapes and bodies of water. If the paintings made during a particular time period show a large number of frozen landscapes of locations that seldom freeze now, researchers can conclude that the time period was cooler than normal. These lines of evidence are not scientific, but many reports of a particular time period being colder than normal strongly suggest that it actually was colder, even in the absence of scientifically reliable weather records. These lines of evidence apply primarily to Europe, so the Little Ice Age could have been either a strictly European phenomenon or a global phenomenon.

The Maunder Minimum and the Little Ice Age

There is fairly strong evidence that variations in the Sun's luminosity related to sunspot activity caused the Little Ice Age. Sunspots are dark regions on the surface of the Sun caused by solar magnetic activity. Solar magnetic activity also causes bright areas, or faculae, on the Sun's surface. The Sun undergoes an eleven-year cycle regulating the amount of sunspots, faculae, and related solar magnetic activity it experiences. Satellite measurements over the most recent solar cycles show that the Sun's luminosity is a very small amount higher during sunspot maximum than during sunspot minimum. The net effect of the bright areas on the Sun is slightly larger than the net effect of the dark areas, so the Sun is brighter during sunspot maximum.

There are also less well established longer cycles in solar activity. Notably, the Maunder Minimum was a period from about 1645 (possibly as early as 1620) to 1715 when there were very few sunspots. This period corresponds to the coldest portion of the Little Ice Age. If the observation that the Sun emits less energy during periods of minimal sunspot activity holds, then the Sun's lower luminosity during the Maunder Minimum may have caused the coldest portions of the Little Ice Age.

Closer comparison of the sunspot activity and global temperatures over the past thousand years supports this hypothesis. The warm period from 1000 to 1200 corresponds to the Medieval Grand Maximum in sunspot activity: During sunspot maxima, there were many more sunspots than is usual during sunspot maxima. There were also other extended periods of very few sunspots similar to the Maunder Minimum, including the Spörer, Wolf, and Dalton minima. Like the Maunder Minimum, these other minima correspond to the cooler periods of the extended Little Ice Age. It is not proven that variations in the Sun's luminosity caused the Little Ice Age, but it seems to be the most likely explanation.

Context

Global warming, especially since the later half of the twentieth century, has become a serious worldwide concern. Most climate researchers attribute this global warming to anthropogenic causes, particularly increased emissions of carbon dioxide (CO_2) and other greenhouse gases (GHGs). The extremely hot surface temperatures on Venus clearly demonstrate that CO_2 can warm a planet.

If, however, the Little Ice Age resulted from solar luminosity variations related to long-term solar activity cycles, then there is the possibility that similar solar variations are contributing to current global warming. Some, but not all, late-twentieth-century sunspot maxima were higher than normal, suggesting the possibility that Earth is entering another sunspot grand maximum similar to the Medieval Grand Maximum. If this is the case, then global warming may have solar variations as well as increased GHGs as a component cause.

Paul A. Heckert

Further Reading

Eddy, J. A. "The Maunder Minimum." Science 192 (1976): 1189–1192.

Foukal, P., C. Fröhlich, H. Spruit, and T. M. L. Wigley. "Variations in Solar Luminosity and Their Effect on Earth's Climate." Nature 443 (2006): 161–166.

Golub, Leon, and Jay M. Pasachoff. Nearest Star: The Surprising Science of Our Sun. Cambridge, Mass.: Harvard University Press, 2001.

Hoyt, Douglas V., and Kenneth H. Schatten. The Role of the Sun in Climate Change. New York: Oxford University Press, 1997.

Jones, P. D., T. J. Osborn, and K. R. Briffa. "The Evolution of Climate Over the Last Millennium." Science 292 (2001): 662–667.

Maunder, E. Walter. "A Prolonged Sunspot Minimum." Knowledge 17 (1894): 173–176.

Maunder, E. Walter. "The Prolonged Sunspot Minimum, 1645-1715." Journal of the British Astronomical Society 32 (1922): 140.

Miller, Gifford H., Áslaug Geirsdóttir, Yafang Zhong, Darren J. Larsen, Bette L. Otto-Bliesner, Marika M. Holland, David A. Bailey et al. "Abrupt onset of the Little Ice Age triggered by volcanism and sustained by sea-ice/ocean feedbacks." *Geophysical Research Letters* 39, no. 2 (2012). The paper argues that four extremely large volcanic eruptions in the half century before the onset of the Little Ice Age triggered cooling, and that feedback between sea ice and the ocean kept conditions cold long after the direct volcanic effects had dissipated.

Soon, Willie Wei-Hock, and Steven H. Yaskell. The Maunder Minimum and the Variable Sun-Earth Connection. River Edge, N.J.: World Scientific, 2003.

Louisiana coast

Category: Geology and geography

The southern third of Louisiana lies no more than 7.6 meters above sea level, and the extensive salt marshes on the central and western coast average less than 1 meter above sea level. Much of the city of New Orleans is below sea level. This land is highly susceptible to the consequences of rising sea levels and increased storm intensity associated with global warming.

Key concepts

barrier island: an offshore island running parallel to a coastline that protects the coastline from storm surges

bayou: a small, often intermittent, distributary

delta: a network of distributaries through deposited sediments at the mouth of a river

distributary: a branch of a river that removes water from the main river, especially near or in a delta

levee: a natural or human-made raised bank along a river

saltwater intrusion: inland movement of seawater

subsidence: land sinkage due to compaction of underlying material

Background

New Orleans was founded in 1718 on a crescent of land on the east bank of the Mississippi River that was considered high enough to be safe from tidal surges and hurricanes. This high ground, about 5 meters above sea level, was a natural levee produced by sediment deposited during annual floods. Although New Orleans remains the highest land along the entire coast of Louisiana, the extensive salt marshes and river delta that once protected the city have eroded during the past century as a

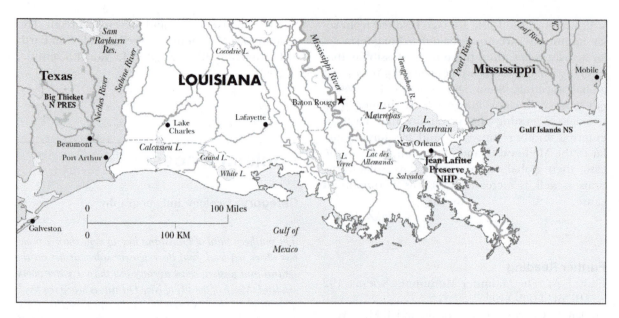

result of "improvements" to the river and development associated with the oil and gas industry.

Natural Mississippi River

Much of the coast of Louisiana was formed by alluvial deposits of the Mississippi River, and even today the Mississippi and its primary distributary, the Atchafalaya River, have a tremendous impact on the Louisiana coast, carrying 1-2 million metric tons of sediment per day to the Gulf of Mexico. During the past five thousand years, six different outlets formed deltas whose remnants can be identified. The earliest followed roughly the course of the present-day Atchafalaya River and emptied into what is now Cote Blanche Bay. Much of the sediment was carried by the westward gulf current toward the Texas border. The next outlet was further east, near present-day Terrebonne Bay. The third moved to its westernmost outlet along the general course of today's Bayou Teche. Once again, sediments moved westward, stranding the earlier beaches, called cheniers, behind extensive mudflats.

The fourth delta, far to the east, formed present-day St. Bernard Parish, most of which was underwater during the flooding caused by Hurricane Katrina. The river again moved west along the present course of Bayou Lafourche, and finally, about six hundred years ago, the river moved to its current course and began to form the bird's-foot delta at its mouth. As each delta was abandoned, sediments began to compact, resulting in local land subsidence. The interaction of sediment deposition and wave action formed a series of barrier islands, from the Chandeleur Islands on the east to Marsh Island on the west.

Managed Mississippi River

Since the time of French settlement, landowners along the Mississippi were required to build and maintain levees, but it was not until major floods in 1849 and 1850 that national concern was raised for controlling the Mississippi River. In 1882, the U.S. Army Corps of Engineers began levee construction. Following the 1927 flood, the federal government committed to a comprehensive flood-control and navigation program that included levees, floodways, channel improvements, and stabilization.

The entire lengths of the Mississippi River and Atchafalaya Basin in Louisiana are confined by levees. Two breaches are designed into the levee system to provide controlled floodways for diverting high flow. The Old River Control Structure, near the natural confluence of the Red River, diverts up to half the flow of the Mississippi during high flood into the Atchafalaya basin. There, it is divided between the Morganza and West Atchafalaya floodways straddling the Atchafalaya River channel. About 24 kilometers above New Orleans,

Wetland on the Mississippi Delta, Louisiana. Even a small rise in sea level could send salt water into this area and replace this vegetation with salt-tolerant species. (© Steven I. Dutch)

a second floodway, the Bonnet Carre' Spillway, can divert more than 7,000 cubic meters per second from the Mississippi River into Lake Pontchartrain to relieve pressure on the New Orleans levees.

The Atchafalaya River formed from the lower Red River in the 1500's, when a new bend in the Mississippi River captured the Red River and most of its flow. Over the years, the Atchafalaya gradually broadened and deepened until it began to capture Mississippi River water even during normal flow. By 1953, 30 percent of the flow moved down the Atchafalaya, and there was concern that the Mississippi would shift course and strand Baton Rouge and New Orleans. The Old River Control Structures, completed in 1963, regulate flow from the Mississippi. In 1973, part of the structure nearly failed during a major flood. It is likely the Mississippi will change course if and when such a failure occurs.

Marshes

The broad alluvial plain of the Louisiana coast supports a sequence of four marsh types categorized primarily based on elevation above sea level. They range from 24-32 kilometers wide on the west side of the state to more than 80 kilometers wide south of New Orleans. The first 1 to 24 kilometers from the shore, a total of 3,640 square kilometers, is saline marsh dominated by salt-tolerant species. The salt marsh merges gradually into brackish marsh covering 4,850 square kilometers. Another 2,830 square kilometers are intermediate marshes, which grade into nearly 4,850 square kilometers of freshwater marsh. Because of the levees, annual floods no longer cover the marshes with fresh sediments, and erosion and subsidence exceeds land building along the entire coast, except for the mouth of the Atchafalaya River, which has a growing delta.

Oil and Gas

Although most of Louisiana's oil and gas production is now offshore in the Gulf of Mexico, the first productive well was about 40 kilometers north of the coast, near the town of Jennings. The easiest way to access well sites was to dredge canals and float equipment to the site. Virtually all of the state's marshland is laced with service canals running from the coast, rivers, bayous, or the commercial Intracoastal Waterway. South of the Intracoastal Waterway, there are more than 7,240 kilometers of canals and 11,590 kilometers of bayous. These waterways permit saltwater intrusion into the heart of freshwater marshes, killing intolerant plant species. They also provide water courses through which storm surges can move far inland, eroding the fragile marshes.

Context

As a result of human "improvements" in south Louisiana, approximately 155 square kilometers of coastline is lost every year, and what remains is even more susceptible to storm surges that are expected to increase as a result of global warming. New floodways have been opened through the Mississippi River levees south of New Orleans in an effort to flood the marshes with new sediments so they will grow again. However, the region remains increasingly vulnerable to warming-related weather patterns, and if the sea level rises, it will only become more vulnerable still.

Marshall D. Sundberg

Further Reading

Barbier, Edward B., Ioannis Y. Georgiou, Brian Enchelmeyer, and Denise J. Reed. "The value of wetlands in protecting southeast Louisiana from hurricane storm surges." *PloS one* 8, no. 3 (2013): e58715. http://journals.plos.org/plosone/article?id=10.1371/journal.pone.0058715 [8 July 2016] Modeling of storm surges suggests that storm surges are most affected by continuity of wetlands and vegetation roughness, that is, presence of larger plants.

Dunne, Mike. *America's Wetland: Louisiana's Vanishing Coast.* Photographs by Bevil Knapp. Baton Rouge: Louisiana State University Press, 2005. Heavily illustrated study of land loss on the Louisiana coast.

Kelley, Joseph T., Orrin H. Pilkey, and Alica Kelley. *Living with the Louisiana Shore.* Durham, N.C.: Duke University Press, 1983. Describes the geologic history of the shoreline, particularly the influence of the Mississippi River, and provides a variety of contemporary perspectives from the barrier islands on the east to Holly Beach (destroyed in 2005 by Hurricane Rita) on the west.

Mathur, Anuradha, and Dilip da Cunha. *Mississippi Floods: Designing a Shifting Landscape.* New Haven, Conn.: Yale University Press, 2001. Combines environmental history and engineering to look at human attempts to control the river and the river's drive to control itself.

See also: Barrier islands; Coastal impacts of global climate change; Coastline changes; Flood barriers, movable; Floods and flooding; Levees; New Orleans; Sea-level change; Soil erosion; Tropical storms; Wetlands.

Maldives

Category: Nations and peoples

Key facts
Population: 379,174 (July, 2008, estimate)
Area: 298 square kilometers
Gross domestic product (GDP): $1.738 billion (purchasing power parity, 2008 estimate)
Greenhouse gas (GHG) emissions in millions of metric tons of carbon dioxide equivalent (CO_2e): 0.6 in 2004
Kyoto Protocol status: Ratified December, 1998

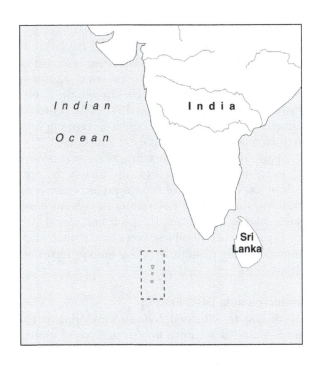

Historical and Political Context
Maldives is an archipelago grouped into nineteen atolls located on the Indian Ocean, south-southwest of India. It is an Islamic republic and became independent from England on July 26, 1965. The majority of its people are Sunni Muslims who were originally Buddhists. The official language is Dhivehi, an Indo-European language related to Sinhala, the major language of Sri Lanka.

Maldives became a republic under the 1968 constitution, which established executive, legislative, and judicial branches of government. The Majilis (legislature) is composed of fifty members who serve five-year terms. The judicial system is derived mainly from traditional Islamic law and administered by secular officials, a chief justice and judges on each of the nineteten atolls. Ibrahim Nasir became the president in 1968 and was succeeded in 1978 by Maumoon Abdul Gayoom.

Maldives has faced no external threat since independence, but has witnessed two internal threats. President Gayoom disclosed in 1980 details of an abortive coup involving Ibrahim Nasir, the first president. The other threat occurred in 1988, when Abdullah Luthufi, a Maldivian businessman, led a group of Tamil militants who invaded the nation in an effort to overthrow the government. Gayoom asked for help from the Indian government, which sent troops to put down the invasion. By 1990, the government had embarked on a democratization program, and the country's security improved.

In 2005, members of the Majilis voted unanimously to legally recognize political parties, and a year later the government introduced a roadmap for reform. A first-ever presidential election under a multiparty system was held in October, 2008. The Maldivian Democratic Party led by Mohamed Nasheed and his running mate, Mohammed Waheed Hassan, defeated incumbent president Abdul Gayoom. On November 11, 2008, Mohamed Nasheed was sworn in as president and Dr. Waheed Hassan as vice president of Maldives.

Impact of Maldivian Policies on Climate Change
Climate change and rising sea levels are a great concern to Maldives, which is less than 2.5 meters above sea level at its highest point. The country's entire existence is in jeopardy, as global warming causes the polar ice caps to melt and sea levels to rise. Scientists predict that Maldives could sink beneath the ocean surface by 2100 if global warming continues. As a result, the government of Maldives is doing all it can to fight global warming.

The nation was among the first to raise climate change as a serious issue at the United Nations. President Gayoom raised the alarm in 1987, when most nations had not recognized the problem. In 1989, Maldives played a major role in calling and hosting the first small states conference on sea levels. The conference issued the Male' Declaration on global warming and sea-level rise and urged intergovernmental action on the issue. Maldives was also one of the first countries to sign the Kyoto Protocol, written in Japan in 1997. The protocol aimed at tackling the issue of global warming and reducing greenhouse gas (GHG) emissions.

Apart from the county's involvement in several international conferences on climate change, it has also implemented several projects aimed at resisting the rising sea by raising the elevation of some islands. For example, it has completed the construction of a flood-resistant island named Hulhumale, made possible by a $60 million Japanese grant.

Maldives as a GHG Emitter

According to a United Nations development report, Maldives accounts for 0.0 percent of global emissions, with an emission rate of 2.5 metric tons of carbon dioxide (CO_2) per person in 2004. Maldives' economy is based mostly on tourism and fishing and hence it has little or no industries that pollute. However, that rate was only 0.7 metric ton per person in 1990. As a result, Maldives is not bound by a specific target for GHG emissions. The developed countries, led by the United States, are the greatest emitters of GHGs. With 15 percent of the world's population, they account for almost half of all emissions.

Summary and Foresight

Tourism is the largest industry in Maldives, taking advantage of the islands' beautiful resorts and beaches. Government revenues from import duties and tourism-related taxes may be threatened by rising sea levels. For example, the 2004 Asian tsunami battered Maldives, forcing the evacuation of thirteen of its two hundred inhabited islands and causing millions of dollars in damage to its resorts and beaches.

Corals are the major ecological component of coral reefs, and elevated sea temperature causes corals to bleach. In 1998, temperature-induced bleaching killed over 95 percent of shallow-water corals in Maldives. The elimination of coral reefs is beginning to affect Maldives, since reefs protect coastlines from storm damage, erosion, and flooding by reducing wave action.

The beaches and shorelines of Maldives are vanishing. They are being washed into the sea, a consequence of global warming. These tiny islands cannot combat this threat alone. They need support from the world community, especially the developed nations that are the major contributors to global warming.

Femi Ferreira

Further Reading

Gayoom, Maumoon Abdul. *The Maldives—a Nation in Peril: Speeches by Maumoon Abdul Gayoom, President of the Maldives, on the Challenge of Environmental Instability to the Maldives and the Global Community.* Maldives: Ministry of Planning Human Resources and Environment, 1998. Maldives president Gayoom discusses tourism, sustainable development, and the environment in relation to his island nation.

Goreau, Thomas, Wolf Hilbertz, and Azeez A. Hakeem. "Maldives Shorelines: Growing a Beach." New York: Global Coral Reef Alliances, 2004. Addresses the death of corals around the world caused by global warming by focusing on countries such as Maldives, Tuvalu, Bangladesh, and others.

Kothari, Uma. "Political discourses of climate change and migration: resettlement policies in the Maldives." *The Geographical Journal* 180, no. 2 (2014): 130-140. Migration in the face of sea level rise may be inevitable for low lying islands, but concern over environmental risks also coincides with long-standing plans to resettle population more densely for economic reasons.

Naseer, Abdulla. *Coral Reefs and Impacts of Climate Change in the Maldives.* Bangkok, Thailand: Workshop on Climate Science and Policy, 2008. Discusses the importance of coral reefs on the formation of beaches. Argues that the destruction of coral reefs in Maldives is responsible for the continued destruction of its beaches.

United Nations Environment Programme. *Maldives:* Post-tsunami Environmental Assessment. Geneva, Switzerland: Author, [2005?]. Examines the environmental impact of the 2004 Indian Ocean tsunami on the Maldives.

See also: Alliance of Small Island States; Coastal impacts of global climate change; Islands; Reefs; Sea-level change.

Malthus, Thomas Robert

English demographer and economist

Born: February 13, 1766; The Rookery, near Dorking, Surrey, England
Died: December 23, 1834; St. Catherine, near Bath, Somerset, England

Malthus hypothesized that, while food supplies increase only in an arithmetical progression, populations increase in a geometric progression. Thus, as both entities increase, the food supply would become insufficient for the population, resulting in famines and increased poverty. Malthus's predictions failed to materialize but may become relevant in the context of global warming.

Life

The British demographer and political economist Thomas Robert Malthus was born in 1766. Malthus was homeschooled until he entered Jesus College, Cambridge, in 1784. He studied English, Latin, and Greek, but his main interest was math. Receiving a master's degree in 1791, he was elected a fellow of Jesus College and ordained an Anglican minister in 1797. Malthus's training as a church leader served him well in describing his ideas about poverty, disease, and famine. These issues and his schooling led Malthus to develop a unique philosophy about God's rule and how populations could be controlled.

Malthus formulated his theory of population growth in the pamphlet *An Essay on the Principle of Population, As It Affects the Future Improvement of Society* (1798), creating a public flurry. Malthus researched many methods of containing catastrophic population growth. He understood that humankind had natural devices, such as crime, disease, and war, for controlling population. The heart of Malthus's doctrine is that an increasing population puts an ever-increasing pressure on natural resources. He was searching for an answer that would allow a balance between population growth and the increase in the world's supply of food.

Climate Work

Malthus's *An Essay on the Principle of Population* caused immediate debate in intellectual circles. It proposed that population growth would outpace food production, with devastating consequences. The technologies that have allowed food growth to increase in step with population growth may only be an intermediate solution to the problems posed by Malthus. The concept of climate change did not exist in Malthus's world or in his theories, but the effects of climate change today may be more than technology can deal with. Malthus also recognized that human activities could reduce the fertility of natural resources and further impair people's abilities to supply the essential goods needed to survive.

Notable intellectuals have called Malthus's principles of population growth a theory of doom and gloom, yet his work may reveal the need for equal and just allocation and use of resources. Malthus's theory ignited scholarly debate about the balanced use of natural resources in order to keep the climate in harmony with the universe. A new factor in the race between population growth and increasing food supplies is the effect of climate change on food production. Climate change is attributed to many causes, primarily human related, including increasing carbon dioxide (CO_2) levels and increasing ozone levels. Malthus's principles of population growth are an important factor in understanding the effects of change and how these changes affect human beings and their climate. The very process of providing the basics of human existence alters the environment that provided these basics, positively or negatively, and may affect Earth's climate.

In spite of living with the ideas and advances of the Enlightenment, Malthus overlooked the technological innovations that were being developed. These innovations were changing lives, altering the environment, and allowing people to survive longer. The science of agriculture made many advances in the 1800's that allowed farmers to make more

The Premature Death of the Human Race

In An Essay on the Principle of Population *(1798), Thomas Robert Malthus expressed his belief that Earth's limited resources would place a limit on the potential of human societies to thrive or even subsist on the planet.*

The power of population is so superior to the power of the earth to produce subsistence for man, that premature death must in some shape or other visit the human race. The vices of mankind are active and able ministers of depopulation. They are the precursors in the great army of destruction, and often finish the dreadful work themselves. But should they fail in this war of extermination, sickly seasons, epidemics, pestilence, and plague advance in terrific array, and sweep off their thousands and tens of thousands. Should success be still incomplete, gigantic inevitable famine stalks in the rear, and with one mighty blow levels the population with the food of the world.

productive use of their lands. The development of effective contraception also had the potential to check population growth. These events illustrated the flaws in Malthus's theory.

Many developing nations have not adopted improved farming techniques or methods of contraception, however. In the case of these nations, Malthus's theory seems to be coming true. Overpopulation, famine, and war continue to devastate these nations, yet the economic successes of industrializing nations tend to counter Malthus's theory. Economists of the nineteenth and twentieth centuries have criticized Malthus's theory on the grounds that technological advances, the expansion of the market economy, the division of labor, and capital goods have offset large population increases. These developments have made possible the competition between nations for wealth, power, glory, and prestige. This competition, while increasing food and resource supplies, has also resulted in environmental degradation. This environmental abuse has led many environmentalists to claim that continuing resource depletion and waste will lead to a Malthusian end to the recognizable world.

It is sometimes asserted that Malthus was incorrect, because the population grew more slowly than he predicted, while resources have grown at a much faster rate. What is important to his argument is that the population always expands to the limits imposed upon the society. He points out that a growth in population forces an increase in productivity, which causes a larger growth in the population. The same increase in productivity fuels changes to the environment that may cause changes in climate as well.

Loralee Davenport

Further Reading

Coleman, David, and Robert Rowthorn. "Who's afraid of population decline? A critical examination of its consequences." *PoPulation and develoPment review* 37, no. s1 (2011): 217-248. Many people express concerns over population decline such as impacts on productivity, economic growth and military security. Offsetting such concerns are reductions in environmental harm and congestion.

Ehrlich, Paul, and Anne Ehrlich. *The Population Explosion.* New York: Simon and Schuster, 1990. Explains how economists and biologists can work together to understand human beings and how they survive. Illustrates Malthus's role in recognizing that human behavior and money are major factors in population growth.

Eiseley, Loren. *Darwin's Century: Evolution and the Men Who Discovered It.* New York: Doubleday, 1961. Eiseley describes Darwin's study of Malthus's theory of the "struggle for existence." Darwin also gives credit to Malthus for stimulating his work on the theory of natural selection.

Eldredge, Niles. *Time Frames: The Rethinking of Darwinian Evolution and the Theory of Punctuated Equilibria.* New York: Simon and Schuster, 1982. Addresses the concepts of Darwinian natural selection. Explains that Malthusian populations naturally increase geometrically and that there are natural factors controlling population growth.

Fulford, Tim. *Apocalyptic Economics and Prophetic Politics: Radical and Romantic Responses to Malthus and Burke.* Boston: Boston University, 2001. Describes the Romantic and radical views of Malthus's theory of population growth. Explores the impact of rising prices and falling wages during the 1700's on Malthus's views of the impact of population growth.

Hardin, Garrett. *Population Evolution and Birth Control.* San Francisco, Calif.: W. H. Freeman, 1964. Examines Malthus's theory and his understanding of the human elements of love, sex, and contraception. Explains that Malthus believed that humans would not use contraceptives until they were forced to and that overpopulation would therefore occur rapidly.

Hasian, Marouf. *Legal Argumentation in the Godwin-Malthus Debates.* Farmington Hills, Mich.: American Forensic Association, 2001. Details the relationship between Malthus and his contemporary interlocutors, including his father, William Godwin, and Jean-Jacques Rousseau.

Heer, David M. *Society and Population.* Englewood Cliffs, N.J.: Prentice-Hall, 1968. Discusses Malthus's ideas concerning hunger, epidemic, and war as methods of population control.

Malthus, Thomas. *Population: The First Essay.* Foreword by Kenneth E. Boulding. Ann Arbor: University of Michigan Press, 1959. In this volume, Malthus contends that people have a duty to remove evil from themselves and that without the evils of society they could control population growth.

_____. *Principles of Political Economy Considered with* a View to Their Practical Application. London: William Pickering, 1836. This work can aid a researcher in understanding Malthus's views on population growth and his application of mathematical principles.

Pingle, Mark. *Introducing Dynamic Analysis Using Malthus' Principles of Population.* Farmington Hills, Mich.: Heldref, 2003. Illustrates the roots of Malthus's principle of population theory in economics and static models.

See also: Anthropogenic climate change; Energy resources and global warming; Industrial Revolution and global warming; Human behavior change; Population growth; Sustainable development.

Manabe, Syukuro
Japanese meteorologist

Born: September 21, 1931; Shingu-Mura, Uma-Gun, Ehim e-Ken, Japan

Award-winning meteorologist Manabe pioneered the use of computer models to simulate the impact of GHGs on climate change.

Life

Syukuro Manabe was born in Japan, earning a doctorate in meteorology from Tokyo University. After moving to the United States in 1958, he worked as a research meteorologist at the U.S. Weather Bureau's General Circulation Research Section, eventually becoming a senior scientist in the Geophysical Fluid Dynamics Laboratory (GFDL) of the National Oceanic and Atmospheric Administration (NOAA). At NOAA, Manabe contributed his expertise to both national and international panels exploring the global climate. He also held an appointment as professor in the Program in Atmospheric and Oceanic Sciences at Princeton University. In 1997, Manabe returned to Tokyo as the director of the Global Warming Research Program of the Frontier Research System for Global Change. In 2002, he returned to Princeton.

Climate Work

Manabe concentrates on modeling the atmosphere in order to predict global climate change. He has summarized his work on his Princeton University Web page:

> In the early 1960's, we developed a radiative-convective model of the atmosphere, and explored the role of greenhouse gases such as water vapor, carbon dioxide and ozone in maintaining and changing the thermal structure of the atmosphere. This was the beginning of the long-term research on global warming. . . .

Manabe pioneered the use of atmosphere-ocean circulation models in the analysis of climate disturbances involving greenhouse gases (GHGs). With the aid of a computer, modelers construct a system of mathematical equations that mimics a particular segment of nature. Manabe's equations center on the heat balance of Earth's atmosphere in order to mimic Earth's climate. The goal is to use data from previous years to predict global weather patterns. Even an approximate model allows its users to find the long-term effects of short-term environmental atmospheric inputs.

In 1967, Manabe and Richard Wetherald demonstrated that increased carbon dioxide (CO_2) in the atmosphere would increase the altitude at which the atmosphere reradiated heat into space, thereby lending supporting evidence for the greenhouse theory. In 1969, he and Kirk Bryan issued the first climate models that combined atmospheric and oceanic models. Using these models, Manabe demonstrated how Earth's climate might respond to increasing GHGs in the atmosphere, and his work has formed the basis for many of the assessments issued by the Intergovernmental Panel on Climate Change.

Kenneth H. Brown
Updated by: Christina J. Moose

See also: Climate models and modeling; Climate prediction and projection; General circulation models; Intergovernmental Panel on Climate Change.

Mangroves

Category: Plants and vegetation

Definition

Mangroves are both a genus of tropical and subtropical trees, the *Rhizophora,* and the family of plants to which the genus belongs, the Rhizophoracae. The term most commonly refers to an assemblage of mangrove trees and other associated trees and shrubs that includes more than one hundred species. Such an assemblage may also be called a mangrove swamp, a mangrove forest, or a mangal.

Mangroves occur in shallow, protected coastal waters, such as flats in intertidal zones, bays, and estuaries in the tropics and subtropics. They cannot survive freezing or even consistently cold water. They are usually canopied forests up to about 10 meters tall, although in rare cases old-growth mangroves can reach 40 meters in height.

Mangroves survive in a difficult niche. They must be salt-tolerant plants (halophytes), and some can thrive in water of twice the salinity of seawater. Mangrove roots can filter salt from water intake, and those growing in the most saline areas excrete salt from their leaves. Mangrove forests shade from the most salt-tolerant species on their seaward side, to the least salt-tolerant species on their landward side, and they shade to conventional forests or freshwater plants.

Mangrove mudflats tend to be oxygen poor (hypoxic), to contain toxic levels of sulfides, to be subject to periodic flooding, and to be very weak for holding trees. Consequently, mangroves have a maze of roots, some of which (pneumataphores) reach up out of the water to get oxygen, while some function as stilts to help brace the trees.

The mangrove niche between freshwater, dry land, and ocean is very small, perhaps only 0.1 percent of the Earth's surface. However, it is very important for three reasons: The maze of roots catches nutrient flows from the land and holds them for gradual release to the sea rather than short, polluting surges. Conversely, the roots protect the land from ordinary erosion and disasters such as hurricanes and tsunamis. Finally, the maze of roots holding leaves and nutrients from shore provides habitat and food for sea life. Young from as many as three-quarters of tropical deepwater fish species live in mangroves.

Significance for Climate Change

Theoretically, mangroves would benefit from global warming, because their tropical climate area would extend farther toward the poles. More mangroves would increase the capture (sequestration) of carbon from atmospheric carbon dioxide (CO_2) and help reduce the greenhouse effect. Moreover, increased mangrove buffering of nutrient surges washing off the land would release a steadier gentle flow of dissolved organic compounds, allowing

Brown pelicans sit in a group of mangrove trees on the Florida coast. (U.S. Fish and Wildlife Service)

plankton and other marine plants to capture more CO_2 from the air and to increase the oceans' reflectivity. Thus, mangroves would create a significant negative feedback retarding global warming.

In practice, however, the decline of mangroves due to human development may be a significant positive driver of global warming. The percentage of Earth's surface covered by mangroves has been cut in half (from 0.2 percent to 0.1 percent) in the last century, mostly as a result of human development. For instance, sand dredged from Biscayne Bay, Florida, buried a mangrove swamp and built a city—Miami. Recent increases in shrimp aquaculture have also caused major destruction of mangroves.

Major warming would also raise sea levels, causing mangroves to retreat from the deeper water and attempt to colonize new inland swampy areas. However, people would likely erect dikes to protect existing structures and agricultural land. Hence, the effective surface area available for mangroves would be constricted even further. This reduction in surface area would decrease the area of ocean fertility, the corresponding amount of carbon sequestration, and the buffering of the land against ocean disasters. Indeed, the damage to New Orleans caused by Hurricane Katrina in 2005 can be partly attributed to reduced swamplands around the city, which left the dikes more vulnerable (although the major causes of bayou losses in New Orleans were subsidence and navigation channels).

A number of groups have tried to replenish mangroves and even establish them in new areas. The best-known of these is the Manzanar Mangrove Initiative, which is backed by the Eritrean government on the East African coast. (It was largely sparked by Gordon Sato, who was interned during World War II at Manzanar, California.) The Manzanar Mangrove Initiative uses small amounts of fertilizer in bags for slow release to vastly increase the area of mangroves and resulting silage for livestock.

An extension of the Manzanar Mangrove Initiative would pump seawater into desert areas to vastly increase mangrove areas, food production, and capture of CO_2 from the atmosphere. However, the required pumping would require major investments in either some form of alternate energy or fossil-fuel-fired power for the pumping. Moreover, skeptics of the Manzanar Mangrove Initiative worry that overzealous fertilization of the mangroves might damage coral reefs further out to sea.

Roger V. Carlson

Further Reading

Donato, Daniel C., J. Boone Kauffman, Daniel Murdiyarso, Sofyan Kurnianto, Melanie Stidham, and Markku Kanninen. "Mangroves among the most carbon-rich forests in the tropics." *Nature geoscience* 4, no. 5 (2011): 293-297. In addition to their roles as habitat and in shoreline protection, mangroves are extremely carbon rich, and half of that carbon is stored in organic soils. Mangrove deforestation accounts for about 10% of global deforestation emissions, even though mangroves account for less than 1% of tropical forests.

Hogarth, Peter J. *The Biology of Mangroves and Seagrasses.* 2d ed. New York: Oxford University Press,

2007. University of York Lecturer Peter Hogarth provides an excellent introduction to mangroves.

Lieth, Helmut, Maximo Garcia Sucre, and Brigitte Herzog, eds. *Mangroves and Halophytes: Restoration and Utilisation*. New York: Springer, 2008. This collection of articles is focused on restoration and use of mangroves and other halophytes in the Caribbean area, especially Venezuela.

Saenger, P. *Mangrove Ecology, Silviculture, and Conservation*. New York: Springer, 2003. Summarizes pertinent mangrove issues, from global biological patterns to threats to methods for restoration.

Sato, Gordon, et al. "A Novel Approach to Growing Mangroves on the Coastal Mud Flats of Eritrea with the Potential for Relieving Regional Poverty and Hunger." *Wetlands* 25, no. 3 (September, 2005): 776-779. Briefly summarizes the Manzanar Mangrove Initiative, which seeks to develop methods of culturing mangroves along virtually any tropical coast and of developing agricultural products from mangroves.

See also: Coastal impacts of global climate change; Coastline changes; New Orleans; Plankton; Sequestration; Soil erosion.

Marginal seas

Category: Oceanography

Definition

Marginal seas are ocean regions that connect coastal zones to the open ocean. Marginal seas are found as indentations in the continental landmasses and are often separated from the open ocean by an archipelago or peninsula. They vary considerably in size, depth, the nature of their connection to the ocean, and the circulation of water within them.

The Mediterranean Sea is connected to the Atlantic Ocean through the Strait of Gibraltar, which is only 14 kilometers wide. In contrast, the Gulf of Mexico is connected to the Atlantic Ocean via the Caribbean Sea through the Yucatan Channel (271 kilometers wide) and the Florida Straits (180 kilometers wide). The largest of the marginal seas is the South China Sea, with an area of 2.97 million square kilometers, while the Irish Sea is one of the smallest, with an area of approximately 90,000 square kilometers. The Caribbean Sea has the greatest average depth, at 2,400 meters, and the shallowest marginal sea is the Persian Gulf, at 24 meters deep.

Water circulation in marginal seas varies depending on bathymetry, the nature of the connection with the open ocean, riverine input, and local climate. In seas such as the Black Sea, where riverine input exceeds evaporation, surface waters are less saline than is average ocean water. In other seas, such as the Mediterranean Sea, where evaporation exceeds riverine inflow and precipitation, salinity can exceed that of average ocean water. These salinity variations have important implications for the vertical circulation of water in the sea and therefore the chemistry of the bottom waters. Because marginal seas are relatively shallow and small in area compared to the open ocean, they are strongly affected by variations in river runoff, local climate, and direct human impacts, such as nutrient runoff. For example, the sediments of the Arabian Sea are characterized by a regular banding that reflects the annual development of monsoons.

Significance for Climate Change

Marginal seas represent the buffer between land and the ocean system. They have a major economic impact on those nations that adjoin them as a result of their effects upon tourism, fishing, and the transport of goods. Marginal seas in many regions of the world suffer negative impacts from direct human activities. For example, runoff of fertilizers from agriculture triggers algal blooms in the surface water of many marginal seas. When the algae die and settle to the bottom of the sea, decay of the organic matter consumes almost all of the sea's dissolved oxygen, creating regions devoid of sea life known as "dead zones." A dead zone up to 18,000 square kilometers in area develops every summer in the Gulf of Mexico at the mouth of the Mississippi River. The development of dead zones could be exacerbated, assuming societal activities remain unchanged, in a warmer climate. Increased river runoff and warmer temperatures would lead to a situation in which greater amounts of nutrients are delivered to marginal seas, which generally have lower amounts of dissolved oxygen than does the open ocean.

Marginal seas are also sensitive to rises in sea level, such as those that may accompany global warming. Such a sea-level rise could inundate land that is currently populated. Rises in sea level coupled with changes in runoff to the ocean will affect the hydrologic balance of many marginal seas, with potentially negative ecosystem consequences. For example, decreased freshwater flow into the Black Sea could cause upwelling of the oxygen-depleted, sulfidic bottom waters onto the shelf, causing widespread loss of marine life.

Marginal seas in the Arctic and Antarctic are sites of concern because melting sea ice could contribute to sea-level rise and perhaps also to the shutdown of thermohaline circulation in the Atlantic Ocean. The collapse of the Larsen B ice shelf, in the northwestern Weddell Sea, in 2002, remains a dramatic example of the effects of global warming. The loss of sea ice in marginal seas will have major impacts on many aspects of the Earth system, including ocean circulation, sea level, radiative forcing, biological productivity, and community structure.

Anna M. Cruse

Further Reading

Black, Kenneth D., and Graham B. Shimmield. *Biogeochemistry of Marine Systems.* Oxford, England: Blackwell/CRC Press, 2003. Discusses the impacts of a wide range of changes—including climate change—on various ocean systems. Each chapter discusses biogeochemical reactions to changes in specific oceanic environments. Individual chapters are devoted to the Mediterranean, Arctic, and Arabian Seas. Figures, tables, index, references.

Dai, Minhan, Zhimian Cao, Xianghui Guo, Weidong Zhai, Zhiyu Liu, Zhiqiang Yin, Yanping Xu, Jianping Gan, Jianyu Hu, and Chuanjun Du. "Why are some marginal seas sources of atmospheric CO_2?." *Geophysical Research Letters* 40, no. 10 (2013): 2154-2158. In contrast to river-dominated marginal seas, ocean-dominated marginal seas receive carbon dioxide from external sources.

Mee, Laurence. "Reviving Dead Zones." *Scientific American* 295, no. 5 (November, 2006): 78-85. The formation of dead zones could become more widespread with a warmer climate, if current societal practices continue. This article, written by a leader in marine conservation issues, discusses the formation of dead zones and steps that can be taken to reverse and restore dead zones once they exist. Figures.

Pew Oceans Commission. *America's Living Oceans—Charting a Course for Sea Change: A Report to the Nation.* Arlington, Va.: Pew Oceans Commission, 2003. Provides a review of the state of U.S. coastal waters and marginal seas, including the potential direct and indirect consequences of climate change. Figures, illustrations, maps, tables, index.

Yanko-Hombach, Valentina, et al. *The Black Sea Flood* Question: Changes in Coastline, Climate, and Human Settlement. Dordrecht: Springer, 2007. Provides views of expert scientists on the effects of climate change upon the Black Sea, a major marginal sea, and the consequences of those changes for early human societies. Figures, graphs, tables, references, bibliography.

See also: Arctic; Atlantic heat conveyor; Freshwater; Hydrologic cycle; Maritime climate; Mediterranean Sea; Ocean-atmosphere coupling; Ocean dynamics; Ocean life; Sea ice; Sea-level change; Sea surface temperatures; Thermohaline circulation.

Maritime climate

Categories: Oceanography; meteorology and atmospheric sciences

Definition

Maritime climates are generally considered to be those that are moderated by the sea. However, a true maritime climate is in most cases a climate in which it is neither very warm nor very cold, with adequate rainfall throughout most of the year. A typical maritime climate occurs on the coast of Oregon and Washington in the United States, in many parts of New Zealand, in Tasmania, and in much of western Europe. Although surrounded by the sea, tropical islands are not normally considered to have a maritime climate but rather a tropical climate. True maritime climates usually have winter daytime temperatures of about 15° Celsius and winter nighttime temperatures of about 5° Celsius. In the summer, daytime temperatures average about 25° Celsius and nighttime temperatures average about 15° Celsius. Rainfall in a true maritime climate occurs throughout the year, with no pronounced wet or dry season, and averages 10 centimeters per month.

Despite the moderate nature of the true maritime climate, extremes do occur. In many maritime climates, daytime temperatures can reach above 35° Celsius, and occasionally even above 40° Celsius, while nighttime temperatures may reach 0° Celsius, and occasionally as low as –10° Celsius. Sunshine is generally more than adequate for plant growth, and annual bright sunshine hours of two thousand to twenty-five hundred hours are the norm. Rainfall, although generally adequate for plant growth, can vary: Maritime climates can have periods of up to six weeks without any appreciable rainfall, and there is at least one known instance of no rain at all falling during almost every month on the calendar. In contrast, monthly rainfalls of over 20 centimeters in a month are reasonably common, but monthly rainfalls of over 50 centimeters are not impossible. Although a daily rainfall of more than 0.3 centimeter is uncommon, at times, the rainfall in a twenty-four-hour period may exceed 20 centimeters. In summary, a true maritime climate is generally an easy climate to live in for people, plants, and animals, but relatively extreme events do occur.

Significance for Climate Change

Although the true maritime climate has milder winters, generally cooler summers, and greater temperature ranges from nighttime to daytime than do continental climates, there are many variations from place to place. For example, the climate of New York City, although situated on a coast, has a climate quite different from that of Vancouver, on the coast of British Columbia. Alterations in the true maritime climate brought about by global climate change are likely to be relatively small, and in most places plants and animals should be able to adapt to the change. However, if there is an overall increase in temperatures in the middle latitudes of both hemispheres, where most of the true maritime climates occur, summers are likely to become warmer, albeit not as warm as current continental climates, and winters are likely to have fewer days below freezing. Sunshine hours are unlikely to change, but rainfall extremes, especially on a daily basis, are likely to increase somewhat.

New Zealand is considered to have a typical maritime climate, although there are some areas in New Zealand that are quite dry, with an average annual rainfall of only 30 centimeters. Nevertheless, most of New Zealand, as well as much of western Europe, has a typical maritime climate. Questions arise as to the effects of global warming on the temperature and agricultural production of such a climate. Considering New Zealand as an example, vineyards over the past one hundred years have flourished in many parts of the country, but as yet, in the far south of New Zealand in the Southland district, there are no commercial vineyards. Instead, the district is dairy country, covered with very green pastures and grazing cattle. The Southland District Council and the Southland Regional Council might consider the probable effects on the region of changes predicted in the Intergovernmental Panel on Climate Change's 2007 report. If they did, they might advertise the prospects of the Southland district to become a major grape-growing region in the twenty-first century.

W. J. Maunder

See also: Continental climate; Mediterranean climate; Polar climate; Tropical climate.

Marsh, George Perkins

American statesman

Born: March 15, 1801; Woodstock, Vermont
Died: July 23, 1882; Vallombrosa, Italy

In his seminal work, Man and Nature, Marsh argued that humans, from their very beginning, have been active agents of environmental change.

Life

George Perkins Marsh was born in Woodstock, Vermont, in 1801. In his boyhood years, he took careful note of how this forested region was changing before his eyes as a result of extensive settlement. These early impressions of his cultural geography and his firsthand experiences with the mountains and valleys of Vermont, he would later write, set the tone for his perspective on the impact that humans can have on their environment.

Marsh had a keen sense for the environment and enjoyed the outdoors. His academic interests drew him to Dartmouth College, where he studied languages and law. After moving back to Burlington, Vermont, he set up a law practice, ultimately finding his way into politics and serving as a U.S. representative from Vermont between 1843 and 1849. His interests in science and learning in general provided him with the skills to help organize the establishment of the Smithsonian Institution. His fondness for languages served him well, as he later became a U.S. diplomat to Turkey in 1849 and to Italy in 1861.

Marsh was an avid traveler and, as a nineteenth century naturalist, was well versed in the process of collecting specimens and observing his surroundings. These skills helped hone his understanding of the relationship between humans and their environment. Although

he was a prolific writer on a wide range of subjects, it was while he was in Italy that he began his most popular literary work, *Man and Nature*, published in 1864. Emerging from an eclectic scientific mix of geography, statesmanship, and common sense, the text, through an exhaustive collection of local case studies, attempted to increase public consciousness of the impact that humans have on their environment. A later edition included the subtitle *Or, Physical Geography as Modified by Human Action*. The book acquired international acclaim, and Marsh would come to be considered

George Perkins Marsh. (Library of Congress)

the founder of the American conservation movement. He is credited with establishing the first scientific approaches to forest management in the United States.

Climate Work

Marsh suggests in *Man and Nature* that the destruction of Earth's forests could have an influence on climate change. Citing scientists from both sides of the question, he puts forth the notion that deforestation, by virtue of its potential to raise soil temperatures and lower humidity, could generate a change in at least regional climate. Extensive historical evidence for the connection of forests and precipitation is presented. He cites some writers who suggest more precipitation is evident over forested lands. Marsh concludes that forests, within their regional setting, influence the humidity of the atmosphere around them.

Marsh goes on to say that forests influence the climates of continental interiors just as the sea influences coastal climates. He cites sources that argue not only that tropical rain forests and other forests increase rainfall and moderate climate but also that their destruction can be linked to the development of hotter regions. Marsh's linking of forests and precipitation may lead readers to conclude that no place on Earth is free of the potential impact of human actions.

Marsh alludes to the impact of humans on climate by introducing a common nineteenth century forestry management practice, now known as clear-cutting. He suggests that regions and countries that clear their forests subject their soils to drying and could see that effect go on indefinitely. Additionally, fallen trees can dam up water courses, creating an environment conducive to higher humidity. These impacts, he concludes, could lead to seasonal temperature changes in both the atmosphere and the soil.

As a naturalist, Marsh had interests in the whole of the environment, not just a singular component of it. His unique consideration of humans as agents of environmental change suggests that they can affect their environment in positive or negative ways. For his time, Marsh's work embraced an unusual environmental paradigm. It represented a drastic departure from the homesteading Romanticists of the 1860's and the notion that humans had a right of dominion over the land such that the rain and the blessings of God would follow the ploughs of settlement. *Man and Nature* is considered to be one of the earliest influential writings on conservation and the environment. Its insights and advocacy of restraint in the use of resources is a clarion call to any society concerned for the future of its natural landscape.

M. Marian Mustoe

Further Reading

Koelsch, William A. "The legendary "rediscovery" of George Perkins Marsh."*Geographical Review* 102, no. 4 (2012): 510-524. Contrary to many narratives, Marsh's work was cited by numerous other workers prior to his alleged rediscovery in the 1930's.

Marsh, G. P. *Man and Nature*. 1864. Reprint. Edited by David Lowenthal. Foreword by William Cronon. Seattle: University of Washington Press, 2003. The complete text, including extensive case studies on the impact of human actions on the landscape. This publication also includes notes from Marsh's biographer, David Lowenthal.

Sauer, C. O., ed. *Land and Life: A Selection from the Writings of Carl Ortwin Sauer*. Edited by J. Leighly. Berkeley: University of California Press, 1969. Outlines some of Marsh's themes in the context of historical geography.

Trombulak, Stephen C., ed. *So Great a Vision: The Conservation Writings of George Perkins Marsh*. Hanover, N.H.: Middlebury College Press, 2001. An introduction to the conservationist philosophy of George Perkins Marsh.

See also: Axelrod, Daniel; Bennett, Hugh Hammond; Carson, Rachel; Conservation and preservation; Deforestation; Desertification; Ecosystems; Elton, Charles Sutherland; Environmental movement; Forestry and forest management; Thoreau, Henry David.

Mass balance

Category: Cryology and glaciology

Definition

Generally, mass balance refers to the net sum between the input and loss of mass to a reservoir. In glaciology, the term refers to the balance between the loss and gain of ice from a glacier, ice cap, or ice sheet. When more ice is gained than lost, the result is a positive mass balance and the overall growth of the ice body. A negative mass balance occurs when more ice is lost than is gained and the ice body shrinks.

Several other terms are used to specify the gain or loss of glacial ice. "Total mass balance" refers to the total net gain or loss of ice from an entire glacier over a single hydrologic cycle. "Specific mass balance," by contrast, refers to such loss or gain at a particular point on the surface of a glacier. Total and specific mass balances can diverge in extremely large ice bodies, such as the Antarctic ice sheets, whose mass balances vary in different geographic locations. The average mass balance per unit area of an ice body is called the "mean specific mass balance." It can be used to compare the amount of ice gained or lost by ice bodies of different sizes.

Significance for Climate Change

Glaciers and other ice bodies are highly sensitive to climate change and are viewed by many scientists as early warning detectors for global warming or global cooling. Ice bodies do not form in a single year or in response to a single snow event. Instead, they form over a sustained period of time, as temperatures remain cool enough that not all of the snow that falls in a given winter melts in the subsequent summer. It is this long-term accumulation of snow that leads to the development of ice and, eventually, a glacier or other ice body. Ice bodies are dynamic, much like rivers. Ice can accumulate at the head of a glacier at the same time that it is being lost from the toe of the same glacier through ablation or iceberg calving. Also, ice bodies can change in size and shape throughout the course of a hydrologic cycle. Thus, to determine whether a glacier is growing or shrinking in size, it is necessary to consider the mass balance, or net sum, of ice that accumulates and is lost throughout an entire hydrologic cycle, rather than simply examining the size of a glacier at two points in time.

Anna M. Cruse

See also: Antarctica: threats and responses; Arctic; Deglaciation; Glaciations; Glaciers; Greenland ice cap; Ice shelves.

Mauna Loa Record

Category: Meteorology and atmospheric sciences

Definition

The Mauna Loa Record (or Mauna Loa Carbon Dioxide Record) is the longest continuous record of atmospheric carbon dioxide (CO_2) concentrations. The level of CO_2 in the Earth's atmosphere is one of the primary determinants of global temperature, and higher levels of CO_2 correspond with warmer global temperatures. The CO_2 data collected at the Mauna Loa Observatory in Hawaii since 1958 has been called the most important geophysical record on Earth.

Significance for Climate Change

In the 1950's, continuous monitoring of CO_2 levels represented a break from the conventional view that occasional measurements in selected places would provide enough data. Despite this, scientists at the Mauna Loa Observatory, founded by the U.S. Weather Bureau, decided to maintain continuous records of the CO_2 levels at its location. The observatory is located at an elevation of 3,353 meters on the island of Hawaii, and its CO_2 measurements have been supported by the Weather Bureau and the National Oceanic and Atmospheric Administration (NOAA) ever since. Within only a few years of the project's founding, an annual increase in atmospheric CO_2 concentration was measured.

A carbon dioxide monitoring station at Mauna Loa, Hawaii. (NOAA)

In 1974, NOAA installed a new CO_2 analyzer that operated parallel to the original analyzer installed by the Scripps Institution of Oceanography. The Scripps analyzer was replaced with a newer model in 2006. The two-analyzer system provides redundancy to ensure the quality of the Mauna Loa Record and the integrity of the global CO_2 measurement network. In the 1960's and 1970's, Scripps added several measurement stations ranging from the Arctic to Antarctica. However, these stations collect flask samples that are analyzed in a lab rather than collecting on-the-spot data via an analyzer, as the Mauna Loa Observatory does.

Despite the project's importance, it has faced numerous budget and staff cuts over its existence, as well as questions of whether it constituted basic research or routine monitoring. Even today, such long-term studies are rare and often poorly supported. Despite the obstacles it has faced, the Mauna Loa project has continued, producing large amounts of valuable data that have led to several breakthroughs in climate science.

A simple graph of the Mauna Loa Record data (called the Keeling Curve after Scripps scientist Charles Keeling, who cofounded the project) shows a steady increase of 2 parts per million (0.53 percent) per year in atmospheric CO_2 since the observatory began collecting data in 1958. According to NOAA's Earth System Research Lab, 63 percent of global warming due to greenhouse gases is due to CO_2. The record has provided crucial evidence that human activity has changed and is changing the composition of the atmosphere. Climate scientists have used the Mauna Loa data to model the state of the Earth's future climate, affecting political actions such as the drafting of the Kyoto Protocol, a multinational agreement to limit carbon emissions.

The Mauna Loa data also demonstrated for the first time the annual fluctuation in atmospheric CO_2 attributable to biological activity. CO_2 levels reach their maximum in May and their minimum in October. Both of these averages, however, have increased every year since the project began. While the curve is small, it indicates an exponential increase rather than a linear one—the average CO_2 level increases by slightly more each year than the previous year. Atmospheric CO_2 has increased by more than 35 percent over amounts recorded before the Industrial Revolution, and over 6 percent between 1990 and 2009. Comparison with ice-core records suggests that current CO_2 concentrations are unprecedented in the past 650,000 years.

In 1983, Keeling and other scientists began addressing how El Niño weather events affected atmospheric temperature and CO_2 concentration, using the data from Mauna Loa. They were able to identify approximately ten-year fluctuations in temperature correlated with CO_2 fluctuations, leading to a better understanding of long-term global temperature variance.

Skeptics have put forth critiques of the Mauna Loa data, citing the Mauna Loa Observatory's location on an active volcano and increasing traffic near the observatory. However, the project has attempted to minimize contamination through careful sampling and by normalizing to negate local contamination. Atmospheric CO_2 samples from other stations worldwide corroborate the Mauna Loa data, and the Mauna Loa Record continues to form a base data set for crucial research into climate change.

Melissa A. Barton

Further Reading

Field, Christopher B., and Michael R. Raupach, eds. *The Global Carbon Cycle: Integrating Humans, Climate, and the Natural World.* Washington, D.C.: Island Press, 2004. Explores scientific knowledge of the global carbon cycle from a multidisciplinary viewpoint, including coverage of carbon-climate-human interactions. Discusses research into the history of atmospheric CO_2 and the factors that have influenced past changes, as well as the gas's role in the modern carbon cycle. Includes discussion of the Mauna Loa Record within a global context. Tables, figures, color photographs, index.

Gillis, Justin. "Heat-trapping gas passes milestone, raising fears." *New York Times* 11 (2013).In 2013, carbon dioxide measurements on Mauna Loa exceeded 400 parts per million for the first time.

Keeling, Charles D. "Rewards and Penalties of Monitoring the Earth." *Annual Review of Energy and the Environment* 23 (November, 1998): 25-82. A review of the Mauna Loa Record project and its implications for science by one of its founders; autobiographical. Also discusses the history and challenges of the project. Figures, references.

Keeling, C. D., et al. "Atmospheric CO_2 and CO_2 exchange with the Terrestrial Biosphere and Oceans, from 1978 to 2000: Observations and Carbon Cycle Implications." In *A History of Atmospheric CO2 and Its Effects on Plants, Animals, and Ecosystems*, edited by J. R. Ehleringer, T. E. Cerling, and M. D. Dearing. New York: Springer Verlag, 2005. Report on atmospheric carbon research, including the Mauna Loa Record. Figures, references.

See also: Carbon dioxide; Climate reconstruction; Greenhouse effect; Greenhouse gases and global warming.

Mean sea level

Category: Oceanography

Definition

Sea level is the height of the surface of the ocean at any given location. Sea level is highly variable and can undergo very rapid changes due to such events as tides, tsunamis, changes in barometric pressure, wind-generated waves, and even freshwater floods. While these events can produce changes in sea level of several meters, they are local in scale and of a very short duration, generally lasting only for hours. Mean sea level is the average, global height of the sea surface, independent of these local, short-term changes. Changes in mean sea level are on the order of a few millimeters per year.

Mean sea level at specific locations can be calculated using tide gauge records and subtracting the effects of annual changes in atmospheric pressure and long-term changes in tidal ranges, which are driven by astronomical factors. Changes in global mean sea level can be calculated using satellite-based radar altimetry, such as with the TOPEX/Poseidon satellite. The radar altimeter measures the height of the satellite above the ocean, based on the time it takes for a radio signal to travel from the satellite to the sea surface and back. Since the actual altitude of the satellite is known, any changes in the altimeter measurement reflect changes in the height of the sea surface itself.

Significance for Climate Change

On short timescales (decades to centuries), mean sea level is a function of the amount of water stored as ice in glaciers and ice sheets. As global temperature rises, less water is stored as ice, contributing to a rise in mean sea level. A rise in mean sea level in response to global warming has important societal consequences. First, such a rise contributes to a loss of land, as coastal areas are slowly inundated by water. This is a concern for certain low-lying island nations such as the Maldives or Tuvalu. The Maldives is a nation made up of twelve hundred islands in the Indian Ocean, which has a maximum elevation of only 2.5 meters above current sea level. Thus, the Maldivian population of approximately 380,000 people is highly vulnerable to even a slow rise in mean sea level. For other nations, a rise in mean sea level is also a concern because of increased hazards from flooding during high tides—especially spring tides—and storms. A rise in mean sea level provides a higher baseline upon which tidal fluctuations build. According to the Fourth Assessment Report released by the Intergovernmental Panel on Climate Change, from 1993 to 2003, mean sea level rose approximately 3.1 millimeters per year.

Anna M. Cruse

See also: Coastal impacts of global climate change; Floods and flooding; Glaciers; Islands; Maldives; Sea ice; Tuvalu.

Media

Category: Popular culture and society

The battle over climate change has played out publicly in the media. Conservatives have charged the media with bias against questioning anthropogenic climate change. Liberals have argued that there is a scientific consensus and the media are irresponsible to portray the issue as subject to debate.

Key concepts

bloggers: people who publish information or commentary known as blogs on Web sites

mainstream media (MSM): an often derogatory term for major journalistic media outlets, including major television networks and newspapers

media bias: valuations made or endorsed by information sources that are supposed to strive for objectivity

Background

Modern media climate coverage dates to the 1890's. In the late nineteenth century, *The New York Times* warned of a possible return to an ice age. That coverage continued until the 1920's and 1930's, when media outlets cautioned about a warming trend. The coverage again shifted in the 1950's to global cooling, which lasted in some form into the 1990's. The global warming focus, though, began in the 1960's and escalated with publication of Al Gore's book *Earth in the Balance* (1992). Since the release of Gore's movie *An Inconvenient Truth* (2006), the warming debate has become a politicized argument, in which media coverage is scrutinized as much as climate science.

Media Coverage History

As far back as February 24, 1895, *The New York Times* was warning of the return of an ice age. The paper pointed to scientific concerns about a second glacial period following increases in northern glaciers. Fear spread through the print media over the next three decades. On October 7, 1912, both *The New York Times* and *The Los Angeles Times* cited the worries of Cornell University professor Nathaniel Schmidt about a new ice age. Arctic expeditions added to media coverage and public concern about a cooling climate, but by the time *The Atlantic* was reporting on cooling in 1932, many other outlets had started reporting a warming trend.

The Post discussed a warming Earth in 1930. *The New York Times* told readers in 1933 that America was experiencing the longest warming spell since 1776. Both the *Quarterly Journal of the Royal Meteorological Society* and *Monthly Weather Review* printed articles about humankind having a role in making the planet warmer. Major news media continued coverage of warming into the 1950's, when the trend shifted again.

In 1954, *Fortune* magazine reported that the Earth was growing colder. *Science News* described cooling as a major threat in 1969 and again six years later. The 1975 magazine cover depicted a city in a snow globe, as the magazine ranked the threat of a new ice age as high as nuclear war for potentially harming human life.

The New York Times wrote about global warming in 1969, but there was still little coverage of the topic in the 1970's in major news media. Toward the end of the decade, global warming and the greenhouse effect began appearing in print more than a dozen times in the top newspapers. In the 1980's, that number increased to more than one thousand stories on global warming. *The Post, The New York Times, The Los Angeles Times, The Wall Street Journal,* and *USA Today* all told readers of a looming threat from a warming Earth.

That trend continued through the 1990's. By the end of 1999, *The New York Times* alone had published more than sixteen hundred stories on global warming and *The Post* had published more than thirteen hundred. Throughout the decade, other outlets also devoted broadcast time or pages to the issue. The Public Broadcasting System (PBS), *Newsweek,* and others depicted global warming as an issue of national concern.

Climate reporting continued to gain momentum into the new millennium, but the coverage began to change significantly. More and more, the news stories were criticized for either including or excluding scientists who disputed some aspect of anthropogenic climate change. Reporting on the issue increased in virtually all major news outlets.

Gore and warming appeared everywhere from *People* magazine to *Saturday Night Live* when *An Inconvenient Truth* opened in theaters in 2006. The added attention focused the debate not just on the science but on whether a debate even existed. That trend continued, as television networks from the Cable News Network (CNN) to Fox News to the American Broadcasting Company (ABC) focused their efforts on daily coverage as well as news specials about climate. More than two hundred stories on the issue appeared on the Big Three broadcast networks in just the second half of 2007.

Spinning the Debate

The two primary sides of the global warming debate agree that the issue is spun by the media, but they do not agree in what way. Environmental groups and liberal media critics claim that the traditional view of journalist as neutral observer has muddled the debate. They embrace a position often repeated by Gore that the debate is over. Conservatives point to numerous examples of activist journalism ignoring scientific disagreement. They complain that reporters hype weather stories as climate change and call for expensive global warming solutions without giving any time to the opposition.

Mainstream Media

News outlets are the primary source of information about climate. Newspapers have lost their dominance as the primary news source, but newspaper Web sites have gained in popularity at the same time. Thus, newspapers have joined television news and the Internet as the major news media in the early twenty-first century.

Discussion of climate change became easy to find in major news outlets. It was featured repeatedly on the cover of *Time* magazine, on the front page of newspapers, and as the lead story of national network news broadcasts. In most cases, it was reported as an imminent crisis with potentially devastating results. Hurricanes, floods, drought, and a host of minor threats were linked in the press to global warming.

Skeptical scientists and public policy groups roundly criticized what they perceived as media bias in these stories. Journalists who covered the topic, from ABC's Bill Blakemore to *The New York Times'* Andrew Revkin, were often criticized for bias in their reporting. Prominent media watchdog groups and individual bloggers analyzed reporting, while scientists would dissect scientific claims in major news stories.

Other critics attacked any journalist who disseminated stories about climate change skeptics. A conference organized by the Heartland Institute brought together numerous scientists and public policy experts who raised questions about climate consensus in 2008. The event was a metaphor for the debate. The skeptical view received little coverage, and what coverage it did receive was often critical. CNN, for example, compared those who were skeptical to flat-earthers after the world was discovered to be round.

Alternative Media

The alternative media—including talk radio and nonmainstream Web sites—have had a field day with climate change. Bloggers focus much of their effort on monitoring the news media coverage of the climate. Numerous individuals and groups on the Left and the Right maintain media blogs, and much of their focus has been on environmental coverage. The most famous of these, the Drudge Report, regularly links to climate stories. Whether the topic is snowfall during a global warming hearing in Washington or Gore's own carbon footprint, Drudge and others have driven a significant news agenda. The biggest of such stories can cross over into the mainstream media.

Think tanks, environmental groups, and politicians have all participated in the news media debate on climate change. The ability to link from one to another helped further blur the lines of traditional journalism, as advocacy organizations left and right targeted the other side's positions. The diverse voices also allowed readers and viewers to self-select the information they received. That caused ordinary information consumers to harden positions along ideological lines.

Talk radio also has a significant hand in the global warming debate. In 2008, much of private talk radio remained conservative. Criticism of Gore or climate science was commonplace. Often, talk radio would highlight a story made popular by bloggers such as Matt Drudge or would address a topic that, in turn, would drive the blogosphere. Liberal talk radio, including National Public Radio and Air America, took an opposite approach. Environment and climate stories were prime topics of concern for hosts and listeners alike.

Entertainment Media

Film and television played a big role in the climate debate. With the rise of global warming as an issue in the 1980's, Turner Broadcasting responded with a cartoon called *Captain Planet*. The cartoon's superheroes protected the Earth from evils such as pollution and global warming.

Several popular movies featuring global warming themes followed. The made-for-television movie *The Fire Next Time* aired in 1993. Kevin Costner's 1995 disaster picture *Waterworld* depicted a world awash in a flood caused by warming. *The Day After Tomorrow* (2004) was one of the most controversial of these films. When it opened, the climate debate was in full force. The movie's title was reminiscent of that of the antinuclear made-for-television film *The Day After* (1983), and it depicted a climate

apocalypse brought on by global warming. In the film, the changing climate results in a rapid cooling of the Earth and the onset of a new ice age. The movie was criticized by conservatives for characterizations of a president and vice president similar to George W. Bush and Dick Cheney. It also showed Americans fleeing the freeze being stopped at the Mexican border as illegal immigrants.

Also in 2004, science-fiction thriller author Michael Crichton released a novel critical of the environmental movement. *State of Fear* portrayed murderous environmentalists altering the Earth's climate to force humans into eco-friendly behavior. The novel also included extensive footnotes to raise objections to the idea of anthropogenic global warming. Crichton's book was criticized by scientists such as the National Aeronautics and Space Administration's James E. Hansen and praised in Congress by Republican Senator James Inhofe of Oklahoma.

Film and television portrayals of climate change escalated in 2006. Gore's *An Inconvenient Truth* became one of the top-grossing documentaries of all time. The movie version of his PowerPoint presentation was lauded by environmentalists and widely criticized by conservatives. Either way, it prompted widespread discussion of the topic. The film is shown widely in school systems, but a British court ruled that it contained nine significant errors.

A string of climate documentaries followed Gore's film in 2007 and 2008. They included actor Leonardo DiCaprio's *Eleventh Hour*, *Arctic Tale*, *The Great Global Warming Swindle*, and *Everything's Cool*. Each drew predictable criticism from opponents. Even the SciFi network included global warming as one of ten potentially lethal threats to humankind in its *Countdown to Doomsday* in 2007.

Context

Despite almost countless news stories, the climate debate remains a major issue in the mainstream media. Polls show a significant number of Americans remain skeptical both about media coverage and about climate science. Because of this, media coverage has become a major topic of concern. Environmentalists continue to criticize skeptics and challenge the media to disregard such voices. Conservatives are joined by a growing and vocal group of scientists who publicly challenge what eco-groups call a climate consensus. Each side tends to complain about the tenor of media coverage of both themselves and their interlocutors.

Dan Gainor

Further Reading

Gore, Al. *An Inconvenient Truth: The Planetary Emergency of Global Warming and What We Can Do About It.* Emmaus, Pa.: Rodale, 2006. Discusses the dangers of global warming and links it to human production of carbon dioxide emissions. An appropriate companion to the movie of the same name. Illustrations, graphs.

Hanitzsch, Thomas, Folker Hanusch, Claudia Mellado, Maria Anikina, Rosa Berganza, Incilay Cangoz, Mihai Coman et al. "Mapping journalism cultures across nations: A comparative study of 18 countries." *Journalism Studies* 12, no. 3 (2011): 273-293. A cross cultural survey shows that basic journalistic ethical standards like impartiality and factual accuracy are universal but the degree to which journalists should be advocates differs among cultures.

Horner, Chris. *The Politically Incorrect Guide to Global Warming (and Environmentalism).* Washington, D.C.: Regnery, 2007. Chris Horner, attorney and counselor of Cooler Heads Coalition, exposes the misconceptions surrounding the global warming hype. Explains that possible climate remedies may actually be worse than the effects of any theoretical warming. References, graphs, index.

McKibben, Bill. *Fight Global Warming Now: The Hand*book for Taking Action in Your Community. New York: Henry Holt, 2007. McKibben, a one-time writer for *The New Yorker*, describes the use of grassroots campaigns to fight for global warming legislation. The book teaches an activist approach, including ways to maximize use of the media.

Singer, S. Fred, and Dennis T. Avery. *Unstoppable Global Warming: Every Fifteen Hundred Years.* Rev. ed. Blue Ridge Summit, Pa.: Rowman & Littlefield, 2008. Avery, director of the Center for Global Food Issues, and Singer, an atmospheric physicist, explain that global warming is a part of Earth's natural cycle. They elaborate on the reasons for warming and why it is not as perilous as the media suggest. References, glossary, index.

See also: Catastrophist-cornucopian debate; Conservatism; *Day After Tomorrow, The*; Environmental movement; Gore, Al; Heartland Institute; *Inconvenient Truth, An*; Journalism and journalistic ethics; Liberalism; Popular culture; Pseudoscience and junk science; Skeptics.

Medieval Warm Period (MWP)

Category: Climatic events and epochs

Definition

The Medieval Warm Period (MWP) is a term used to describe a period of several centuries that preceded the Little Ice Age (LIA). The MWP (also the Medieval Warm Epoch or the Little Climatic Optimum) was proposed in 1965 by Hubert Horace Lamb, a British meteorologist and groundbreaking climate historian, who believed it lasted from roughly 900 to 1300 C.E. During the MWP, Lamb believed, the North Atlantic and northern and western Europe experienced warmer conditions on average. He presented evidence drawn primarily from historical documentary data such as the expansion of agriculture to higher-altitude fields in mountainous regions, shifts in the cultivation of certain crops (such as wine production in the British Isles), changes in tree lines, and reports of weather and weather-related events (such as floods and droughts) in historical writings. This period also coincided with the Viking settlement of Greenland and excursions to Labrador, as well as a population boom in Europe, which grew from roughly 35 million to 80 million people between 1000 and 1347 C.E. All of these events seemed consistent with a milder climate and improved agricultural production.

Authentic Viking recreation, Newfoundland, Canada. (Dylan Kereluk/Wikimedia Commons)

In the decades following Lamb's assertion, temperature proxies were examined to better define the MWP, the transition to the LIA, and the extent to which they were regional or global phenomena. For example, measurements of oxygen isotope ratios in marine sediments from a Sargasso Sea core indicated an ocean temperature around 1100 that was about 1° Celsius warmer than present levels. This was followed by a nearly 2° Celsius decrease between 1100 and 1700.

Dendroclimatology, or the study of climate through tree-ring growth, conducted in the Sierra Nevada and the Great Basin of the western United States suggests periods of increased warmth and later periods of severe drought during the MWP. Similar tree-ring studies in northern Sweden and the Polar Ural Mountains provide evidence of increased temperatures from 971 to 1100 and from 1110 to 1350, respectively. Ice core and borehole studies in Greenland indicate a warm period peaking around 1000, followed by a 3° Celsius decrease into the LIA. Studies of glaciers and their moraines suggest that the MWP generally coincided with glacial retreat and the shift to the LIA with glacial advance. Other types of proxy studies include examining lake sediments, speleotherm (stalactite/stalagmite) growth, and coral growth. Many, but not all, proxy studies of sufficient length do indicate a temperature peak during the MWP and a decrease into the LIA, but with considerable variability in the details.

Significance for Climate Change

The MWP is a central issue in the debate over climate change, as it is the most recent period of climatic warmth assumed to be free of anthropogenic factors. Likewise, the transition to the LIA is the most recent significant temperature shift before the warming of the twentieth century. The initial estimates by Lamb suggest MWP temperatures were about 0.5° Celsius warmer than those of the late twentieth century. This peak subsequently decreased, as more Northern Hemisphere data sets were combined. While many proxy data sets showed a temperature peak in the MWP, the heterogeneity of the peak locations, magnitudes, and durations meant that averaged temperature changes were reduced from those seen at individual sites.

Even with multidecadal filtering to extract long-term trends during the MWP from year-to-year fluctuations, interludes of cold have appeared in periods of relative warmth. The 2001 report of the Intergovernmental Panel on Climate Change (IPCC) largely dismissed the existence of a distinct, global MWP and LIA, epitomized by the famous hockey stick graph that showed a small, nearly linear decrease in average temperature from 1000 to the late nineteenth century with no distinct MWP to LIA transition. After some controversy, the next IPCC report in 2007 revived the MWP and LIA, using eight proxy-based temperature reconstructions to define a mild Northern Hemisphere temperature peak between 950 and 1100 that was 0.1°–0.2° Celsius cooler than the mean global temperature between 1961 and 1990 and 0.3°–0.4° Celsius warmer than the coolest LIA period.

Attempts to measure the MWP illustrate the challenges of using temperature proxies rather than instrumental temperature readings. Proxies are influenced by effects other than temperature—for example, growth patterns in tree-ring studies can reflect precipitation, diseases, and atmospheric carbon dioxide concentration. Studies that combine multiple proxy data sets must accurately calibrate their results against a common temperature standard, but different researchers use different approaches. Only a limited number of proxy sites extend back to the MWP, with few in the Southern Hemisphere and unbalanced coverage in the Northern. Overall, there is considerable potential error in each reconstruction, such that current views of the MWP are tentative and will likely change with future data.

Interest in the cause of the MWP has been muted by the debate over its nature. If the event was only regional, then changes in regional meteorological features such as the North Atlantic Oscillation might provide a sufficient explanation. If the MWP was more global, solar activity may account for it. Sunspot records do not exist for the MWP, but concentrations of the cosmogenic isotopes carbon 14 and beryllium 10 in tree rings and ice cores suggest that there was stronger than normal solar radiation around 1000 and again between 1100 and 1250. These isotope concentrations are also consistent with evidence of weaker radiation during much of the LIA.

Raymond P. LeBeau, Jr.

Further Reading

Diaz, Henry F., Ricardo Trigo, Malcolm K. Hughes, Michael E. Mann, Elena Xoplaki, and David Barriopedro. "Spatial and temporal characteristics of climate in medieval times revisited." *Bulletin of the American Meteorological Society* 92, no. 11 (2011): 1487. Global analysis of proxy data show that medieval warm and cool periods were regional in extent, with the medieval warm period wetter over western Europe but drier across much of North America.

Fagan, Brian. *The Great Warming: Climatic Change and the Rise and Fall of Civilizations.* New York: Bloomsbury Press, 2008.

Hogan, C. Michael. "Medieval Warm Period." *Encyclopedia of Earth.* Boston U, 1 Jan. 2013. Web. 23 Mar. 2015.

Hughes, Malcolm, and Henry F. Diaz, eds. *The Medieval Warm Period.* Berlin: Springer-Verlag, 1994.

Kaufman, Darrell S., et al. "Continental-Scale Temperature Variability during the Past Two Millennia." Nature Geoscience 6 (2013): 339–46. PDF file.

Lamb, Hubert H. *Climate, History, and the Modern World.* 2d ed. London: Routledge, 1995.

Singer, S. Fred, and Dennis T. Avery. *Unstoppable Global Warming: Every Fifteen Hundred Years.* Rev. ed. Blue Ridge Summit, Pa.: Rowman Littlefield, 2008.

Mediterranean climate

Category: Meteorology and atmospheric sciences

Definition

Climate is the long-term average weather of a particular location. A Mediterranean climate is characterized by wet winters and dry summers, with mild winter and hot summer temperatures. This type of climate covers just under 2 percent of Earth's land area. It is found only in the middle latitudes (around 35° to 45° north or south latitude) and near an ocean or the Mediterranean Sea. Most of the area included in this climatic zone is the Mediterranean Sea basin, for which it was named. Other Mediterranean climate regions include the coastal regions of central and southern California, central Chile, the west side of the tip of South Africa, and parts of southern Australia, particularly the southwest.

The summer dryness in Mediterranean climates is caused by stable atmospheric high-pressure systems preventing storm systems from entering the area. These semipermanent high-pressure systems move away from the area and toward the equator during the fall, allowing rain-producing low-pressure systems to move in. When high pressure is reestablished in the spring, low-pressure areas can no longer move through the area, and rain-free conditions are reestablished for a number of months.

Because this climatic zone is always near a large body of water and such water bodies moderate temperatures (water heats and cools more slowly than does land), Mediterranean winter temperatures rarely reach the freezing point. Summer temperatures are greatly affected by the relative warmth or

Sunset view on the Mediterranean Sea towards Cannes from Juan-Les-Pins near Nice, France. (Túrelio/Wikimedia Commons)

coolness of the body of water adjoining the area. The Mediterranean Sea is a warm body of water, causing the Mediterranean basin to be relatively warm in the summer. Other areas, such as California, have relatively cool summer temperatures near the coast, because the Pacific Ocean, for example, is cooler than the Mediterranean Sea. The dryness of summer allows large temperature fluctuations during the day/night cycle. It also contributes to Mediterranean areas experiencing a relatively large number of wildfires, both because the air contains less moisture to resist fire and because the vegetation native to dry areas is itself drier and more prone to burn than that of wetter climates.

Significance for Climate Change

Mediterranean climates depend for their stability upon a combination of specific atmospheric conditions and the presence of a large body of water. Thus, if the atmospheric conditions are altered by climate change, or if they cease to exist near large bodies of water, Mediterranean climates will disappear. On the other hand, new Mediterranean climates could be established near different water bodies under the right circumstances.

Areas such as California or Chile are on north-south coastlines, and it is possible that climate change could move the climatic zone farther north or south. If the change is not overly abrupt, Mediterranean flora and fauna could migrate along with the climate zone. However, in the Mediterranean basin, a substantial increase in global warming or another climatic change could destroy the ecosystem altogether, because there is no large body of water to the north to which the climate, flora, and fauna could shift. Within Europe, there could be some movement north, as seems to be taking place in the British Isles. The little egret, commonly found in the Mediterranean Sea area, was unknown in England prior to 1996. As southern England has begun to develop a more Mediterranean climate, it was the second most populous heron in the United Kingdom by 2008. While birds have the ability to move from the European mainland to Great Britain, however, plants and land animals do not.

Climatic changes have been documented in several Mediterranean and nearby regions. These changes raise particular concerns regarding biodiversity, which is more affected by climate than by other environmental factors. The Iberian Peninsula, for example, has the greatest animal diversity of any region in Europe, and many Iberian species are endangered by global warming. Average summer temperatures in the region are generally rising, which is putting a substantial heat stress upon both plants and animals. Humans are not immune to such factors, and a study documenting a 200 percent increase in the number of extremely hot days on the peninsula concluded that frail people's lives were at risk from the increase.

Desertification is also taking place in the Mediterranean, as rainfall decreases and higher temperatures increase the rate of evaporation. A large number of droughts in the early twenty-first century have contributed to the encroachment of the desert from the south. Although between the mid-twentieth and early twenty-first century, there was generally a decrease in the amount of rainfall, in some specific locations there was an increase in precipitation.

It is estimated that up to 20 percent of plant species live in Mediterranean regions. In some areas, such as California, population growth has played a major factor in decreasing the number and diversity of plants, but changes in the precipitation patterns have also been a contributing factor, as has the resulting increase in wildfires. Even in areas where climate change has not substantially altered the total rainfall, the number of days on which it rains has changed. Many areas have fewer but harder rains than normal, resulting in less moisture being absorbed by the soil. An Intergovernmental Panel on Climate Change study has indicated that Mediterranean climates will very probably experience an increase in severe droughts, heat waves, and wildfires as a result of the changing climate.

Donald A. Watt

Further Reading

Bolle, Hans-Jurgen, ed. Mediterranean Climate: Variability and Trends. Berlin: Springer, 2003.

Giupponi, Carlo, and Mordechai Shechter, eds. Climate Change and the Mediterranean: Socio-economic Perspectives of Impacts, Vulnerability, and Adaptation. Cheltenham, Gloucestershire, England: Edward Elgar, 2003.

Kassam, Amir, Theodor Friedrich, Rolf Derpsch, Rabah Lahmar, Rachid Mrabet, Gottlieb Basch, Emilio J. González-Sánchez, and Rachid Serraj. "Conservation agriculture in the dry Mediterranean climate." *Field Crops Research* 132 (2012): 7-17. Conservation agriculture, including minimum soil disturbance and no-tillage planting, can be very useful in dry Mediterranean climates. Better moisture retention and infiltration help smooth the effects of erratic rainfall.

Lionello, P., ed. The Climate of the Mediterranean Region: From the Past to the Future. Burlington: Elsevier Science, 2012. Print.

McDade, Lucinda. "Plant Communities and Climate in Southern California." Rancho Santa Ana Botanic Garden. Rancho Santa Ana Botanic Garden, n.d. Web. 23 Mar. 2015.

Moreno, José, and Walter C. Oechel, eds. Anticipated Effects of a Changing Global Environment in Mediterranean-Type Ecosystems. New York: Springer-Verlag, 1995.

Mediterranean Sea

Categories: Oceanography; geology and geography

Definition

Bounded by Africa to the south, Asia to the east, and Europe to the north and west, the Mediterranean Sea is virtually landlocked. It is approximately 2,510,000 square kilometers in area, has an average depth of 1,501 meters, and reaches a maximum depth of 5,092 meters. It is believed that the Mediterranean was created as a result of interactions over millions of years between the Eurasian and African continental plates.

The major source of the Mediterranean's water is a surface current flowing inward from the Atlantic Ocean at the Strait of Gibraltar in the west. This current increases in salinity as it progresses, eventually sinking in the eastern Mediterranean and reversing course to flow outward along the seabed through the strait. Other important sources of water are the Black Sea (considered by some to be an arm of the Mediterranean) to the northeast, several large rivers (including the Nile), and rainfall. Because of its semilandlocked nature, the Mediterranean has limited tides, and its high rate of evaporation makes it saltier than the Atlantic. The narrow, human-made Suez Canal links the Mediterranean to the even saltier Red Sea in the southeast.

In general, the Mediterranean region is dry and hot during the summer and wet and cool during the winter. It is positioned between rainy, temperate Central Europe and arid North Africa and is sensitive to variations in these regions' normal weather patterns.

Significance for Climate Change

Climatologists and meteorologists predict that global warming will have a marked negative effect on the Mediterranean Sea and the countries lying on its shores. Existing problems such as flooding, heat extremes, water shortages, and desertification are likely to grow worse.

Global warming may raise the level of the Mediterranean Sea as much as 1 meter by the end of the twenty-first century, covering beaches and inundating coastal communities that are home to millions. The World Bank has warned that the Nile delta, in which one-tenth of Egypt's population lives and in which nearly half of the country's crops are grown, is in particular danger. Some have suggested that the Mediterranean could grow stagnant and that the flow of surface water through the Strait of Gibraltar might even reverse. The sea's resources are already threatened by overfishing, and warmer, saltier water threatens other native species such as corals and sponges, while favoring invasive species that have entered the Mediterranean from the Red Sea by way of the Suez Canal.

According to the United Nations' Intergovernmental Panel on Climate Change (IPCC), annual mean temperatures will probably increase throughout Europe, and the Mediterranean region is expected to experience hotter summers. This increase could have particularly serious

Mediterranean Sea and Maritime Alps, near Cannes, France. (© Steven I. Dutch)

consequences along the coast of North Africa. Heat waves are likely to increase in number and intensity, as they did throughout the second half of the twentieth century. The region may suffer from three to six times as many dangerously hot days in the twenty-first century, with France at greatest vulnerability. Rising temperatures will probably also result in increasingly intense storms, and although hurricanes rarely touch Europe, a 2007 study suggests that heat-induced atmospheric instability coupled with warmer sea temperatures could lead to tropical storms forming in the Mediterranean.

Overall, annual precipitation is likely to decrease in all but the northernmost parts of the Mediterranean region. Summers in particular are expected to be drier, with some studies suggesting that summer precipitation over southern Europe may decline by as much as 80 percent. Only in the areas dominated by mountain ranges such as the Alps is precipitation likely to rise, and those areas may lose a greater proportion of moisture due to higher temperatures. Droughts are likely to increase in frequency and severity, especially in the western Mediterranean.

Water usage in the Mediterranean region doubled in the second half of the twentieth century, and lack of water, already a serious problem in many parts of North Africa and the Middle East because of growing populations, will intensify. Rising sea levels will result in the salinization of freshwater needed for agriculture and human consumption.

Historically, the lands of the Mediterranean region have suffered from anthropogenic deforestation, and the resulting desertification is likely to spread in many areas. The increase in the number of forest fires forecast by the European Spatial Planning Observation Network in 2006 will add to the severity of the problem and will contribute to a loss of biodiversity.

Under conditions of global warming, death rates in the Mediterranean region from storms, heat stroke, air pollution, lack of water, and even malaria can be expected to rise. Many Mediterranean destinations will become too hot for tourists, undermining one of the region's major sources of revenue.

Grove Koger

Further Reading

Giorgi, Filippo, and Piero Lionello. "Climate Change Projections for the Mediterranean Region." *Global and Planetary Change* 63 (2008): 90-104. Review of projections suggesting that the region is particularly vulnerable to global warming and is likely to experience decreased precipitation, pronounced warming, and increasingly frequent heat waves.

Hughes, J. Donald. *The Mediterranean: An Environmental History.* Santa Barbara, Calif.: ABC-CLIO, 2005. Chronological survey of the sea's environmental development, including considerations of the possible effects of global warming. Maps, photographs, glossary, annotated bibliography.

Marquina Barrio, Antonio, ed. *Environmental Challenges in the Mediterranean, 2000-2050.* Boston: Kluwer, 2004. Papers presented at a North Atlantic Treaty Organization (NATO) research workshop addressing climate change along with other problems such as population growth, desertification, and pollution. Maps, graphs, bibliographies.

Soloviev, Sergey L., Olga N. Solovieva, Chan N. Go, Khen S. Kim, and Nikolay A. Shchetnikov. *Tsunamis in the Mediterranean Sea 2000 BC-2000 AD.* Vol. 13. Springer Science & Business Media, 2013. Although tsunamis are popularly associated with the Pacific, the Mediterranean also has seen some 300 events in the period covered by this catalog.

Ulbrich, U., et al. "The Mediterranean Climate Change Under Global Warming." In *Mediterranean Climate Variability*, edited by Piero Lionello, Paola Malanotte-Rizzoli, and R. Boscolo. Oxford, England: Elsevier, 2006. Overview of

contemporary research, including studies based on regional climate models and atmospheric general circulation models. Maps and graphs.

See also: Coastal impacts of global climate change; Desertification; Floods and flooding; Health impacts of global warming; Marginal seas; Ocean dynamics; Ocean life; Rainfall patterns; Sea-level change.

Megacities

Category: Economics, industries, and products

Megacities are vast population centers with equally great effects upon local climate. They both concentrate and require significant amounts of resources, posing unique challenges for sustainability and environmentalism generally.

Key concepts

megalopolis: a megacity that sprawls over a large area, rather than being concentrated spatially in the manner of traditional cities

urban heat island: an urban region that is significantly warmer than the surrounding rural areas

urbanization: the process of concentration of the human population in cities

Background

A megacity is a city with a population greater than 10 million. When cities such as Mexico City, Los Angeles, or Hong Kong get this large, it becomes difficult to determine their precise boundaries or true population. The U.S. Census Bureau and the United Nations often disagree in nonsystematic ways about the population of some of the world's largest cities, and the discrepancies between their estimates can represent several million people.

The bureaucracy and infrastructure of megacities can be as complex as those of small nations, and resource allocation within them is particularly difficult. In 1950, New York City was arguably

A remarkably clear view of the Los Angeles metropolitan area. The Los Angeles urban area extends from Long Beach on the cost in the foreground to Riverside and San Bernardino almost 100 km (60 miles) inland, and has a population of about 18 million (2015 estimate). (© Steven I. Dutch)

the diversity of human cultures. In their vast urban landscapes, millions of people live and die in sad and tragic conditions of poverty and low life expectancy. At the same time, others live lives of almost unfathomable wealth and freedom to travel about the globe. These urban environments simultaneously represent both a pinnacle of human achievement and a shameful failure to realize human potential.

Change Challenges and Opportunities

Cities almost always depend on a hinterland beyond their spatial extent to provide food, water, energy, and raw materials to sustain the lives of their citizens. They also increasingly depend on this hinterland to absorb their sewage, solid waste, and greenhouse gas (GHG) emissions. Historically this hinterland was predominantly nearby. Increasingly, however, hinterlands are farther and farther away from megacities, and, in regard to GHG emissions, the global atmosphere itself may be considered part of the hinterland. The hinterland of Los Angeles is global, as the city receives oil from the Middle East, water from the eastern Sierra, and food from Mexico, Europe, and Asia. Mexico City has built vast tunnels to divert sewage to distant hinterlands. Denver, Colorado, uses a network of tunnels to divert water from the western slope of the Rocky Mountains that would normally flow into Mexico's Sea of Cortez.

Almost all of these processes relate to climate change forcing factors in direct or indirect ways. Nonetheless, a fundamental and primary impact of megacites with respect to climate change is the energy used by these cities to provide electricity (often provided by coal-fired power plants) and the energy used to provide transportation (predominantly generated by fossil-fuel combustion). These urban areas are the most densely populated areas of the world. This density is an opportunity for numerous efficiencies with respect to energy consumption for

Earth's only megacity. In less than one human lifetime there were at least twenty megacities on the planet. One of the most challenging aspects of this dramatic increase in the number of megacities is that they are increasingly appearing in some of the poorest nations of the world. In 1950, only three of the world's most populous cities were in the developing world. In 2005, about 75 percent of the world's largest cities were in the developing world. The rise of these megacities represents a profound development in the history of humanity, and the challenges they present with respect to climate change may also be among the greatest opportunities to address that change.

Reality on the Ground

A megacity is a complex structure consisting of a sophisticated built environment that shelters and sustains millions of human agents. These agents, living in close proximity to one another, often represent extremes of human experience, as megacities juxtapose great wealth and poverty, as well as electricity, transportation, and the myriad other related needs of urban residents that require electricity and transportation. Leveraging the energy efficiency opportunities that these densely populated areas represent will be of paramount importance with respect to humanity's collective response to the challenges of climate change.

Paul C. Sutton

Further Reading

Andrade, Laura Helena, Yuan-Pang Wang, Solange Andreoni, Camila Magalhães Silveira, Clovis Alexandrino-Silva, Erica Rosanna Siu, Raphael Nishimura et al. "Mental disorders in megacities: findings from the Sao Paulo megacity mental health survey, Brazil." *PloS one* 7, no. 2 (2012): e31879. Survey of mental health problems in the megacity of Sao Paulo found up to 10% of the population affected, with most cases untreated. Anxiety disorders and depression were the most common problems.

Burdett, Ricky, and Deyan Sudjic. *The Endless City: The Urban Age Project by the London School of Economics and Deutsche Bank's Alfred Herrhausen Society.* New York: Phaidon, 2007. This brilliantly illustrated text provides detailed case studies of the realities, prospects, and challenges facing six major cities of the world (New York City, Shanghai, London, Mexico City, Johannesburg, and Berlin).

"Cities." *Science* 319 (February 8, 2008). This special journal section includes numerous articles on various urban issues such as poverty, sewage treatment, population growth, and transportation challenges.

Hardoy, Jorge E., Diana Mitlin, and David Satterthwaite. *Environmental Problems in an Urbanizing World.* Sterling, Va.: Earthscan, 2001. Focuses on human health issues related to environmental problems in the urban areas of the developing world. Relates the political and economic realities and challenges of urban environments to the growing social and environmental justice issues of the developing world.

National Geographic, November, 2002. Edited by Erla Zwingle. This special issue titled "Cities: Challenges for Humanity" includes a pull-out map of megacities that illustrates their increase through time and projections into the future. Features many photographs and a good general discussion of contemporary urbanization trends around the world.

See also: Energy efficiency; Energy resources and global warming; Population growth; Urban heat island; Water resources, global.

Meridional overturning circulation (MOC)

Category: Oceanography

Definition

Meridional overturning circulation (MOC) is an oceanographic term for water flows in the plane defined by the vertical and meridional (or north-south) axes. It is calculated by averaging those north-south, up-down flows from east to west across the width of an ocean basin.

Most discussions of the MOC focus on the deep, overturning circulations that connect the ocean abyss to the surface. Deepwater formation (the "sinking branches" of the deep MOC) occurs in two broad regions of the global ocean: the high-latitude North Atlantic Ocean, predominantly in the Labrador Sea and the Nordic Sea, and the Southern Ocean, near Antarctica. Water's density increases when it is cold and salty. Thus, the densest surface waters occur in the polar regions. Temperatures there are low, and brine-rejection during ice-formation increases the salinity of surface ocean water.

Dense polar water sinks to form deepwater masses that spread out horizontally along the meridional axis to fill most of the global deep oceans. The return branch of this deep MOC is more diffusely distributed. Deep water becomes more buoyant, and it will return back toward the surface if it is heated or made less saline. Vertical mixing with less dense water higher in the water column will decrease the density of deep water, producing a

Thermohaline Circulation

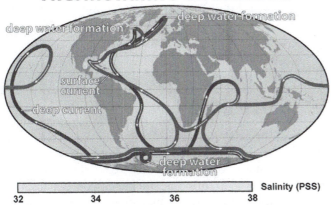

This map shows the pattern of thermohaline circulation also known as "meridional overturning circulation." This collection of currents is responsible for the large-scale exchange of water masses in the ocean, including providing oxygen to the deep ocean. (Robert Simmon, NASA)

return flow toward the surface. The deep MOC is sometimes called the "thermohaline circulation," although that term refers to such movement in all oceans, not just the Atlantic. The term "thermohaline circulation" is meant to evoke the idea that vertical motions are caused by changes in the temperature and salinity of seawater.

Shallow, wind-driven overturning circulations exist closer to the ocean's surface, the most prominent such feature being the subtropical cells in the Atlantic and Pacific Oceans. The winds blowing over the subtropical oceans force a convergence of surface waters, which pushes surface water downward in a process known as "Ekman pumping." This water travels at depth toward the equator, where the pattern of winds forces surface waters to diverge, bringing the water that was pumped downward in the subtropics back to the surface. Surface winds then force the surface waters back toward the subtropics, completing the subtropical cell.

Significance for Climate Change

The MOC plays a crucial role in maintaining Earth's climate. The earth is heated primarily in the tropics. Warm ocean currents move a large amount of tropical heat to higher latitudes in the subtropical cell and the deep MOC. This transfer of heat helps keep the high latitudes warm. The poleward heat flux is particularly strong in the North Atlantic Ocean. Heat brought poleward in the North Atlantic is then advected by large-scale winds eastward, where it warms Europe. Scientists have suggested that the North Atlantic MOC may slow down or shut down completely as a result of global warming; if this happens, the heat flux associated with the MOC would decrease, and Europe would become colder.

A climatic change that warms the high latitudes will introduce freshwater into the high-latitude North Atlantic Ocean as ice sheets melt and precipitation increases. This added freshwater would decrease the salinity of surface waters, reducing their tendency to sink. If polar water stops sinking, the warm surface current that brings lighter water to replace the sinking water would be disrupted, so the flow of heat to the North Atlantic and Europe would diminish or cease entirely.

Geologic evidence supports the argument that the introduction of freshwater into the North Atlantic can weaken heat transport by the Atlantic MOC. During the last ice age, ice sheets several kilometers thick covered a large portion of North America and northern Europe. Around 14,000 years ago, glaciers in the Northern Hemisphere retreated as a result of astronomically forced changes in Earth's orbit. Temperatures rose, and the ice sheets began to melt. Then, approximately 12,800 years ago, temperatures dropped rapidly back into the glacial range and ice sheets returned for another 1,300 years. This rapid drop in temperatures, known as the Younger Dryas, is believed to have been caused by the rapid input of freshwater into the North Atlantic from the melting North American ice sheet, which dramatically decreased the Atlantic MOC. While there is no large ice sheet on North America today, scientists are concerned that global warming could cause freshwater melt from Greenland to trigger analogous processes.

Alexander R. Stine

Further Reading

Kuhlbrodt, T., et al. "On the Driving Processes of the Atlantic Meridional Overturning Circulation." Reviews in Geophysics 45 (April 24, 2007).

"The Ocean Conveyor." Woods Hole Oceanographic Institution. Woods Hole Oceanographic Inst., 2014. Web. 23 Mar. 2015.

Richardson, P. L. "On the History of Meridional Overturning Circulation Schematic Diagrams." Progress In Oceanography 76 (2008): 466–486.

Schmittner, A., John C. H. Chiang, and Sidney R. Hemming, eds. Ocean Circulation: Mechanisms and Impacts. Geophysical Monograph 173. Washington, D.C.: American Geophysical Union, 2007.

Siedler, Gerold, John Church, and John Gould, eds. Ocean Circulation and Climate. London: Academic Press, 2001.

Srokosz, M., M. Baringer, H. Bryden, S. Cunningham, T. Delworth, S. Lozier, Jochem Marotzke, and Rowan Sutton. "Past, present, and future changes in the Atlantic meridional overturning circulation." *Bulletin of the American Meteorological Society* 93, no. 11 (2012): 1663-1676. That Atlantic meridional circulation is unusual in that it transports heat northward. Changes in the circulation can have abrupt impacts on lands bordering the North Atlantic.

Mesosphere

Category: Meteorology and atmospheric sciences

Definition

The mesosphere is the region of the atmosphere extending from the top of the stratopause (about 50 kilometers high) to the bottom of the mesopause-lower thermosphere boundary (MLT), about 100 kilometers high. It is the coldest region of the atmosphere. The warmest temperatures in the mesosphere are found just above the stratopause, where air temperatures may be as high as -5 °Celsius. Mesospheric temperatures decrease with increasing altitude, with the lowest temperatures, around -125 °Celsius, occurring during summer at the mesopause.

Significance for Climate Change

The natural source of water in the mesosphere is oxidation of methane. Atmospheric concentrations of the greenhouse gas (GHG) methane have increased dramatically since the beginning of the Industrial Revolution and may have influenced the appearance of noctilucent clouds. Water vapor emitted by rockets and the space shuttle has been observed to form noctilucent clouds. Exhaust from one shuttle mission may increase the appearance of noctilucent clouds by as much as 20 percent. Carbon dioxide (CO_2) in the mesosphere releases heat to space. Solar proton events have been observed to cool the lower mesosphere by causing photochemical reactions leading to ozone depletion, causing an estimated temperature drop of up to 3° Celsius.

Gravity waves are the dominant form of motion in the mesosphere, with wavelengths of around 10 kilometers. Wind speeds on the order of 100 meters per second may occur in the MLT. Gravity wave flux is lowest at solar maximum. Direction of atmospheric transport is from the summer hemisphere to the winter hemisphere.

In the mesosphere, clouds can form only at the coldest temperatures, within about thirty days of the summer solstice at latitudes above 50° north and below 50° south. Because they form at 82 kilometers altitude, the thin filamentous noctilucent clouds (sometimes called polar mesospheric clouds, or PMCs) are illuminated only when the Sun is between 6° and 16° below the horizon.

There has been speculation that noctilucent clouds are a recent phenomenon resulting from anthropogenic activity, as they were not reported prior to 1885, when scientists began studying the colorful twilight displays that followed the eruption of Krakatoa. Soon after the discovery of noctilucent clouds, speculations began that the particles composing the clouds were of extraterrestrial origin. In 1962, rockets launched to investigate noctilucent clouds determined that the cloud particles had nuclei of iron or nickel dust of extraterrestrial origin, which were coated by water ice. One of the most surprising aspects of noctilucent clouds is their strong radar reflectivity. After some study, an explanation was advanced in 2008 that molecules of sodium and molecules of iron form a thin metallic film over the tiny ice crystals. This film reflects radar waves in a special, amplified way that makes the clouds more prominent on radar than if they were a cloud of disordered metal dust. Noctilucent clouds are postulated to remove about 80 percent of the sodium and iron from their environment in the mesosphere.

Water is formed in the mesosphere as a result of the oxidation of methane gas, rather than arising from Earth's surface. Because methane is released as a by-product of incomplete combustion, as well as through coal-mining activity and anaerobic production from animals, swamps, and rice paddies, its concentration has been rising steadily as a result of anthropogenic activity. The concentration of methane in the atmosphere has increased 150 percent since the industrial age began, around 1750.

After rockets are launched through the mesosphere, their exhaust trails may form artificial noctilucent clouds. These noctilucent trails can form in the mesosphere in temperate latitudes, but ground observations of them are made only when the Sun is between 6° and 16° below the horizon.

After the launch of the STS-107 shuttle mission in January, 2003, it was determined that vaporized iron and water from its exhaust traveled to Antarctica at 110 kilometers altitude in about two days. This exhaust plume went on to form noctilucent clouds in the southern polar summer. Data from this event indicate that a single space shuttle's exhaust might cause a 10 percent to 20 percent increase in the appearance of seasonal noctilucent clouds.

Anita Baker-Blocker

Further Reading

Arnold, Neil. "Solar Variability, Coupling Between Atmospheric Layers, and Climate Change." *Philosophical Transactions of the Royal Society of London A* 360, no. 1801 (December, 2002): 2787-2804. Reprinted in *Advances in Astronomy*, edited by J. M. T. Thompson. Vol. 1. London: Imperial College Press, 2005.

Bellan, Paul M. "Ice Iron/Sodium Film as Cause for High Noctilucent Cloud Radar Reflectivity." *Journal of Geophysical Research D* 113 (2008): 16,215-16,218.

Brasseur, Guy, and Susan Solomon. *Aeronomy of the Middle Atmosphere: Chemistry and Physics of the Stratosphere and Mesosphere.* 3d rev. ed. Dordrecht: Springer, 2005. eBook Collection (EBSCOhost). Web. 20 Mar. 2015.

Glickman, Todd S., ed. *Glossary of Meteorology.* 2d ed. Boston: American Meteorological Society, 2000.

Schroeder, Wilfried, and Karl-Heinrich Wiederkehr. "Johann Kiessling, the Krakatoa Event, and the Development of Atmospheric Optics After 1883." *Notes and Records of the Royal Society of London* 54, no. 2 (May, 2000): 249-258.

Siingh, Devendraa, R. P. Singh, Ashok K. Singh, Sanjay Kumar, M. N. Kulkarni, and Abhay K. Singh. "Discharges in the stratosphere and mesosphere." *Space science reviews* 169, no. 1-4 (2012): 73-121. Long reported by pilots and ground observers. but only recently well documented, electrical discharges in the upper atmosphere are related to lightning ischarges at lower altitudes. This paper contains a description of the different types and numerous illustrations.

Meteorology

Category: Meteorology and atmospheric sciences

Arguably the most important aspect of climate change is its effects on the physical and chemical properties of Earth's atmosphere. Meteorologists can assess those effects and their consequences, informing climate policy decisions.

Key concepts

climate: long-term average weather conditions

mesoscale: the atmospheric scale between microscale and synoptic scale, ranging from a few kilometers to hundreds of kilometers *microscale:* the smallest scale of atmospheric motion, ranging from meters to kilometers

planetary scale: the largest scale of atmospheric motion, covering the entire globe

synoptic scale: the typical scale of weather maps, showing such features as high- and low-pressure systems, fronts, and jet streams over an area spanning a continent

weather: a particular atmospheric state at a given time and place

Background

Meteorology is the scientific study of the atmosphere, weather, and climate. It combines most of the basic scientific disciplines, such as mathematics, physics, chemistry, statistics, and computer science, and applies them to Earth's atmosphere and its phenomena. Thus, meteorology is a branch of Earth science and of physical science.

Meteorology provides the core knowledge about climate change and global warming. It can be used to analyze planetary climate patterns as well as continental, regional, and local patterns. Meteorologists measure weather and other specific atmospheric phenomena and abstract from those measurements to determine the climate, the long-term average conditions in a given location.

Dynamic Meteorology

Dynamic meteorology is the core discipline of the atmospheric sciences. It employs dynamics, fluid mechanics, and classical mechanics, coupled with rigorous mathematics, to study atmospheric motion and evolution. Dynamic meteorology treats the atmosphere as a fluid continuum, applying Newtonian principles to atmospheric systems. Modern numerical weather prediction is a result of this approach. Many methods in dynamic meteorology are also extended to study climate systems in the closely related discipline of climate dynamics. Climate dynamics may provide a good tool for studying climate change. Various global circulation models (GCMs) are examples of this application.

Physical Meteorology

In addition to its kinetic properties and dynamic evolution, the atmosphere possesses many other physical properties, such as its thermal content, its humidity, its electrical and optical properties, and so forth. To study these physical properties of the atmosphere, meteorologists incorporate the principles and approaches of physics. A wide range of subjects is helpful in studying such atmospheric phenomena, including cloud physics, thermodynamics, precipitation physics, boundary-layer meteorology, thermal convection, atmospheric electricity, and atmospheric optics.

Many atmospheric physical properties are directly related to global climate change. For example, clouds are an important factor in global warming. Clouds play a dual role in the global climate system. On one hand, they contribute significantly to Earth's albedo, reflecting a large amount of solar energy back into space and producing a global cooling effect. On the other hand, they absorb long-wave radiation from the Earth's surface and re-emit it, thereby heating the atmosphere and surface. Increased atmospheric humidity due to global warming will increase cloud cover and influence severe weather patterns as well. Physical meteorology can provide a detailed understanding of these aspects of climate change.

Applied Meteorology

As meteorologists' understanding of the complexity of Earth's climate system has increased, new applied meteorological specializations have emerged. Thus, the field now includes satellite meteorology, radar meteorology, statistical meteorology, agricultural micrometeorology, and climatology. These new subdisciplines are fundamentally interdisciplinary. They not only help translate meteorological concepts and research methods to other scientific disciplines but also strengthen meteorology by incorporating the technologies and methods of its sister sciences. For example, modern technologies, such as radar and satellites, add fresh content to meteorology and provide new observational tools for studying the atmosphere. These new areas are also important for global climate studies. For example, satellites can provide a global view of the global warming effect.

Atmospheric Chemistry

Traditional meteorology is mostly concerned with the physical aspects of the atmosphere. Climate researchers, however, have found that atmospheric chemistry is just as important for understanding climate change. Greenhouse gases are central to global climate change, and other aspects of atmospheric composition may play similarly important roles. Ozone depletion in the stratosphere is both an atmospheric dynamics and atmospheric chemistry problem that concerns climate change. In addition, Earth's carbon cycle is an important area for climate study.

With an increasing level of global industrialization and urbanization, environmental conservation and protection become more concerned issues. Air pollution and air quality are central to these environmental problems. Acid rain and environmental acidification are also concerns for environment protection and conservation. For all these problems, atmospheric chemistry can provide fundamental understanding.

Context

Meteorology is one of the primary sciences employed in the study of Earth's climate system. Earth's climate is a complex system, however, that includes five components: atmosphere, hydrosphere, lithosphere, cryosphere, and biosphere. It is the interaction of all five of these components that determines Earth's climatic environment. Therefore, meteorology, although providing a core understanding of global climate and climate change, must be combined with knowledge from other scientific disciplines, such as oceanography, geology, hydrology, chemistry, biology, ecology, astronomy, and glaciology, to address global climate change.

Chungu Lu

Further Reading

Ahrens, C. Donald. *Meteorology Today.* 11th ed. Pacific Grove, Calif.: Thomson/Brooks/Cole, 2015. Updated version of a widely used standard textbook.

Geer, Ira W., ed. *Glossary of Weather and Climate.* Boston: American Meteorological Society, 1996. Collection of terms, concepts, and definitions related to weather and climate.

Glickman, Todd S., ed. *Glossary of Meteorology.* 2d ed. Boston: American Meteorological Society, 2000. Collection of most meteorological terms, concepts, and definitions, and is useful for a quick lookup.

Lutgens, Frederick K., and Edward J. Tarbuck. *The Atmosphere.* 10th ed. Upper Saddle River, N.J.: Pearson Prentice Hall, 2007. Elementary textbook covering a wide range of atmospheric sciences.

See also: Atmosphere; Atmospheric chemistry; Atmospheric dynamics; Atmospheric structure and evolution; Average weather; Climate and the climate system; Climate change; Climatology; Weather vs. climate.

Methane

Category: Chemistry and geochemistry

Methane is among the six GHGs restricted under the Kyoto Protocol. Its global warming potential is more than twenty times that of CO_2.

Key concepts

alternative fuel: clean or renewable fuel that can replace traditional fossil fuels

archaea: a taxonomic group of prokaryotic, single-celled microorganisms similar to bacteria, but evolved differently

energy from waste: technologies that are designed to produce energy and reduce or eliminate waste at the same time

fuel: an energy source that is burned to release energy

fuel alternative: replacement energy source that can be used instead of fuel

greenhouse gas (GHG): a gas in the atmosphere that traps heat on Earth that would otherwise radiate into space

Background

Methane is a colorless, odorless gas with the molecular formula CH_4. It is the main chemical component of natural gas (accounting for 70-90 percent of such gas). Natural gas makes up up to 20 percent of the U.S. energy supply. Methane was discovered by the Italian scientist Alessandro Volta, who collected it from marsh sediments and demonstrated that it was flammable. He called it "combustible air."

Methane as a Greenhouse Gas

As with all greenhouse gases (GHGs), methane in the atmosphere acts similarly to glass in a greenhouse. It allows light energy from the Sun to reach Earth's surface, but it traps heat energy radiated back from the surface in the form of infrared radiation. Since the beginning of the Industrial Revolution in the mid-eighteenth century, methane concentrations have more than doubled in the atmosphere, causing nearly one-quarter of the planet's anthropogenic global warming. Continuous release of methane into the atmosphere causes rapid warming, because methane's contribution to the greenhouse effect is much more powerful than that of carbon dioxide (CO_2).

Global warming itself may trigger the release of methane trapped in tundra permafrost or ocean deposits, thereby accelerating climate change in a positive feedback loop. The release of large volumes of methane from such geological formations into the atmosphere has been suggested as a possible cause for global warming events in the past. Methane oxidizes to CO_2 and therefore remains in the atmosphere for a shorter time-period of nine to fifteen years, compared to CO_2, which may remain in the atmosphere for one hundred years.

Sources of Methane

According to the U.S. Environmental Protection Agency (EPA), about 60 percent of global methane emissions are a direct result of human-related activities. These activities include creating landfills, treating wastewater, animal husbandry (through enteric fermentation and manure production), cultivating rice fields, mining coal, and producing and processing natural gas. For instance, the livestock sector (including cattle, chickens, and pigs) generates 37 percent of all anthropogenic methane. Landfills are the second largest anthropogenic source of methane in the United States.

Natural sources of methane include wetlands, lake sediments, natural gas fields, termites, oceans, permafrost, and methane hydrates. Wetlands are responsible for up to 76 percent of global natural methane emissions. Surprisingly, according to EPA data, termites contribute about 11 percent of global natural methane emissions. In most of these processes, methane is produced by microorganisms called archaea as the integral part of their metabolism. Such microbes are called methanogens, and the route of methane generation is called methanogenesis.

Archaea live in oxygen-depleted habitats, because the presence of oxygen would kill them instantly. For their food source,

AIRS-Retrieved Global Tropospheric Methane for August 2005

parts per billion by volume, ppbv

1687   1723   1760   1797   1833

A satellite image of the global distribution of methane in Earth's troposphere. (NASA-JPL)

Global Methane Emissions by Sector, 2000

Economic Sector	Percent of Total Emissions
Agricultural by-products	40.0
Fossil fuel retrieval, processing, and distribution	29.6
Waste disposal and treatment	18.1
Land use and biomass burning	6.6
Residential, commercial, and other sources	4.8
Power stations	0.9

Data from the Netherlands Environmental Assessment Agency.

methanoges use products of bacterial fermentation such as CO_2 and molecular hydrogen (H_2); different acids such as acetate, pyruvate, or formate; or even carbon monoxide. That is why methanogenic archaea usually exist in consortium with other microorganisms (bacteria). They also live in symbiotic relationships with other life forms, such as termites, cattle, sheep, deer, camels, and rice crops.

Methane as a Fuel

In the 1985 science-fiction film *Mad Max Beyond Thunderdome* starring Mel Gibson, a futuristic city was run on methane generated by pig manure. In reality, methane can be a very good alternative fuel. It has a number of advantages over other fuels produced by microorganisms. It is easy to make and can be generated locally, obviating the need for long-distance distribution. Extensive natural gas infrastructure is already in place to be utilized. Utilization of methane as a fuel is a very attractive way to reduce wastes such as manure, wastewater, or municipal and industrial wastes. In local farms, manure is fed into digesters (bioreactors), where microorganisms metabolize it into methane. Methane can be used to fuel electrical generators to produce electricity.

In China, millions of small farmers maintain simple, small, underground digesters near their houses. There are several landfill gas facilities in the United States that generate electricity using methane. San Francisco has extended its recycling program to include conversion of dog waste into methane to produce electricity and to heat homes. With a city dog population of 120,000, this initiative promises to generate significant amounts of fuel with a huge reduction of waste at the same time.

Methane was used as a fuel for vehicles for a number of years. Several Volvo car models with bi-fuel engines were made to run on methane, with gasoline as a back up. Methane is more environmentally friendly than are fossil fuels. Burning methane results in production of CO_2 and contributes to global warming, but with less impact on Earth's climate than methane itself would have in the atmosphere, for a net benefit. Even though the use of methane as an energy source releases CO_2, the process as a whole can be considered CO_2 neutral, in that the released CO_2 can be assimilated by archaea.

Methane Removal Processes

The natural mechanism of methane removal from the atmosphere involves its destruction by the hydroxyl radical (OH). Significant amounts of methane are also consumed by microorganisms called methanotrophs, which use the methane for energy and biosynthesis. These bacteria are prevalent in nature and potentially could be used for methane mitigation.

Context

Since methane is a powerful contributor to global warming, any efforts to reduce methane emissions will have a rapid impact on Earth's climate. One way to avoid methane release into the atmosphere is to turn it into a fuel. Supply of fossil fuels, particularly oil, is limited and does not satisfy world energy demands, which consistently increase. The extensive use of fossil fuels causes global warming. Methane utilization in place of fossil fuels as an energy source can provide significant environmental and economic benefits. In the future, landfills and wastewater treatment facilities can possibly be redesigned to optimize methane production. However, further research is needed to better understand archaean-bacterial methanogenic communities in landfills and wastewater treatment facilities in order to improve methane generation.

Some technical obstacles exist to efficiently converting landfill wastes that primarily contain plant lignocellulosic material. Lignocellulose is a combination of lignin, cellulose, and hemicellulose that strengthens plant cell walls. Microbial communities in landfills cannot utilize lignin. Creating efficient methane-producing facilities that are also capable of reducing waste is a feasible option for sustainable development to provide fuel to heat homes, run cars, generate electricity, and eliminate powerful GHG and health hazards.

Sergei Arlenovich Markov

Further Reading

Archer, David. *Global Warming: Understanding Forecast.* Malden, Mass.: Blackwell, 2007. Devotes a chapter to discussion of methane and its greenhouse effects.

Madigan, Michael T., et al. *Brock Biology of Microorganisms.* 12th ed. San Francisco, Calif.: Pearson/Benjamin Cummings, 2009. Introductory microbiology textbook. Chapter 17 describes methane production by archaea.

National Academy of Sciences. *Methane Generation from Human, Animal, and Agricultural Wastes.* Washington, D.C.: Author, 2001. Describes the role of disparate sources in methane production.

Nebel, Bernard J., and Richard T. Wright. *Environmental Science: Towards a Sustainable Future.* Englewood Cliffs, N.J.: Prentice Hall, 2008. Several chapters describe methane as a GHG and an alternative fuel.

Osborn, Stephen G., Avner Vengosh, Nathaniel R. Warner, and Robert B. Jackson. "Methane contamination of drinking water accompanying gas-well drilling and hydraulic fracturing." *proceedings of the National Academy of Sciences* 108, no. 20 (2011): 8172-8176. Methane in drinking water is non-toxic and its principal danger is building up to explosive concentrations in enclosed spaces. Fluids used in fracking, however, can constitute a health risk and can contaminate ground water through the same avenues as methane.

See also: Biofuels; Carbon dioxide; Clean energy; Energy from waste; Fossil fuel emissions; Fossil fuels and global warming; Greenhouse effect; Greenhouse gases and global warming.

Mexico

Category: Nations and peoples

Key facts

Population: 109,955,400 (January, 2009, estimate); 121,736,809 (July 2015 estimate.)

Area: 1,922,550 square kilometers

Gross domestic product (GDP): $1.559 trillion (purchasing power parity, 2008 estimate); $2.227 trillion (purchasing power parity, 2015 estimate).

Greenhouse gas (GHG) emissions in millions of metric tons of carbon dioxide equivalent (CO_2e): 686 in 2000, 723.85 in 2012

Kyoto Protocol status: Ratified 2000

Historical and Political Context

Mexico has experienced political strife throughout its history, including major wars and executions of leaders. The Spaniards invaded the region in 1519, conquering the Aztec and Mayan native cultures. Mexico attempted to declare independence from Spain in 1810, resulting in war and the execution of a number of leaders from Mexico. This war of independence lasted until 1821, when Spain finally granted Mexico its autonomy. The Mexico of 1821 included most of present-day Central America and the southwestern United States.

There was constant strife in Mexico until 1867, especially between a group supporting the centralized federal government required by the 1824 constitution and a group that supported a more localized government. During this time, the constitution was suspended, which resulted in civil war. The Republic of Texas, among others, declared independence from Mexico and was able to defeat the Mexican forces. Later, Mexico lost a war with the United States for control of Texas, and in addition lost what is now the southwestern United States.

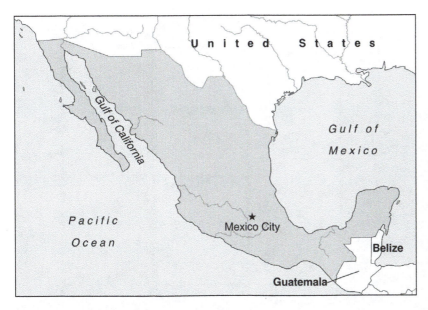

Party politicians were elected. In 1989, the first noninstitutional Revolutionary Party governor of a state was elected. It was suspected, however, that the Institutional Revolutionary Party changed the election results in the 1988 presidential elections so that its candidate, Carlos Salinas, won. This infuriated many people. Salinas signed the North American Free Trade Agreement and helped control inflation. By 1994, however, the economy in Mexico collapsed. The United States helped spur the Mexican economy, so that it had rebounded by 1999. In 2000, Vincente Fox became the first noninstitutional Revolutionary Party candidate to be elected president.

In 1867, Benito Juárez restored the republic of Mexico as a democracy, and he began to modernize the country. His acts reduced the power of the Catholic Church over Mexican politics, required equal rights for all people, and brought the army under civilian control. Porfirio Díaz was the ruler of Mexico from 1876 to 1911. He helped invest in the arts and sciences, and he improved Mexico's economy, reducing economic inequality and political repression. Díaz resigned in 1911 after an election fraud was found, and this event caused another Mexican revolution.

The government entered a chaotic period. A number of elected leaders were overthrown and assassinated. Finally, in 1929, Plutarco Calles founded the Institutional Revolutionary Party, which became the dominant party in Mexican politics until the end of the twentieth century. Mexico became more stable under the party. The economy grew significantly from 1940 to 1980. The government took over mineral rights, including nationalizing the petroleum industry into the organization called PEMEX. The Institutional Revolutionary Party, however, gradually became oppressive and authoritarian. For example, the government in 1968 killed many protestors.

The Institutional Revolutionary Party began to lose its control over Mexican politics in the 1970s, as some noninstitutional Revolutionary

Impact of Mexican Policies on Climate Change

Mexico has a free-market economy with the eleventh highest gross domestic product (GDP) in the world. Mexico has improved railroads, the distribution of natural gas, airports, and the generation of electricity. It has also become the largest producer of cars and trucks in North America. The economy of Mexico tends to be linked to that of the United States, as economic downturns and upturns in the United States have been reflected in Mexico. The country has a large middle class, although there are still significant income disparities among the Mexican people. Many persons live in poverty, especially in rural areas.

Mexico has made a definite commitment to energy conservation to reduce pollutants and greenhouse gases (GHGs) in the atmosphere and to promote energy efficiency. The nation has seven major metropolitans areas with populations over one million. The Mexico City metropolitan area is the largest, with over twenty-two million inhabitants.

The use of refining fuels such as gasoline and diesel, the two most used liquid fuels, has been growing steadily in Mexico. For instance, the use

The border crossing between Tijuana, Mexico and San Ysidro, California (a suburb of San Diego) is the busiest in the world. In this view, automobiles and pedestrians wait to pass through border controls to enter the United States. In addition to legal border crossings, large numbers of people enter the U.S. illegally along the 2,000-mile (3200-kilometer) border. (© Steven I. Dutch)

of refining fuels in Mexico grew from 141,000 kiloliters per day in 1988 to 206,000 kiloliters per day in 2000. Lead was eliminated from gasoline, and sulfur in gasoline and diesel fuels was drastically decreased in the 1990s. Fuel oil (4 percent sulfur) and national diesel (2 percent sulfur) have been replaced by gas oil (2 percent sulfur), industrial fuel (1 percent sulfur), and PEMEX Diesel Industrial (0.05 percent sulfur). These improvements in fuel have helped greatly to reduce air pollution, especially from lead and sulfur, in the Mexico City metropolitan area. The metropolitan area lies in a valley, with mountains surrounding much of the region. Pollution settles in this valley for much of the year because of an atmospheric inversion layer that forms over the valley trapping them. This results in a multitude of health effects on the city's inhabitants. The adoption of catalytic converters has also improved air quality. Mexico City's carbon monoxide emissions decreased by 454,000 metric tons per year from 1989 to 1994, reducing the amount of carbon monoxide in the atmosphere by 67 percent. In addition, ozone concentrations decreased by 36 percent from 1991 to 2003.

The use of fossil fuels will likely continue to increase with time in Mexico. Thus, other means need to be found to further reduce fossil fuel emissions. For example, public transport needs to be increased so fewer persons drive their own vehicles, and all vehicles need to burn fuel more efficiently. Mexico City has attempted to reduce pollution by imposing restrictive driving rules, with only limited success.

Mexico as a GHG Emitter

The GHG emissions of Mexico have grown steadily since 1990. For instance, carbon dioxide (CO_2) emissions were estimated to be 385 million metric tons in 2000 and 438 million metric tons in 2007. During this period, Mexico produced about 1.6 percent of the world's CO_2 emissions. The largest CO_2 emissions are from burning of fuels from industry and in the home (about 32 percent) and from transportation (about 15 percent). Mexico's CO_2 emissions will likely continue to increase as the country remains dependent on fossil fuels and its population grows, unless a more effective means to reduce emissions can be found.

Mexico signed the Kyoto Protocol in 2000. The goal of the protocol was to stabilize GHG emissions (especially those of CO_2, methane, nitrous oxide, and sulfur hexafluoride) to prevent climate change. The industrialized countries agreed to reduce their GHG emissions by 5.2 percent compared to those of 1990. The Kyoto Protocol was put into effect in February, 2005. The treaty divided countries into Annex I (industrialized) countries and non-Annex I (developing) countries. Under the treaty, an Annex I country can invest in projects to help reduce GHG emissions in a non-Annex I country. The Annex I country will earn credit for reducing the other country's GHG emissions that it can use to offset its own emissions in excess of its treaty obligations. For instance, Japan could invest in developing solar energy in Mexico to be used for electrical generation. Japan would then be given credits that it could either use to gain permission to emit more GHGs or sell to a third party.

The major commitment of Mexico as a non-Annex I country is to examine its GHG emissions with time to help limit their emissions. One approach to limiting its GHG emissions might be to add a tax on fossil fuels to increase the costs of the fuels and to reduce the Mexican government's subsidies on fossil fuels. The higher cost of the fossil fuels would reduce the use of the fuels and reduce GHG emissions. For example, it was predicted that if a tax of $10 per ton (in U.S. dollars) had been levied on Mexican fossil fuels starting in 1987, this tax would have resulted in a reduction of about 943 kilograms per person in CO_2 emissions. It would also have increased the Mexican government's revenues by $772 million, or about 3.2 percent. The petroleum, mining, chemical, and construction industries of Mexico would likely stand to lose the most from such a tax. Also, a tax on fossil fuels would encourage the development of more energy-efficient motors in vehicles and the use of more energy-efficient fuels.

Bioenergy sources, such as wood fuels, grain ethanol and other farming fuels, and cattle residues, have been suggested to replace some fossil fuels and reduce GHG emissions. For instance, if Mexico used biofuels (especially ethanol, biodiesel, and electricity generated from biological materials) to replace 16 percent of fossil fuels, then CO_2 emissions could be reduced by 79 million metric tons of CO_2 by 2030.

Summary and Outlook

The economy of Mexico in the latter half of the twentieth century went from an agrarian economy to a more industrialized economy, increasing national consumption of fossil fuels and production of hard goods such as cars and trucks. Correspondingly, GHG emissions spiraled upward as well. The Mexican government has become more stable in the early twenty-first century than it was for much of its past, as political and police corruption have decreased. Nevertheless, there still appear to be a large number of corrupt government officials, including police officers, who are ready to take bribes.

Traffic in illegal drugs is a steadily increasing problem in Mexico. The so-called drug lords have been killing the police, army officials, one another, and many innocent civilians. Some civilians appear to have been held for ransom to obtain more money for the drug lords and some persons appear to have been killed to provoke fear in the general population. Also, some local areas in Mexico have been taken over by those involved with drugs. Targets of violence include rival cartels, police, elected officials and journalists. Estimates of casualties in the drug war range from 60,000 dead to twice that level, with thousands more missing. There may be an imminent danger that much of Mexico could be destabilized if these actions continue, and an unstable society will be much less equipped to institute climate policy initiatives to respond to global warming. Thus, Mexico must curtail these problems to be able to progress as a stable, democratic society. The efforts of Mexico to reduce the GHG emissions could fail if money is channeled elsewhere to stop the activity in drugs.

Robert L. Cullers

Further Reading

Segarra, Paulina and Ajnesh Prasad. How corruption is hurting Mexico City's efforts to tackle air pollution. The Conversation (May 5, 2016) https://theconversation.com/how-corruption-is-hurting-mexico-citys-efforts-to-tackle-air-pollution-57517 [10 July 2016] Problems include bribery at enforcement centers, entrenched private transportation interests, poorly conceived remedial laws and resource embezzlement.

Boyd, Roy, and Maria E. Ibarraran. "Costs of Compliance with the Kyoto Protocol: A Developing Country Prospective." *Energy Economics* 24 (2002): 21-39. Describes a potential method for Mexico to limit GHG emissions by imposing an energy tax on the burning of fossil fuels.

Islas, Jorge, Fabio Manzini, and Omar Masera. "A Prospective Study of Bioenergy Use in Mexico." *Energy* 32 (2007): 2306-2320. Describes the use of wood, farming fuels (such as alcohol produced to run cars), and municipal waste as alternative fuels to reduce CO_2 emissions. Numerous tables and figures.

McKinley, Galen, et al. "Quantification of Local and Global Benefits from Air Pollution Control in Mexico City." *Environmental Science and Technology* 39 (2005): 1954-1961. Summarizes information about air pollution in Mexico City and potential means for improving air quality. Includes seven data tables.

Mexico Air Quality Management Team. "Improving Air Quality in Metropolitan Mexico City: An Economic Evaluation." Washington, D.C.: World Bank Latin America and the Caribbean Region, 2002. Evaluates air pollution in Mexico City and argues that reducing air pollution by certain amounts could affect the health of the residents. Tables and graphs.

Nova, M., J. Gasca, and U. Gonzalez. "The Energy Demand and the Impact by Fossil Fuels Use in the Mexico City Metropolitan Area from 1988 to 2000." *Energy* 31 (2006): 3381-3390. Compares contemporary and predicted future use of fossil fuels in Mexico generally to that of the Mexico City Metropolitan Area. Discusses the city's influence on the air quality in Mexico. Numerous tables and figures.

See also: Kyoto Protocol; United Nations Framework Convention on Climate Change; U.S. energy policy.

Microwave sounding units

Categories: Science and technology; meteorology and atmospheric sciences

Definition

A microwave sounding unit sounds (produces a profile of) the temperature and the moisture levels in the atmosphere. The first microwave sounding units were placed on National Oceanic and Atmospheric Administration (NOAA) satellites in 1978. In 1998, the Advanced Microwave Sounding Units replaced the older units. The satellites are polar-orbiting and synchronous with the sun. The satellites will make about sixteen orbits in twenty-four hours. The instruments make a scan every eight seconds, during which time the point on the surface below the satellite will have moved by 45 kilometers. During six of the eight seconds, the AMSU-A instrument is making thirty observations about 3° apart from 48° on one side to 48° on the other. The next scan will be 45 kilometers farther along its orbit. After the thirty observations, the instrument makes an observation of a warm calibration target and an observation of cold space before returning to its original starting position. The AMSU-A instrument is determining temperature of the different layers of the atmosphere. Another instrument, AMSU-B, is scanning in a similar mode but makes three times as many observations and determines water vapor concentration. In 2005, the AMSU-B instrument was replaced by a Microwave Humidity Sounder (MHS). In 2009 or 2010, the Advanced Technology Microwave Sounder (ATMS) is slated to replace the AMSU-A.

The AMSU-A is detecting thermal emissions from atmospheric oxygen. Those emissions are in the microwave region of 23 to 89 gigahertz. The microwave radiation is low-energy and high-frequency. The AMSU-A instrument has several channels. Different channels record different wavelengths of radiation. Groups of channels emphasize different layers of the atmosphere. Combined together a profile of the temperature of the different layers of the atmosphere is produced. Channels 2 through 4 emphasize the surface, channels 5 through 8 emphasize the mid-troposphere, and channels 9 through 14 emphasize the lower stratosphere.

Significance for Climate Change

The data obtained from AMSU-A includes temperature and water vapor profiles of the atmosphere, snow and ice coverage, cloud water content, and rain rate. With data from other instruments, scientists can produce not only the temperature and water vapor profiles but also a measurement of the ozone, properties of clouds, and the amount of infrared radiation emitted by the Earth in areas not covered by clouds. The AMSU data are used in weather prediction. The more quickly the data get to weather prediction centers, the better the accuracy of the predictions.

It would appear that the data on temperature profiles that have been acquired since 1958 by balloons and since 1978 by MSU would allow for a determination of the increase or decrease of the temperature of the atmosphere. It has not proven to be so simple. Because of a change in equipment, the balloon data have a discontinuity that makes it much less valuable in determining temperature change. The data from MSU are not temperatures but thermal radiation from oxygen molecules.

Different scientists use different methods to calculate the temperatures from the MSU data and consequently obtain different temperatures. Two of the groups doing calculations of this type are the Remote Sensing System (RSS) and the University of Alabama in Huntsville (UAH). The lower troposphere temperatures, in which stratospheric cooling has been eliminated, has been measured by RSS with an increase of 0.156° Celsius per decade and by UAH with an increase of 0.13° Celsius per decade. Other scientists have used the RSS data and the UAH data and have derived increases from 0.050° Celsius per decade to 0.20° Celsius per decade. Climate models indicate that the troposphere should warm about 20 percent more than the surface over the entire world and 50 percent more in the tropics. The surface temperature has increased by 0.17° Celsius per decade since 1979. Reconsideration of the data from UAH and RSS has led to corrections that have brought the data into fair agreement and into a not inconsistent agreement with the surface data. Probably the best statement about the data was made in a report by the National Research Council. The report states that whether the two sets of data agree with each other or with surface data, the fact is that there is a warming trend. The amount of warming may be less important than the fact that it exists.

C. Alton Hassell

Further Reading

Chahine, Moustafa T. *AIRS/AMSU/HSB—The Atmospheric Infrared Sounder, with Its Companion Advanced Microwave Sounding Unit and Humidity Sounder for Brazil: Providing New Insights into Earth's Weather and Climate.* Greenbelt, Md.: National Aeronautics and Space Administration, Goddard Space Flight Center, 2001. Describes the infrared equipment used to determine the temperature and humidity of the atmosphere. Illustrations.

Kegawa, Seiichiro, and Tsan Mo. *An Algorithm for Correction of Lunar Contamination in AMSU-A Data.* Washington, D.C.: U.S. Department of Commerce, National Oceanic and Atmospheric Administration, 2002. Technical report that covers microwave remote sensing and radiometers, in addition to the errors caused by the Moon in the AMSU data. Illustrations, biography.

Kegawa, Seiichiro, and Michael P. Weinreb. *An Algorithm for Correction of Navigation Errors in AMSU-A Data.* Washington, D.C.: U.S. Department of Commerce, National Oceanic and Atmospheric Administration, 2002. Technical report detailing radiometry that demonstrates the change of radiant energy into mechanical work. Illustrations, biography.

Mo, Tsan. *NOAA-L and NOAA-M AMSU-A Antenna Pattern Corrections.* Washington, D.C.: U.S. Department of Commerce, National Oceanic and Atmospheric Administration, 2000. Technical report that also covers microwave remote sensing, radio meteorology, microwave antennas, and satellite meteorology. Illustrations, biography.

Wang, Likun, Cheng-Zhi Zou, and Haifeng Qian. "Construction of stratospheric temperature data records from stratospheric sounding units." *Journal of Climate* 25, no. 8 (2012): 2931-2946. A fairly technical discussion of how raw satellite data is converted into temperature records.

See also: Geographic Information Systems in Climatology; Meteorology; Weather forecasting; Weather vs. climate.

Middle East

Category: Nations and peoples

The Middle East is largely desert, with limited annual precipitation and limited usable freshwater sources. Increased consumption of Middle Eastern petroleum and related fossil fuels appears to be increasing atmospheric CO_2 concentrations, and reliance upon those resources causes the relative political instability of the region to have disproportionate effects on global socioeconomic stability.

Key concepts

desalination: Removal of soluble salts from water, usually to make it potable or suitable for irrigation

dew point: The temperature at which airborne water vapor condenses

Fresnel lenses: Small plastic sheets with very small, patterned, concentric circles stamped or milled into them that concentrate sunlight

orographic precipitation: Rainfall caused by rising topography that cools moisture-laden marine air

paleoclimatology: The scientific study of ancient climates

palynology: The study of relict pollen from ancient pollen traps

water rights: The rights of sovereign states or other legal entities to contested water supplies

Background

The Middle East comprises the sovereign states of the Eastern Mediterranean and North Africa, as well as those states along the Persian Gulf north of the Indian Ocean. Political definitions of the Middle East normally include the states of Syria, Lebanon, Israel, Egypt, Jordan, Saudi Arabia, Iraq, Yemen, Kuwait, Qatar, Bahrain, the United Arab Emirates, and Iran and the quasi-state territories of Palestine. These nations often act as a geopolitical bloc, because many of them are unified by their interests in the global oil market and their largely Muslim populations. The region is more than 92 percent Muslim, and it contains over 60 percent of known global oil reserves, as well as over 40 percent of natural gas reserves.

Petroleum Production

Middle Eastern oil reserves from both conventional and unconventional sources combined are estimated at over 750 billion barrels. In 2008, the Middle East supplied over 60 percent of all global oil production. Middle Eastern oil reserves may be divided into three categories, in a formula known as 3P: what is provable, what is probable, and what is possible. These are all based on measurable analytic estimates and hypotheses or directly known from instruments used in petroleum exploration.

Saudi Arabia is arguably the most progressive state in the Middle East in the sense that it has declared a long-term goal of preparing for a future without oil. This plan requires Saudi Arabian society to be restructured to find alternative sources of

With the end of cheap, abundant oil in sight, some Middle Eastern nations are seeking to reinvent themselves as financial and business hubs. Shown here is Dubai, United Arab Emirates, from the 456-meter (1495 foot) viewing level of the Burj Khalifa, tallest building on earth. (© Steven I. Dutch)

economic wealth and diversification to replace lost oil revenues. The Saudis seek to become self-sufficient in their ability to provide themselves with food and water, among other resources. Nonetheless, Saudi Arabia remains the largest oil producer in the world, producing over 10 million barrels per day, 12 percent of global output.

Climate change represents a significant threat to the peoples and nations of the Middle East. Warming in the region would be likely to spread desertification, and the already marginal vegetation cover would become even more threatened. Ironically, much of the regional Middle Eastern economy is overwhelmingly dependent on oil production, which is seen as one of the major contributors to greenhouse gas (GHG) emissions and global warming, because burning oil releases carbon dioxide (CO_2) into the atmosphere.

Water Desalination

The peoples of the arid Middle East have been working for decades on increasing freshwater supplies by developing desalination technology. The Middle East accounts for 75 percent of global water desalination. Saudi Arabia—the world's largest producer of desalinated water—has thirty desalination plants that supply 70 percent of national drinking water needs for a 2009 population of 29 million people, although this population is increasing. The Red Sea facility at Shoaiba, for example, a multiflash distillation operation, uses intense hot steam from a local power plant to boil out the salts from water. Its estimated freshwater production is 150 million cubic meters for 2009 and 3 billion cubic meters by 2012. Thus, satisfying the Middle East's demand for freshwater contributes directly to GHG emissions.

Israel is also a high-tech leader in desalination, with a volume of 100 million cubic meters of fresh-water produced in 2006 at the Israeli Ashkelon desalination facility alone, which uses seawater reverse osmosis (SWRO) technology. This facility, the world's largest SWRO operation, provides 13 percent of Israel's domestic water needs. Overall, the world's largest multiflash distillation operation is at Jebel Ali in the United Arab Emirates, which produces 300 million cubic meters of freshwater per year from the Persian Gulf.

If the Middle East could increase its collective water desalination efforts by 1,000 percent, it could theoretically produce enough freshwater to alter the region's climate. Such a quantity of freshwater would make it possible to sustain significant vegetation and forestation programs that could cause increased humidity and give rise to a sustainable regional hydrologic cycle. If this cycle were achieved, it could transform the desert into a sustainable agricultural region.

Major obstacles stand in the way of such a program, however. The necessary desalination technologies would be extremely energy intensive, practically requiring that they be powered by renewable energy sources such as solar and wind power to avoid depleting even the Middle East's energy reserves. Because the Middle East enjoys high levels of insolation, solar-powered desalination is not beyond the realm of theoretical possibility, and many global pioneering efforts in solar energy technologies are funded by Middle Eastern venture capital.

Water Rights

Gudea of Lagash, ruler of Neo-Sumeria circa 2100 B.C.E., made a statement four millennia ago that may resonate during the twenty-first century: "He who controls water controls life." This statement is increasingly true in the Middle East, where marginal water sources are increasingly sources of conflict and the United Nations is encountering increasing difficulties in mediating disputes.

Sovereign Middle Eastern states involved in water rights disputes include Israel, Lebanon, Syria, and Jordan, where mountain ranges of the Amana and Hermon Mountains and the Golan Heights lie in the territories of multiple countries. The Amana Mountains are primarily in Lebanon, but their Mediterranean coastal rain shadow extends to Syria. The Hermon Mountain massif is shared between Syria and Israel. The contested Golan Heights are shared between Israel and Syria.

The Jordan River watershed, issuing from the Sea of Galilee, has become a highly charged geopolitical battleground: Jordan has long claimed that the technologically adept Israel supports its agricultural needs by siphoning off at least 70 percent of the watershed above the Jordanian border, leaving little water for Jordan's own agriculture and reducing its potential self-sustenance while increasing its need to import food. If climate change renders the Middle East increasingly arid, water rights will become an even more acrimonious issue, potentially further destabilizing an already tense region where the Jewish state of Israel and its predominantly Muslim neighbors coexist uneasily, and where tensions between Sunni and Shiite Islam and between secular and religious factions and states also cause unrest.

Increasing desertification as a result of climate change is likely, unless growth in regional desalination can compensate for lowering water tables combined with higher demand for rainwater. Overconsumption of water resources (for example, by denizens of the Jordan Valley in the Levant) results in smaller bodies of water (for example, the Sea of Galilee and the Dead Sea), which increases desertification by reducing the available watershed and increases evaporation by increasing the overall land temperature.

Potential Reforestation

Portions of the eastern Mediterranean landscapes of the Levant now composing Israel were covered in antiquity by dense forests that were lost over time. Palynological studies of remnant pollen verify that this ancient hill forest comprised oak and other hardwoods, among other species. Photographs from the late nineteenth and early twentieth centuries show, however, that the coastal hills were completely denuded of trees in an arid landscape. Systematic, extensive planting of thousands of hardy pine trees, mostly fast-growing Aleppo pines acclimated to aridity, began in Israel in the 1950s.

Meteorological records kept since 1948 show that the annual average Israeli rainfall in the 1950s was around 25 to 30 centimeters. As the new pine forests matured, they cooled the surrounding air and mitigated surface temperatures, drastically altering rainfall. Moisture-laden air off the Mediterranean had previously risen over these hot hills and kept going, since the dew point was too high to cause precipitation. With cooler temperatures from forest cover, there was marked increase in orographic precipitation at lower elevations, as the dew point lowered significantly. Thus, rainfall in Israel increased dramatically over the fifty years between 1950 and 2000, reaching around 1 meter annually. This reforestation practice is thus now known to be effective and could potentially change microclimates all over the Middle East, especially where moisture-laden winds from bodies of water could be "harvested" to produce rainfall for agriculture.

Solar Energy

Middle Eastern states such as Saudi Arabia and Israel are pioneering solar energy research and experimental projects. The amount of sunlight in the Middle East exceeds that of most other global regions, so solar energy has an extremely high potential there. Near Ashdod and in the Negev area, for example, where 330 sunny days per year are normal, Israel has planned or is building some of the world's most progressive solar energy facilities, with many hectares of solar collectors placed on solar "farms," some using rotating dishes made from mirrors. One Negev site alone is expected to cover 400 hectares when completed in 2012.

Rotating-mirror solar collection can harness 75 percent of incoming sunlight, which is about five times the proportion harvestable using traditional solar panels. The technology also reduces the quantity of photovoltaic cells needed by a factor of about one thousand. Israel is planning to produce about 65 percent of its energy by such means within twenty years. Saudi Arabia also has comparable technology and goals, embodied in its Saudi Solar Village Project, where model solar villages have solar units that use Fresnel lenses to concentrate sunlight. This renewable solar energy source has great potential to reduce dependence on carbon-emitting fossil fuels and also to power desalination for agricultural needs.

Context

The Middle East is not only a geographic but also a geopolitical bloc that often operates on ideological commonalities reinforced by shared language, culture, and religion. It is predominantly Arabic in language and Islamic in culture. Additionally, its shared climate zone is overwhelmingly arid, bordering on desert, with overall annual precipitation under 20.8 centimeters. Only Lebanon and parts of Syria exceed this annual precipitation average and usually by only a small margin.

The arid Middle East as a geographic and geopolitical unit will be highly influenced by any increased global warming, because it is already under climatic stress. That the region contains a majority of the world's known petroleum reserves only complicates this economic and climatic problem, because its economic health is likely to diminish as a result of decreased global reliance on fossil fuels.

Although models by which one can infer climatic relationships are becoming increasingly sophisticated, one of the most difficult tasks ahead may be to differentiate between correlation and causation in global warming, especially given the increasing fossil fuel carbon footprint. The Middle East, with its enormous but nonrenewable energy reserves, is vital to the future of anthropogenic climate change.

Patrick Norman Hunt

Further Reading

Atlas of Israel: Cartography, Physical and Human Geography. New York: Macmillan, 1985. Documents and maps all geophysical data available in Israel from over forty years of quantitative measurements from meteorology, oceanography, demography, and other parameters.

BP [British Petroleum]. Statistical Review of World Energy, June, 2008. London: Author, 2008. Reveals that regional and global oil production fell during 2008 for the first time since 2002, mostly as a result of changing crude oil costs tied to economically driven conservation.

Hyne, Norman J. Nontechnical Guide to Petroleum Geology, Exploration, Drilling, and Production. 2d ed. Tulsa, Okla.: PennWell, 2001. Explains key concepts of geology and applied engineering relevant to the petroleum industry globally; provides useful statistics on Middle Eastern oil production.

Kemp, Geoffrey. *The East Moves West: India, China, and Asia's Growing Presence in the Middle East.* Brookings Institution Press, 2012. China and India are increasingly looking to secure energy resources in the Middle East and also expand their military and political influence.

"Middle Eastern Oil Consumption Shows Strong Growth." Bahrain Tribune, April 9, 2009. Reports that global consumption of Middle Eastern oil rose to 6.2 million barrels per day in 2007, representing a 4.4 percent increase over 2006 consumption. Examines this growth from the perspectives of both GHG emissions and Middle Eastern economic health.

"Saudi Arabia's Prince Nayef: A Rising but Enigmatic Prince." *The Economist* 391, no. 8625 (April 4–10, 2009): 53. Suggests potential trends in current Saudi Arabian succession for the royal Saudi family, acknowledging that Saudi Arabia contains 25 percent of the known global oil reserves; discusses how political succession might impact those reserves.

See also: Desertification; Iran; Kyoto Protocol; Saudi Arabia; Security; United Nations Framework Convention on Climate Change.

Milanković, Milutin
Serbian mathematician

Born: May 28, 1879; Dalj, Slavonia (now in Croatia)
Died: December 12, 1958; Belgrade, Serbia

Milanković produced a curve demonstrating the variation in intensity of summer sunlight over the past 600,000 years. Once established, the curve enabled scientific exploration of past and future relationships between solar radiation, eccentricity, and climate.

Life

Milutin Milanković is credited with developing the concept of astronomical cycles that affect global climate and the timing of ice ages. In December, 1904, he received a doctorate of technical science from the Vienna University of Technology. Late in 1909, he was offered a professorship at the University of Belgrade, where for almost a half century, he would lecture on rational mechanics, theoretical physics, and celestial mechanics. Around 1915, he began to study the astronomical explanation for the Pleistocene epoch ice age. This work examined long-term solar radiation at various latitudes and seasons to determine the influence of the astronomical input to Earth's climate. The research was completed in the 1930's, culminating in the 1941 publication of *Kanon der Erdbestrahlung* (*Canon of Insolation and the Ice Age Problem*).

Climate Work

Milanković proposed that if incoming solar radiation cyclically varies, then those variations of solar energy may be correlative with glacial and interglacial ages, which may arise from the orbital geometries between the Sun and Earth. About every 100,000 years, Earth's orbit around the Sun changes from a near-circular orbit to a slightly elliptical orbit, a variable known as eccentricity. Circular orbits are said to exhibit low eccentricity (around 0); more elliptical orbits have a high eccentricity (around 0.07).

An additional orbital cycle is the tilt (obliquity) of Earth's rotational axis relative to the ecliptic. The tilt varies from 22° to 25° over a period of forty-one thousand years and increases seasonal cycles at high latitudes, whereas lower latitudes undergo a reduced seasonal effect. The greater the obliquity, the more warmth high latitudes receive in summer, and the less they receive in winter. Only the eccentricity cycle affects the amount of solar radiation reaching Earth, so when Earth's eccentricity is at a maxim (causing less solar input) and its obliquity is at a minimum (resulting in less warmth in high latitudes), Milanković proposed that the onset of glacial conditions would occur.

An additional cyclic variation is the precession of the equinoxes, which has two components that regulate the seasonal change of the distance between Earth and the Sun. One component is elliptical precession, which relates to Earth's rotation around one of the foci of the orbit, having a periodicity of twenty-six thousand years; the other component is the "wobble" of the Earth on its rotational axis. Recall the slowing motion of a stationary top: It spins around its rotational axis, but the top of the top also wobbles (orbits). The effects of combining these two components of precession are present at the low latitudes (with periodicities of nineteen thousand and twenty-three thousand years).

With these three major variations in operation, the essence of Milanković's theory is demonstrated: Cooler summers would retard the melting of winter snow and ice; in turn, the relative mildness of winter at low latitudes would lead to sizable evaporation, which would produce abundant snowfall at middle and high latitudes. As snow and ice accumulate on land and remains, the icy surface reflects more energy back into space; over time, this snow and glacial ice can develop into continental ice sheets and mountain glaciers. Also, due to reduced insolation in the cooler, middle- and higher-latitude oceans, less greenhouse gas (GHGs) is available. Less GHG allows more long-wave radiation to leave the Earth, further cooling the planet.

Until the 1960's, Milanković's astronomical explanation for the presence or absence of glaciers was disputed because of a lack of geologic evidence. However, a complex study using climate-sensitive microorganisms in deep-sea sediments was undertaken to establish a chronology of temperature changes over the past one-half million years. The microorganisms used were foraminifera ("forams"), a protozoa that secretes a shell and whose shell, upon death, is deposited on the ocean floor, mixing with other sediments. The foram shell contains specific percentages of oxygen 18 (O^{18}) and oxygen 16 (O^{16}). Since ordinary oxygen (O^{16}) is atomically lighter than heavy O^{18}, during evaporation of ocean water O^{16} would go into ice formation in ice sheets and O^{18} would remain in ocean waters. Thus, when forams with shells rich in O^{18} were plentiful in ocean sediments, Earth was in a phase of glaciation.

Using isotope analysis, temperature variations over time can be compared. The established climate-time scale was then compared to the astronomical calculations of eccentricity, obliquity, and precession to determine whether there was a correlation. The conclusion of Milanković's mathematically complex research is that major variations in climate are closely linked to periods of obliquity, precession, and orbital eccentricity.

Research into future climate regimes investigates the relationship between solar radiation and eccentricity. Studies have demonstrated that for the next twenty-five thousand years, insolation will increase only 25 watts per square meter as received at 65° north latitude in June. Eccentricity will approach 0 for the same period, which will subdue the variations of precession. With these values (which do not consider anthropogenic changes to climate), the present interglacial cycle may continue into the future for twenty-five thousand years. Climate modeling that considers CO_2 input from anthropogenic sources and insolation variation over the next 100,000 years suggests that CO_2 concentrations over 220 parts per million volume will lead to protracted interglacial periods—about fifty thousand years into the future. An overview of most climate models confirms that future global climates will be similar to the warmest portions of the last few tens of millions of years. If the models are correct in predicting long, warm interglacial periods, the additional heat of global warming may strongly modify Earth processes and severely stress living organisms.

Mariana L. Rhoades

Further Reading

Berger, A., and M. F. Loutre. "An Exceptionally Long Interglacial Ahead?" *Science* 297 (August 23, 2002): 1287-1288. Summarizes Milanković cycles as they relate to climate. Describes the current status of those cycles and projects future glaciations.

Eyles, Nick, and Andrew Miall. *Canada Rocks: The Geologic Journey.* Markham, Ont.: Fitzhenry & Whiteside, 2007. Provides textual accounts and useful charts of Milanković variables, sea-level change, the oxygen isotope record and its correlation to the geomagnetic polarity timescale, Pleistocene glaciation, and sediments.

Hays, J. D., J. Imbrie, and N. J. Shackleton. "Variations in the Earth's Orbit: Pacemaker of the Ice Ages." *Science* 194 (December 10, 1976): 1121-1132. Geologic investigation of Milanković cycles using deep-sea foraminifera to document variations in obliquity and precession.

Petrovic, Aleksandar. "Canon of Eccentricity: How Milankovic Built a General Mathematical Theory of Insolation." *Climate Change: Inferences from Paleoclimate and Regional Aspects* (2012): 131-138. Milankovich worked at a university whose goal is fairly common today but exceptional at that time: interdisciplinarity. The university's stated goal was "acquiring unity among the sciences." Without a uniquely encouraging academic environment, he might never have made his discovery.

See also: Earth motions; Solar cycle; Sun.

Military implications of global warming

Category: Nations and peoples

Climate change is an emerging global threat that may aggravate conflicts and undermine security in many regions of the world. Climate-related impacts and shocks could trigger military responses and absorb resources for risk reduction, disaster mitigation, and conflict resolution.

Key concepts

conflict: Discord between opposing or incompatible values, interests, or actions of individual or collective actors

crisis: A potential turning point to the worse in a difficult situation

disaster: A drastic disruption in the functioning of a system with severe implications

instability: A state in which one or more normally constant, essential features of a system are influx

risk: an indicator used to evaluate possible dangers that combines possible damages and the likelihood of their occurrence

vulnerability: Susceptibility of a social system or group to pressures and stresses

Background

Global warming not only affects the lives of individual human beings but may also have larger societal consequences. By triggering a cycle of environmental degradation, economic decline, social unrest, and political instability, climate change may become a crucial issue of geopolitical security and conflict. In some parts of the world (notably in Africa, Asia, and Latin America), the erosion of social order, state failure, and violence have traditionally gone hand in hand. In the most susceptible regions, conflicts that begin within one state or between two states may spread to other neighboring states—for example, through refugee flows, ethnic links, environmental resource flows, or arms exports. Such spillover effects can destabilize entire regions and expand the geographical extent of a crisis, overstretching global and regional governance structures.

The German Advisory Council on Global Change concluded in its 2007 comprehensive analysis that, without resolute counteraction, climate change will overwhelm many societies' adaptive capacities in the near future, resulting in a level of instability that could jeopardize national and international security. In a 2008 report, the Office of the European Union High Representative and the European Commission suggested that "climate change acts as a threat multiplier, worsening existing tensions in countries and regions which are already fragile and conflict-prone." In particular, the warming of the Arctic region opens up new avenues for potential cooperation, but also for possible territorial disputes.

Policy Debate on Climate Security

In 2007, the United Nations Security Council held its first discussion on the security risks of climate change, and the U.N. secretary general warned that climate change may pose as much of a danger as war. With its 2007 joint award of the Nobel Peace Prize to Al Gore and the Intergovernmental Panel on Climate Change (IPCC), the Nobel Prize Committee emphasized that extensive climate change could result in "increased danger of violent conflicts and wars, within and between states." A panel of experts that included, among others, former U.S. director of central intelligence James Woolsey and Nobel Laureate Thomas Schelling asserted that climate change "has the potential to be one of the greatest national security challenges that this or any other generation of policy makers is likely to confront."

Water Stress and Conflict

Climate change will likely exacerbate water scarcity for hundreds of millions of people. Uneven water distribution may induce migration or the quest for resources from neighboring regions. Individual case studies suggest that water scarcity undermines human security and heightens competition for water and land resources. According to Peter Gleick's chronology, water has been a factor in at least forty-two violent conflicts since the beginning of the twentieth century, but only in a few cases was the exchange of fire involved. Incontrast, the Basins at Risk project at Oregon State University has concluded that water scarcity does not increase the likelihood of interstate conflicts but rather strengthens cooperation through transboundary water agreements and institutions. Quite likely, the outcome depends on the severity of the conflict, the stability of the neighboring states, and their willingness to cooperate.

In the Middle East, water scarcity is intertwined with the region's general conflicts. The arid climate, the imbalance between water demand and supply, and the ongoing confrontation between key political actors exacerbate the water crisis of the Nile, Euphrates, and Jordan Rivers. Global warming—together with population growth, overexploitation, and pollution—is projected to increase the likelihood and intensity of droughts in the region, undermining the conditions for peace and human security. However, statements on "Water Wars" in the Middle East have been questioned. The region's conflicts are largely determined by

political differences, where hydrological matters represent an additional dimension of conflict as well as cooperation. Further progress of the water talks is connected to the fate of the Middle East peace process.

Central Asia is another region vulnerable to water conflicts, and the IPCC projects a sharp temperature rise in that region. Agriculture largely relies on irrigation and accounts for 20 to 40 percent of the gross domestic product of most central Asian nations. Electricity in the region is based almost completely on hydroelectric power, which depends on glacier melt water from mountain ranges. Some of the glaciers in the region have already declined in the past decades, and by 2050 about 20 percent of the glaciers in some mountains may disappear. The states of central Asia are characterized by largely closed markets, extreme social disparities, and weak state structures, making them unable to cope with these changes. Struggles over land and water resources have already played a major role in this region, and they have been aggravated by ethnic disputes, separatist movements, and religious fundamentalist groups.

Land-Use Conflicts and Food Insecurity

Reduction of arable land, water shortages, diminishing food and fish stocks, increased flooding, and prolonged droughts already threaten food security in many parts of the world, and climate change will aggravate this trend. With global warming, a drop in agricultural productivity is anticipated that will be reinforced by desertification, soil salinization, and water scarcity. More frequent extreme weather events may trigger regional food crises and further undermine the economic performance of weak and unstable states, thereby exacerbating destabilization, the collapse of social systems, and violent conflicts.

Particularly vulnerable will be Africa's food production, which has been in decline per capita for more than twenty years. Around 34 percent of the African population lives in arid regions, and about one-third of the population in sub-Saharan Africa is malnourished or undernourished. By 2020, in some African countries, yields from rain-fed agriculture could decline by as much as 50 percent, severely compromising agricultural production and access to food. Food crises impair the livelihoods of subsistence farmers and increase unemployment and migration for millions of people. As a result of migration from rural to urban areas, slums in African cities grow and become breeding grounds for crime and violence. Growing numbers of marginalized people could join riots and armed rebel groups, possibly culminating in civil war and ethnic conflict.

An example of ethnic conflict aggravated by resource scarcity is the 1994 genocide in Rwanda. In Rwanda, soil degradation, population growth, and unequal land distribution gave radical forces an opportunity to escalate ethnic rivalries into a political power struggle. Elsewhere in Northern Africa, a series of droughts caused Arab herders to move into the more fertile areas of Darfur, where grazing cattle trampled farmers' fields, contributing to existing clashes and tensions. The Sudan Post-Conflict Environmental Assessment of the U.N. Environmental Programme in 2007 concluded that Darfur is a "tragic example of the social breakdown that can result from ecological collapse."

Natural Disasters

The IPCC projects that many areas of the globe will experience more frequent and intense extreme weather events and natural disasters, including droughts, heat waves, wildfires, flash floods, and storms. Disasters have dramatic impacts on human lives, generate rising economic and social costs, cause large numbers of fatalities, and temporarily impair or collapse state functions. Regions at high risk from storm and flood disasters generally have weak economic and political capacities, making adaptation and crisis management more difficult. Storm and flood disasters along the densely populated east coasts of India and China could cause major damage and trigger large migration processes. Developed countries are also vulnerable to natural disasters, as was seen during the 2003 heatwave in Europe, when more than thirty-five thousand people died and agricultural losses reached $15 billion.

The record hurricane season of 2005 demonstrated that even the world's most powerful nation is vulnerable and unable to cope with natural disasters. When Hurricane Katrina hit the U.S. Gulf Coast with wind speeds of up to 230 kilometers per hour, it left a trail of destruction over an area as large as the British Isles. In the Gulf of Mexico, 90 percent of oil refinery capacity had to be shutdown. When New Orleans was flooded, over fifteen hundred people lost their lives, and hundreds of thousands fled their homes. The Earth Policy Institute in Washington, D.C., called this outflux "the first documented mass movement of climate refugees." The city's entire infrastructure was devastated, including water, food, energy, transportation, communications, and sanitation. Public order broke down. Most vulnerable were those living in poor-quality housing in high-risk areas and having few financial resources and no insurance to cope with disasters.

Environmental Migration

In response to environmental degradation and weather extremes, or their indirect consequences such as economic decline and conflict, people will be forced to leave their homelands for other regions. Most vulnerable are high-risk climate hot spots, especially coastal and riverine areas and areas whose economies depend on climate-sensitive resources. Although most of the affected people in the Southern Hemisphere will remain within their national borders, industrialized regions face substantially increased migratory pressure—Europe from sub-Saharan Africa and the Arab world, and North America from the Caribbean and Central and South America.

The potential pressure on China to resettle large populations from flooded coastal regions or dry areas may put migration pressure on neighboring countries, including Russia. Migration of people can increase the likelihood of conflict in transit and target regions where migrants have to compete with the resident population for scarce resources such as land, accommodation, water, employment, and basic social services. Immigrants are perceived as competitors who change the "ethnic balance" in the region. Beginning in 2015, Europe experienced a large influx of refugees from Africa and the Middle East fleeing warfare and economic hardship. Globally, the number of displaced persons reached almost 60 million, the largest number since World War II. The influx of refugees contributed to a rise in internal political tensions within Europe, especially the rise of political parties demanding an end to immigration and even their countries seceding from the EU.

Populated mega-deltas in southern and eastern Asia will be at greatest risk due to increased flooding from the sea and, in some mega-deltas, from rivers as well. Climate change would significantly aggravate human insecurity in Bangladesh, one of the poorest and most densely populated countries of the world. Since 1960, about 600,000 persons have died as a result of cyclones, storm surges, and floods. Improved warning systems and shelters have drastically lowered the number of such deaths in recent years. The impacts of projected sea-level rise could be disastrous, threatening the Bangladeshi economy and exacerbating insecurity.

A 1-meter increase in the sea level could inundate about 17 percent of Bangladesh and displace some 40 million people, according to the World Bank. On several occasions, the migration of impoverished people has already caused violent clashes within Bangladesh and between emigrating Bangladeshi and tribal people in northern India, where several thousand people have died. The complex interaction of both anthropogenic and natural trends and their socioeconomic and political implications may further lead to situations of political instability and violent clashes, undermining young democratic institutions.

Context

The security risks of climate change are determined by the causal links among climate stress factors, human impacts and responses, and societal instabilities. Whether societies are able to cope with the impacts and restrain the risks depends on their responses to change and their ability to solve and moderate associated instabilities and conflicts. While a gradual temperature rise of several degrees will already severely affect national and international security, abrupt and large-scale climate change beyond critical tipping points (for example, the collapse of the North Atlantic thermohaline circulation, the loss of the Amazon rainforest or of the South Asian monsoon, and the melting of the Greenland or the western Antarctic ice sheets

with several meters of consequent sea-level rise) will likely have catastrophic consequences that could be comparable to major wars. Addressing the problem will require integrated approaches that combine climate and security policy in a mutually enforcing way. Adaptive strategies for mitigation and adaptation are needed that minimize security risks and mitigate conflicts by strengthening institutions, economic well-being, energy systems, and other critical infrastructures.

Jürgen Scheffran

Further Reading

Campbell, K. M. The Age of Consequences: The Foreign Policy and National Security Implications of Global Climate Change. Washington, D.C.: Center for Strategic and International Studies, 2007. Provides a U.S. perspective on emerging national security threats, which are evaluated in the framework of three scenarios of global temperature change.

Scheffran, J. "Climate Change and Security." Bulletin of the Atomic Scientists, May/June, 2008, 19–25. Provides an overview of contemporary and historical perspectives on the climate-security link. U.N. Development Programme. Human Development Report, 2007: Fighting Climate Change. New York: Author, 2007. Covers the link between climate shocks and human development and discusses cooperative strategies to avoid dangerous climate change.

Scheffran, Jürgen, and Antonella Battaglini. "Climate and conflicts: the security risks of global warming." Regional Environmental Change 11, no. 1 (2011): 27–39. Military conflicts likely to arise as a result of global warming will revolve around the classic causes of war: resources and territory. They include water stress, land use and food security, natural disasters and environmental migration. Conflict is likely to become more frequent and severe as climate stresses increase.

See also: Agriculture and agricultural land; Displaced persons and refugees; Drought; Extreme weather events; Famine from global warming; Floods and flooding; Human migration; Tropical storms; Water resources, global; Water resources, North American; Water rights.

Minerals and mining

Category: Economics, industries, and products

Definition

Minerals are naturally occurring inorganic substances with definite chemical compositions and characteristic physical properties. A rock is an aggregate of minerals. Minerals range in composition from native elements and metals, sulfides, halides, and carbonates to complex silicates, which are the most common rock-forming substances. Contemporary society depends on the availability of mineral resources: Metallic minerals such as iron, copper, aluminum, lead, and zinc; nonmetallic minerals such as salt, clay, gypsum, soil, and water; and energy resources such as coal, petroleum, natural gas, uranium, and palladium have all become necessary to the functioning of human civilizations.

With few exceptions, minerals are generally nonrenewable (exhaustible), since mineral resources generally have required long expanses of geologic time to develop and thus are present in effectively fixed amounts in the Earth. Minerals are extracted from the Earth by a variety of mining methods when the resource is economically viable. As world population increases exponentially and that population aspires to and achieves middle-class socioeconomic status, an escalating resource crisis has begun to develop that propels the mining of nonrenewable mineral resources. This industrial mining is coupled with many environmental concerns.

It is believed that Earth may have reached its maximum capacity to absorb environmental degradation related to mining. Potential environmental impacts of mineral mining depend on factors such as mine waste management, mining procedures, local hydrology, climate, rock types, size of operation, and related factors. Mining disrupts the landscape and can instigate mine subsidence, disrupting biological and water resources. Underground mining is generally more hazardous as a result of poorer ventilation and visibility and slope instability along mine walls. Moreover, the dust and toxic gases in mines lead to severe respiratory problems, and the possibility of exposure to radiation poses serious health threats.

Through mining processes, large amounts of material accumulate as waste that needs to be disposed of. Many copper mines, for example, extract ore that contains less than 1 percent copper. For many nonferrous metals, almost all of the mined ore becomes waste. Artisanal mining, such as alluvial mining for gold and diamonds, often has impacts on landscape, which is disrupted by trenches. These activities can lead to erosion and localized destruction of river banks. The waste also contains dangerous substances, such as heavy metals, which leach into the soil and result in the generation of acid or alkaline mine drainage.

The sulfide-containing minerals in metal mining get oxidized in the presence of air and react with water to form sulfuric acid. The acid mine drainage water contaminates surface and underground water. With the proliferation of the petroleum industry, numerous large-scale oil spills are becoming usual industrial accidents, and oil percolates through the soil and pollutes groundwater. The fuels and chemicals used in the mining industry are potential pollutants too. These chemicals left behind by explosives are usually toxic, and they increase the salinity of water. Small-scale artisanal mining may also affect water where mercury is used to process gold. Excavation and removal of raw ore are only the initial stage in the mining process. The ores are processed at refineries, and the valuable portion is extracted by flotation, gravity, or chemical methods. The by-products of mineral refining are sulfur, arsenic, and radioactive substances that are dangerous if they are released into the environment.

Significance for Climate Change

The mineral mining industries contribute to the global climate. Fossil fuels are used to generate the energy required for moving mining equipment, mining procedures, ore processing and drying, transportation, and building operations. Burning them generates greenhouse gases (GHGs). The mining industry consumes large amounts of electricity to transport material by huge vehicles or extensive hoisting systems for underground mines. The underground mines become very warm with time, and cooling of deep underground mines is energy intensive. Refining metal ores by smelting also requires large amounts of energy. Surface mines pollute the air through blasting operations. Coal mines release methane, which is a primary GHG. However, methane can be captured through expensive processes to reduce the enhanced greenhouse effect.

Some members of the mining industry use ozone-depleting gases such as chlorofluorocarbons, hydrochlorofluorocarbons, and hydrofluoro-compounds for cooling. Similar to coal mining, the petroleum industry also has severe impacts on climate. The production and use of oil and natural gas makes a significant contribution to global warming by increasing atmospheric carbon dioxide (CO_2) concentrations. Additionally, mining operations cause extensive deforestation, one of the major changes in landscape that lead to increased CO_2 concentration in the atmosphere and promotes warming.

All these impacts can have long-term environmental and socioeconomic consequences and will be extremely difficult and expensive to address through remedial actions. Therefore, the mining industries are moving toward meeting standards of air and water quality set by Environmental Protection Agency (EPA) and other government agencies. At present, wastes from the extractive industries are properly managed in order to ensure the long-term stability of disposal facilities and to minimize water and soil contamination arising from acid or alkaline drainage and leaching of heavy metals.

The coal industries are participating in the Clean Air Acts (1963-1990) and controlling methane emissions, although this is not economically feasible in many locations. Methane is removed from coal mines through degasification systems or mine ventilation systems during mining activities, or after mining has occurred. Over the 1990-2000 time period, the EPA reports that recovery of coal mine methane resulted in a reduction of methane emissions by 30 percent or by the equivalent of approximately 25 million metric tons of CO_2 per year. The terrestrial and geological sequestration of carbon and improvement of carbon uptake of soils are areas of research in mining industries that can reduce anthropogenic CO_2 significantly. Moreover, the mining industries

are promoting clean coal technologies to refine emissions reduction. The final goal is the development and implementation of zero-emissions mining industries. This requires both time and research dollars, and several organizations are participating and investing in this cutting-edge research endeavor, along with the potential for emissions savings.

Arpita Nandi

Further Reading

Caldeira, Ken, et al. "A Portfolio of Carbon Management Options." *Journal of Geophysical Research* 110 (2005). Discusses the options available to help stabilize the radiative forcing from GHGs and other atmospheric constituents.

Hoffert, Martin I., et al. "Advanced Technology Paths to Global Climate Stability: Energy for a Greenhouse Planet." *Science* 298, no. 5595 (November, 2002): 981-987. Report of a group of researchers that surveyed possible future energy sources, evaluating them for their capability to supply massive amounts of carbon-emission-free energy and for their potential for large-scale commercialization.

Raupach, Michael R., et al. "Global and Regional Drivers of Accelerating CO_2 Emissions." *Proceedings of National Academy of Science* 104 (2007). Focuses on trends in emissions and their demographic, economic, and technological drivers, using annual time-series data on national emissions, population, energy consumption, and gross domestic product. Understanding the magnitudes and patterns of CO_2 emissions will help scientists predict future climate changes.

Valero, Antonio, Andrés Agudelo, and Alicia Valero. "The crepuscular planet. A model for the exhausted atmosphere and hydrosphere." *Energy* 36, no. 6 (2011): 3745-3753. "Crepuscular" means "pertaining to twilight," and refers to an attempt to model the earth after all resources have been extracted.

See also: Arctic seafloor claims; Coal; Energy resources and global warming; Fossil fuels and global warming; Lithosphere.

Modes of climate variability

Categories: Climatic events and epochs; meteorology and atmospheric sciences

Modern technologies, particularly satellites and supercomputers, have allowed scientists to view climate from a global perspective, revealing recurrent patterns of climate parameters that affect large areas of the planet over varying periods of time, often years or decades.

Key concepts

El Niño-Southern Oscillation (ENSO): a coupled oceanic/atmospheric seesaw that occurs in the equatorial Pacific but often has global climatic consequences

modes: phases of a climatic seesaw; for example, El Niño is the warm mode of the ENSO seesaw, whereas La Niña is the cold mode

North Atlantic Oscillation (NAO): a seesaw in pressure between the Azores and southwestern Iceland, thought by some scientists to be an expression of the Northern Annular Mode

Northern Annular Mode (NAM) and Southern Annular Mode (SAM): also called, respectively, the Arctic and Antarctic Oscillations, seesaws in pressure between the latitudes near 45° northern (or southern) latitude and the North (or South) Pole

Pacific Decadal Oscillation (PDO): a temperature, pressure, and wind seesaw in the Pacific Ocean

Pacific-North American (PNA) pattern: a seesaw between northern Pacific and North American pressures

regimes: another word for "modes," fitting in with meteorological metaphors such as "fronts"

seesaw: a change in opposite directions, such as high pressure in one region and low pressure in the other

teleconnection: a connection between two widely separated regions of the planet that have highly correlated changes in some climatic parameter, usually resulting from a seesaw

Background

A college town with 2,000 residents that hosts 10,000 students for eight months every year has an average population of 8,666, but it will almost never have that number present. In "academic year" mode, it has a population of 12,000, and in "vacation" mode, it has 2,000. The climate also has modes, and its behavior varies dramatically between them. Temperatures, rainfall, winds, and other climatic phenomena during El Niño are very different from those during La Niña. Recognizing these modes is essential to understanding how the climate operates.

Seesaws

Modes of climate variability are often referred to as "seesaws" because what is missing from one region (such as warmth, atmospheric pressure, or precipitation) is found in excess in the other region. Such seesaws result from the fact that the atmosphere is a finite body of gas that obeys the laws of physics.

By 1924, a number of seesaws had been identified by Sir Gilbert Walker. The data set he presented in 1932 had 183 stations widely spaced across the globe, with multiyear records that permitted statistical analysis. He put his North Atlantic Oscillation (NAO) and North Pacific Oscillation (NPO) on a statistical footing, detailing the strength of correlations between what he called "action centers." He also established the existence of the Southern Oscillation, which he defined in terms of a pressure seesaw. By the 1960's others had shown that this coincided with a pattern of sea surface temperature fluctuations called El Niño, and so it is now known as the El Niño-Southern Oscillation (ENSO).

Additional workers found more seesaws, and often an index was determined by combining the values of some climatological variable at two or more locations in a simple algebraic way: The Southern Oscillation Index was obtained by subtracting the sea-level barometric pressure at Darwin, Australia, from that at Tahiti; the North Atlantic Oscillation Index was obtained by subtracting the sea-level pressure at Iceland from that at the Azores.

Statistical Identification of Seesaws

Since Walker's work, the data series from many of his stations have been extended by more than seventy years, hundreds of new stations have been established, and satellite and other remote-sensing techniques have contributed immense amounts

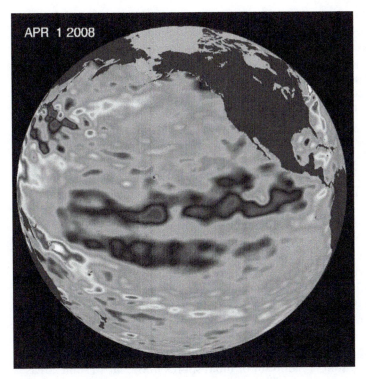

APR 1 2008

The dark area above reveals a cooling trend corresponding to La Niña and the cool phase of the Pacific Decadal Oscillation. (NASA-JPL)

Context

Teleconnections show that Earth's climate is not entirely random. Spatial patterns exist, and the polarity of the seesaws within these patterns alternates, often with far-reaching consequences. What is less well known is the temporal behavior of these patterns, what causes them to switch polarity, and how they interact.

ENSO is the shortest and best-known seesaw, having an average period of four years, but this period can vary from two to seven years. Efforts to explain why its period should change, or to predict how long a particular El Niño or La Niña will last, have so far been unsuccessful.

The PDO has effects that are geographically similar to those of the ENSO but has a period of twenty to fifty years. Cool before 1924, then warm until 1947, cool again until 1976, and warm again until at least 1998, it has had ambiguous behavior since then. In addition to not knowing why it reverses or when it might reverse again, climatologists do not know whether the PDO causes increased ENSO fluctuations or is caused by them.

Scientists' understanding of the modes of climate variability is incomplete, but most climate scientists agree that they play an important role over periods of years to decades. Our ability to interpret climate data correctly depends on being able to place them in context with respect to these modes. Predictions of climate change would improve if it were possible to predict when these modes will reverse and how strong they would be.

Otto H. Muller

of climate information. New statistical techniques involving eigenvector analysis have been developed to analyze these data, particularly Principal Component (PC)/Empirical Orthogonal Function (EOF) analysis.

These techniques take data sets, which can be enormous, and rearrange them into separate, independent components that reveal how the data points are linked. As an example, consider the pressure data for points in the Northern Hemisphere, but outside the tropics (that is, at latitudes greater than 20° north). Eigenvector analysis finds that much of the variability in these data can be explained by two regional seesaws, EOF1, and EOF2. A map of which regions are controlled by each EOF shows that EOF1 corresponds to the NAM, and EOF2 corresponds to the PNA. Because this representation of PNA uses criteria that differ from its original index-based definition, it is often referred to as PNA.

Further Reading

Alley, R. B., et al. *Abrupt Climate Change: Inevitable Surprises.* Washington, D.C.: National Academy Press, 2002. In the section on modes of climate variability (pages 48-69), this book presents an excellent overview of what is known, and what remains to be learned, about teleconnections and their influence on twentieth century climate.

Büntgen, Ulf, Willy Tegel, Kurt Nicolussi, Michael McCormick, David Frank, Valerie Trouet, Jed O. Kaplan et al. "2500 years of European climate variability and human susceptibility." *Science* 331, no. 6017 (2011): 578-582. Although there are many suggestive coincidences between historic and climatic events, establishing causality is difficult because the climatic data are not detailed enough.

Collier, Michael, and Robert H. Webb. *Floods, Droughts, and Climate Change.* Tucson: University of Arizona Press, 2002. Written for the layperson, this book provides a very accessible explanation of many teleconnections, with an emphasis on how they affect people.

Rohli, Robert V., and Anthony J. Vega. *Climatology.* Sudbury, Mass.: Jones & Bartlett, 2008. With excellent images and clear descriptions, this textbook offers an outstanding examination of the various modes of climate variability. The treatment of statistical methods, particularly those involving eigenvalues and principal components, provides a good feel for how they work without requiring the reader to become enmeshed in equations and linear algebra.

See also: Atlantic multidecadal oscillation; Climate and the climate system; Ekman transport and pumping; El Niño-Southern Oscillation; La Niña; North Atlantic Oscillation (NAO); Ocean-atmosphere coupling; Rainfall patterns; Thermohaline circulation.

Mold spores

Category: Plants and vegetation

Definition

Mold spores are reproductive structures of filamentous fungi (molds). A single microscopic mold filament, called a hypha, forms a mat that is called a mycelium. Mycelia are visible without a microscope. Molds are very common organisms and can be found where there is moisture, oxygen, and food they need. Molds can be seen on bread, cheese, or fruit. Hot spots of mold growth can be found in basements and bathrooms (especially shower stalls), house plants, and even air conditioners. Molds grow on fallen leaves, rotting logs, certain grasses, and weeds. They also can be found in barns, dairies, bakeries, and greenhouses.

The mold mycelium produces reproductive branches above the surface of the mold. These branches carry spores called conidia that function in distribution of mold by air, water, and animals. Among different molds, spores—employed in asexual reproduction—vary in size, shape, and color. Each spore can germinate to start a new mold, which in turn produces million of spores. Spores are very tough structures: they are resistant to drying, freezing, heating, and some chemicals.

The majority of the mold spores are disseminated by air. A sample of air may contain up to 2 million spores per cubic meter, but on average, about 10,000 spores inhabit one cubic meter of air. The amount of mold spores in the air in some areas is greater than the amount of pollen. Certain types of mold spores can cause various allergic reactions in humans, such as irritations of the eyes, nose, and throat. About 20 to 30 percent of the population develops allergic responses after exposure to these mold spores. The most common allergenic spores in the United States are *Alternaria cladosporium, Aspergillus, Fusarium, Mucor, Rhizopus,* and *Penicillium.*

In some people, exposure to mold spores leads to asthma. Some mold spores, if they reach lungs, can cause infections called mycoses. Systemic mycoses are the most serious category of mold infection. The host becomes infected by inhaling spores that germinate in the lungs. In the United States, two of the most common mold infections of that type are coccidioidomycosis, caused by *Coccidioides immitis,* and histoplasmosis, caused by *Histoplasma capsulatum.* Mild coccidioidomycosis may go unnoticed or produce symptoms similar to those of pneumonia or tuberculosis. The human immune system normally destroys mold spores and neutralizes mold infections. In a small number of cases, however, more serious coccidioidomycosis develops and lesions of the skin, bones, joints, internal organs, and brain (meningitis) occur. Progressive histoplasmosis symptoms include lung cavities, sputum production, night sweats, and weight loss.

Significance for Climate Change

The weather and mold-spore distribution are closely related. Spore count is usually higher in temperate and tropical regions than in the polar and northern regions. In colder climates, molds are present in the air during the period between late winter and late fall. In warmer climates, mold spores are found throughout the year. It is likely that warmer temperatures due to global warming will result in an increase and even abundance of mold spores and, therefore, in considerable increase of allergic reactions. Repeated exposure to a massive amount of mold spores (100 million per cubic meter) can cause serious allergy-related health problems, including chills, fever, dry cough, breathlessness, weight loss, and even permanent lung damage.

Global warming is believed to be a major factor in the explosion of mold-related asthma and mold infections. For instance, the causative agent of coccidioidomycosis can be found in geographical areas with high summer temperatures and mild winters. In the southwestern regions of the United States, where this climate prevails, an estimated 80 percent of inhabitants are currently infected. Infectious disease specialists suggest that global warming will cause the further expansion of the geographic ranges of coccidioidomycosis infection.

Scientists predict that climate change could also increase the spread of histoplasmosis, which at present afflicts about 500,000 people annually. Another example of mold-spore infection spreading as a result of climate change is the infection caused by *Cryptococcus gattii*. Though previously it was only seen in Australia and other subtropical regions, this mold is spreading in Canada's Vancouver Island and the Pacific Northwest. It can cause serious human infection of the lungs and brain.

In addition, molds are the cause of numerous plant diseases. The increase of plant fungal diseases due to global warming may have a negative impact on plants' ability to take up carbon dioxide (CO_2), a greenhouse gas, thereby increasing the CO_2 concentration of the atmosphere and contributing to further warming. Managing mold plant infections may also require pesticides whose production consumes fossil fuels and generates even more CO_2 emissions.

There is another indirect relationship between mold spores and climate change. In nature, many molds are capable of decomposing woody plants such as trees. Cellulose and lignin in these trees are the biological molecules most resistant to decomposition. Molds, however, use cellulose and lignin from woody plants as their source of energy and carbon, and they release CO_2 in the process. Trees function as carbon sinks, retaining carbon for the duration of their lives and sequestering it from the atmosphere. As global climate change increases the amount of mold spores and, eventually, molds, the CO_2 released by decomposition of woody plants by those molds will also increase. Eliminating mold spores is impractical. Therefore, the only solution to keep molds under control is to control global warming.

Sergei Arlenovich Markov

Further Reading

Beggs, Paul J. "Impacts of Climate Change on Aeroallergens: Past and Future." *Clinical and Experimental Allergy* 34, no. 10 (October, 2004): 1507-1513. Describes the link between mold spores, climate change, and asthma.

Madigan, Michael T., et al. *Brock Biology of Microorganisms.* 12th ed. San Francisco, Calif.: Pearson/Benjamin Cummings, 2009. Introductory microbiology textbook. Chapter 18 describes molds.

Moore-Landecker, Elizabeth. *Fundamentals of the Fungi.* 4th ed. Upper Saddle River, N.J.: Prentice Hall, 1996. Comprehensive textbook about fungi and molds.

Shah, Rachna, and Leslie C. Grammer. "An overview of allergens." In *Allergy and Asthma Proceedings*, vol. 33, no. 3s, pp. S2-S5. OceanSide Publications, Inc, 2012. Survey of how allergens act chemically and physiologically. Reactions depend on the avenue of ingestion and the mode of transport.

See also: Carbon 4 plants; Carbon 3 plants; Composting; Mangroves; Sequestration.

Monsoons

Category: Meteorology and atmospheric sciences

Definition

Monsoons are seasonal changes in surface wind and precipitation patterns over the tropical and subtropical continents and surrounding oceans. These changes are due to the differences in thermal properties between land and ocean that give rise to different responses to seasonal changes in insolation.

The specific heat of land is typically less than that of ocean. That is, it requires less thermal energy to raise the temperature of a given amount of land by a given amount than it does to raise the temperature of the same amount of ocean by the same amount. As a result, the summer Sun warms the land more quickly and to a greater extent than it warms the oceans, which remain relatively cool. Typically, low-pressure weather systems form over warm surfaces and high-pressure systems form over cool surfaces. Thus, in the summertime, landmasses often become centers of low pressure, whereas the adjacent oceans become centers of high pressure. Wind blows from high-pressure centers to low-pressure centers, so in summer winds typically blow from the ocean to the land. This phenomenon is called "wind convergence" over land.

When wind converges over land, convection causes clouds to form. Precipitation typically follows the development of these clouds. Therefore, summer monsoons are characterized by winds blowing from the ocean to the land, over which clouds then form and rain falls.

In the winter, the entire process reverses. The lower specific heat of land, which caused it to heat more quickly than the oceans in the summer, causes it to cool more quickly than the oceans in the winter. As insolation decreases, the land quickly loses heat, becoming cold relative to the oceans, which retain much of their heat. As a result, the oceans become centers of low pressure, landmasses become centers of high pressure, and winds begin to blow from the land to the oceans. Clouds and precipitation also move from land to the oceans.

The contrast between the thermal properties of land and seawater is the basic cause of monsoons, but certain topographic features can enhance this effect. Larger landmasses and higher-altitude land surfaces increase the thermal and pressure differentials between land and water. As a result, the world's strongest monsoons are all related to the world's largest mountain ranges: The East Asian and South Asian monsoons are related to the Tibetan Plateau, the North American monsoon is related to the Rocky Mountains, and the South American monsoon is related to the Andes Mountains.

While monsoons are thus caused by contrasts between land and ocean, the atmosphere plays a crucial role in the system as well. The atmosphere couples with both land and sea to mediate and react to the thermal differences between them. It forms different pressure systems over the two bodies based on their temperature differential, reversing the direction of the winds. It also transports large amounts of water from sea to land, first by absorbing evaporated water vapor and then by precipitating the water onto the land.

Significance for Climate Change

Monsoons form a central part of many regional climate systems. In many parts of the world, monsoonal precipitation constitutes a major rainfall system, providing water resources for regional ecosystems. Many tropical and subtropical climates are dependent upon the monsoonal rains.

Any significant climate change will inevitably affect global monsoon circulations, causing changes in the patterns and intensity of monsoonal rainfall. However, it is a matter of debate among climatologists as to whether global warming will intensify or weaken the monsoons. On one hand, generally warmer climates may reduce the temperature contrast between land and ocean, resulting in weaker monsoons. Evidence to support this hypothesis exists in studies of Himalayan ice cores. Data for the last three hundred years indicate that for every 0.1° Celsius increase in temperature in the Himalayas, there was a decrease of about 100 millimeters in monsoonal rainfall.

On the other hand, global warming may enhance evaporation from the ocean's surface, because warmer air holds more water vapor than does cooler air. An increase in the amount of atmospheric water vapor would likely lead to greater monsoonal rainfall and stronger monsoons. Data from the Tibetan Plateau—where Earth's strongest monsoon is located—indicate that over the last fifty years, total atmospheric water vapor content and total surface rainfall both increased.

Some scientists argue that, rather than strengthen or weaken monsoons, global warming might simply redistribute rainfall without significantly altering average amounts. Such redistribution would entail more severe events at both extremes, as heavy precipitation and flooding would occur in some areas and drought would occur in others. Thus, life in monsoonal areas is likely to change signficantly if global warming continues, but the nature of those changes remains uncertain.

Chungu Lu

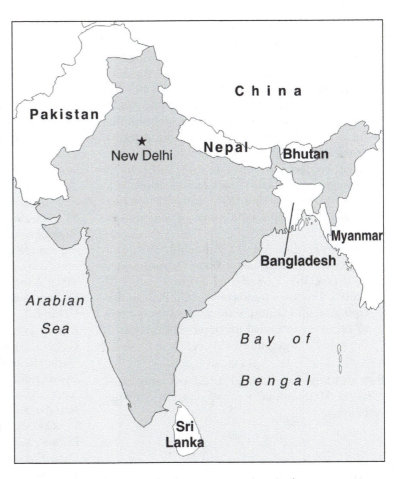

The monsoons determine the beginning and end of rainy seasons in many parts of the world, but particularly in the Indian subcontinent. The map above illustrates the progression of rains northward during a typical year, as the summer monsoon moves onto the subcontinent, bringing moist air from the south.

Further Reading

Ahrens, C. Donald. *Essentials of Meteorology: An Invitation to the Atmosphere.* 5th ed. Belmont, Calif.: Thomson Brooks/Cole, 2008. Widely used introductory textbook on atmospheric science; covers a wide range of topics on weather and climate.

Cherchi, Annalisa, Andrea Alessandri, Simona Masina, and Antonio Navarra. "Effects of increased CO2 levels on monsoons." *Climate dynamics* 37, no. 1-2 (2011): 83-101. Global warming increases atmospheric water vapor but that does not necessarily increase precipitation. In fact, in a series of modeling experiments, it did not.

Duan, Keqing, Tandong Yao, and Lonnie G. Thompson. "Response of Monsoon Precipitation in the Himalayas to Global Warming." *Journal of Geophysical Research* 111 (2006). Examines historical ice core data and finds that Indian monsoon precipitation decreased in response to an increase in surface temperature.

Intergovernmental Panel on Climate Change. *Climate Change, 2007—Synthesis Report: Contribution of Working Groups I, II, and III to the Fourth Assessment Report of the Intergovernmental Panel on Climate Change.* Edited by the Core Writing Team, Rajendra K. Pachauri, and Andy Reisinger. Geneva, Switzerland: Author, 2008. Comprehensive overview of global climate change published by a network of the world's leading climate change scientists under the auspices of the World Meteorological Organization and the United Nations Environment Programme.

Kripalani, R. H., A. Kulkarni, and S. S. Sabade. "Indian Monsoon Variability in a Global Warming Scenario." *Natural Hazards* 29 (2003): 189-206. Examines Indian monsoon variability under global warming conditions and suggests that there is no clear evidence of changes in Indian monsoon rainfall.

Lutgens, Frederick K., and Edward J. Tarbuck. *The Atmosphere.* 10th ed. Upper Saddle River, N.J.: Pearson Prentice Hall, 2007. Introductory textbook that covers a wide range of atmospheric sciences.

Xu, Xiangde, Chungu Lu, Xiaohui Shi, and Shouting Gao. "World Water Tower: An Atmospheric Perspective." *Geophysical Research Letters* 35 (2008). Discusses trends in Tibetan Plateau precipitation and atmospheric water vapor content in response to global warming. The plateau is the world's largest monsoon region.

See also: Average weather; El Niño-Southern Oscillation; Meteorology; Rainfall patterns; Weather vs. climate.

Montreal Protocol

Category: Laws, treaties, and protocols
Date: September, 1987

The Montreal Protocol initiated a process to restore Earth's ozone layer, which protects the planet's inhabitants from dangerous ultraviolet rays. It has been ratified by most of the world's nations.

Participating nations: *1988:* Belarus, Belgium, Canada, Denmark, Egypt, European Community, Finland, France, Germany, Greece, Ireland, Italy, Japan, Kenya, Luxembourg, Malta, Mexico, Netherlands, New Zealand, Nigeria, Norway, Portugal, Russian Federation, Spain, Sweden, Switzerland, Uganda, Ukraine, United Kingdom, United States; *1989:* Australia, Austria, Burkina Faso, Cameroon, Fiji, Ghana, Guatemala, Hungary, Iceland, Jordan, Liechtenstein, Malaysia, Maldives, Panama, Singapore, Sri Lanka, Syrian Arab Republic, Thailand, Trinidad and Tobago, Tunisia, United Arab Emirates, Venezuela; *1990:* Argentina, Bahrain, Bangladesh, Brazil, Bulgaria, Chile, Ecuador, Gambia, Iran, Libyan Arab Jamahiriya, Poland, South Africa, Zambia; *1991:* Botswana, China, Costa Rica, Malawi, Philippines, Togo, Turkey, Uruguay; *1992:* Algeria, Antigua and Barbuda, Barbados, Croatia, Cuba, Cyprus, El Salvador, Guinea, India, Indonesia, Israel, Republic of Korea, Kuwait, Mauritius, Niger, Pakistan, Papua New Guinea, Paraguay, Saint Kitts and Nevis, Samoa, Slovenia, Swaziland, Zimbabwe; *1993:* Bahamas, Benin, Bosnia and Herzegovina, Brunei Darussalam, Central African Republic, Colombia, Côte d'Ivoire, Czech Republic, Dominica, Dominican Republic, Grenada, Guyana, Honduras, Jamaica, Kiribati, Lebanon, Marshall Islands, Monaco, Myanmar, Namibia, Nicaragua, Peru, Romania, Saint Lucia, Saudi Arabia, Senegal, Seychelles, Slovakia, Solomon Islands, Sudan, Tanzania, Turkmenistan, Tuvalu, Uzbekistan; *1994:* Bolivia, Chad, Comoros, Congo, Democratic Republic of the Congo, Ethiopia, Gabon, Lesotho, Macedonia, Mali, Mauritania, Mozambique, Nepal, Vanuatu, Viet Nam; *1995:* Democratic People's Republic of Korea, Latvia, Lithuania, Micronesia, Morocco; *1996:* Azerbaijan, Estonia, Georgia, Liberia, Madagascar, Republic of Moldova, Mongolia, Qatar, Saint Vincent and the Grenadines, Yemen; *1997*: Burundi, Suriname; *1998:* Belize, Kazakhstan, Lao People's Democratic Republic, Tajikistan, Tonga; *1999:* Albania, Armenia, Djibouti, Oman; *2000:* Angola, Haiti, Kyrgyzstan; *2001:* Cambodia, Cape Verde, Nauru, Palau, Rwanda, Sao Tome and Principe, Serbia, Sierra Leone, Somalia; *2002:* Guinea-Bissau; *2003:* Cook Islands, Niue; *2004:* Afghanistan, Bhutan; *2005:* Eritrea; *2006:* Equatorial Guinea, Montenegro; *2008:* Holy See, Iraq; *2009:* Andorra, San Marino

Background

In September, 1987, representatives of most of the world's nations met in Montreal, Quebec, to draft a protocol that called for a cessation in the production of chlorofluorocarbons (CFCs) and hydrochlorofluorocarbons (HCFCs). By July, 1996, this protocol had been subscribed to by representatives of 158 nations. The Montreal Protocol grew out of a concern among most of the world's nations about the depletion of the ozone layer, a thin invisible band in the stratosphere ranging from 10 to 40 kilometers above the Earth's surface. The ozone layer is crucial to the survival of life on Earth, because it filters out most of the deadly ultraviolet rays the Sun produces, permitting moderate temperatures and allowing visible light to get through. Without this natural filtration system, Earth would not be habitable.

Scientists have studied closely the four levels of the atmosphere—the troposphere, the stratosphere, the mesosphere, and the thermosphere—since the 1960's, using weather balloons and satellites to determine what gases are present. The ozone in the atmosphere was first measured in 1956 by scientists working out of Halley Bay in Antarctica.

The earliest weather satellites were launched in the 1970's. In 1978, the United States launched Nimbus 7, the first satellite designed to measure the ozone in the stratosphere in order to ascertain how profusely distributed the ozone atoms are and to determine the extent of the ultraviolet rays the ozone is filtering out. Using scientific data from these experiments, Joseph Farman, a British scientist, concluded in 1985 that the ozone layer had become extremely thin. People began to refer to an ozone "hole," a somewhat inaccurate designation. By this time, earlier researchers had discovered that CFCs, widely used as propellants in aerosol cans, were causing substantial damage to the ozone layer. The presence of CFCs and HCFCs in the atmosphere tripled between 1976 and 1987.

Halons, which contain bromine rather than chlorine, are produced in smaller quantities than CFCs and HCFCs, but they are extremely threatening to the atmosphere, because they destroy ozone at ten times the rate that these other pollutants do. Halons are widely used for extinguishing vehicular fires.

By 1985, after scientists had determined categorically that the depletion of the ozone layer was much more advanced than they had originally feared, many nations banned the continued manufacture and use of CFCs and other such substances. Reexamining information gathered by Nimbus 7, researchers were startled to find that a serious depletion of the ozone layer began in the 1970's and had continued apace since then with a steady thinning in the ozone layer.

Another source of pollution came from the refrigerants used in refrigerators and in air conditioning systems. The improper disposal of refrigerators released dangerous toxins into the air, so an effort was launched to make sure that refrigerators were disposed of safely. The refrigerants in them began to be reused in new refrigerators, employing a process developed by scientists in Germany. By far the greatest offender in pollution from refrigerants is the automobile, because many air conditioned vehicles have leaking rubber tubes from which refrigerants can escape.

By 1985, the hazards the ozone layer was experiencing from various sorts of pollution were considered so critical that immediate action was essential. This pollution compromised the safety of the entire planet, although it was most pronounced over the Northern Hemisphere. There, the ozone layer was shrinking above industrialized regions, where harmful chemicals were regularly released into the atmosphere as by-products of manufacturing. By the end of the decade that began in 1970, ozone in the stratosphere was found to be disappearing at twice the rate it previously had.

Summary of Provisions

One of the earliest efforts to control the depletion of the ozone layer occurred in March, 1985, when a group of concerned scientists met in Vienna and drafted the Framework for the Protection of the Ozone Layer. The provisions of this document, signed by representatives of twenty-one countries, set a deadline of August 1, 1988, for the signatory countries to cease manufacturing CFCs and called for bans on the use of these chemicals as propellants in aerosol and spray cans, which were major factors in compromising the ozone layer.

This meeting in Vienna marked an important beginning but did not accomplish the sort of widespread action that the urgency of the situation

Preamble to the Montreal Protocol

The Parties to this Protocol,

Being Parties to the Vienna Convention for the Protection of the Ozone Layer,

Mindful of their obligation under that Convention to take appropriate measures to protect human health and the environment against adverse effects resulting or likely to result from human activities which modify or are likely to modify the ozone layer,

Recognizing that world-wide emissions of certain substances can significantly deplete and otherwise modify the ozone layer in a manner that is likely to result in adverse effects on human health and the environment,

Conscious of the potential climatic effects of emissions of these substances,

Aware that measures taken to protect the ozone layer from depletion should be based on relevant scientific knowledge, taking into account technical and economic considerations,

Determined to protect the ozone layer by taking precautionary measures to control equitably total global emissions of substances that deplete it, with the ultimate objective of their elimination on the basis of developments in scientific knowledge, taking into account technical and economic considerations and bearing in mind the developmental needs of developing countries,

Acknowledging that special provision is required to meet the needs of developing countries, including the provision of additional financial resources and access to relevant technologies, bearing in mind that the magnitude of funds necessary is predictable, and the funds can be expected to make a substantial difference in the world's ability to address the scientifically established problem of ozone depletion and its harmful effects,

Noting the precautionary measures for controlling emissions of certain chlorofluorocarbons that have already been taken at national and regional levels,

Considering the importance of promoting international co-operation in the research, development and transfer of alternative technologies relating to the control and reduction of emissions of substances that deplete the ozone layer, bearing in mind in particular the needs of developing countries,

HAVE AGREED [to adopt this protocol.]

demanded. In May, 1985, reliable scientific journals published findings clearly pointing to the existence of a substantial depletion of the ozone layer over Antarctica. Because of the speed at which this depletion was taking place, it was argued that the future of life on Earth was endangered.

By July, 1996, representatives of 158 countries had signed the protocol that grew out of the Vienna meeting. Although this meeting was not initially a notable success, it was significant for highlighting a critical problem and bringing it to the attention of the world's governments, which, despite their marked political and ethnic differences, realized that no time could be wasted in moving toward a tenable solution.

The Montreal Protocol on Substances That Deplete the Ozone Layer called for a 50 percent reduction in the production of CFCs by the year 2000. It spelled out how the gradual reduction in the production and use of chemicals that threaten the ozone layer must occur.

The protocol defines procedures for controlling the production of CFCs and specifies limits on their export and import. Many serious, seemingly unresolvable disagreements arose among the many nations participating in the drafting of the agreement. Representatives from industrialized nations were pressured by industrial lobbyists to resist approving any actions that might harm industry. In the United States, the Ronald Reagan administration, which was fiercely probusiness, viewed the protocol as a dangerous infringement on free enterprise. Most of Europe's industrial nations had vested interests that ran counter to the provisions of the protocol.

Representatives from countries in the Southern Hemisphere had other concerns, with many representatives of developing countries that teetered on the brink of insolvency resisting measures that might disturb the delicate equilibrium that characterized their economies. Special provisions and alterations in deadlines as they pertained to developing countries were enacted. Provision was made in the protocol to establish the Ozone Fund, which was subsidized by the industrialized nations for the benefit of developing nations. By 1996, this fund exceeded $500 million.

Despite the numerous quibbles that the organizers of the Montreal meeting had to address, the development of the Montreal Protocol moved forward. Annual meetings of those who participated in the Montreal Protocol were held in various venues, including Helsinki, Copenhagen, Beijing, Nairobi, Bangkok, and Vienna. The Montreal Protocol, signed in September, 1987, went into effect in 1989. It created mechanisms to regularly review parties' adherence to the protocol, taking into consideration the changes that scientific, environmental, technical, and financial factors might dictate in the future.

In 1990, at a meeting in London, the parties to the protocol voted to eliminate controlled substances that endanger the world's atmosphere gradually. At a meeting of the group in Copenhagen in 1992, the schedule for phasing out ozone-depleting elements was accelerated. Specified substances included CFCs, halons, carbon tetracholoride, methyl chloroform, HCFCs, hydrobromofluoro-carbons (HBFCs), and methyl bromide.

Parties to the Montreal Protocol organized panels to discuss scientific issues relating to the depletion of the ozone layer, to consider the effects this depletion had on the environment worldwide, and to deal with technical and economic problems that provisions of the Montreal Protocol created. Some of these meetings resulted in amendments to the original protocol aimed at strengthening earlier provisions. As information about new substances dangerous to the ozone layer—such as CFCs, methyl chloroform, and carbon tetrachloride—came to be known, such substances were added to the list of those already banned or controlled.

Significance for Climate Change

Because of the danger of ultraviolet rays emitted by the Sun, Earth's earliest inhabitants lived under water. Over billions of years, the oxygen released by aquatic plants rose and over eons created the ozone layer. With such a layer to protect Earth from ultraviolet rays, life eventually was able to emerge from the oceans and, over considerable time, to establish itself on Earth. Ozone absorbs all ultraviolet rays with wavelengths less than 290 nanometers and some with wavelengths between 290 and 320 nanometers, but it absorbs almost none with wavelengths above 320 nanometers. It is the latter rays, known as UV-C rays, that are deadly.

Ozone is made continuously in nature, originating mostly above the tropics and then carried by winds toward the Arctic and Antarctic, where ozone concentrations are greatest. Thus, once the phenomenon of the depletion of the ozone layer was recognized, scientists sought to explain this depletion. They determined that such substances as bromine, chlorine, the hydroxyl radical, and nitric acid threaten the ozone layer. It thus became clear that industrial pollutants were the culprits and that the depletion of the ozone layer could cause ultraviolet irradiation of Earth's surface, endangering human and nonhuman life. The actions demanded by the Montreal Protocol offer humankind's most effective hope of averting this irradiation and preserving Earth's ozone layer.

R. Baird Shuman

Further Reading

Calhoun, Yael, ed. *Climate Change.* New York: Chelsea House, 2005. Section B of this study, "Greenhouse Gases and the Ozone Layer," addresses the depletion of the ozone layer and details attempts being made to reverse this depletion. The introduction to this volume discusses the Montreal Protocol.

DiMento, F. C., and Pamela Doughman, eds. *Climate Change: What It Means for Us, Our Children, and Our Grandchildren.* Cambridge, Mass.: MIT Press, 2007. Chapter 5, "Climate Change: How the World Is Responding," written by the editors, is particularly relevant. Richard A. Matthew's contribution, "Climate Change and Human Security," focuses productively on dealing with the problem.

Grundmann, Reiner. *Transnational Environmental Policy: Reconstructing Ozone.* New York: Routledge, 2001. Chapters 5 and 6, "The Road to Montreal" and "The Montreal Protocol and After," are valuable for their clear overviews of the Montreal Protocol.

Hillman, Mayer, with Tina Fawcett and Sudhir Chella Rajan. *The Suicidal Planet: How to Prevent Global Climate Catastrophe.* New York: St. Martin's Press, 2007. Chapter 7, "The Blueprint for Survival," deals directly with means of replenishing the ozone layer.

Labitzke, Karin G., and Harry van Loon. *The Strato-sphere: Phenomena, History, and Relevance.* Berlin: Springer, 1999. The material presented in chapter 5, "The Ozone Layer," is valuable for its discussion of measures directed at reducing the emissions of chlorine atoms that attach themselves to ozone atoms and destroy them.

Parson, Edward A. *Protecting the Ozone Layer: Science and Strategy.* New York: Oxford University Press, 2003. A thorough account of the Montreal Protocol. Especially valuable for its discussion of the protocol's provisions for developing countries.

Velders, Guus JM, Akkihebbal R. Ravishankara, Melanie K. Miller, Mario J. Molina, Joseph Alcamo, John S. Daniel, David W. Fahey, Stephen A. Montzka, and Stefan Reimann. "Preserving Montreal Protocol climate benefits by limiting HFCs." *Science* 335, no. 6071 (2012): 922-923. The Montreal Protocol was highly successful in halting degradation of the earth's ozone layer, but continued agreement on limiting ozone-depleting chemicals is necessary to preserve the gains already achieved.

See also: Aerosols; International agreements and cooperation; Kyoto Protocol; Ozone; Ultraviolet radiation; United Nations Framework Convention on Climate Change.

Motor vehicles

Categories: Economics, industries, and products; energy

Passenger vehicles in the United States by 2007 accounted for 40 percent of the country's oil consumption and 10 percent of the world's consumption. Cars and trucks used in the United States burned 15 percent of the world's annual oil production.

Key concepts

congestion charge: a tax levied for driving in the urban core of a large city

Golden Quadrilateral: India's first superhighway system, linking the country's east and west coasts with Delhi

miles per gallon (MPGs): a fuel efficiency standard for motor vehicles, measuring the average distance traveled on one gallon of fuel

Target Neutral: a carbon-offset plan for drivers in Great Britain

Background

Motor vehicles consume one-third of the world's oil production. The number of automobiles has been increasing more quickly than population, especially in countries that have just begun to experience widespread industrial development, such as India and China. The tailpipes of motorized vehicles also emit 47 percent of North America's nitrous oxide, a primary component of ozone in the lower atmosphere that also retains heat as a greenhouse gas (GHG).

U.S. Cities

Cities in the United States and elsewhere have been remade to accommodate automobiles. The U.S. cities that reached maturity after the 1930's—the last decade in which the average person did not own a car—have become energy-intensive, sprawling regions of suburbs and freeways, in which most residents are required to use automobiles for transportation. Outside of a few major eastern cities (New York and Boston being prominent examples), the automobile has become an everyday necessity for nearly everyone in the United States. After the year 2000, urban planners began to deemphasize the automobile in order to reduce GHG emissions. Even General Motors (GM) during 2007 entered a partnership to develop residences in once-abandoned buildings along the Detroit waterfront. One of the selling points for these lofts was the fact that people who live in them will be able to walk to work at GM's headquarters.

In New York City, Mayor Michael Bloomberg proposed a congestion charge for the crowded southern half of Manhattan Island, roughly from 86th Street southward. Under Bloomberg's plan, on weekdays between 6:00 a.m. to 6:00 p.m., trucks would be charged $21 per day and cars would be charged $8 per day to drive in the city. This tax would be in addition to the premium parking fees charged by city-owned and private lots. In 2008, only 5 percent of the people who worked on Manhattan Island and lived outside the island commuted by car. According to Bloomberg's proposal, drivers traveling only within the congestion zone would pay half price; taxis and livery cabs would be exempt, with uncontrolled, free access to the area.

Studies have shown that vehicle speeds on southern Manhattan Island average from 4 to 6 kilometers per hour. Many times, walking is almost as fast as driving. Subways are generally much faster means of transportation in the city. The value of the time lost to congestion delays in New York City has been estimated at $5 billion per year. When the cost is adjusted to include wasted fuel, lost revenue, and increasing costs of doing business, the total rises to $13 billion per year. Bloomberg's plan was defeated by the New York State Assembly in July of 2007.

Existing Congestion Charges

Singapore was the first city to introduce a congestion charge. London introduced such a charge in 2003, followed by Milan, Italy. In London, vehicle speeds have since risen by 37 percent and carbon dioxide (CO_2) emissions have fallen by 15 percent. London's mayor, Ken Livingstone, a major proponent of the charge, was easily reelected in 2004, and in 2006 two-thirds of London residents supported the charge. During January, 2007, London's congestion zone was expanded westward to include most of Kensington, Chelsea, and Westminster. By early 2007, London's congestion charge had reduced private automobile use 38 percent and CO_2 emissions 20 percent in the congestion charge zone.

London also improved its public transport system to offer its residents easier alternatives; many commuters had complained that the buses were slow and expensive. By the time the London bus fleet was upgraded, more than six million people used it daily. The number of people commuting by bicycle in London soared by 80 percent after the congestion charge was implemented.

In the meantime, BP proposed that every motorist in Great Britain sign up for a plan called Target Neutral. Drivers can fund ventures that offset the amount of CO_2 that their driving adds to the atmosphere. Drivers register at a Target Neutral Web site, which calculates the estimated amount of CO_2 that may be produced by their driving over the coming year. Drivers then pay offsets based on the estimate. The typical family car, traveling 16,100 kilometers per year, is likely to cost about £20 ($39) to offset.

Stockholm

Congestion charging in downtown Stockholm was a controversial issue in the Swedish general election during the late summer of 2006. A congestion charge of up to $7 per day was narrowly approved by 52 percent in a referendum September 17, 2006. The Stockholm congestion charge reduced auto traffic by 20 to 25 percent, while the use of trains, buses, and Stockholm's subway system increased. Emissions of CO_2 declined 10 to 14 percent in the inner city and 2 to 3 percent in Stockholm County. The project also increased the use of environmentally friendly cars (such as hybrids), which are exempt from congestion taxes. As in London, commuting by bicycle also increased. Following a trial period, the Stockholm congestion tax became permanent during August of 2007.

Context

While some cities restrict automobile use, developing countries' fleets are increasing quickly. In China, by 2008, there were 20 personal vehicles per 1,000 people of driving age; in India, there were 18 per 1,000. In the United States, the same figure stood at 1,148 per 1,000. Between 2000 and 2006, sales of heavy trucks in China increased 800 percent. Sales of passenger cars increased 600 percent. Sales of new passenger vehicles in India tripled (from 500,000 per year to 1.5 million) between 1998 and 2008, a period during which the country built its first interstate highway system, the Golden Quadrilateral, which links Mumbai (Bombay), Delhi, Kolkata (Calcutta), and Bangalore.

If people in India and China drove at half the rate of those in the United States, world oil consumption, 86 million barrels per day in 2006, would balloon to more than 200 million. If drivers in India and China used cars as Americans do, world oil consumption would more than triple, with attendant impacts on its price, as well as GHG emissions.

Bruce E. Johansen

Further Reading

Brown, Lester R. *Plan B 3.0: Mobilizing to Save Civilization.* 3d ed. New York: W. W. Norton, 2008. Provides a detailed analysis of motor vehicles' role in global warming.

Cline, William R. *The Economics of Global Warming.* Washington, D.C.: Institute for International Economics, 1992. An early attempt to sketch the costs and benefits of global warming, including the role of motorized transportation.

Ozguner, Umit, Tankut Acarman, and Keith Redmill. Autonomous ground vehicles. Artech House, 2011. An overview of the engineering of autonomous cars. A brief perusal of the table of contents shows just how complex the problem is. Not only does the vehicle have to navigate and sense the external world, but also has to sense itself during braking, turning and acceleration.

Steinman, David. *Safe Trip to Eden: Ten Steps to Save Planet Earth from Global Warming Meltdown.* New York: Thunder's Mouth Press, 2007. A solutions-oriented work that includes alternatives to motor vehicles.

See also: Automobile technology; Carbon dioxide; Carbon footprint; Catalytic converters; Fossil fuel emissions; Fossil fuel reserves; Fossil fuels and global warming; Hubbert's peak; Transportation.

Mount Pinatubo

Category: Geology and geography

Definition

Mount Pinatubo is an explosive volcano located near the tropics at latitude 15.1° north and longitude 120.4° east, in the Philippine Islands in the western Pacific Ocean. The volcano erupted on June 15, 1991, producing the second largest eruption of volcanic material in the twentieth century. About 10 billion metric tons of magma (molten rock plus suspended crystals) were brought to the surface, along with abundant volcanic gases, including about 20 million metric tons of sulfur dioxide. This was the largest amount of sulfur dioxide injected into the atmosphere since the eruption of the Krakatoa (Krakatau) volcano near Java in 1883.

Part of the hot gas, volcanic ash, and larger rock fragments spewed by the eruption tore down the valleys of Mount Pinatubo as pyroclastic flows. Great quantities of volcanic ash and gases rose soon after the eruption to heights of over 34 kilometers into the atmosphere. Much of the island of Luzon was completely dark during the day, as the dense ash cloud covered around 125,000 square kilometers around the volcano. Volcanic ash covered everything, and many people died when their roofs collapsed under the weight of the ash. Much of the ash became saturated with water from a nearby typhoon that produced large volcanic mudflows. Many previously abundant organisms such as foraminifera were greatly depleted in the nearby oceans after the eruption.

Significance for Climate Change

The eruption of Mount Pinatubo gives some evidence as to how a big volcanic eruption at equatorial latitudes can change the climate. Larger volcanic eruptions that occurred through geologic time should have had an even more drastic effect on climate than did the eruption of Mount Pinatubo. Based on this eruption, volcanic ash quickly settles to the ground, so its effect on climate is short-lived. Only large eruptions such as that of Mount Pinatubo can eject volcanic gases (mostly water vapor, carbon dioxide, and sulfur dioxide) into the stratosphere to affect climate for more than a few days. The sulfur dioxide gas ejected from Mount Pinatubo circled the globe in twenty-two days. Sulfur dioxide gas ejected from erupting volcanoes at higher latitudes generally takes less time to circle the Earth.

Sulfur dioxide injected into the atmosphere by the eruption was rapidly oxidized to sulfuric acid and mixed with water vapor. This reduced the amount of heat absorbed by the atmosphere from the Sun by about 10 percent. Much of the sulfuric acid stayed in the atmosphere for over a year. This

Mount Pinatubo, as it spews volcanic ash and steam during its 1991 eruption. (USGS/USAF/R. Batalon)

appears to have reduced the average temperature close to the Earth's surface by about 0.5° Celsius.

This cooling reversed the trend of global warming for several years after the eruption of Mount Pinatubo. For example, the ice sheet in Greenland did not melt as much as usual during this time. This cooling was not uniform, however, as parts of North America, Siberia, and Europe experienced higher-than-normal temperatures during this time. The warming trend in those regions was due to circulation changes in the atmosphere that are not completely understood, although several climate models to explain these changes were successful.

The generally cooling temperature of the lower atmosphere reduced the temperature of the ocean at the surface by about 0.4° Celsius for several years after the eruption, especially at midlatitudes. This cooling slightly reduced the evaporation rate of the ocean, so there was on the average less precipitation on the land's surface. The slow rise in sea level that occurred before the eruption was somewhat reduced as well, presumably because there was less evaporation from the oceans and less melting of the glaciers.

Various species of chlorine (such as Cl, ClO, HCl, HOCl, and $ClONO_2$), bromine, and iodine catalyze the removal of ozone in the stratosphere. The eruption of Mount Pinatubo liberated a large amount of chlorine species. Thus, the amount of ozone in the atmosphere dropped significantly after the eruption. For instance, ozone in the atmosphere of the tropics was reduced by about 15 percent after the eruption. The ozone hole in the Antarctic became much larger after the eruption.

Robert L. Cullers

Further Reading

Church, John A., Neil J. White, and Julie M. Arblaster. "Significant Decadal-Scale Impact of Volcanic Eruptions on Sea Level and Ocean Heat Content." *Nature* 438 (November 3, 2005): 74-77. Estimates the drop in ocean heat content due to aerosols being injected into the atmosphere after the Mount Pinatubo eruption.

Fiocco, Giorgio, Daniele Fu'a, and Guido Visconti, eds. *The Mount Pinatubo eruption: Effects on the atmosphere and climate.* Vol. 42. Springer Science & Business Media, 2013. One of the largest eruptions of the 20ᵗʰ century produced globally observed climatic effects.

Grainger, R. G., and E. J. Highwood. "Changes in Stratospheric Composition, Chemistry, Radiation, and Climate Caused by Volcanic Eruptions." In *Volcanic Degassing*, edited by C. Oppenheimer, D. M. Pyle, and J. Barclay. Geological Society Special Publication 213. Bath, Somerset, England: Geological Society, 2003. Provides an overview of the effects of volcanic ejecta and gases upon the composition of the atmosphere and the climate. Some of the discussion is moderately technical.

Kump, Lee R., James F. Kasting, and Robert G. Crane. *The Earth System.* Upper Saddle River, N.J.: Prentice Hall, 2003. Describes global environmental changes. Includes global warming, atmospheric shifts, and the relation of volcanic activity to climate. Appendixes, tables, figures, glossary.

Robock, Alan. "The Climatic Aftermath." *Science* 295 (February 15, 2002): 1242-1244. Describes the climate variations over the Earth for several years after the Mount Pinatubo eruptions and some of the reasons why this occurred.

See also: Chlorofluorocarbons and related compounds; Global dimming; Ozone; Sulfur cycle; Volcanoes.

Mount Toba Eruption

Category: Geology and Geography

Mt. Toba, on the Island of Sumatra, erupted explosively 74,000 years ago, ejecting 2,800 cubic kilometers of magma in the largest volcanic eruption since Homo sapiens emerged on the planet. The eruption caused short-term cooling and possibly accelerated the developing Wisconsin glaciation. A combination of climate deterioration and direct damage due to ashfall are thought to be responsible for a population bottleneck early in the history of the human race.

Key Concepts

Supervolcano: A volcano drawing upon a large magma chamber, capable of producing a VEI-7-8 explosive eruption ejecting more than five hundred cubic kilometers of magma in a single eruption.

Volcanic veiling: Blocking of the sun's rays by fine ash and sulfuric acid aerosol – one mechanism by which volcanoes produce global cooling.

Population bottleneck: An event, inferred from genetic studies of modern human groups, which reduced the human population to a few thousand individuals approximately 80,000 years ago.

Tephra: Fragmental material produced by a volcanic eruption, including ash, lapilli or cinders, and bombs (boulder-sized rocks).

Volcano Explosivity Index (VEI): A logarithmic scale used to rank eruptions based on the amount of material ejected. VEI-5 (1-10 km^2); VEI-6 (10-100 m^2) VEI-7 (100-1,000 km^2) VEI-8 (>1,000 km^2)

Background

The scenic waters of Lake Toba, on the Island of Sumatra, conceal a violent volcanic past that dwarfs any catastrophic eruption of historic times. The lake occupies the caldera of a supervolcano, Mount Toba, one of six volcanoes worldwide that have produced eruptions with a volume of ejected magma greater than 500 km^2 in the last million years. The Toba eruption is the largest. Supervolcanoes form above a very large magma chamber, which fills over an extended period of time. There are several theories about what triggers a catastrophic eruption, including magma buoyancy increasing internal pressure and earthquakes creating cracks that cause initial release of magma, followed by "unzipping' along the entire margin of the chamber. As the chamber is exhausted, its roof collapses, creating a caldera. The length of time between major eruptions correlates with the size of the magma chamber. Volcanoes such as Vesuvius experience a major eruption every few hundred years, producing 1 to 10 km² of ejecta. The eruption interval for Toba and Yellowstone is on the order of 400,000 years. No supervolcano is believed likely to endanger civilization with a climate-changing megaeruption in the foreseeable future.

The Eruption

Mt. Toba lies near the Sunda megathrust fault, source of the devastating 2004 earthquake and tsunami, and atop the parallel Sumatra fault, so an immediate earthquake trigger is likely. One theory postulates a general climate–volcano feedback mechanism, whereby changes in sea level associated with continental glaciation or deglaciation cause stresses on the earth's crust, increasing earthquake frequency. Earthquakes can trigger massive volcanic eruptions, and those eruptions temporarily reduce global temperatures, which in turn leads to increased glaciation. The Toba eruption, which occurred when the Wisconsin glaciation was well underway, is used in support of the feedback hypothesis.

The main eruption, which must have occurred in July or August judging from the direction of prevailing winds, occurred at multiple points along the rim of what is now the caldera. The main ash producing phase lasted only nine to fourteen days. Fountains of ash rose high into the stratosphere. Fountain collapse gave rise to pyroclastic flows that devastated the width of Sumatra, reaching both coasts and burying the northern half of the island in fifty or more meters of lava and ash. Pyroclastic flows gave rise to secondary ignimbrite fountains and this was the source of much of the fine ash that fell at remote locations. The ash cloud travelled west across the Indian Ocean, blanketing Malaysia and India. Ash deposits in India average 100 cm but in places are up to 2 meters deep. The

immediate ecological impact and long term climatic effects of this eruption would have been greater had not the bulk of the ash fallen in the ocean.

Within days to weeks after the onset of the main eruption, the roof of the depleted magma reservoir collapsed, forming a caldera 2 km deep. Subsequent activity was confined within the caldera. There followed a long period – perhaps 40,000 years – of dormancy. Refilling of the magma reservoir produced small eruptions and a resurgent dome that formed Samosir Island. This dome building phase has long since subsided.

Effect on Climate

Massive volcanic eruptions cause abrupt cooling by injecting fine ash and sulfur dioxide into the upper atmosphere, blocking incident radiation. Ash rapidly precipitates out; however, sulfur dioxide reacts to form a smog of minute droplets of sulfuric acid, which takes one to three years to dissipate and is responsible for most of the veiling observed in historic eruptions. Assuming the same ratio of SO_2 to ash for other Indonesian volcanoes, it is estimated that this eruption could have caused a drop of as much as 6 degrees centigrade in summer temperatures at high latitudes, but only for one to three years. Acid rain would have had a major impact on vegetation. That volume of sulfur dioxide can also be expected to cause ozone layer depletion. The environmental effects of this eruption were decidedly negative to life on earth over a wide geographical area.

At one time it was suggested that the Toba eruption caused the Wisconsin glaciation, but a careful examination of the timing of glacial advances and of major volcanic events in the Pleistocene demonstrates that the effect is of secondary importance. The advance and retreat of glaciers follows a regular pattern correlated with variations in the earth's orbit (Milankovitch cycles), with supereruptions occurring after glacial advance is well underway. At most they provide a booster effect. Volcanic ash can also contribute to increased planetary albedo, as it is nearly as reflective as snow. Since the Toba ashfall occurred predominantly at sea, the albedo cooling effect was probably not great, but the large land area and vegetation destruction involved in the two Yellowstone megaeruptions may well have been significant, and of far longer duration than volcanic veiling.

Significance to Human Evolution

Based on genomic sequencing, rates of gene substitution, and comparison of modern human populations, scientists studying human evolution have postulated a population bottleneck around 80,000 years BP, when only a few thousand individuals survived. The Toba eruption has been proposed as a cause. However, investigations of Lake Malawi sediments indicate minimal cooling in Africa, and human archaeological sites below and above the ash layer in India demonstrate cultural continuity. It is difficult to project how a few "years without a summer" would have affected human hunter gatherers who were already adapting to colder conditions in Europe and East Asia as the last Pleistocene glaciation progressed. Clearly Toba is significant, but its central role is questionable.

Martha A. Sherwood

Further Reading

Ambrose, S.H. 1998. Late Pleistocene human population bottlenecks, volcanic winter, and differentiation of modern humans. *J. Human Evol.* 34:623-665. Scholarly article arguing for a pivotal role of Toba.

Bindeman, Ilya N. The Secrets of supervolcanoes. *Scientific American* 294(6). 2006. Eruption dynamics, with a focus on Yellowstone..

Jones, Morgan, Sparks, R.S.J., and Valdes, Paul J. 2007. The climatic impact of supervolcanic ash blankets. *Climate Dynamics* 24(6)553-564. Scholarly article on albedo effect.

Rampino, M.R., S. Self, and R.W. Fairbridge.1979. Can rapid climatic change cause volcanic eruptions? *Science*, 206, 826-829. A concern of current global warming.

http://www.volcanocafe.org/the-toba-supereruption/

See also: Climatology; Volcanoes.

National Center for Policy Analysis

Category: Organizations and agencies
Date: Established 1983
Web address: http://www.ncpa.org

Mission

An American nonprofit conservative think tank, the National Center for Policy Analysis (NCPA) promotes an environmental regulation program that seeks to solve problems by relying on a competitive private sector rather than on government control and regulation. The NCPA maintains offices in Dallas, Texas, and Washington, D.C. Well over half of its funding comes from foundations, with the rest from corporations and individuals. Acting as an organizer for other conservative groups, the NCPA also conducts its own free-market-oriented analysis of various issues.

Significance for Climate Change

The NCPA's E-Team focuses on environmental policy. The individuals who form the E-Team are climate-change skeptics who opposed the Kyoto Protocol and also oppose any greenhouse gas (GHG) regulation. NCPA scholars hold that the causes and consequences of the current global warming are unknown. Since it would be very expensive to reduce carbon dioxide (CO_2) emissions substantially, they believe doing so would result in economic decline and increased environmental destruction with little or nothing accomplished to prevent global warming, whatever its cause.

A typical project of the group was an analysis of the 2007 Fourth Assessment Report of the Intergovernmental Panel on Climate Change (IPCC). Citing the purpose of the IPCC as being to provide a

> comprehensive, objective, scientific, technical and socio-economic assessment of the current understanding of human-induced climate change, its potential impacts and options for adaptation and mitigation,

the NCPA concludes that, despite dire predictions of world temperature changes that could result in a global sea level rise, tropical disease spread, accelerated rate of loss of glaciers and ice caps, and increased severity of drought and flooding, forecasts in Chapter 8 of the report violate basic forecasting principles and are therefore invalid.

Because the NCPA is principally concerned with opposing government regulation in favor of promoting private alternatives, the group has concluded, for example, that global warming regulation is a key portfolio risk for state and local pension funds and has recommended that pension fund administrators not promote global warming legislation unless they can demonstrate how such regulation will benefit their portfolios.

The 2007 NCPA publication *A Global Warming Primer* maintains that predictions by some scientists that global warming will cause such things as droughts, floods, and hurricanes of greater intensity are not valid. Given its stance, the group recommends "focused adaptation," taking steps to adapt to warmer conditions, rather than implementing measures that it feels would have more negative economic impact than is justified given the facts that the group accepts as valid regarding global warming.

Victoria Price

See also: Cato Institute; Conservatism; Cooler Heads Coalition; Heartland Institute; Intergovernmental Panel on Climate Change; Nongovernmental International Panel on Climate Change; Skeptics.

National Climate Program Act

Category: Laws, treaties, and protocols
Date: Signed into law September 17, 1978

The National Climate Program Act created an interagency program that conducts climate research, provides climate information, and supports policy decisions in the United States.

Background

Following a period of reduced global average temperatures in the early 1970's, and growing out of years of effort by groups of climatologists, several bills were introduced in the U.S. Congress to coordinate climate research, prediction, and planning. One such bill, the National Climate Program Bill of 1975, failed to pass. Two years later, U.S. representative George Brown of California introduced the National Climate Program Bill of 1977, but it also failed, largely because of disagreements between the American Association of State Climatologists and the National Oceanic and Atmospheric Administration (NOAA) over funding priorities. Finally, in 1978, Congress passed the National Climate Program Act, and it was signed by President Jimmy Carter on September 17, 1978.

Summary of Provisions

The National Climate Program Act (Public Law 367 of the Ninety-Fifth Congress), called for the establishment of the National Climate Program (NCP), as well as the Climate Program Advisory Committee and the Climate Program Policy Board. These entities are to issue periodic reports and plans to "assist the Nation and the world to understand and respond to natural and human-induced climate processes and their implications." The act required the secretary of commerce to establish a National Climate Program Office that would coordinate efforts and develop a series of research and climate services, drawing together the strengths of NOAA and other governmental agencies. These responsibilities were delegated to NOAA. The Department of the Interior and its U.S. Geological Survey are among the other agencies assigned specific roles under the NCP.

Significance for Climate Change

The NOAA Climate Program conducts research and monitoring related to climate, climate change, and climate impact. It gathers and manages data from surface, marine, upper-air, and satellite observations; issues monthly and seasonal predictions of temperature, precipitation, and other weather indicators; predicts the impact of climate fluctuations on water resources, including fisheries, crop irrigation, and energy demands; and conducts new research. Five divisions of NOAA contribute to these efforts: the National Environmental Satellite, Data, and Information Service; the National Marine Fisheries Service; the National Ocean Service; the National Weather Service; and the Office of Oceanic and Atmospheric Research.

Several climate projects under NOAA have yielded important results. Under the direction of the NOAA administration, the United States is part of the Group on Earth Observations, an international organization developing the Global Earth Observation System of Systems (GEOSS), which will collect and manage data around the world. NCP awards grants and fellowships for outside research on the Arctic, on atmospheric composition and climate, on the global climate cycle, and other topics. It also operates the Regional Integrated Sciences and Assessments program, a partnership with American universities to connect with local and regional researchers and policy makers. The Climate Program Office operates separate divisions for climate observations, research, climate assessments and services, planning, and communications and education.

Although the 1978 act established the National Climate Program Office, the office received only a few million dollars of funding, so for its first five years the program accomplished little. No climate-related bills were introduced in the year after passage of the act, and in 1980 most of the funding for climate research was canceled as a result of a budget crisis. By 1984, pilot programs and new structures, including a strongly linked network of regional monitoring centers, enabled the NCP to produce and disseminate useful climate data. These data were essential in the growing national and international understanding of the causes and the effects of global warming.

As policy makers became more interested in global warming, they were unable to make use of much of the pure science that NCP was conducting, and they pressed for more information in forms that would help them draft policy. In response, in 1990 Congress created the United States Global Change Research Program to increase understanding of and response to global warming through research presented by NCP. Several amendments to the National Climate Program Act have been proposed since the 1980's to provide more funding, to solidify the various roles of the various agencies involved, or to require specific actions based on the data. In 2007, the Climate Change Adaptation Bill was introduced in Congress to amend the National Climate Program Act. It would require the president to draw up a strategic plan to address the impacts of global warming in the United States, and to establish a national climate service within NOAA. The bill was reported out of comittee in June, 2008, but no further action was taken by the 110th Congress.

Cynthia A. Bily

Further Reading

Dessler, Andrew Emory, and Edward Parson. *The Science and Politics of Global Climate Change: A Guide to the Debate.* New York: Cambridge University Press, 2006. An introductory overview explaining the science and the politics of global warming.

Gerrard, Michael. *Global Climate Change and U.S. Law.* Washington, D.C.: American Bar Association, 2007. An examination of existing U.S. law related to global warming, including a historical overview of governmental involvement in the scientific research.

Harris, Paul G. *Climate Change and American Foreign Policy.* New York: Macmillan, 2000. Analyzes U.S. climate change policy, including important federal agencies and legislation, from the Jimmy Carter administration to Kyoto.

Henderson, Gabriel. "Governing the Hazards of Climate: The Development of the National Climate Program Act, 1977-1981." *Historical Studies in the Natural Sciences* vol. 46 p. 207-242. (2016). The need for a program to address climate issues was originally conceived as service-oriented, but conflicts between State climatology bureaus and Federal agencies hampered the effectiveness of the Act at first.

Mooney, Chris. *Storm World: Hurricanes, Politics, and the Battle Over Global Warming.* Orlando, Fla.: Harcourt Trade, 2008. Examines scientific disagreements over the ties between hurricanes and global warming; illuminates differing views within NOAA.

See also: Clean Air Acts, U.S.; Energy Policy Act of 1992; U.S. energy policy; U.S. legislation.

National Research Council (NRC)

Category: Organizations and agencies
Date: Established 1991

Mission

The National Research Council (NRC) is the operating arm of the United States National Academies of Science. It was established in 1916 and made permanent by President Woodrow Wilson in 1918. The National Academies of Science include the National Academy of Sciences (NAS), the National Academy of Engineering (NAE), and the Institute of Medicine (IOM). These are private, nonprofit organizations chartered to provide policy advice to the federal government on science, technology, and medicine. They provide this policy guidance through the six divisions of the NRC. For global warming questions, most of these activities are carried out through the Division of Earth and Life Studies.

The NRC initiates studies at the request of the White House, a department of the federal government, or the Congress by calling together a committee of experts from the academies and from the nation. The committee studies the issue and publishes a public report on its findings, including policy recommendations. The NRC publishes more than two hundred reports and other documents each year.

Significance for Climate Change

The NRC has been studying climate change since 1992. In 2002, it was charged with providing strategic advice to the US Climate Change Science Program (CCSP), which coordinates the climate change activities of thirteen federal agencies. In addition to the annual advisory reports for CCSP, twenty-five other reports regarding global warming have been issued. The highlights of these reports have been compiled and published as Understanding and Responding to Climate Change, 2008 Edition.

The report indicates that available scientific data clearly show that the Earth is warming, and

The headquarters of the National Academies in Washington, DC. (Matthew G. Bisanz/Wikimedia Commons)

most of this temperature increase is likely due to human causes. Temperatures reconstructed by several different methods indicate that the planet's surface temperatures since the middle of the twentieth century have been higher than in any comparable period since about 1500 and have increased at the same rate as has the emission of greenhouse gases (GHGs). The report states, "Climate change will affect ecosystems and human systems—such as agriculture, transportation, and health infrastructure—in ways we are only beginning to understand." It concludes, "The increasing need for energy is the single greatest challenge to slowing climate change." As a result, the main action that should be taken is to reduce the amount of carbon dioxide and other GHGs that are released into the atmosphere: The world must work together to make use of alternative energy sources and prepare its populations for the effects of higher temperatures during the next decades.

Raymond D. Cooper

Further Reading

Climate Change: Evidence, Impacts, and Choices. Washington, DC: Natl. Academy of Sciences, 2012. PDF file.

"Climate Intervention Is Not a Replacement for Reducing Carbon Emissions; Proposed Intervention Techniques Not Ready for Wide-Scale

Deployment." National Academies. Natl. Academy of Sciences, 10 Feb. 2015. Web. 25 Mar. 2015.

"New Report Calls for Attention to Abrupt Impacts from Climate Change, Emphasizes Need for Early Warning System." National Academies. Natl. Academy of Sciences, 3 Dec. 2013. Web. 25 Mar. 2015.

Roston, Eric. "Geoengineering. The Bad Idea We Need to Stop Climate Change." Bloomberg Business. Bloomberg, 10 Feb. 2015. Web. 25 Mar. 2015.

"Strong Evidence on Climate Undersores Need for Actions to Reduce Emissions and Begin Adapting to Impacts." National Academies. Natl. Academy of Sciences, 19 May 2010. Web. 25 Mar. 2015.

Natural Resources Stewardship Project

Category: Organizations and agencies
Date: Established October 12, 2006
Web address: http://www.nrsp.com

Mission

As a skeptic organization, the Natural Resources Stewardship Project (NRSP) casts doubt on groups that view global warming as an issue that demands immediate attention, calling this stance a "hypothesis" and arguing that "current climate change is within natural variations." The NRSP is a Canadian, federally incorporated, nonprofit, nonpartisan organization that claims to promote responsible environmental stewardship through media and public relations; consumer education; promotion of private property rights; market-based approaches; and efficient and sensible regulatory and legislative frameworks, particularly at the federal level.

Significance for Climate Change

The NRSP was established when there was a perception that caring for the natural environment had changed from being about individual responsibility to being about nonaccountable actions of transnational and nongovernmental organizations (NGOs), with negative economic consequences. This perceived change resulted in lessening citizens' initiatives and undermining private property rights. Members felt that many governmental and NGO initiatives followed an agenda not based on science or on rational economics. Overall, they felt that the situation was damaging the economy more than it was helping the environment.

The NRSP defines "responsible environmental stewardship" as prudent use of all resources, minimizing unnecessary pollution, transforming waste into resources, and improving material conditions; formulating practical environmental policies based on logic, scientific objectivity, and an understanding of risk; individual rather than governmental action as the preferred means to achieve goals; an understanding that private property encourages private responsibility; a recognition that regulation of resources is best at the most local level possible; and an understanding that more economic freedom allows more responsible individual action.

A top-priority initiative of the NRSP entitled "Understand Climate Change" used proactive grassroots groups to campaign against the Kyoto Protocol and other greenhouse gas emission reduction schemes and to promote sensible climate change policy. Another project, "The Science Centre," will establish a credible independent auditing mechanism to review scientific studies before they are used as a basis for widespread environmental decisions.

The NRSP addresses climate issues differently from less skeptical groups. They applaud the Canadian government's moving toward reducing greenhouse intensities over the short term rather than in terms of absolute emission caps. Believing that carbon dioxide is "almost certainly not a significant driver of global climate change," they will devote time and effort to studying natural factors, such as changes in the Sun's output. Funds will be used to find ways to adapt to the natural phenomenon of climate change rather to reduce carbon dioxide, and the group will oppose establishing further emission standards except at the most local level possible.

Victoria Price

See also: Canada; Skeptics.

Netherlands

Category: Nations and peoples

Key facts
Population: 16,715,999 (July, 2009, estimate), 17,000,059 (2016 estimate)
Area: 41,526 square kilometers
Gross domestic product (GDP): $670.2 billion (purchasing power parity, 2008 estimate); $856.3 billion (purchasing power parity, 2016 estimate)
Greenhouse gas (GHG) emissions in millions of metric tons of carbon dioxide equivalent (CO_2e): 213 in 1990; 174 in 1999; 217 in 2004, 205.4 in 2012.
Kyoto Protocol status: Ratified 2002

Historical and Political Context
By the seventeenth century, the Netherlands played a major role in the world's economy. The creation of the Dutch East India Company established the country as one of the major seafaring and trading powers. The mercantile class became extremely influential in all aspects of the country's life. The Netherlands' major trade rival was England and the competition between the countries led to the Dutch Wars, which were resolved in 1667 by the Treaty of Breda. England recognized the Netherlands' right to the Dutch East Indies. During this period, the Netherlands' capital city of Amsterdam became the site of the first stock exchange and was recognized as the wealthiest trading city at the time. In 1652, the Netherlands had become a republic under Jan de Witt. In 1672, the French invaded the country and killed de Witt. This marked the beginning of a decline in the economic prosperity of the country, which lasted throughout the eighteenth century. From 1795 to 1815, the Netherlands was under French control as part of Napoleon's Empire.

The Industrial Revolution of the nineteenth century did not bring about rapid changes in the Netherlands, because the country relied heavily on waterways for transportation and on wind power for energy. Up to the time of World War II, the Netherlands maintained a neutrality and independence from its neighboring countries that adversely affected its economic prosperity. After the war, the country made a dramatic change in policy and began interacting with its neighbors; this new policy brought about renewed economic prosperity. The Netherlands became an important founding member of major international organizations, including the Benelux, the North Atlantic Treaty Organization, and the European Coal and Steel Community. As one of the fifteen founding members of the European Union, the Netherlands plays an important role in Europe's economy and welfare.

Impact of Dutch Policies on Climate Change
Sea level, rising tides, and the potential of flooding have always been of major concern to the Netherlands since approximately 27 percent of its land lies below sea level. Throughout its history, the country has intervened to protect its land and inhabitants from encroachment by the sea and from flooding caused by overflowing rivers and rising sea levels. The industrialization of the Netherlands has adversely affected the country's situation. With

the exception of the natural gas fields near Sloctern, the country has few natural resources and the economy depends on manufacturing and processing for a considerable amount of its wealth. Petroleum refining, food processing, and chemical processing, along with the manufacture of electrical machinery, all increase greenhouse gas (GHG) emissions. This adds to global warming, which may raise sea levels and increase erratic weather patterns that may contribute to the overflow of rivers and flooding. This situation is worsened by the Netherlands' use of intensive agriculture and horticulture, which produce more GHGs.

The coastline of the Netherlands has undergone considerable changes over the centuries. In 1134, a severe storm created the archipelago of Zeeland. In 1421, the Saint Elizabeth flood caused serious damage to the Netherlands. To combat these problems, the Dutch built polders and dikes to control the water levels and prevent disasters. There are three different kinds of polders, which are tracts of lowlands enclosed by dikes. The first is land reclaimed from a lake or the sea, the second is an area prone to flooding that is protected from the sea by dikes, and the third is a drained marsh separated from the surrounding water by dikes. Windmills are used to pump the excess water. Water bodies or home councils, which are groups independent of any government control, supervise the maintenance of the flood-prevention systems.

In 1953, the Netherlands experienced one of its worst floods. The country put into effect the Delta Works, which raised 3,000 kilometers of outer sea dikes and 10,000 kilometers of canal and river dikes to a flood-prevention level and closed off the Zeeland sea estuaries.

Netherlands as a GHG Emitter

As a member of the European Union, the Netherlands ratified the Kyoto Protocol in 2002. According to the data gathered by the European Environment Agency, the Netherlands emitted 213 million metric tons of GHGs in the base year of 1990. In 2006, the Netherlands had reduced its emissions to 207.5 million metric tons, ranking seventh among the EU27 and sixth among the EU15 as an emitter of GHGs. The burden-sharing target of the Netherlands under the Kyoto Protocol is

−6 percent, to an annual average of 200.3 million metric tons of emissions between 2008 and 2012. From 2002 to 2006, the Netherlands produced an average of 213.6 million metric tons of GHG emissions, representing an increase of 0.3 percent. The country's 2006 emissions, at 207.5 million metric tons, were 3 percent below its base year emissions but were still above its treaty target. The Netherlands has projected an increase in its emissions to a level just 2 percent below the 1990 levels, as present policies continue in energy supply and use, as well as in transport and agriculture. 2012 emissions, 205.42 million metric tons, were still above its Kyoto target.

Summary and Outlook

The Netherlands believes that it will exceed its target of 6 percent below 1990 levels by 2012. By implementing carbon sink activities and using Kyoto mechanisms, specifically that of providing funds for projects reducing GHG emissions in other countries under the clean development mechanism, the Netherlands projects a level of GHG emissions 8 percent below those of 1990.

In view of the prospects of continued global warming, the Netherlands is embarking upon a major flood-control project. Projected to continue to the year 2100, the project is estimated to cost about one billion Euros per year. At the core of the project is the planned raising of dikes and reinforcing of storm barriers. Many different approaches are being considered, from amelioration of the protection from the sea at major ports, especially Rotterdam, to extending the coastline of the North Sea as much as 1 kilometer by dumping millions of metric tons of sand into the ocean. The Netherlands is also working to use technology to protect the country from floods. A system of sensors to determine the stability of the dikes is being developed to replace the human volunteers who carry out the inspection. The Netherlands is also working with International Business Machines to create software that can analyze weather, provide early warning of flood threats, and help coordinate plans for evacuation. The law approving funding for the massive and costly project and maintaining its funding over the century-long period has yet to be passed, but the Netherlands is convinced that the sea level will rise and significant preventive measures must be in place before 2100.

Shawncey Webb

Further Reading

Abboud, Leila. "Before the Deluge." *The Wall Street Journal*, March 9, 2009, p. R10. Good account of the Netherlands' plans for future flood control.

Clapp, Jennifer, and Peter Dauvergne. *Paths to a Green World: The Political Economy of the Global Environment.* Cambridge, Mass.: MIT Press, 2005. Good discussion of environmental issues and global warming with respect to trade and the world economy.

Delta Project. *The Delta Project: Preserving the Environment and Securing Zeeland Against Flooding.* Goes, Netherlands: Florad Marketing Group, 2002. Provides details of Netherlands' flood protection projects.

Katsman, Caroline A., A. Sterl, J. J. Beersma, H. W. Van den Brink, J. A. Church, W. Hazeleger, R. E. Kopp et al. "Exploring high-end scenarios for local sea level rise to develop flood protection strategies for a low-lying delta—the Netherlands as an example." Climatic change 109, no. 3-4 (2011): 617–645. A series of modeling experiments to test the Netherlands' flood control measures against future threats.

Hoeksema, Robert J. *Designed for Dry Feet: Flood Protection and Land Reclamation in the Netherlands.* Reston, Va.: American Society of Civil Engineers, 2006. Excellent study of flood threats and water control covering the reclamation of the Zuiderzee and the coasts, sea defenses, and the Delta Project.

Shogren, Jason F. *The Benefits and Costs of the Kyoto Protocol.* Washington, D.C.: AEI Press, 1999. Excellent overview of the environmental and economic effects of the Kyoto Protocol.

See also: Europe and the European Union; Floods and flooding; Kyoto Protocol; Levees; Sea-level change; United Nations Framework Convention on Climate Change.

New Orleans

Category: Nations and peoples

After Hurricane Katrina devastated New Orleans in 2005, the city's experience became a pivotal point for increasing interest in understanding and mitigating the contribution of global warming to future coastal disasters.

Key concepts

Atlantic Multidecadal Oscillation: Cyclical increases and decreases in sea temperatures that last for decades

Power Dissipation Index: A measure of the total annual energy output of all hurricanes in a region

Saffir-Simpson scale: A scale that ranks hurricanes from category 1 through category 5, based on their sustained wind velocity

Storm Surge: Rising water that is caused when a hurricane's winds push the ocean's surface inland

U.S. Army Corps of Engineers: A government agency whose duties include flood control

Background

New Orleans, the only major American city below sea level (average 1-2 feet below, range about -7 to +20 feet), depends on a system of levees and natural buffers, such as coastal wetlands and barrier islands, for flood protection. When Hurricane Katrina's storm surge hit and the levees failed in August, 2005, 80 percent of New Orleans flooded. Post-storm queries sought to determine how to assess and reduce the probability of future New Orleans disasters by better understanding the conditions that caused Katrina's devastation.

Rising Sea Levels and Hurricane Strength

Whether New Orleans will suffer future catastrophes that are similar or worse in magnitude than the one caused by Katrina depends on a number of factors, including the current and projected effects of climate change. There is a fairly strong consensus among scientists that sea levels have increased because of global warming and will rise another 0.3 to 1 meter over the next one hundred years. Additionally, some scientists have hypothesized that hurricanes high winds, rain, and tornados push water vapor up into the atmosphere, creating more sea surface heat that increase the power of storms.

A study conducted at the Massachusetts Institute of Technology combined hurricanes' duration and wind speed to create the Power Dissipation Index to measure their intensity. The research report, published a month before Katrina, examined 1,557 Pacific and 558 Atlantic hurricanes over a thirty-year period, and found a doubling of storm power, with a strong correlation between rising water temperatures and increasing hurricane strength. Research conducted at the Georgia Institute of Technology similarly showed that, since the mid-1990's, the number of category 4 or 5 hurricanes on the Saffir-Simpson scale doubled. A report issued in 2007 by the United Nations Intergovernmental Panel on Climate Change (IPCC) asserts that climate change has created the potential for more intense hurricanes that could threaten the Gulf Coast for the next one hundred years or more.

These and similar studies' conclusions are not accepted by all scientists. Some meteorologists, including National Oceanic and Atmospheric Administration researchers, take the position that the Atlantic Multidecadal Oscillation (AMO) has created decades-long increases and decreases in sea temperatures, and the AMO is more responsible for the recent upsurge in hurricane strength than is global warming. These critics concede, however, that the AMO will cause decades of more powerful hurricanes.

Loss of Natural Protections

Historically, Southeast Louisiana's coastal wetlands and barrier islands were natural buffers that reduced hurricane damage to New Orleans (the U.S. Army Corps of Engineers asserts that 4.3 kilometers

View of New Orleans looking east. The French Quarter is in the center of the picture. Much of the area beyond the French Quarter is below sea level and was flooded by Hurricane Katrina in 2005. (© Steven I. Dutch)

of wetlands dissipates 0.3 meter of storm surge). Continuous silt deposits from the Mississippi River counteracted soil subsidence. Both human development and climate change create the potential for extensive damage from future storms. Flood control and navigation projects stopped the land buildup from the Mississippi River in both the wetlands and the New Orleans area, while the channels created by the oil industry sped up the pace of coastal erosion.

The combined effects of sea level rise, land subsidence, and erosion have resulted in the loss of thou-sands of square kilometers of Louisiana's coastal land, a 1-meter increase in "relative sea level" for New Orleans over the past one hundred years (one-third due to sea rise and two-thirds due to soil compaction), and concerns that New Orleans may be on the coast in another one hundred years. The destruction of barrier islands has also increased (Katrina destroyed half of the Chandeleur Islands east of New Orleans). Many experts claim that current and proposed coastal restoration efforts would only minimally reduce annual losses.

Government Responses and Calls for Action

Whether New Orleans can avoid future disasters of Katrina's magnitude will depend on the effectiveness of measures that protect the city. The federal government began increasing the height and strength of the levees shortly after Katrina, and spent millions of dollars on coastal restoration programs. Currently, a 133-mile chain of levees, flood walls, gates and pumps are in place.

The protections include two 50-foot lift gates that can be lowered to block the waters of Lake Pontchartrain. Also, 95-foot navigation gate with 220-ton curved sides can swing open, which will be most of the time, to allow easy boat traffic. When a storm threatens, however, they can swing shut to seal off the canal from the devastating storm surge that occurred in Hurricane Katrina. Further, a two-mile Great Wall with steel support piles that extend as far as 200 feet underground can seal off the channel from Lake Borgne to the east, or the billion-dollar west closure complex, which features the world's largest pumping station. Finally, new pumping stations and closure structures have

safe houses for workers built to withstand 250 mph winds, with air-conditioning and enough provisions to last a week.

Critics argue that these measures, funded by $14.5 billion from Congress, cannot counteract the ongoing effects of global warming. They propose that a comprehensive approach that includes coastal restoration and local, national, and global efforts to significantly reduce greenhouse gas emissions will be necessary to ensure New Orleans's future survival.

Context

Some scientists assert that concerns about that city's fate should be a wake-up call that draws attention to a much broader threat. They posit that, if the future effects of global warming occur as currently projected, all of America's coastal areas, where half of the nation's population resides, would be in danger from catastrophes caused by coastal erosion, higher sea levels, and strong storms. Global warming could result in cities like New York and Miami depending on levees for protection. The IPCC report asserts that, during the next century, coastal communities all over the world will be at risk of disasters like Katrina's destruction of New Orleans. This report also states that, beyond the hundred-year projections, current rates of ocean temperature increases could cause much higher sea levels as huge Antarctic and Greenland ice sheets melt.

Jack Carter
Updated by: Lillian M. Range

Further Reading

Gratz, Roberta Brandes. *We're Still Here Ya Bastards: How the People of New Orleans Rebuilt Their City.* New York, NY: Nation Books, 2015. A journalist and urbanist who moved to New Orleans after Hurricane Katrina shares stories of people who returned after being dislocated. Book reveals her love of New Orleans, and insights about urban planning and community activism.

Mooney, Chris C. *Storm World: Hurricanes, Politics, and the Battle Over Global Warming.* Orlando, Fla.: Harcourt, 2007. A New Orleanian examines the debate regarding the effect of global warming on increased hurricane threats, including the impact of leading authorities, interest groups, politics, and the media. Appendixes, bibliography, and recommended reading.

Sargent, William. *Just Seconds from the Ocean: Coastal Living in the Wake of Katrina.* Hanover, N.H.: University Press of New England, 2007. Uses analyses of historical and recent coastal disasters to posit that shortsighted development has increased the effects of global warming and undermined natural safeguards such as barrier islands and wetlands. Argues for more effective environmental and coastal regulatory policy initiatives.

Strauss, Ben, Kulp, Scott, & Levermann, Anders. Carbon choices determine U.S. cities committed to futures below sea level. *Proceedings of the National Academy of Sciences.* Lists U.S. cities that would be impacted by sea level rises resulting from carbon emissions. Argues that with aggressive cuts, many of these cities could be spared.

Tidwell, Mike. *The Ravaging Tide: Strange Weather, Future Katrinas, and the Coming Death of America's Coastal Cities.* New York: Free Press, 2006. The author, who predicted Hurricane Katrina in a 2003 publication, asserts that cities everywhere will experience similar catastrophes absent a concerted, comprehensive effort to address the causes of global warming, which will otherwise continue to raise sea levels and produce stronger hurricanes.

See also: Atlantic multidecadal oscillation; Coastal impacts of global climate change; Louisiana coast; Tropical storms.

Nitrogen cycle

Categories: Chemistry and geochemistry; environmentalism, conservation, and ecosystems

Definition

The nitrogen cycle is a natural process by which nitrogen in the air moves into the soil, is utilized by living organisms, and returns to the air. Organisms need nitrogen to make macromolecules, including amino acids and nucleic acids. Air is 79 percent nitrogen gas (N_2). N_2 has a triple bond, making it relatively inert and unable to be used by most organisms. Nitrogen fixation, converting N_2 to NH_4+ (an ammonium ion), is carried out by *Rhizobium* bacteria, which live in root nodules of host plants of the legume family such as peas, beans, and clover. Bacteria get carbohydrates from the plant, and the plant uses some of the ammonium the bacteria fix.

Free-living cyanobacteria also fix nitrogen. Lightning fixes smaller amounts of nitrogen. Plants absorb nitrate or ammonium ions from the soil through root hairs. Ammonium is converted into organic nitrogen compounds by bacteria and by plants. Other organisms get organic nitrogen from what they eat. Organic nitrogen is converted to ammonia as microorganisms decompose dead matter. Soil bacteria, *Nitrosomona*, perform nitrification where NH_4 is converted to NO_2^- (nitrite). NO_2^- is converted to NO_3^- (nitrate) by *Nitrobacter*. Completing the nitrogen cycle, nitrites are converted to N_2 and N_2O (nitrous oxide) in anaerobic conditions by bacteria *Pseudomonas* and *Clostridium*.

Significance for Climate Change

Humans have more than doubled the annual transfer of nitrogen gas into biologically available forms of nitrogen. This has occurred through burning of fossil fuels, manufacture of synthetic nitrogen fertilizers, and widespread cultivation of legumes (soy, alfalfa, and clover). Burning fossil fuels causes formation of oxides of carbon (carbon monoxide, carbon dioxide). The burning of such fuels occurs at elevated temperatures that cause nitrogen and oxygen molecules in air to react to form oxides of nitrogen (NO, N_2O, and NO_2). NO_2 forms smog and mixed with water forms nitric acid (HNO_3), contributing to acid rain.

The Haber-Bosch process fixes N_2 using hydrogen, high temperature, and pressure to form ammonia. Synthetic nitrogen fertilizers (ammonia, ammonium nitrate, and urea), applied directly to the soil, have led to a huge increase in agricultural productivity. The applied fertilizer not utilized by plants leaches out of soil and accumulates in water. N_2O and CO_2 are greenhouse gases (GHGs) that contribute to global warming by absorbing energy from the Earth's surface, stopping the loss of this energy, and raising the Earth's temperature. NO_2 is not as abundant as CO_2, yet it is an important, stable GHG that absorbs infrared energy about 270 times more strongly than does CO_2. Nitrogen cycle influence on global warming is complex, as an increase in biologically active nitrogen stimulates plant growth, which increases uptake of CO_2.

Susan J. Karcher

See also: Carbon cycle; Hydrologic cycle; Nitrogen fertilization; Nitrous oxide; Phosphorus cycle; Sulfur cycle.

Nitrogen fertilization

Categories: Plants and vegetation; Economics, industries, and products; Chemistry and geochemistry

Definition

Nitrogen fertilization is a phenomenon in which plant growth is unusually stimulated by the addition of reactive nitrogen compounds, such as ammonia and nitrate. Nitrogen compounds are usually applied as fertilizers to soil for uptake by plant roots. They are also often applied through air deposition from the atmosphere to foliage for uptake through leaves and to surface soil for plant-root uptake. Nitrogen fertilizers are usually manufactured through chemical processes, such as the Haber-Bosch process, to produce ammonia. This ammonia is applied directly to the soil or used to produce other compounds, including ammonium nitrate and urea. Reactive nitrogen compounds in the atmosphere are side-products of fossil fuel combustion.

Nitrogen is an essential element of all proteins, enzymes, and metabolic processes for the synthesis and transfer of energy. Nitrogen is also an element in chlorophyll, the green pigment of plants that facilitates light harvest and photosynthesis. Thus, nitrogen is essential for plant survival, growth, and reproduction. Many experiments have shown that photosynthesis is linearly correlated with nitrogen concentration in leaves. Increased nitrogen availability via fertilization or deposition generally increases photosynthetic carbon fixation and stimulates plant growth.

Significance for Climate Change

Nitrogen is very abundant in the Earth's atmosphere but mostly not active for plants to use. Most plants take up only reactive nitrogen compounds, mostly from soil, for photosynthesis and growth. In natural ecosystems, inactive nitrogen in the atmosphere is converted to biologically useful forms mainly via nitrogen fixation by lightning or by a limited number of plant and microbial species. Anthropogenic activities, such as manufacture of nitrogen fertilizer, cultivation-induced nitrogen fixation, and combustion of fossil fuels, have accelerated the addition of reactive nitrogen to ecosystems, to 5 times the natural level.

Nitrogen addition, particularly through fertilization and deposition, has profound implications for climate change and global warming. Nitrogen deposition has been suggested to be a major mechanism underlying terrestrial ecosystem carbon sequestration. Extensive experimental evidence supports the theory that plant growth is limited by nitrogen in almost all ecosystems, and nitrogen addition to ecosystems often stimulates plant biomass growth and increases carbon storage in plant pools. The nitrogen limitation is usually persistent in ecosystems, largely because of the transient nature of biologically available forms of nitrogen. Biologically available nitrogen enters natural ecosystems primarily by biological fixation, but it is highly susceptible to loss by leaching and volatilization.

Although nitrogen stimulates biomass growth and carbon storage in plant pools, the effects of nitrogen fertilization on soil carbon storage are controversial. In several studies, nitrogen addition did not significantly affect soil carbon storage. It caused significant increases in soil carbon content in European and North American forests, and it stimulated substantial carbon loss from soil in other ecosystems. Nitrogen addition not only stimulates plant biomass growth and carbon input to the soil systems but also potentially stimulates microbial decomposition of litter and soil organic matter. Particularly

Severnside fertilizer works. This plant is owned by Terra Nitrogen UK Ltd. and manufactures ammonia and ammonium nitrate. (Sharon Loxton/ Wikimedia Commons)

in forests, most litter and soil organic matter have high carbon contents and low nitrogen contents.

Microbial activities are strongly limited by nitrogen availability. Nitrogen deposition can relieve nitrogen stress and stimulate microbial activities. As a consequence, soil carbon content can decrease with additional nitrogen input from deposition. In an ecosystem with low carbon content and high nitrogen content in litter and soil organic matter, nitrogen-induced increases in plant biomass growth and carbon input can result in increases in soil carbon sequestration. Since carbon in terrestrial ecosystems is mostly stored in soil, the inconsistent, often opposite responses of soil carbon storage to nitrogen addition suggest that stimulation of carbon sequestration in terrestrial ecosystems by nitrogen deposition may be minor.

Nitrogen fertilization and deposition may indirectly affect land-surface energy balance and then feedback to climate change. Nitrogen addition usually results in increases in leaf chlorophyll content and greenness of plant canopy and land surface. Green land surface has low albedo, so it readily absorbs solar radiation, and its high transpiration levels influence local and regional water vapor dynamics and air circulation.

Nitrogen fertilization and deposition can cause emissions of nitrous oxide (N_2O), ammonia gas (NH_3), and nitrogen oxides to the atmosphere, leading to greenhouse effects. The production and application of nitrogen fertilizers also consume energy and emit carbon dioxide (CO_2) into the atmosphere. Reactive nitrogen in the atmosphere can influence tropospheric aerosols and stratospheric ozone, resulting in cooling effects. Moreover, nitrogen-induced increases in tropospheric aerosols can reduce plant carbon uptake and ecosystem carbon storage.

Nitrogen fertilization and deposition can result in degradation of ecosystems. Runoff with nitrogen-rich compounds from high-deposition regions and fertilized fields causes eutrophication in rivers and lakes and hypoxia in coastal zones.

Yiqi Luo

Further Reading

Bhattacharyya, P., K. S. Roy, S. Neogi, T. K. Adhya, K. S. Rao, and M. C. Manna. "Effects of rice straw and nitrogen fertilization on greenhouse gas emissions and carbon storage in tropical flooded soil planted with rice." *Soil and Tillage research* 124 (2012): 119-130. Rice cultivation is a major source of methane. Fertilization experiments showed that a mix of organic and inorganic fertilization was most effective in sequestering carbon.

Dybas, Cheryl, and Layne Cameron. "How Much Fertilizer Is Too Much for Earth'S Climate?." National Science Foundation. NSF, 9 June 2014. Web. 23 Mar. 2015.

Galloway, J. N., et al. "Transformation of the Nitrogen Cycle: Recent Trends, Questions, and Potential Solutions." Science 320 (2008): 889–892.

Hungate, B. A., et al. "Nitrogen and Climate Change." Science 302 (2003): 1512–1513.

LeBauer, D. S., and K. K. Treseder. "Nitrogen Limitation of Net Primary Productivity in Terrestrial Ecosystems Is Globally Distributed." Ecology 89 (2008): 371–379.

Reay, D. S., et al. "Global Nitrogen Deposition and Carbon Sinks." Nature Geoscience 1 (2008): 430–437.

Nitrous oxide

Category: Chemistry and geochemistry

Definition

A colorless gas with a sweet odor, nitrous oxide is one of the many oxides of nitrogen. Sometimes called laughing gas, funny gas, nitrogen suboxide, or dinitrogen monoxide, it was first manufactured by Joseph Priestley, an English chemist, in 1775. It can be used to render a person insensible to pain in short surgical procedures; if inhaled for too long a period, it will cause death. It is also used as a propellant in aerosals and as an oxidizer to increase a racing-car engine's power output.

Significance for Climate Change

Nitrous oxide is a trace gas produced by human industrial activity that affects the environment as do the other greenhouse gases (GHGs) and, as such, is limited by the Kyoto Protocol. (The other nitrogen oxides are not considered to be GHGs.) Another source of nitrous oxide is nitrogen fertilizers. Nitrous oxide also occurs naturally as a by-product of the bacteria in soil and water, particularly in tropical forest areas, and in animal waste products. Nitrous oxide may have slightly less effect than the halocarbons, as nitrous oxide reacts naturally with soil and water, so is less likely to make its way into the atmosphere. However, one molecule of N_2O in the atmosphere has two to three hundred times the warming effect of one molecule of CO_2.

As light from the Sun enters the atmosphere, some of that light gets scattered by molecules in the air or gets reflected from clouds back into space.

Some of the light that reaches Earth is reflected back into space, such as light that bounces off snow or ice. Much of the light that reaches Earth is absorbed into the Earth and retained as heat. The Earth's surface warms and emits infrared photons, which make several passes between the Earth and the atmosphere, warming the atmosphere and the Earth as they go back and forth. Eventually, these infrared photons return to space.

The GHGs, including nitrous oxide, all have three or more atoms. They are able to absorb infrared photons as they pass, transferring the energy from the photon to the molecule and affecting the way these infrared photons are able to warm the Earth by trapping them. Eventually, this absorption of energy affects the net change in the Earth's energy balance. The effect caused depends on the radiative force associated with the gas.

This balance is also affected by the global warming potential (GWP) of the gas. Some gases stay in the atmosphere much longer than others before any natural process is able to remove them. Some gases can stay in the atmosphere for hundreds or thousands of years. Gases with long lifetimes continue to affect the warming of the Earth and its atmosphere. This warming, in turn, affects changes in weather, sea levels, and land.

These gases can be removed from the atmosphere through condensation and precipitation or by chemical reactions in the atmosphere. However, due to their long lives in the atmosphere, they are, generally speaking, accumulating more quickly than they can be dispersed.

Because of their effects on the global climate, the retention of these gases in the atmosphere could lead to melting of glaciers and polar ice caps, flooding and droughts becoming more severe, rising sea levels, increases in the salinity of freshwater, more devastating tropical cyclones and tidal waves, and erosion of beaches on the coasts.

These effects on the global climate could also help increase food production. As the Earth becomes warmer, growing cycles lengthen, more land becomes available for food production, and more and different varieties of food are able to be grown.

Climate change from the greenhouse gases, including nitrous oxide, could also lead to more insect-borne disease spreading further throughout

Global Nitrous Oxide Emissions by Sector, 2000

Economic Sector	Percent of Total Emissions
Agricultural by-products	62.0
Land use and biomass burning	26.0
Industrial processes	5.9
Waste disposal and treatment	2.3
Residential, commercial, and other sources	1.5
Transportation fuels	1.1
Power stations	1.1

Data from the Netherlands Environmental Assessment Agency.

the world. As mosquitoes and other pests are able to survive in more and different areas, malaria, dengue fever, and cholera could spread further.

Marianne M. Madsen

Further Reading

Cowie, J. *Climate Change: Biological and Human Aspects.* New York: Cambridge University Press, 2007. Reviews past and projects future climate changes due to GHG release.

Emanuel, Kerry. *What We Know About Climate Change.* Cambridge, Mass.: MIT Press, 2007. Discusses how GHGs, including nitrous oxide, have increased in the atmosphere and the climate changes they may cause.

Fagan, B. M. *The Little Ice Age: How Climate Made History, 1300-1850.* New York: Basic Books, 2001. Reviews the history of climate change over time, discussing whether the current changes are part of a natural cycle.

Flannery, Tim. *We Are the Weather Makers: The Story of Global Warming.* Rev. ed. London: Penguin, 2007. Argues that global warming is real, providing definitions and models.

Hoffman, Jennifer, Tina Tin, and George Ochoa. *Climate: The Force That Shapes Our World and the Future of Life on Earth.* London: Rodale Books, 2005. Explains how human actions have caused a shift in the climate; includes photos, graphs, diagrams, and maps.

Marshall, John, and R. Alan Plumb. *Atmosphere, Ocean, and Climate Dynamics.* Boston: Elsevier Academic Press, 2008. Discusses changes in the atmosphere and how those changes affect the oceans.

Reay, Dave S., Eric A. Davidson, Keith A. Smith, Pete Smith, Jerry M. Melillo, Frank Dentener, and Paul J. Crutzen. "Global agriculture and nitrous oxide emissions." *Nature Climate Change* 2, no. 6 (2012): 410-416. Increasing use of chemical fertilizers contributes to nitrous oxide emissions. Global emissions are reasonably well known but local emissions are hard to measure.

See also: Enhanced greenhouse effect; Greenhouse effect; Greenhouse gases and global warming; Nitrogen cycle; Nitrogen fertilization.

Noctilucent clouds

Category: Meteorology and atmospheric sciences

Definition

At very high latitudes, usually farther than $50°$ from the equator, very thin, extremely high-altitude clouds sometimes form during the summer. These clouds are very difficult to observe directly from Earth. They are easiest to see at the end of twilight, when darkness has settled on the surface of the Earth but sunlight is still shining at very high altitudes: The ultra-high-altitude clouds appear to shine after dusk, so they are called noctilucent clouds (German for "night-shining" clouds).

Noctilucent clouds form in the mesosphere, an upper layer of Earth's atmosphere, at an altitude of about 85 kilometers. At that altitude, the air is exceedingly thin and cold. Dust and tiny ice

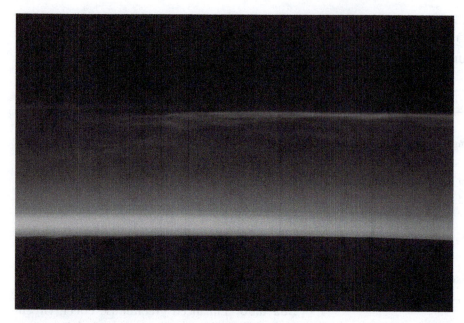

Noctilucent clouds are very high altitude (80 km) clouds, so called because they remain sunlit long after the ground and surrounding sky are dark. In this very unusual photo from the International Space Station, we see them from space in their true relationship to the surface. Note the slight curvature of the horizon. (NASA)

engines, have also been implicated in the formation of noctilucent clouds. Despite over a century of research, the nature and actual cause of noctilucent clouds is still not known.

Significance for Climate Change

Because noctilucent clouds were not reported until 1883, some have speculated that they did not exist before then, but that claim is unsubstantiated. Many subtle phenomena were first studied scientifically in the nineteenth century, but that does not mean that they did not exist beforehand. Noctilucent clouds have also been more commonly reported over the course of the twentieth century, leading to claims that these clouds are becoming more common. That claim, too, is disputed. Noctilucent clouds have been more widely publicized over the course of the last century, so there may have simply been more people looking for them or recognizing and reporting them when they saw them. Concerted and in-depth scientific investigations of noctilucent clouds have been fairly recent, so there is little long-term data.

Not only have noctilucent clouds been observed more frequently in high latitudes, but they have also been observed at lower latitudes than before, with reports of noctilucent clouds from as far south as Colorado. The apparent increase in noctilucent cloud reports appears to correlate with the reported increases in global temperatures over the last century. Since noctilucent clouds require extremely cold mesospheric temperatures, sighting these clouds at lower latitudes may suggest a cooling of the mesosphere. This has led to claims that noctilucent clouds are in some manner related to global

crystals form the clouds, but the ice crystals require exceedingly cold temperatures, typically colder than $-120°$ Celsius, which occur during the summer months in the polar regions (when surface temperatures are warmest). As a result, the technical name for noctilucent clouds is polar mesospheric clouds.

The first known reports of noctilucent clouds followed the 1883 explosive eruption of Krakatoa, when the clouds were extensively studied by Jesse Otto. Initially, it was believed that the clouds were simply a by-product of the volcanic eruption. However, the clouds persisted long after the ash from the eruption had dissipated. Otto and subsequent researchers were able to determine very little about the clouds. They are too high for balloon studies and have prompted satellite missions to study them, including the Aeronomy of Ice in the Mesosphere (AIM) mission.

Though little is known about noctilucent clouds, scientists have observed that they are somewhat more frequent after meteor showers. Some rocket exhausts, such as those of the Space Shuttle main warming, which may also relate to cooling of the upper atmosphere. However, these claims are difficult to substantiate, because the nature of the noctilucent clouds is still not understood, nor is the mechanism by which these clouds form.

Some theoretical models indicate that certain greenhouse gases, such as methane, may provide hydrogen to the mesosphere that can oxidize to form water from which the clouds can form. One of the mysteries about noctilucent clouds is the source of the water in the mesosphere, one of the driest parts of Earth's atmosphere. An increase in methane could be one explanation. However, other sources for the water in the mesosphere could also result in noctilucent clouds. Major volcanic eruptions or impacts from icy bodies from space could also impart water to the mesosphere, but impacts should occur at fairly consistent rates, and there have been no major volcanic eruptions of the order of Krakatoa in the last century.

In recent years, it has been discovered that noctilucent clouds could be monitored from the ground by radar even when they were not visible. The ice crystals forming the clouds seem to act as attractors for iron and sodium ions in Earth's thermosphere. These metal ions are believed to come from meteoroids burning up in Earth's upper atmosphere. Satellite detection of noctilucent clouds is also now possible. These new methods of study may answer questions about the origins of the clouds and any possible link between them and global climate change.

Raymond D. Benge, Jr.

Further Reading

Dalin, P., et al. "Ground-Based Observations of Noctilucent Clouds with a Northern Hemisphere Network of Automatic Digital Cameras." *Journal of Atmospheric and Solar-Terrestrial Physics* 70, nos. 11/12 (August, 2008): 1460-1472. Extensive technical report of scientific investigations of noctilucent clouds.

Dalin, P., N. Pertsev, and V. Romejko. "Notes on historical aspects on the earliest known observations of noctilucent clouds." *History of Geo- and Space Sciences* 3, no. 1 (2012): 87-97. One mystery of noctilucent clouds is the suddenness of their appearance. They were first recorded in 1885. The authors discuss historical evidence for their discovery and the possible connection with the Krakatoa eruption of 1883.

Petersen, Carolyn Collins. "Noctilucent Clouds from Rocket Exhaust." *Sky and Telescope* 106, no. 3 (September, 2003): 26. Illustrated summary of research linking noctilucent clouds to rocket exhaust.

Schroder, Wilfried. "Otto Jesse and the Investigation of Noctilucent Clouds 115 Years Ago." *Bulletin of the American Meteorological Society* 82, no. 11 (November, 2001): 2457-2468. Details the early history of noctilucent cloud studies.

Stevens, Michael H. "Heavenly Harbingers." *Smithsonian* 32, no. 2 (November, 2001): 20. A brief firsthand account of studies of noctilucent clouds, with illustrations.

Witze, Alexandra. "Enigmatic Clouds Illuminated." *Nature* 927, no. 7172 (2007): 927. Some of the early findings of the AIM mission are summarized.

See also: Atmospheric structure and evolution; Clouds and cloud feedback; Mesosphere; Polar stratospheric clouds.

Nongovernmental International Panel on Climate Change

Category: Organizations and agencies
Date: Established 2007
Web address: http://www.sepp.org

Mission

The Nongovernmental International Panel on Climate Change (NIPCC) is an international panel of nongovernmental scientists and scholars assembled to address the causes and consequences of global climate change. The NIPCC was established by the Science and Environmental Policy Project

(SEPP). The SEPP was founded in 1990 by eminent atmospheric physicist (and former director of the United States Weather Satellite Service) S. Fred Singer,

> on the premise that sound, credible science must form the basis for health and environmental decisions that affect millions of people and cost tens of billions of dollars every year.

NIPCC set out to produce an independent evaluation of the available scientific evidence on the causes of climate change. Motivation for this grew out of widespread dissatisfaction with the global climate assessment reports of the United Nation's Intergovernmental Panel on Climate Change (IPCC), in particular the Fourth Assessment Report (2007).

SEPP brought together an international panel of nongovernmental scientists and scholars who were not predisposed to believe that climate change is caused mostly by human greenhouse gas emissions. The organization's report, *Nature, Not Human* Activity, Rules the Climate (2008), focused on evidence that the NIPCC felt that the IPCC had ignored. The report stems from an international climate workshop in Vienna in April, 2007, organized by the NIPCC.

Significance for Climate Change
The NIPCC's 2008 report aims to provide an independent, nongovernmental second opinion on the global warming issue. NIPCC claims

> the central problems for policymakers in the debate over global warming are (a) is the reported warming trend real and how significant is it? (b) how much of the warming trend is due to natural causes and how much is due to human-generated greenhouse gases? and (c) would the effects of continued warming be harmful or beneficial to plant and wildlife and to human civilization?

The report presents evidence that helps provide answers to all three questions.

The NIPCC could find no convincing evidence or observations of significant global climate change from other than natural causes. The authors sum up their findings as follows:

This NIPCC report falsifies the principal IPCC conclusion that most of the reported warming (since 1979) is "very likely" (that is, 90-99 percent certain) caused by the human emission of greenhouse gases. In other words, increasing carbon dioxide is not responsible for current warming. Policies adopted and called for in the name of "fighting global warming" are unnecessary.

C R de Freitas

See also: Intergovernmental Panel on Climate Change.

Nongovernmental organizations (NGOs) and climate change

Categories: Environmentalism, conservation, and ecosystems; Ethics, human rights, and social justice

Definition
Nongovernmental organizations (NGOs) are broadly defined as private, philanthropic, and nonprofit organizations that aim to provide services for needy people and to challenge sociopolitical inequalities at local, national, or international levels. Oxfam, Greenpeace, the Red Cross, Save the Children, and Doctors Without Borders are a few high-profile NGOs that have been active in shaping climate change policies.

Significance for Climate Change
An increasing number of NGOs have showed their growing interest and influence in putting the environment on the global agenda and pressuring governments, business, and international organizations to take climate change seriously. Climate change affects both developed and developing countries, but many NGOs pay extra attention to the needs of vulnerable groups in developing countries

US Air Force General Douglas Fraser, commander of US Southern Command, listens to members of nongovernmental organizations at Ancien Aeroport Militaire in Port-au-Prince, Haiti, March 6, 2010. (MCC Spike Call/Wikimedia Commons)

since the climate-change-induced destruction falls disproportionately on them. These impoverished people rely heavily on natural resources, such as land and sea, for survival, but climate change would cause more floods, droughts, and extreme weather, and that will destroy their livelihoods and worsen poverty. However, the adaptive capacity of developing countries is low, and their governments are not strong enough to respond effectively to the challenges posed by climate change. Some NGOs, such as Oxfam, take an environmental justice perspective, arguing that rich countries control access to carbon and are capable of protecting themselves from the ravages of climate change. In contrast, poor countries are least responsible for greenhouse gas (GHG) emissions, but they pay the price for industrial growth in rich countries. They highlight the implicit power imbalance between rich and poor countries and between carbon haves and have-nots.

NGOs have comparative advantages in global environmental politics. First, global warming is a crisis on a global scale and requires global coordination to provide a solution. Most international NGOs have dense networks, and they often work together in coalitions and coordinate their lobbying efforts in order to maximize the impact. Second, donors do not always trust governments in the developing countries. They prefer offering development aid through NGOs, which are seen as transparent, accountable, and trustworthy. Third, NGOs can respond more quickly than many governments to the needs of local communities. They are also sensitive to gender and power dynamics.

Therefore, NGOs try to build bridges for local people, governments, and international communities. Apart from raising public awareness and engaging affected communities in the decision-making process, NGOs are keen to advocate mitigation and adaptation policies. They launch campaigns to cut GHG emissions in order to reduce global average temperature. They also help poor countries to develop alternative livelihoods, to improve global humanitarian systems, to increase emergency aid, and to reduce risks of disaster by building long-term social protection. They build local capacity by focusing on disaster preparedness and recovery plans. They also help identify new funding sources for the Adaptation Fund, the largest potential source of funds for climate adaptation in poor countries. Furthermore, they build partnerships with other organizations to provide essential services, such as water and sanitation, to local people.

It needs to be noted that NGOs are a diverse group with varied goals, structures, and motivations. These differences mark heterogeneous policy design, priority, and response to climate change. Different targets, priorities, and strategies show the disparities in history, resources, leadership, expertise, networks, and visions. While some focus more on targets for national CO2 emissions, financial mechanisms, and technology transfer, others are more concerned with humanitarian work and disaster preparedness. For example, Doctors Without Borders focuses on the health implications of higher temperature on disease patterns, while the International Red Cross, in contrast, emphasizes community-based self-reliance, early-warning operation, risk-mapping, and vulnerability assessment. Oxfam takes a rights-based perspective, arguing that the excessive GHG emission of rich countries has violated the human rights of poor people in developing countries.

NGOs have faced a number of challenges. First, NGO relationships with local governments in developing countries can be rough. Some NGOs suffer from excessive government interference. Second, NGOs rely on financial support from donors. Donors can make unreasonable demands and put their own agendas above the needs of recipient communities. Third, the legitimacy and governance of NGOs are sometimes questioned because the public wants to know how funds are allocated. Finally, local NGOs feel cash-rich international NGOs imposing their agendas on them, without building their capacity for dealing with local issues. Local NGOs are concerned that overreliance on foreign NGOs will create a culture of dependence, and that is not conducive to the healthy development of NGOs in developing countries.

Sam Wong

Further Reading

Bernauer, Thomas, and Carola Betzold. "Civil society in global environmental governance." *The Journal of Environment & Development* 21, no. 1 (2012): 62-66. Non-governmental organizations can often do things more effectively than central governments, but may suffer from perceived lack of legitimacy and accountability.

Magrath, John, with Andrew Simms. Africa: Up in Smoke 2. London: New Economics Foundation, 2006.

Revi, Aromar. "Climate Change Risk: An Adaptation and Mitigation Agenda for Indian Cities." Environment and Urbanisation 20, no. 1 (2007): 207–229.

World Bank. Not If but When: Adapting to Natural Hazards in the Pacific. Washington, D.C.: Author, 2006.

Nonlinear processes

Category: Science and technology

Definition

In science, a process is a series of changes of one or more variables that represent properties or states of an object or a system. Examples of physical, chemical, and biological processes include evaporation, land erosion, oxidation, cell division, germination, dispersion, growth, accumulation, and global warming. In addition to the processes themselves, the study of the relationship between causes and effects of changes accounts for a major portion of scientific research. There are many ways to classify a process, based on properties of the process itself or the cause-effect relationship behind it. For example, a process can be continuous or discrete, stable or unstable, convergent or divergent, and linear or nonlinear.

A process is nonlinear when its effect is not simply proportional to its cause. For example, water evaporation is a nonlinear process, because water being boiled will not vaporize until the temperature reaches the critical threshold of 100° Celsius, causing a change in water's state from liquid to gas.

Significance for Climate Change

The global climate system includes a variety of nonlinear processes that are subject to positive feedbacks, as well as complex interrelations between numerous factors affecting the climate. Such complexity exposes the Earth to a high risk of abrupt climate changes.

A positive feedback loop worth noting is the Arctic permafrost melt, which can speed up the cycle between the accumulation of GHGs in the atmosphere and temperature growth. The greenhouse effect contributes to global warming and higher temperature leads to melting frozen soils in the Artic region. The ice melting can release vast amounts of carbon dioxide (CO_2) and methane trapped in the permafrost soils. Estimates show that billions of metric tons of methane—a greenhouse gas twenty-one times more potent than is CO_2—can be emitted into the atmosphere and amplify the greenhouse effect.

Another important positive feedback loop is the Arctic albedo change. The albedo of a surface is the percentage of incident light that it reflects back into space. Since ice masses have high albedos, the ice covering the Arctic Ocean and land surfaces can help the Earth absorb only a small fraction of solar energy. As the Arctic ice melts due to global warming, the uncovered surface assimilates more solar energy and as a consequence intensifies the warming effect.

Besides permafrost melt and the Arctic albedo change, rain-forest decline, water scarcity, land degradation, ocean decline, and persistent toxins have the potential to cause abrupt climate change. Studies of the effects of the nonlinear processes, as well as their interactions with other elements of the climate system, are in progress to resolve uncertainty about abrupt and irreversible climate changes.

To N. Nguyen

See also: Albedo feedback; Climate feedback; Clouds and cloud feedback.

North Atlantic Oscillation (NAO)

Category: Oceanography

Definition

The North Atlantic Oscillation (NAO) is the dominant pattern of atmospheric circulation in the North Atlantic region ranging from central North America to Europe and northern Asia. The NAO is usually developed in the winter and is caused by fluctuations in atmospheric pressure between a subpolar, low-pressure center near Iceland and a subtropical, high-pressure center near the Azores-Gibraltar region. The NAO is generally described by the North Atlantic Oscillation index, which is a weighted measurement of the difference between the subpolar low-pressure zone and the subtropical high-pressure center during the winter season of the North Atlantic region.

The positive NAO index phase corresponds with time periods when a stronger subtropical high-pressure center and a deeper-than-normal subpolar low-pressure zone exist in the North Atlantic region, increasing the atmospheric pressure gradient in this region. During positive-NAO-index years, the western subtropical North Atlantic Ocean is warm. Strengthened westerly winds blow warmth and moisture into north-central Europe. The warm, moisture-bearing winds arriving from the subtropical Atlantic Ocean make Europe warmer and wetter. In the meantime, northern Canada

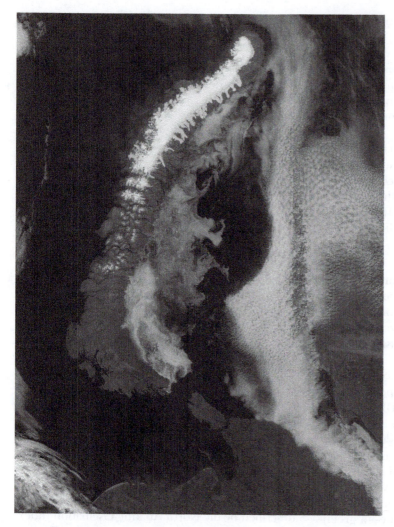

North of the western Russian mainland lies the island archipelago of Novaya Zemlya. The northern island is glacier covered and is the site of ongoing research into the effects of the North Atlantic Oscillation and climate change on the glaciers. (NASA)

pressure gradient in the North Atlantic region. As a result, fewer and weaker winter storms occur in this region. More moist air is brought to the Mediterranean, and cold air is brought to northern Europe. Northeastern Canada and Greenland experience mild and wet winters and the eastern United States undergoes a cold and dry winter season.

Significance for Climate Change

The NAO index varies from year to year and has evidenced a cyclicity of decadal scales over the past 150 years. The NAO index was persistently positive in the early 1900s, negative in the 1960s and 1970s, and considerably more positive during the 1980s and early 1990s. Since the heat capacity of the ocean is much greater than that of a continent, the NAO accounts for approximately one-third of the changes in average winter surface temperatures in the northern hemisphere. Variations in the NAO have significant impacts on many aspects of North Atlantic societies and the environment, such as agricultural harvests, water resources, fishery yields, industrial energy production, and ecosystems. Significant changes in the NAO may in turn influence climatic changes, including changes in sea surface temperature (SST), ocean circulation patterns, and Arctic sea-ice coverage.

Many mechanisms have been proposed to account for NAO index variability, including atmospheric response to changes in SST, variability of atmospheric convection in the tropics, internal and nonlinear dynamics of the extratropical atmosphere, and anthropogenic forcing caused by greenhouse gas (GHG) emissions and ozone depletion. Tropical heating has been proved to influence the atmospheric circulation over the North Atlantic region. Since tropical convection is sensitive to the underlying SST distribution, recent

and Greenland experience cold and dry winters. Cooler temperatures occur off the west coast of Africa. Strong trade winds send more dust out across the ocean toward the Caribbean Sea. The eastern United States undergoes a mild and wet winter season.

The negative NAO index phase corresponds with time periods when both the subtropical high-pressure center and the subpolar low-pressure zone are weakened, which would reduce the atmospheric warming of the tropical oceans may lead to persistently positive values for the NAO index.

Some scientists think that changes in atmospheric circulation associated with the NAO index contributed to the winter warming of the Northern Hemisphere. Statistical evidence has demonstrated that the forcing of increased GHG concentration in the atmosphere may have affected the long-term variability of the NAO. Recent comparisons of NAO index records between the 1800s and the late twentieth century demonstrate that global warming may cause the increased variability of the NAO. Even though studies have linked climate change to the NAO, however, the mechanism of the NAO is still not fully understood. The NAO needs to be further investigated to advance understanding of the linkages between anthropogenic forcing and NAO variability.

Yongli Gao

Further Reading

Appenzeller, C., T. F. Stocker, and M. Anklin. "North Atlantic Oscillation Dynamics Recorded in Greenland Ice Cores." Science 282 (1998): 446–449.

Goodkin, N. F., K. A. Hughen, S. C. Doney, and W. B. Curry. "Increased Multidecadal Variability of the North Atlantic Oscillation Since 1781." Nature Geoscience 1 (2008): 844–848.

Holland, D. M., et al. "Acceleration of Jakobshavn Isbrae Triggered by Warm Subsurface Ocean Waters." Nature Geoscience 1 (2008): 659–664.

Hurrell, J. W., et al., eds. The North Atlantic Oscillation: Climate Significance and Environmental Impact. Washington, D.C.: American Geophysical Union, 2003.

Osborn, Timothy J. "Winter 2009/2010 temperatures and a record-breaking North Atlantic Oscillation index." *Weather* 66, no. 1 (2011): 19-21. Although 2009/2010 was a very warm year globally, Britain and western Europe had one of the most severe winters in decades. The regional severity seems linked to an unusually strong North Atlantic Oscillation.

Paeth, H., et al. "The North Atlantic Oscillation as an Indicator for Greenhouse-Gas Induced Regional Climate Change." Climate Dynamics 15, no. 12 (1999): 953–960.

North Korea

Category: Nations and peoples

Key facts

Population: 24.983,205 (2015 estimate)
Area: 120,538 square kilometers
Gross domestic product (GDP): $40 billion (purchasing power parity, 2014 estimate)
Greenhouse gas (GHG) emissions in millions of metric tons of carbon dioxide equivalent (CO_2e): 71.8 in 2004; 94.5 in 2010
Kyoto Protocol status: Ratified 2005

Historical and Political Context

In the aftermath of World War II in 1945, the Korean peninsula was divided into two countries in accordance with a United Nations arrangement. North of the 38th parallel, the North Korea, or the Democratic People's Republic of Korea (DPRK), was formally established in 1948. North Korea established a centralizing system and nationalized all properties. Economic activities were conducted under state control. In June, 1950, North Korea launched the Korean War against South Korea, and the war lasted until July, 1953. After the war, North Korea emphasized heavy industry, along with a centralized, planned economy.

North Korea developed a political ideology, *Juche* (self-reliance), whose goal was to strengthen the country economically and militarily and finally to make North Korea immune to foreign invasion and capitalist intervention. This ideology has prevented the nation from keeping technological pace with other industrialized countries and has made North Korea one of the poorest, most isolated nations in the world.

In July, 2002, North Korea started to open its border and to adopt capitalism and set up a free-trade zone near its border with China. However, North Korea sought to set back these economic reform policies from 2005 to 2009, and it currently maintains a firm political control of the economy and allows only limited access to foreign capital. In 2013, North Korea announced a new policy calling for the simultaneous development of its nuclear weapons

program and its economy. By January 2016, North Korea has conducted four underground nuclear detonation tests which drew multiple sanctions from the United Nations Security Council.

Impact of North Korean Policies on Climate Change

After the Korean War, North Korea moved toward a command economy and all economic activities have been conducted under state control. The centralized, planned economy grew significantly in industrial production and made significant progress in social infrastructure. North Korea has consistently promoted economic development policies that place top priority on heavy industry, including electricity production, steel production, and machine building. In turn, unreasonably one-sided promotion of heavy industry has deepened structural imbalances between North Korean industries. This "heavy industry first" policy has made North Korea unable to catch up to new technological developments or to phase out its coal-and-steel-based economy. North Korea also pursues military strength while striving to develop its economy.

In the 1970s, environmental pollution became a serious issue after two decades of industrialization emphasizing heavy industry and reckless development of natural resources. However, North Korea did not regulate air pollution until the Environmental Protection Law was established in 1986. North Korea established the State Environmental Protection Bureau to implement the Environmental Protection Law in 1996. North Korea has also participated in the United Nations Environment Programme (UNEP) since 1992.

In spite of North Korea's efforts to restore its environment, environmental degradation in the country is getting worse. This is primarily because of the government's policies emphasizing development in areas of heavy industries and an increase in food production, rather than environmental protection, as well as the lack of investment in developing energy-saving equipment and renewable energy alternatives due to the weak economy.

North Korea as a GHG Emitter

The World Bank estimated that North Korea emitted 94.5 million metric tons of greenhouse gas (GHG) of carbon dioxide (CO_2) equivalent in 2010. The GHG emission as a percent of global CO_2 production is about 0.2 percent. GHG emissions of North Korea plummeted in the early 1990s as its economy collapsed. Then, the emission sharply decreased in the late 1990s and it stabilized in recent years. In 2000, GHG emission is primarily from the energy sector (92.6 percent), as well as Industrial Processes and Product Use sector (6.1 percent) and waste sector (1.4 percent).

Because of its self-reliance policy, North Korea avoids dependency on foreign energy sources, such as petroleum and natural gas, but it relies heavily on its own coal as a primary source of energy (70.1 percent in 2002). Coal with high sulfur content emits not only carbon dioxide but also a large amount of other air pollutant emissions such as SO_x. North Korea's use of coal is projected to increase fivefold from 2005 to 2020. In addition to that, the use of obsolete industrial technology with poor energy efficiency, lack of exhaust gas purification technology, lack of renewable energy alternatives, and poor electricity transmission infrastructure induce a great deal of GHG emission.

To cooperate in international efforts to reduce GHG emission, North Korea has signed several international environmental agreements and treaty. North Korea ratified the Kyoto Protocol in 2005. At the 2015 Paris Climate Conference, North Korea pledged to active engage in global environment efforts. North Korea declared war on deforestation and put forward a large-scale tree-planting project over the next decade. North Korea pledged reduction of GHG emission at 37.4 percent compared with the levels of the 1990s by 2030.

Summary and Outlook

A lack of liquid energy sources has forced North Korea to use abundant but poor-quality coals to meet its energy demands. In addition, North Korea's high GHG emission levels result from the governmental enforcement of unbalanced economic policies such as putting heavy industry first and engaging in the parallel development of military strength and the economy. Furthermore, its self-reliance ideology rendered North Korea unable to carry out reforms of its closed economy.

North Korea attempts to strengthen cooperation with international societies such as the United Nations Development Programme (UNDP) to acquire economic support and to lift its economy. North Korea also emphasizes the need for foreign trade and economic cooperation with other nations. North Korea needs to make internal changes to facilitate and acquire international support. International organizations and community also need to respond to North Korea's effort to reduce GHG by promoting energy infrastructure, such as clean coal combustion, purification technologies, and renewable energy alternatives.

North Korea has also demonstrated its willingness to engage with the global community such as the United Nations Framework Convention on Climate Change (UNFCCC) and United Nations Environment Programme (UNEP) to reduce GHG emission and to protect its environment. Through the Clean Development Mechanism (CDM) of the UNFCCC, North Korea intends to upgrade its energy sector which is responsible for most of its GHG emissions. North Korea has registered several CDM projects in hydroelectricity, methane reduction program, reforestation, and energy efficiency with the UNFCCC. However, its commitment to the global climate change mitigation projects through collective international actions remains uncertain.

Jongnam Choi

Further Reading

Hong, Soon-jick. "Environmental Pollution in North Korea: Another South Korean Burden." *Asian-Pacific Economic Literature* 13, no. 2 (November, 1999): 199-214. Describes the causes of pollution in North Korea and discusses approaches that South Korea should take to prevent pollution on the Korean Peninsula.

Park, Song-dong. "The Situation Regarding the North Korean Environment and the Long-Term and Short-Term Tasks." In *Unified Economy*. Seoul, South Korea: Hyundai Research Institute, 1997. Discusses South Korea's long- and short-term strategies to preserve the environment in North Korea and promote good relations between the two Koreas.

United Nations Environment Programme. *DPRK: State of the Environment, 2003*. Nairobi, Kenya: Author, 2003. The first comprehensive survey of North Korean environmental issues, including water resources, air and soil pollution, deforestation, and biodiversity.

World Bank, 2016, World Development Indicators: Energy Dependency, Efficiency and Carbon Dioxide Emissions.

See also: Industrial greenhouse emissions; Kyoto lands; Kyoto mechanisms; Kyoto Protocol; United Nations Environment Programme; United Nations Framework Convention on Climate Change.

Northwest Passage

Categories: Geology and geography; oceanography

Definition

In the 1400's and early 1500's, Portuguese and Spanish navigators developed routes to the rich oriental spice trade going south of Africa and South America, respectively. Their rivals, England and France, hoped to find another, shorter route. They sought a way west at the north end of North America—the Northwest Passage. Many expeditions failed, often tragically, to find an ice-free northwest passage through the maze of islands west of Greenland in the Canadian north. Their attempts pitted them against the polar cold and floating pack ice of the Arctic Ocean.

Finally, in 1903, Roald Amundsen allowed his small ship, *Gjoa,* to be locked in ice and drift westward for three years, until it melted free on the Alaska side. That was an exploring feat, but it had no commercial potential. Likewise, the 1969 passage of the supertanker *Manhattan* was a nautical triumph but a commercial failure, because it required assistance from two icebreakers and repairs afterwards.

Since then, two things have changed: The climate has warmed, and the Arctic's energy resources have become better known and more sorely desired. The Little Ice Age reached its coldest point in the early 1800's, and world temperatures have moved generally higher ever since (with cooling retreats such as the 1880's to through about 1910 and the early 1940's through the mid-1970's). Beginning in the mid-1970's, temperature increases resumed and increased more steeply. This increase has been ascribed to the greenhouse effect of global warming caused by higher levels of carbon dioxide (CO_2), which slows the escape of heat through the Earth's atmosphere.

By the mid-1980's, cruise ships began making summer cruises through many Northwest Passage waters. In 2007, the summer melt reduced the extent of Arctic ice to its smallest area since comprehensive satellite data were collected beginning about 1970. The Northwest Passage was largely ice free for several weeks before the cold returned. Moreover, those open summer waters replace the near total solar reflectivity of polar ice with the near total solar absorption of dark, open water. Several climate models suggest that all of the Arctic pack ice may melt in summers by 2100, or even much sooner.

Beside climate change, development is coming to the entire Arctic region, because world demand for petroleum and natural gas may soon exceed production both because of continued economic growth and exhaustion of a number of existing deposits. Meanwhile, about a quarter of the world's hydrocarbon deposits ring the Arctic from the Norwegian Sea through Siberia, Alaska, and northern Canada.

Significance for Climate Change

The fabled Northwest Passage may soon become a functional reality because of the confluence of climate warming, the resource wealth of the Arctic, and the Arctic's central position between Europe and East Asia. These factors could make a navigable Arctic Ocean "the Mediterranean of the north." However, there is still doubt about a functioning Northwest Passage. A minority group of climatologists suggests that most of the warming in the Arctic since the early 1800's is due to natural cycles. They posit that the mechanisms of global change are still only dimly understood and that computer modelers are applying mere decades of firm data against processes that operate over millennia. Consequently, the recent melting could be replaced by cold and increased pack ice comparable to the 1800's or beyond any in recorded history.

Still, the general pattern of two centuries has been of warming climate and decreased ice. This suggests that offshore drilling and even coal mining will progress northward into the various seas of the Arctic Ocean. It suggests that thousands of ships will ply that waterway, either through Canadian waters or through a longer but serviceable route north of Russia (the Northeast Passage). Once in place, it may be a factor for increased global warming for four reasons.

First, incidental pollution and accidental spills are inevitable. Pollution tends to melt ice. Soot on ice or snow increases melting by increasing absorption. Any major petroleum spills would result in black, absorbing areas on, in, and beneath the ice for decades, because petroleum-digesting bacteria function slowly at near freezing temperatures.

Second, major industrial development also brings thermal pollution that could significantly increase melting of the pack ice. Moreover, there is a major limitation to cooling waters compared to temperate regions. The Arctic Ocean has an unusual thermal structure, with warm salty Gulf Stream water reaching the Arctic Ocean but diving under a major layer of less-dense freshwater. The deeper waters are often a degree above freezing, and they are more saline, which hinders them from freezing. Thus, bringing cooler waters from the depths is not an option.

Third, industries and governments making sizeable investments in ports, icebreaking ships, and mining facilities would not gladly surrender those treasures to returning ice. The first two mechanisms could be applied deliberately to melt ice. Fourth, activities to enforce sovereignty claims, such as building bases, would also contribute to warming.

The sheer volume of Arctic ice makes it difficult to predict the full climatological consequences of its melting. One climate theory has held that an ice-free Arctic in winter would provide sufficient water vapor for vastly increased snowfall on surrounding lands, perhaps enough to generate a new ice age.

Roger V. Carlson

Further Reading

Easterbrook, Gregg. "Global Warming: Who Loses—and Who Wins?" *The Atlantic* 299 (April, 2007): 52-64. Science journalist Easterbrook focuses on the politics of who owns the Arctic, especially the Northwest Passage.

Hall, Sam. *The Fourth World: The Heritage of the Arctic and Its Destruction*. New York: Alfred A. Knopf, 1987. Journalist Sam Hall describes the interrelated Arctic issues of climate, resources, people, and sovereignty.

Heide-Jørgensen, Mads Peter, Kristin L. Laidre, Lori T. Quakenbush, and John J. Citta. "The Northwest Passage opens for bowhead whales." *Biology letters* 8, no. 2 (2012): 270-273. Bowhead whales, formerly believed to consist of two separate populations separated by sea ice, have now been observed to mingle.

Moss, Sarah. *The Frozen Ship: The Histories and Tales of Polar Expeditions*. New York: Blue Bridge, 2006. Historian Sarah Moss provides not just historical accounts but also a survey of the literature and beliefs surrounding polar expeditions.

Open University Course Team. *Case Studies in Oceanography and Marine Affairs*. New York: Pergamon Press, 1991. Although dated, this book has an extensive chapter on the key Northwest Passage issues of resources, pollution, sovereignty, trade, and climate change.

See also: Albedo feedback; Arctic; Arctic peoples; Arctic seafloor claims; Greenhouse effect; Oil industry.

Nuclear energy

Category: Energy

Nuclear energy may offer a partial alternative to fossil fuels. Nuclear power generates no GHGs, but adequate measures to dispose of nuclear waste have yet to be developed, and the long lead time required to construct a nuclear power plant precludes using nuclear energy as a global replacement for GHG-emitting energy sources in the short term.

Key concepts

fission: the splitting of a heavy nucleus (such as in a uranium atom) into two, with a release of a relatively large amount of energy

fossil fuels: fuels, including coal, gas, and oil, based on compressed, ancient organic materials

fusion: the combination of two light nuclei to produce a heavier nucleus; if accomplished in a contained setting, this is infinitely sustainable

greenhouse gases (GHGs): gases emitted by burning fossil fuels that trap heat in Earth's atmosphere

high-level waste: extremely radioactive fission products and transuranic elements such as plutonium, often produced by fission processes

Background

Since the 1950's, countries such as the United States, the Soviet Union (now Russia), and France have relied on nuclear power as an energy source. Electric power is generated in nuclear power plants by fission reactors that heat water to turn turbines.

Nuclear fission produces no greenhouse gases (GHGs). Mining uranium to power the reactors causes some environmental degradation, but the most significant drawbacks of nuclear energy are the high cost of reactor construction, the long lag time required to build reactors, and the difficulty of safely disposing of the nuclear waste generated by the reactors.

Nuclear Energy as a Power Source

When several countries began to build nuclear reactors to generate electric power in the 1950's, nuclear energy was hailed as the energy of the future. Uranium was and still is in plentiful supply, and supporters of nuclear power indicated it would provide inexhaustible electric power in the future. A few countries, most notably France, engaged in extensive reactor construction, and approximately 80 percent of France's electric power now comes from nuclear power. Other countries, such as the United States, initially built several reactors. Support for nuclear power declined in the United States, however. No new power reactors were ordered after 1978. Approximately 20 percent of the electric power produced in the United States in the first decade of the twenty-first century came from nuclear energy.

Worldwide, approximately 16 percent of all electric power is generated by nuclear energy. In the first decade of the twenty-first century, some 440 nuclear power reactors were in operation in thirty countries. Six different types of power reactor were in operation, with research underway to expand this number. The most common reactor was the pressurized water reactor (268 reactors), followed by the boiling water reactor (94 reactors).

In spite of its initial problems, nuclear power may offer some advantages in dealing with climate change generated by burning fossil fuels. The operation of fission reactors does not generate any GHGs. Uranium is in plentiful supply, both from mines throughout the world and as material reprocessed from old nuclear weapons. Nuclear reactors can be constructed to fit existing needs, with electric power capacities ranging from nearly 2 gigawatts per reactor to much smaller capacities to serve small communities. Once in operation, a

power reactor may be expected to operate for forty to sixty years.

Although challenging, it may be possible to generate as much as 1 terawatt of electric power per year with nuclear energy by 2050. If this goal is accomplished, it will be possible to achieve a substantial reduction in carbon emissions by mid-century. The technology to build nuclear reactors already exists and does not have to be invented; new advances, however, can lead to increased reactor efficiency. The construction of nuclear power reactors is a feasible alternative to fossil fuel power plants.

One criticism that has sometimes been levied against nuclear power is that it is not safe. Some critics indicate that reactors may leak radioactive matter. Power reactors in Western countries such as the United States, the United Kingdom, and France have had minimal leakages with no health risks. Other critics indicate that power reactors may have major accidents, including core meltdowns, pointing to accidents at the Three Mile Island Nuclear Power Plant in the United States in 1979 and the Soviet reactor at Chernobyl in 1986. The Three Mile Island accident led to a radiation leak of less than 50 curies of radiation, and no health issues have been traced to this incident. The Chernobyl accident was far more severe, with a release of 100 million curies of radiation, numerous deaths, and continuing health problems resulting from the accident, as well as lingering health problems in the region. Western countries do not use reactors of the Soviet design, which had limited containment around the reactors. The safety claims regarding most Western power reactors are generally overstated. In many ways, nuclear power produces fewer health problems than do coal-fired power plants.

All existing nuclear power reactors are fission reactors. The potential exists, however, eventually to construct fusion reactors. Such reactors would produce much less radioactive waste and operate more efficiently than fission reactors. Although advocates over the last thirty years have repeatedly indicated that fusion would become the energy source of the future, the technology remains relegated to the laboratory for the near term. No means of sustaining the temperatures required for fusion (about 100 million° Celsius) has yet been found. In addition, questions involving the high costs of construction and the production of the tritium gas necessary for a fusion reaction must be addressed.

Several environmentalists and other advocates of nuclear energy argue that the United States and other Western countries should reconsider its use. They point to the negative impacts of coal and oil plants, such as acid rain and GHG emissions. In addition, some economists predict that oil will become increasingly expensive, making nuclear energy more cost-effective.

Drawbacks and Liabilities of Nuclear Energy

Most of the twentieth century problems with nuclear energy remain, and some new issues have been raised as well. As fears of terrorism and so-called suitcase devices have increased, the need to control even relatively small amounts of nuclear material has become more urgent, and even well-protected nuclear power plants pose security risks. Although terrorists might prefer more high profile targets than reactors and upgrading nuclear fuel is difficult, the security threat may sway public opinion against expanding nuclear energy.

Moreover, there are no ideal methods of nuclear waste disposal. Wastes with low levels of radioactivity are currently being stored underground in the United States and elsewhere. The greatest source of concern is high-level radioactive waste, such as spent fuel rods, piping, and the like. France has long followed a policy of reprocessing and reusing spent fuel rods. This is an attractive solution, but the process produces weapons-grade plutonium as a by-product. The United States and most other countries have not adopted this practice, in part out of fear that some plutonium might fall into the wrong hands.

The United States and most other countries that use nuclear energy have not solved the problem of spent nuclear fuel as yet. The United States is developing an underground storage facility at Yucca Mountain in Nevada, but it is controversial and is unlikely to be operational until 2015 at best. Until the Yucca Mountain site becomes operational,

American nuclear power facilities are storing spent fuel on site. This approach is also being followed by most of the other countries that operate nuclear power reactors. The Russian Federation stores spent fuel on site but also engages in a limited amount of reprocessing at Chelyabinsk-65, and another facility at Krasnoyarsk is scheduled to start operation in 2015. The former Soviet government also engaged in extensive injection into the ground of a good deal of its lower-level waste such as cesium 137 and strontium 90. This injection occurred at three Soviet sites, producing several environmental problems.

Nuclear waste storage on site remains an option for nuclear power facilities, but the waste issue will need to be resolved if extensive use is to be made of nuclear energy in the future. The French and possibly the Russians seem to be comfortable with reprocessing. France and the United Kingdom are engaged in a joint project to develop a means to recycle spent nuclear fuel without producing weapons-grade plutonium. Most other countries are trying to develop underground sites, as Finland is doing at Olkiluoto, or shipping their nuclear waste to countries such as the United States for storage. Because of the long period of radioactive decay of some of this waste (several million years in some cases), developing safe, permanent storage facilities remains a challenging issue.

Timetables for Nuclear Deployment

Nuclear energy may provide some help in mitigating global warming, but its impact is unlikely to be felt in the short run. In the United States, for example, the time required to obtain a permit to build a power reactor had increased to three and one-half years by the time the last reactor was permitted in 1978. Once a permit was issued, construction required around ten additional years. Although reactors have been constructed in other countries such as France or Russia in much shorter periods of time, they are complex facilities that cannot be constructed in short order. In some nations—such as Germany or the Scandinavian countries, where power reactors have been shut down—the largely antinuclear public will have to be convinced to allow new nuclear facilities to be contructed before any such project can be considered.

A large-scale expansion of nuclear energy is not likely until the 2020's, if then. In the United States, sixteen utilities had announced plans for potential reactor construction by 2007, but it would be several years before any of these facilities could go online. Elsewhere, the situation was much the same.

Context

Nuclear energy has much to offer as an alternative to fossil fuels. Utilization of nuclear energy can help to provide a middle-term solution to the need for clean energy until ways can be found to enable cleaner, renewable sources such as solar energy to satisfy humanity's energy needs. In addition to environmental factors, the increasing cost of oil will help make nuclear energy more attractive. Fears of terrorism, radiation leaks, and the difficulties surrounding nuclear waste disposal are obstacles to a resurgence of nuclear energy. The length of time required to build an operational reactor negates any advantages of nuclear energy as a short-term solution to energy-related GHG emissions. Public opinion appears to be becoming more favorably inclined toward nuclear energy, but a good deal of opposition remains. As with many issues concerning global warming, the longer the wait before construction of power reactors begins, the more costly the process will be, and the less help the reactors will provide in combating the increasing emission of GHGs.

John M. Theilmann

Further Reading

Caldicott, Helen. *Nuclear Power Is Not the Answer.* New York: New Press, 2006. Argues that the various issues facing nuclear energy are so great as to outweigh the advantages of nuclear power.

Deutch, John M., and Ernest J. Moniz. "The Nuclear Option," *Scientific American,* September, 2006, 75-83. Indicates that a significant expansion of nuclear power could reduce GHG emissions in a cost-effective fashion.

Graves, Christopher, Sune D. Ebbesen, Mogens Mogensen, and Klaus S. Lackner. "Sustainable hydrocarbon fuels by recycling CO 2 and H 2 O with renewable or nuclear energy." *Renewable and Sustainable Energy Reviews* 15, no. 1 (2011): 1-23. With sufficient energy, carbon dioxide and water can be recombined into hydrocarbons. The energy source can either be nuclear or renewable energy, such as wind or solar.

Herbst, Alan M., and George W. Hopley. *Nuclear Energy Now.* New York: John Wiley, 2007. Primarily economic analysis of nuclear energy that advocates its adoption.

Lake, James A., Ralph G. Bennett, and John F. Kotek. "Next Generation Nuclear Power." In *Oil and the Future of Energy.* Guilford, Conn.: Lyons Press, 2007. Examines the evolving technology of nuclear reactor construction.

Kim, Younghwan, Minki Kim, and Wonjoon Kim. "Effect of the Fukushima nuclear disaster on global public acceptance of nuclear energy." *Energy Policy* 61 (2013): 822-828. Not surprisingly, the Fukushima disaster had a negative impact on public perception of nuclear power. Government attempts to affect public opinion are often counterproductive.

Macfarlane, Allison M., and Rodney C. Ewing, eds. *Uncertainty Underground.* Cambridge, Mass.: MIT Press, 2006. Useful collection of essays about the technology of dealing with high-level nuclear waste at Yucca Mountain, as well as public policy issues.

Morris, Robert C. *The Environmental Case for Nuclear Power.* St. Paul, Minn.: Paragon House, 2000. Contrasts nuclear energy favorably with fossil fuels in terms of their environmental and medical effects.

Nuttall, W. J. *Nuclear Renaissance: Technologies and Policies for the Future of Nuclear Power.* New York: Taylor and Francis, 2004. Good treatment of the technical, political, and economic issues surrounding nuclear power.

See also: Clean energy; Coal; Energy resources and global warming; Fossil fuels and global warming; Fuels, alternative; Hubbert's peak; International Atomic Energy Agency; Nuclear winter; U.S. energy policy.

Nuclear winter

Category: Climatic events and epochs

Definition

"Nuclear winter" was the term coined in the 1980's to describe the global climatic effects of a major nuclear weapons exchange between the United States and the Soviet Union. The concept arose from scientific theories attempting to explain the extinction of the dinosaurs. Evidence to suggest an unusual geological event was found at the boundary between the Cretaceous and Tertiary periods (65-70 million years ago). An iridium-rich layer of sediment separated the dinosaur-fossil-bearing layers below from the dinosaur-fossil-lacking layers above. It was theorized that this amount of iridium could only have come from a collision with an asteroid-sized object, at least 10 to 16 kilometers in diameter.

If such an object had hit the Earth, the impact would have had a global effect on the planet's environment. Rock and ash debris would have been thrown high into the atmosphere, and the impact would have ignited huge forest fires. Ash from these fires, combined with the fine dust from the impact, could have remained in the atmosphere for several years and altered the Earth's albedo. Less sunlight would reach the surface, as the dust-laden atmosphere reflected it back into space. Those life forms surviving the initial impact would have to adapt to a rapidly changing environment. Less sunlight would affect photosynthesis, and the food chain would be disrupted as the Earth entered into a long global winter.

Some scientists saw a correlation between this scenario and the possible aftermath of a massive nuclear weapons exchange. The smoke and ash coming from the hundreds of burning cities and forests would pollute the atmosphere and alter weather patterns. Scientists envisioned an Earth enshrouded by dust-laden clouds that would block out sunlight and lower surface temperatures. The summer days would be cloudy, cooler, and wetter, with a short growing season. Winters would be cold and brutal. It was estimated that within three years of the nuclear conflagration, over one-half of the Earth's population would face starvation.

Significance for Climate Change

The end of World War II in 1945 saw the beginning of the Cold War between the United States and the Soviet Union. During the war, the United States and the Soviet Union had been allies fighting against a common enemy in Nazi Germany. Although the two countries had different political ideals and agendas, they put aside their differences to win the war. Once the war was over, each side began to mistrust the other. What made matters worse was the development and use of the atomic bomb.

The United States was the first country to develop a nuclear bomb, which it used twice to end the war with Japan, bombing the cities of Hiroshima and Nagasaki. Conventional bombing raids over Germany and Japan killed more people and damaged more property than a single atomic bomb. However, the psychological effect of one bomb doing the same amount of damage as thousands of conventional weapons made a difference and motivated Japan to end the war. The bomb blasts had unanticipated, lingering effects on the area, however, in the form of radioactive fallout. Tens of thousands of people may have died from the nuclear blast, while perhaps hundreds of thousands more would later die from radiation poisoning.

The nuclear attacks on Hiroshima and Nagasaki demonstrated to the world the horrors of nuclear war. It was hoped that they would serve as a reason for avoiding any further nuclear conflicts. Throughout the 1950's, 1960's, and 1970's, the United States and the Soviet Union each developed sufficient nuclear capability to deter the other from attacking, following a doctrine of mutual assured destruction. It was understood that in the event of a nuclear exchange, the fallout would affect the global community. In an attempt to persuade political leaders to cut back their stockpiles of nuclear weapons, scientists began to study the global environmental effects of a massive nuclear weapons exchange.

Throughout the 1980's and 1990's, scientists and political activists publicized the effects of a nuclear weapons exchange and the threat it would pose to the survival of humankind. Perhaps their efforts played an important role in reducing the stockpiles of nuclear weapons and bringing an end to the

Cold War. In the wake of the Cold War, however, nuclear weapons remain. The threat of an all-out nuclear exchange involving thousands of warheads may be a thing of the past, but several nations have the ability to wage a limited nuclear war and are apparently willing to do so.

The possibility also exists of nuclear weapons falling into the hands of terrorists. Even a relatively small conventional engagement such as the 1991 Gulf War has environmental effects. When Saddam Hussein set fire to hundreds of Kuwaiti oil wells, the intense smoke from these fires polluted the environment. While full-scale nuclear war—and thus full-scale nuclear winter—appears unlikely, even a limited exchange would have severe environmental consequences and could affect the global climate in an unforeseeable manner.

Paul P. Sipiera

Further Reading

Baum, Seth D. "Winter-safe deterrence: The risk of nuclear winter and its challenge to deterrence." *Contemporary Security Policy* 36, no. 1 (2015): 123-148. Although nuclear winter no longer has a high media profile, the danger nevertheless is real. Studies indicate that even relatively small nuclear exchanges might have global effects. This paper explores ways nations might deter attack without risking nuclear winter.

Fisher, David E. *Fire and Ice: The Greenhouse Effect, Ozone Depletion, and Nuclear Winter.* New York: Harper and Row, 1990. A good introduction to the concept of global climatic change and humankind's effects upon the environment.

Greene, Owen, Ian Percival, and Irene Ridge. *Nuclear Winter: The Evidence and Risks.* New York: Polity Press, 1985. One of the first books written to capture the public's attention to the dangers of climatic change induced by nuclear weapons or natural disasters.

Sagan, Carl, and Richard P. Turco. *A Path Where No Man Thought: Nuclear Winter and the End of the Arms Race.* New York: Random House, 1990. Co-author Sagan, a popular scientist, was probably the best-known proponent of the nuclear winter concept.

See also: Mount Pinatubo; Nuclear energy; Volcanoes.

Obama Administration Efforts on Climate Change

Category: Ethics, human rights, and social justice

"Someday, our children, and our children's children, will look at us in the eye and they'll ask us, did we do all that we could when we had the chance to deal with this problem and leave them a cleaner, safe, and more stable world?"—President Barack Obama, June 25, 2013

The earth's climate is changing and human activity is responsible for it, and the price of inaction will be severe – these scientific facts are indisputable. However, the politics of climate change in the United States does not agree with the scientific consensus. As a result action on addressing climate change and its impacts suffers. Despite this obstruction, President Obama, through a series of executive actions has paved the way to prepare the United States for the impacts of climate change and also to lead international efforts to reduce carbon pollution to address global climate change. He believes that no other country is better equipped to lead the world to address this crisis. It is this belief that has led the Obama Administration to do more to combat climate change than ever before.

In June 2013, President Obama using his executive authority put forward America's first Climate Action Plan - a historic moment in U.S. action on climate change to strategically achieve three overarching goals:
- Cut domestic carbon pollution
- Prepare the United States for climate change impacts
- Lead international efforts to address global climate change

The U.S. Environmental Protection Agency (EPA) under the auspices of the Climate Action Plan established the first ever standards regulating carbon pollution from new and existing power plants, a major breakthrough in controlling greenhouse gas (GHG) emissions. Power plants are among the largest emitters accounting for one-third of all US GHG emissions. It is estimated that the new standards for existing power plants will result in a 30 percent reduction in carbon pollution by 2030 and generate climate and health benefits worth an estimated \$55 billion to \$93 billion per year.

To further reduce GHG emissions, the Climate Action Plan details steps to cut emissions of hydrofluorocarbons (HFCs) and methane, two pollutants that are more potent GHGs than carbon dioxide. Given that these are among the fastest-growing GHGs in the world, efforts to curb their production is needed both at the domestic and international fronts. Domestically, US EPA has proposed measures to transition from HFCs to climate-friendly alternatives. The Administration is working with its international partners to phase out the production and consumption of HFCs globally. The Administration's Strategy to Reduce Methane Emissions takes steps to cut methane emissions from landfills, oil and gas, agriculture, and mining sectors. Reaching the goal set for cutting methane emissions is projected to generate enough energy to heat more than 2 million homes for a year.

To promote a clean energy economy, the Department of Interior under the auspices of the Climate Action Plan opened public lands for renewable energy projects for the first time and approved over 50 solar, wind, and geothermal utility scale projects on public and tribal lands. Increased funding for clean energy technology and efficiency improvements under the Plan includes initiatives to expand and modernize the nation's electric grid by establishing energy corridors in western states, expediting the review of transmission projects in non-Western states; and improving the siting, permitting and review of transmission projects.

The Obama Administration recognizes that to create and sustain a clean energy economy, it is not sufficient to just adopt cleaner forms of energy, but that technological advances in the sector are essential. With this in mind, the President under the Department of Energy, created the Advanced Research Project Agency – Energy (ARPA-E) in 2009. ARPA-E helps to advance high-impact energy projects and to develop programs that will keep the United States at the forefront of clean energy research, development, and deployment.

To prepare the US to address climate change impacts, the Obama Administration has put in

place mechanisms for building a Clean Energy Infrastructure. One of the actions under this effort has been to propose fuel economy standards – the toughest in U.S. history to date - for passenger, commercial, and heavy-duty vehicles. These standards are projected to save over 500 million barrels of oil and save vehicle owners and operators an estimated $50 billion in fuel costs. The new fuel economy standards have brought down both U.S. oil demand and oil prices at the pump thus creating savings for consumers. Together with increasing savings and reducing emissions, these standards also decrease US dependence on foreign oil, key to our national security.

To further reduce America's GHG emissions and increase its energy efficiency, the Obama Administration has invested in several initiatives to develop energy conservation standards for appliances and equipment, and for buildings. The President's Better Buildings Challenge advances energy efficiency of billions of square feet of building space. This initiative also ensures that rural America is not left behind and that it is also made energy efficient.

Even before the Climate Action Plan was launched, the President had set aggressive goals to achieve energy efficiency. The investments made under the American Recovery and Reinvestment Act of 2009 have led to a ten-fold increase in solar-generated electricity and tripled wind-generated electricity production. These initiatives besides reducing the impact of climate change are also health protective. Consequently, these measures will not only save lives and prevent illnesses, the investments in clean energy will also help build a robust economy.

To ensure that the US defense system is climate resilient, the Department of Defense (DOD) is also on board the President's Climate Plan. The DOD, the single largest consumer of energy in the United States has released a roadmap in 2012 that transforms energy use in military operations to become more climate friendly and efficient.

The Administration recognizes the role of communities in building the nation's climate preparedness and resilience. To support the disparate local needs of communities from Alaska to Louisiana, from Maine to Texas, the President established a Task Force on Climate Preparedness and Resilience. The Task Force —made up of state, local, and tribal leaders with experience in building climate-resilient communities—is tasked with advising the federal government on how to address the needs of communities across the nation dealing with impacts of climate change.

The experience of rebuilding post Hurricane Sandy was used by The President's Hurricane Sandy Task Force to develop a Rebuilding Strategy, which is being used as a model by communities across the nation recovering from disasters. The Administration is also committed to ensuring that recovering communities are not only rebuilt but rebuilt to be climate resilient to ward off future climate disasters such as sea level rise, extreme storms, fires, and heat waves.

To assess the impacts of climate change in the United States, the Obama Administration relies heavily on scientific tools and evidence for climate-based decision making. In 2014, the Administration released the Third U.S. National Climate Assessment (NCA), the most authoritative and comprehensive source of scientific information ever generated about climate change impacts in the United States. The information presented in the NCA is a valuable resource for decision makers at the national, state, and local levels to become prepared and climate response ready.

According to the NCA, climate change is already impacting the lives and livelihoods of people in the United States. Climate Change is considered the biggest public health threat of this century. Compared to previous assessments, the health impacts of climate change featured prominently in the third NCA for the first time reflecting the growing evidence about the range of health impacts, including asthma and allergies, cardiovascular disease, and others. It also notes that, "there is growing recognition that the magnitude of health 'co-benefits,' like reducing both pollution and cardiovascular disease could be significant, both from a public health and an economic standpoint.

Besides impacting the public's health, climate change also threatens the healthcare infrastructure in the event of extreme weather scenarios. The Department of Health and Human Services, caretaker of the nation's health, has been tasked by the Administration to develop resources and tools to promote hospital resilience. As part of this effort, the Administration developed a best practice guide to promote continuity of care before, during, and after extreme weather events for policy makers, health care providers, and others.

In leading international efforts to address global climate change, the Obama Administration has forged some key bilateral co-operations, specifically with India and China, two countries that alongside the U.S. are among the top GHG emitters. India and the US have pledged to enhance cooperation in clean energy. The US has agreed to support India in its bid to increase its share of renewable energy in electricity generation through the US-India Joint Clean Energy Research and Development Center, a $125 million effort jointly funded by both governments and the private sector. Other initiatives under this joint agreement include launching an Air Quality Cooperation wherein the United States will implement EPA's AIRNow program in India's megacities; the United States will work with India to adopt cleaner fuels, and emissions and efficiency standards in heavy-duty vehicles; help India develop its own Climate Resilience Toolkit. The historic U.S.-China Joint Announcement on Climate Change signed in 2014 reiterates the personal commitment of Presidents Obama and Xinping to address climate change by implementing domestic policies, strengthening bilateral co-operation, and by promoting sustainable and climate-resilient economies, and marks a new era in climate diplomacy. Last but not least, the signing of the Paris Climate Agreement - a treaty dealing with GHG emission mitigation, adaptation, and finances and considered a historic turning point in reducing global warming – by 177 United Nations member countries is considered by many as the cornerstone of President Obama's climate legacy.

President Obama, one of the most liberal Presidents in United States history, also faced one of the most hostile Congresses, who regularly obstructed votes on his proposed programs as well as blocking or refusing to consider nominations to vacant offices. As a result, many of President Obama's accomplishments were not legislated into law but were enacted by executive order.

Banalata Sen

Further Reading

The White House Office of Press Secretary FACT SHEET: The United States and India – Moving Forward Together on Climate Change, Clean Energy, Energy Security, and the Environment https://www.whitehouse.gov/the-press-office/2016/06/07/fact-sheet-united-states-and-india-%E2%80%93-moving-forward-together-climate

The White House Climate Action Plan https://www.whitehouse.gov/sites/default/files/image/president27sclimateactionplan.pdf

National Climate Assessment http://nca2014.globalchange.gov/

The White House Office of the Press Secretary US-China Joint Presidential Statement on Climate Change https://www.whitehouse.gov/the-press-office/2016/03/31/us-china-joint-presidential-statement-climate-change

See also: Climate change, Greenhouse gases and global warming; Climate Project

Ocean acidification

Categories: Pollution and waste; oceanography

Definition

Ocean acidification describes the decrease in pH of the oceans due to increased concentration of dissolved carbon dioxide (CO_2). CO_2 is a trace gas found in the atmosphere. The concentration of dissolved oceanic CO_2 is in equilibrium with the gas's atmospheric concentration, such that an increase in the atmospheric concentration leads to an increase in the dissolved concentration, and vice versa. Throughout geologic time, atmospheric

CO_2 reflected a balance between various sources and sinks found on land and in the oceans—such as photosynthesis, respiration, chemical weathering of rocks and burial of organic and inorganic carbon. Since the onset of the industrial age, human activities, such as fossil fuel combustion and intensification of agriculture, represent a new source of atmospheric CO_2.

These activities are thought to cause the observed increases in atmospheric CO_2, which in turn lead to increased dissolved CO_2 concentrations in the oceans. Dissolved CO_2 combines with water to form carbonic acid (H_2CO_3), which subsequently dissociates to produce bicarbonate (HCO_3^-) and carbonate ions (CO_3^{2-}) and protons (H^+). The balance among these different reactions is such that increased concentrations of dissolved CO_2 lead to increased proton concentrations (acidity). Since pH is an inverse scale of the concentration of dissolved H^+, the higher the proton concentrations, the lower the pH and the greater the acidity. Modeling experiments using different scenarios to project future atmospheric CO_2 concentrations indicate that by 2100, the ocean's pH will drop by up to 0.45 unit.

Significance for Climate Change

The reactions by which CO_2 and water form H_2CO_3, HCO_3^- and CO_3^{2-} control the pH of the world's oceans. Oceanic pH has important ecological consequences, because many plankton (including the *Coccolithophoridae*, which are marine algae, and the *Foraminiferida*, which are planktic protists) and macrofauna (corals, mollusks, brachiopods, and so on) precipitate the minerals calcite or aragonite (which both have the chemical formula $CaCO_3$) to make exoskeletons, shells, and tests in a process called calcification. Under normal conditions, calcite and aragonite are stable minerals in surface waters, because CO_3^{2-} concentrations are naturally maintained at levels that prevent their dissolution. However, increasing atmospheric CO_2 concentrations lead to decreased oceanic pH values and a concomitant decrease in carbonate ion concentration.

This effect is currently observed in a vertical profile of the oceans, where carbonate ion concentrations decrease with increasing depth. At a depth

Effects of Ocean Acidification on Coral Reef Calcification Rate

Atmospheric CO_2 Level[a]	Coral Calcification Rate[b]
280 (preindustrial)	70
380 (current)	55
560	20
840	5

Data from National Oceanographic and Atmospheric Administration.
a. parts per million.
b. millimoles per square meter per day.

known as the lysocline, the rate of carbonate mineral dissolution increases rapidly with decreasing carbonate ion concentrations. Thus, in surface and intermediate waters, carbonate minerals are not dissolved. Below the lysocline, carbonate minerals are readily dissolved. Many laboratory experiments seem to indicate that ocean acidification will prove detrimental to many marine ecosystems. This finding has been recently validated by a study of a benthic ecosystem located near a naturally occurring volcanic vent that delivers CO_2 to the surrounding waters. The full-scale consequences of acidification of the global ocean are likely to be numerous, and could include extinctions as food webs collapse due to a loss of calcareous planktonic primary producers and consequences for the strength of the biological pump.

The biological pump is the process by which CO_2 is actively removed from the atmosphere by primary producing phytoplankton who convert the CO_2 to organic matter via photosynthesis. When these organisms die, their hard parts act as ballast to help the organic matter sink to the ocean floor, where it is buried in sediments. Ultimately, this sequesters the CO_2 in rocks for geologic time scales. However, if ocean acidification leads to dissolution of calcareous tests, organic matter will not be effectively buried because it is not dense enough to settle to the ocean floor. Not all scientists agree with the negative predicted consequences of ocean acidification. They argue that carbonate minerals present

in sedimentary rocks on the ocean floor should, over time, consume the excess H^+ produced from increased atmospheric CO_2 concentrations, causing ocean acidification to slow or even stop.

The increase in atmospheric CO_2 concentrations appears to be linked to an increase in global temperatures. Temperature is a strong control on the conversion of atmospheric CO_2 to dissolved CO_2 in the ocean. Henry's Law states that the dissolved concentration of a gas is proportional to the partial pressure of that gas in the atmosphere in contact with that liquid. Gases are characterized by different Henry's Law constants that give the proportionality of dissolved gas that will be in equilibrium with the overlying atmosphere. For most gases, including CO_2, the value of the Henry's Law constant decreases with increasing temperature—that is, at higher temperatures, the amount of CO_2 that can be dissolved in the ocean will decrease. This temperature dependence provides a negative feedback on the amount of ocean acidification that could occur. However, this also means that a greater proportion of anthropogenic CO_2 would remain in the atmosphere as global temperature rises.

Anna M. Cruse

Further Reading

Doney, Scott C. "The Dangers of Ocean Acidification." *Scientific American* 294, no. 3 (March, 2006): 58-65. Discusses the impact of fossil fuel combustion on the world's oceans. Examines the ecosystem effects of increased ocean acidity.

Hall-Spencer, Jason M., et al. "Volcanic Carbon Dioxide Vents Show Ecosystem Effects of Ocean Acidification." *Nature* 454, no. 7200 (July 3, 2008): 96-99. Presents data from an ecosystem-scale study that used a naturally occurring volcanic CO_2 vent as an analogue for the effects of ocean acidification caused by increased anthropogenic CO_2 concentrations.

Loáiciga, Hugo. "Modern-Age Buildup of CO_2 and Its Effects on Seawater Acidity and Salinity." *Geophysical Research Letters* 33, no. 10 (May 26, 2006). Uses an equilibrium model of carbonate speciation to conclude that increased atmospheric CO_2 concentrations will not cause ocean acidification on time scales of hundreds of years.

Pandolfi, John M., Sean R. Connolly, Dustin J. Marshall, and Anne L. Cohen. "Projecting coral reef futures under global warming and ocean acidification." *science* 333, no. 6041 (2011): 418-422. Corals have survived half a billion years under conditions far more extreme than projected global warming scenarios. However, their response to rapid changes is less well known. There have been a number of sharp declines in corals coinciding with other mass extinctions.

Royal Society. *Ocean Acidification Due to Increasing Atmospheric Carbon Dioxide.* London: Author, 2005. This policy document provides a comprehensive overview of ocean acidification and its possible consequences. Illustrations, figures, maps, references, glossary.

See also: Ekman transport and pumping; Freshwater; Groundwater; Hydrologic cycle; Ocean-atmosphere coupling; Ocean dynamics; Ocean life; Petroleum hydrocarbons in the ocean; pH; Plankton.

Ocean disposal

Category: Pollution and waste

Definition

Ocean disposal ranges from dumping toxic chemicals, sewage, dredge spoils (material removed during dredging), and bilge water to accidental spills from oil tankers. Dumping of waste and dredge spoils has long been a part of the United States' waste process, as it provides an economically viable option for disposing millions of metric tons of waste. Dumping waste into the oceans has increasingly been shown to adversely affect aquatic ecosystems and contribute to global warming. There are strict ocean disposal regulations in the United States, including the Marine Protection, Research, and Sanctuaries Act (MPRSA) of 1972 and the Clean Water Act (CWA) of 1972; nevertheless, illegal ocean disposal still occurs frequently.

Significance for Climate Change

Toxins from pesticides, as well as heavy metals, phosphorus, and nitrogen from dredge spoils, are a few of the harmful wastes dumped into the ocean. The effects of ocean disposal are not fully understood, but ocean disposal has been shown to cause substantial loss of marine life and indirectly enhance global warming. When plants and animals die, the dead organisms decay and release carbon dioxide (CO_2). CO_2 collects in the atmosphere and traps the Sun's heat, increasing the Earth's temperature. Recent consequences attributed to this climate change include the melting of glaciers, rising sea levels, disruption of coral reef habitats, and more extreme hurricanes.

In addition to the evidence of climate change caused by ocean disposal, hazardous waste disposal, such as nuclear refuse, can cause genetic defects, leading to cancer or death in both humans and marine life. Because of this danger, long-term burial of hazardous waste at sea is controversial. From the 1940's through the 1960's, the United States discarded barrels of radioactive waste in the ocean. The Environmental Protection Agency (EPA) then prohibited all dumping as a result of many of the barrels leaking.

Most of the material being dumped into the ocean is dredged waste. One approach to long-term storage of dredge spoils entails drilling holes into the ocean's floor and saturating the holes with the dredged waste. San Francisco, California, houses one such site approximately 91 kilometers offshore. This site is the deepest ocean disposal site in the United States, covering 22 square kilometers at a depth of 2,500-3,000 meters. The consequences of drilling deep into the ocean floor are not well known, raising questions of the short- and long-term effects of dredge spoils deposited in deepwater sediment on deep-sea biodiversity.

Sandie Zlotorzynski and
Kathryn Rowberg

See also: Ocean acidification; Ocean life; Petroleum hydrocarbons in the ocean; Water resources, global; Water resources, North American.

Ocean dynamics

Categories: Meteorology and atmospheric sciences; oceanography

The movement of water in the Earth's oceans is the primary process that transports heat from the tropics to the polar regions, keeps the oceans uniform in chemistry, and transports heat between the ocean's surface and its floor.

Key concepts

gyres: large, rotating loops of ocean current found in all major oceans; they are driven westward by the trade winds near the equator and eastward by the westerlies at high latitudes

meridional: referring to the motion of air or water in a generally north-south direction, that is, along meridians

thermohaline circulation: a vertical circulation in the oceans that is mostly driven by water-density differences, which in turn are governed by temperature and salinity

trade winds: twin wind belts on either side of the equator that generally blow westward

westerlies: belts of wind in midlatitudes that generally blow from west to east

Background

Motions in the Earth's oceans include tides, which are caused by the gravity of the Sun and Moon; surface currents, which are driven by wind; and thermohaline circulation, which is driven by density differences in seawater. All of these motions move large amounts of water from place to place. In so doing, they also transport heat from equator to pole and from the surface to the deep ocean, and they also transport dissolved chemicals, including greenhouse gases. Although the saltiness of ocean water varies from place to place, the relative proportions of dissolved materials—say, the ratio of sodium to chlorine—are extremely uniform. Ocean dynamics keep the oceans thoroughly mixed and profoundly affect Earth's temperature; hence, they are important in climate modeling. Although waves are the most obvious water motion to most people, the water in waves merely oscillates back and forth; that is, waves do not transport water long distances and thus are not discussed in this article.

Tides

Tides are the result of the gravitational attraction of the Moon and Sun. Although the Sun is far more massive than the Moon, its much greater distance means that its tidal effect is only about half that of the Moon. Nevertheless, if the Earth lacked a Moon, it would still have appreciable tides.

The continents prevent water from moving freely, so the actual movement of the tides is very complex. In most ocean basins the high and low tides revolve like the spokes of a wheel or the wave in a drinking glass oscillating in a circle. Tides move water through the connections between oceans and are important in keeping ocean water uniformly mixed. In small bodies of water it can take hours for the tides to progress from one end to the other, so tide predictions have to be based on local observations as well as the positions of the Sun and Moon.

Solar and lunar tides affect each other appreciably. Solar and lunar high and low tides can reinforce each other or partially cancel each other out. When the Earth, Sun, and Moon are in a straight line, at new or full moon, solar and lunar tides reinforce each other. The range between low and high tide is large, a condition called spring tide. When the Sun and Moon are 90° apart, as at first or last quarter moon, solar and lunar tides partially cancel each other out. The range between low and high tide is unusually small, a condition called neap tide.

Ocean Currents

Surface ocean currents are driven by the winds. Generally, water near the equator is pushed west by the trade winds until it strikes a continent. Most of it then is diverted poleward, where it encounters the prevailing westerlies and is pushed east. Once it reaches the eastern side of the ocean, most of the water is diverted toward the equator. Thus, in all the ocean basins there is a large loop, or gyre, rotating clockwise in the Northern Hemisphere and counterclockwise in the Southern.

Scientists in the Argo program deploy a float to monitor ocean temperature and salinity. More than three thousand such floats are deployed worldwide. (NOAA)

Around the Antarctic is a unique geography, a belt of latitude consisting entirely of ocean. With no topography to hinder them, the westerly winds in the Southern Hemisphere, called the Roaring Forties, create some of the roughest seas in the world. They also create a globe-girdling current, the Circum-Antarctic Current, that continuously circles Antarctica and is the principal mechanism for transferring water from one ocean to another.

Thermohaline Circulation

The densest seawater on Earth is found around the Antarctic. The water is dense because it is both cold and salty. The water is salty because freezing of sea ice leaves dissolved salt concentrated in the remaining liquid. This cold, dense water sinks to the bottom and flows northward as a dense layer called Antarctic bottom water. The largest amount of Antarctic bottom water flows beneath the Pacific until it reaches Alaska, where it rises and merges into the surface circulation. It then flows around the North Pacific gyre until it reaches the southwest Pacific,

where it has warmed to over 30° Celsius. Some warm water circulates through Indonesia to the Indian Ocean, and even though the amount of water involved is fairly small, the amount of heat transferred is large. Warm Indian Ocean water rounds Africa, cooling somewhat, then warms again in the South Atlantic. Some warm water crosses into the North Atlantic, travels up the eastern coast of North America as the Gulf Stream, then crosses to Europe. Finally, in the Arctic, the water cools and sinks. It then begins traveling south as North Atlantic deep water. Antarctic bottom water is also creeping northward in the Atlantic, colder and denser than North Atlantic deep water. Thus, in the North Atlantic, Gulf Stream water is moving northward on the surface, Antarctic bottom water is moving north along the bottom, and North Atlantic deep water is moving south just above the Antarctic bottom water. Because much of the water movement is northward or southward, thermohaline circulation is sometimes called meridional overturning circulation.

Context

Large-scale movements of ocean water affect local and global climate by transporting heat. They also affect climate change by transporting dissolved greenhouse gases into the deep ocean for storage and back to the surface for release. One aspect of thermohaline circulation has received particular attention. Many scientists are convinced that a sudden release of freshwater from glacial lakes in North America covered the North Atlantic with a layer of freshwater that prevented the exchange of heat between the ocean and the atmosphere and caused a sharp cooling event, called the Younger Dryas, about twelve thousand years ago. Some have suggested that melting of the Greenland ice cap might have a similar effect, so that warming of the climate might, paradoxically, produce a cooling episode.

Steven I. Dutch

Further Reading

Daniel, Tom, Justin Manley, and Neil Trenaman. "The Wave Glider: enabling a new approach to persistent ocean observation and research." *Ocean Dynamics* 61, no. 10 (2011): 1509-1520. A new device that consists of a submerged glider and instrument platform tethered to a surface float. The glider uses wave energy to propel itself. Test vehicles have survived large storms, transited thousands of kilometers, and remained active for over a year.

Denny, Mark. *How the Ocean Works: An Introduction to Oceanography.* Princeton, N.J.: Princeton University Press, 2008. Summary of oceanography for introductory college students.

Garrison, Tom S. *Essentials of Oceanography.* 4th ed. Belmont, Calif.: Brooks/Cole, 2006. Summary of oceanography for introductory college students.

Pinet, Paul R. *Invitation to Oceanography.* Sudbury, Mass.: Jones & Bartlett, 2009. Discusses geology, biology, chemistry, and physics of the oceans at an introductory college level. Includes a chapter on climate change.

Stewart, Robert H. *Introduction to Physical Oceanography.* College Station: Texas A&M University, 2004. Available at http://oceanworld.tamu.edu/home/course_book.htm. This treatment of oceanography offers extensive discussion of the physics of water movement in the oceans. Intended for upper-level college students.

See also: Atlantic heat conveyor; Atlantic multidecadal oscillation; Climate and the climate system; Ekman transport and pumping; El Niño-Southern Oscillation; General circulation models; Gulf Stream; Gyres; La Niña; Marginal seas; Mean sea level; Mediterranean Sea; Meridional overturning circulation; North Atlantic Oscillation (NAO); Ocean acidification; Ocean-atmosphere coupling; Ocean life; Rainfall patterns; Sea-level change; Sea sediments; Sea surface temperatures; Slab-ocean model; Thermocline; Thermohaline circulation; Younger Dryas.

Ocean life

Categories: Oceanography; animals; plants and vegetation

Oceans, which cover 71 percent of the Earth's surface, respond more slowly than land to global warming. This warming, however, may cause damage to Arctic and Antarctic ecosystems due to melting sea ice, bleaching and die-off of reef corals, disruption of ocean currents, and shifting predator-prey relationships favoring reduced biodiversity. Long-term effects could include massive extinctions due to changes in seawater chemistry.

Key concepts

dead zones: areas of deepwater oxygen depletion due to surface algal blooms or disruption of thermohaline circulation

El Niño-Southern Oscillation (ENSO) events: periodic fluctuation of temperatures and currents in the Pacific Ocean on a four-, ten-, and ninety-year cycle

primary productivity: production of fixed carbon through photosynthesis

thermohaline circulation: the rising and sinking of water caused by differences in water density due to differences in temperature and salinity

Background

Until quite recently, scientists and the general public considered the Earth's oceans to be impervious to anthropogenic degradation. Oceans cover 71 percent of the Earth's surface and account for a little less than half of its primary productivity, that is, the photosynthetic conversion of carbon dioxide (CO_2) into the organic compounds that make up the bodies of living organisms.

The ocean is far from being a uniform habitat; however, with the exception of some near-shore environments, ecological niches cover wide areas and intergrade, meaning that species can readily adapt by shifting their ranges. In consequence, environmental pressures producing elevated extinction rates on land have a less dramatic effect in the open ocean.

Nonetheless, human activity has had an adverse effect on marine life, from phytoplankton to top marine predators such as sharks. While overfishing, pollution, and damming of rivers that serve as spawning grounds for marine fish have all taken their toll, these are only indirectly related to global warming.

Present effects attributable to elevated land temperatures include displacement of currents and upwelling zones and increased runoff in major river systems. Effects attributable to elevated sea surface temperatures include melting sea ice in the Arctic and Antarctic, reducing habitat for polar bears, penguins, and the many humbler species of plants and animals that thrive at the margins of the polar ice caps. In the tropics, higher sea surface temperatures alter coral metabolism, causing corals to bleach when they lose symbiotic algae (zooxanthellae) critical to their growth. Degradation of coral reefs profoundly affects the many organisms restricted to this habitat.

If atmospheric CO_2 continues to increase, altered seawater chemistry will become a concern. An increase in dissolved CO_2 increases acidity, which in turn inhibits production of shells. Coelenterates such as corals, whose skeletons are made up of aragonite, are more susceptible than are mollusks that have calcite shells. Present-day stunting effects attributable to this cause have not yet been observed in nature, but such stunting is suspected in the geologic record, so coelenterates are being closely monitored.

Climate Change and Marine Life in the Geologic Record

Scientists recognize at least five major global mass-extinction events, of which the one at the Permian-Triassic boundary, 251 million years ago, was the most devastating. At that time, 95 percent of marine genera and 70 percent of land genera became extinct in three distinct pulses over a period of about eighty thousand years. At least two respected theories suggest that climate change caused these extinctions.

According to these theories, massive volcanic eruptions in Siberia started the catastrophe in motion. Each eruption caused cooling due to atmospheric volcanic dust and sulfur dioxide, followed by warming due to the longer-lived carbon-dioxide, augmented by heightened decomposition. Over the course of a million years, repeated eruptions, dwarfing anything humans have experienced in their brief tenure on Earth, eventually overwhelmed the Earth's capacity to self-correct.

One theory postulates that the ocean depths became increasingly oxygen-depleted, favoring the growth of bacteria that produce hydrogen sulfide. High pressures and cold temperatures in the abyss allowed that gas to build up, only to be released in a gigantic "burp" of highly toxic fumes. Another theory points to the storage of large quantities of methane in the form of clathrates in deep-sea sediments. It suggests that this methane was abruptly released when warming raised the temperature of the deeper regions of the ocean by 5° Celsius. In addition to being toxic and a powerful greenhouse gas (GHG), methane is explosive at concentrations as low as 5 percent.

Whatever the cause, the extinction, which devastated every group of plants and animals, was extremely abrupt by geological standards. Following the cataclysm, sedimentary rocks are nearly bare of fossils for the first ten million years of the Triassic period.

The Present and Near Future

Unless the most carefully researched models are far off the mark, nothing resembling the devastating geochemical upheavals of the Permian-Triassic period looms in the foreseeable future, even if present levels of fossil fuel consumption persist. These models presuppose that volcanic activity will continue at levels typical of the Holocene and that no asteroids are headed in Earth's direction.

Possible changes due to increasing acidity are being closely monitored, but so far no notable effects on organisms have been observed in nature. Scientists working with a tiny marine snail that is crucial to Antarctic food chains have demonstrated, however, that there is ample cause for concern, because both higher temperatures and increasing acidity inhibit growth and shell production and can be expected to act synergistically. Acidity alone is not expected to reach lethal levels for another fifty years, but temperatures are rising rapidly, and marine organisms in the Antarctic cannot adjust their ranges southward. A wide variety of fish and birds depend on this snail for continued survival.

Polar Regions

In both the Arctic and the Antarctic, chilled surface seawater sinks, allowing nutrient-rich waters to well up from below and support high phytoplankton productivity. The polar seas teem with life. The lower surfaces of ice sheets also support dense growth of attached algae. Global warming near the North Pole causes the most productive zone to retreat northward and contract in extent. This restricts the number of herbivores and carnivores the system can support. Most polar animals are unable to extend their ranges into temperate seas, because their unique adaptations to frigid temperatures make them poor competitors and susceptible to disease in warmer climates. The situation in the Southern Hemisphere is even more acute, as species migrating southward encounter the continental margin.

The plight of polar bears has received considerable attention. These huge carnivores prey almost entirely on seals that they hunt on sea ice; they hibernate on land. The seals are declining in numbers and retreating farther from shore as the ice cap shrinks. Bears are starving and failing to reproduce. Whale populations that had begun to recover from overexploitation by the whaling industry are also declining again as a result of low food supplies. Antarctic penguins also face declining food supplies and an influx of predators, including sharks, which are extending their ranges southward.

Coral Reefs

Reef-building corals, and the numerous species that depend upon them, have a narrow temperature range for optimum growth. They are also vulnerable to changes in sea level due to either global warming or global cooling. During the last Pleistocene glaciation, the resulting drop in sea level exposed much of Australia's Great Barrier reef, restricting this unique ecosystem to isolated pockets. A rapid rise in sea level would damage existing reefs by reducing light levels below those needed by symbiotic algae.

A 2° Celsius rise in surface temperature is sufficient to cause bleaching in corals as the individual polyps eject symbiotic algae. Bleaching initially causes growth to cease and eventually kills the coral colony. In recent years, there have been massive die-offs of corals—80 percent in the Caribbean and 50 percent in the South Pacific—but it is uncertain how much of this is due directly to global warming. The die-off in the South Pacific was associated with a severe El Niño-Southern Oscillation (ENSO) event to which global warming may have contributed. Near-shore pollution also devastates coral reefs in populated areas.

Dead Zones

In a number of parts of the world, extensive areas of ocean have become depleted in oxygen, turning once productive fisheries into wastelands. Most of these dead zones are associated with rivers draining populated areas; one of the largest lies offshore of the mouth of the Mississippi River. This dead zone owes its existence to influxes of nutrient-laden freshwater to the Gulf of Mexico. These nutrients stimulate massive algal blooms. There is an indirect connection to global warming, in that warming generally causes increased precipitation and therefore increased runoff. Dead zones off the west coasts of the United States and South Africa result from disruption of cold currents and associated upwelling zones and thus may be directly related to global warming.

Productivity

While global warming due to elevated CO_2 levels can cause local drops in productivity due to drought on land and disruption of thermohaline cycles in the ocean, the long-term predicted effect, on a global scale, of such warming is a net increase in photosynthesis, with an upper limit that far exceeds any projections based on realistic economic indicators. In the long term, this is good news. If the Earth is producing more food, both the numbers and the diversity of herbivores and predators can be expected to increase.

In the short term, however, such changes lead to the proliferation of weedy species with high reproductive rates and broad ecological ranges, loss of diversity, and generally unstable conditions. Species with specialized ecological requirements become extinct, and natural ecosystems increasingly resemble intentional agriculture or aquaculture. A glimpse of the future may be gleaned from the formerly rich fisheries off the West Coast of North America. These have been in decline for several decades, mainly because of pollution and over-fishing. Warmer waters coupled with a persistent dead zone off the coast of California and Oregon have further reduced stocks of commercial and sport fishes, but they have favored proliferation of the Humboldt giant squid, an aggressive predator adapted to warm temperatures and low oxygen levels. Sport fishermen are now being lured with the prospect of landing a 34-kilogram squid that puts up a mean fight.

Rising temperatures can be expected to reduce areas of high planktonic productivity near the poles while expanding them near the equator and at continental margins, threatening polar species with starvation and extinction while increasing numbers in warmer climates without a corresponding increase in diversity.

There are undoubtedly reef-building organisms ready to replace corals should the seas become inhospitable. During the very warm late Cretaceous, rudists, a group of bivalve mollusks related to clams, were the main reef builders. Several types of algae also have limestone skeletons. If any of these groups were to replace corals, the structural integrity of reefs would be preserved, but the beauty and diversity of the ecosystem would be sadly compromised on any conceivable human time scale.

Context

The main threats to the abundance and diversity of marine life derive from Earth's human population explosion with its concomitant overexploitation and pollution of coastal waters. The exploding population is also a major factor in global warming. In terms of direct threats posed to marine life from rising temperatures, dwindling sea ice in the Arctic and especially the Antarctic is probably the most clear-cut. While anthropogenic warming is undoubtedly a factor in coral reef destruction and the decline of fisheries, it may well not be the

principal cause. As scientists learn more about long-term cycles involving ENSO and analogous oscillating pressure and current systems in other oceans, a better understanding of the relationship of current extreme events to long-term trends should emerge.

Martha A. Sherwood

Further Reading

Benton, Michael J., and Richard Twitchett. "How to Kill (Almost) All Life: The End-Permian Extinction Event." *Trends in Ecology and Evolution* 18, no. 7 (July, 2003): 358-365. Attributes this mass extinction to massive methane release producing a catastrophic breakdown in normal temperature feedback mechanisms.

Bottjer, David J. "Life in the early Triassic ocean." *science* 338, no. 6105 (2012): 336-337. In the early Triassic Period (roughly 252 to 247 million years ago) the oceans were very hot and unusually acidic, with large oxygen poor areas. Thus the early Triassic may provide a model for a greenhouse future earth.

Peters, Robert L., and Thomas Lovejoy, eds. *Global Warming and Biological Diversity*. New Haven, Conn.: Yale University Press, 1992. Includes a chapter on marine biodiversity that emphasizes coral reefs.

Reynolds, Colin S. *Ecology of Phytoplankton*. New York: Cambridge University Press, 2006. Technical discussion of dead zones and their relationship to global warming and of theories of the Permian extinctions.

Saltzman, Barry. *Dynamical Paleoclimatology: Generalized Theory of Global Climate Change*. New York: Academic Press, 2002. Treats interactions between abiotic, biotic, and anthropogenic variables; discusses controversies about the magnitude of human climatological impact.

See also: Dolphins and porpoises; Endangered and threatened species; Extinctions and mass extinctions; Fishing industry, fisheries, and fish farming; Invasive exotic species; Ocean acidification; Ocean disposal; Penguins; Plankton; Polar bears; Reefs; Whales.

Ocean-atmosphere coupling

Categories: Meteorology and atmospheric sciences; oceanography

Definition

Ocean-atmosphere coupling describes the interdependency between the temperatures and circulation of water in the ocean and those of air in the atmosphere. Changes in the surface temperature of ocean water produce changes in the atmosphere above the water, which alters wind patterns and leads to further changes in surface ocean temperature. If these changes are significantly large or long-lived, changes in atmospheric patterns capable of producing changes in global weather patterns can result. The most prominent example of this is the El Niño/La Niña weather cycle, which gained notoriety as the cause of numerous climate disruptions in North America during the late twentieth and early twenty-first centuries.

An image of the point of interface between atmosphere and ocean. (NOAA)

Significance for Climate Change

The specific heat of water—that is, the amount of energy required to alter water's temperature—is extremely high. As a result, Earth's water systems are a significant stabilizing influence on global surface temperatures. The ocean-atmosphere coupling is among the most significant points of distribution of water in the global hydrologic cycle. Water evaporates from the oceans into the atmosphere, which carries water vapor over land, where it precipitates, providing freshwater to terrestrial ecosystems. This cycle, and particularly the interface between air and sea, both directly affect and are affected by global climate patterns, particularly those involving temperature.

The significance of ocean-atmosphere coupling to global warming increased as cyclical meteorological phenomena, such as the El Niño/La Niña cycle in the southern Pacific and the North Atlantic Oscillation patterns affecting weather in northern Europe and North America, began accelerating in the early 1980's. Of particular concern was the alteration of cyclical El Niño periods, triggered by higher sea surface temperatures in the southern Pacific, and La Niña periods, caused by lower sea surface temperatures in this region. Seven El Niño periods took place during the 1980's and 1990's, while only three La Niña periods occurred. Many scientists regarded this phenomenon as possible evidence of the escalation of global warming.

The study of ocean-atmosphere coupling thus intensified during the late twentieth and early twenty-first century, as scientists sought to determine the causes and extent of global warming, as well as its future duration and potential for escalation. Many of these studies sought to determine whether global warming patterns have human causes, while others attempted to forecast changes in global weather patterns to determine the potential for future extreme weather events such as severe droughts, flooding, and drastic changes in temperature. Long-term studies of patterns of climate variability involving ocean-atmosphere coupling have failed to yield definitive answers to these inquiries, leading many scientists to conclude that these patterns must be examined for longer periods in order to determine their implications for global warming and its causes.

Michael H. Burchett

See also: Atmosphere; Atmospheric dynamics; El Niño-Southern Oscillation; Hadley circulation; Inter-Tropical Convergence Zone; La Niña; Ocean dynamics.

Offsetting

Categories: Pollution and waste; economics, industries, and products

Definition

Offsetting is the practice of compensating for pollution from one source or location by reducing or mitigating pollution from another—for example, planting trees to offset carbon dioxide (CO_2) emissions from vehicle use. Individuals have become more conscious of their possible impact on global climate (referred to as their "carbon footprint"), as well as of the difficulties of reducing GHG emissions. Thus, an idea has emerged that increases in greenhouse gas (GHG) emissions from a given source can be mitigated—or "offset"—by parallel activities to decrease emissions from another source or to remove carbon from the atmosphere and store it in sinks.

Significance for Climate Change

Using vegetation to remove carbon from the atmosphere was proposed as early as 1976 by physicist Freeman Dyson. A decade later, in 1988, one of the first projects to provide an offset for carbon emissions—the planting of 52 million trees in Guatemala—was undertaken at the instigation of the Arlington, Virginia, energy company Applied Energy Services to justify a proposed power station. By the late 1990's, celebrity participation began to popularize the idea of individuals offsetting their carbon-emitting activities by paying companies to engage in tree planting and other green activities.

For example, the Rolling Stones in 2003 highlighted "carbon neutral touring," paying the company Future Forests to plant trees to neutralize the environmental impact of their British concerts: The Edinburgh Center for Carbon Management calculated that one tree for every 60 of the band's anticipated 160,000 fans should suffice. (The following year, questions were raised about how Future Forests was spending its money, and the company subsequently renamed itself the Carbon Neutral Company.) In 2006, former vice president and later Nobel laureate Al Gore applied the idea to his influential book, *An Inconvenient Truth*: "By supporting a new Native American wind farm and a new family farm methane energy project through NativeEnergy, this publication is carbon neutral."

The emphasis on newness reflects the so-called additionality requirement—that any carbon decrease produced by an offset be additional to what would otherwise occur. Such counterfactual claims are notoriously uncertain. Though such offsetting does focus attention on the environmental impact of people's lifestyle choices, critics view it as either confused (because of insurmountable problems in definition and measurement) or self-deceptive (because people are questionably encouraged to believe their environmental impact has really been neutralized).

On a larger scale, analogous issues arise about carbon emissions trading among, or within, nations. In agreements such as the Kyoto Protocol's clean development mechanism, many countries have committed to reduce GHGs by certain future dates, but such agreements sometimes permit targets to be reached through credits for activities that remove CO_2 from the atmosphere (carbon sequestration) or that reduce the amount of CO_2 added to the atmosphere (alternative energy), even if emissions are otherwise increasing. Critics of carbon trading question whether pollution is really offset, pointing to scams, misjudgments, and counterproductive efforts. They also challenge the fairness of offsets, criticizing in the name of environmental justice what has been labeled "CO_2 lonialism." (See, for example, the 2004 Durban Declaration against carbon trading.)

"Carbon forestry," Larry Lohmann complained in 1999,

> proposes to lessen the atmospheric effects of the mining of fossil fuels by colonizing still other resources and exerting new pressures on local land and water rights; the community evicted by oil drillers today may find itself displaced by carbon-"offset" plantations tomorrow.

Nearly a decade later, Dyson remarked: "The humanist ethic accepts an increase of CO_2 in the atmosphere as a small price to pay, if worldwide industrial development can alleviate the miseries of the poor half of humanity." The empirical and ethical complexities of the offsetting issue are highlighted by the fact that Dyson and his critics both claim the moral high ground.

Edward Johnson

Further Reading

Dyson, Freeman J. "Can We Control the Carbon Dioxide in the Atmosphere?" *Energy* 2 (1977): 287-291. Originally a 1976 Institute for Energy Analysis occasional paper. Dyson proposes removing CO_2 from the atmosphere through tree cultivation.

_____. *A Many-Colored Glass: Reflections on the Place of Life in the Universe.* Charlottesville: University of Virginia, 2007. Predicts and lauds the power of genetic engineering to provide for the world's food and energy needs while controlling GHG emissions.

Gore, Al. *An Inconvenient Truth: The Planetary Emergency of Global Warming and What We Can Do About It.* Emmaus, Pa.: Rodale, 2006. The print version of Gore's call to action represents itself as "the first book produced to offset 100 percent of the CO_2 emissions generated from production activities with renewable energy."

Lange, Andreas, and Andreas Ziegler. "Voluntary Emission Reductions versus CO2 Offsetting: A Theoretical and Empirical Analysis for the US and Germany." Center of Economic Research at ETH Zurich Working Paper 1261. (2011). Investing in offsets is more likely at higher income levels, but further increases in income result in increases in the polluting activity. Also, as the price of offsets decreases, investment in offsets becomes more likely.

Lohmann, Larry. *The Dyson Effect: Carbon "Offset" Forestry and the Privatisation of the Atmosphere.* Sturminster Newton, Dorset, England: Corner House, 1999. Briefing paper critiquing "the privatization of carbon sinks" inspired by Dyson's proposal.

_____, ed. *Carbon Trading: A Critical Conversation on Climate Change, Privatisation, and Power.* Development Dialogue 48. Sturminster Newton, Dorset, England: Corner House, 2006. Study published by the Dag Hammarskjold Foundation, the Durban Group for Climate Justice, and the Corner House. Detailed, 362-page critical analysis of carbon trading and offsets.

Smith, Kevin, ed. *The Carbon Neutral Myth: Offset Indulgences for Your Climate Sins.* Amsterdam, the Netherlands: Transnational Institute/Carbon Trade Watch, 2007. A critique of the idea that carbon offsets could neutralize environmental impact; Jamie Hartzell's essay, "Offsets and 'Future Value Accounting,'" is particularly useful.

Willey, Zach, and Bill Chameides, eds. *Harnessing Farms and Forests in the Low-Carbon Economy: How to Create, Measure, and Verify Greenhouse Gas Offsets.* Durham, N.C.: Duke University Press, 2007. Collects a dozen essays and twenty-seven appendixes that discuss the practical specifics of establishing offsets.

See also: Baseline emissions; Carbon footprint; Certified emissions reduction; Clean development mechanism; Gore, Al; *Inconvenient Truth, An*; Kyoto Protocol.

Oil industry

Categories: Fossil fuels; economics, industries, and products

The history, globalization, exploration, and geographic distribution of oil reserves, along with their eventual depletion, will affect not only the future of oil markets but also the way humans react to climate change.

Key concepts

alternative energy: fuels and other energy resources that are renewable, nonpolluting, or both and have the potential to replace fossil fuels and other traditional energy resources

fossil fuels: energy resources formed from decayed organic matter under geological pressures over millions of years

greenhouse gases (GHGs): trace atmospheric gases that trap heat, preventing it from escaping into space

peak oil: the point at which oil availability and production reaches its zenith, before the Earth's oil resources begin either to dwindle or to become prohibitively expensive to exploit

Background

The oil industry is a capital-intensive industry, and, therefore, its story generally is one of enormous wealth and, through wealth, one of political power and influence. The modern form of oil industry began when John D. Rockefeller's Standard Oil monopoly was vertically integrated, spanning production, refining, transporting, and retailing operations. The industry's fruits were initially spread abroad by Rockefeller's fleet of kerosene tankers. At the same time, Standard Oil was also securing a tight hold on the U.S. oil market. At the beginning of the twentieth century, it already controlled 87 percent of production, 82 percent of refining, and 85 percent of all petroleum marketing operations in the United States.

In short order, however, a brace of developments turned both the U.S. oil market and the petroleum industry into a competitive, global operation. The application of the Sherman Antitrust Act (1890) and subsequent breakup of Standard Oil into its regional components in 1911 forced some of its newly independent, "oil-short" units (notably Standard Oil of New York and later Mobil Oil) to look abroad for the oil that its gas stations had previously acquired from other parts of the Standard Oil trust. At approximately the same time, the conversion of navy ships to oil prompted Britain and the United States to urge their nascent oil companies to explore abroad for secure sources of oil to service their fleets in remote parts of the world. Soon the ancestor of British Petroleum and Standard Oil of New Jersey (originally Jersey Oil and later Exxon in the United States and Esso in Canada), Rockefeller's core unit, were competing with one another for the status of the world's largest petroleum corporation. That competition would endure throughout the twentieth century.

World War I introduced aircraft, tanks, and ambulances to the battlefield, further underscoring the relationship between national security and a healthy oil industry. Taking advantage of that fact, by the time that Henry Ford introduced the assembly-line technique for making automobiles an affordable part of the average American's life, the U.S. oil industry had already used the war to turn the government in Washington from a trust-busting foe of big oil into one of its biggest supporters. Except for a few minor disruptions, that relationship lasted throughout the twentieth century, manifested in favorable tax laws, support of oil company efforts to stabilize the markets, and—following the rise of the Organization of Petroleum Exporting Countries (OPEC)—a willingness to allow the major oil corporations to undertake mergers akin to those that the Sherman Antitrust Act had been enacted to prevent. However, even before the rise of OPEC and those mergers, U.S. petroleum corporations had remained major players in the U.S. and global economies. On the eve of the 1973 oil crisis, the American petroleum industry was generating 30 percent of all domestic investment and 40 percent of all American investment in the developing world.

The benefits that the U.S. government offered to its smaller oil companies to go abroad and find new sources of oil to meet the growing demand for oil after World War II ultimately undermined the cartel of private oil companies that had stabilized

Rockers like this are familiar features of oil fields. The rocker raises and lowers a rod connected to a piston at the bottom of the well. The piston sucks oil up, and valves within the well prevent the oil from draining back down. (© Steven I. Dutch)

the international price of oil for two generations. Known as "the Seven Sisters," this cartel—composed of Exxon, Mobil, Standard Oil of California (SoCal, later Chevron), British Petroleum, Royal Dutch Shell, Texaco, and Gulf—accounted for 90 percent of all global production outside the United States and Russia, 80 percent of all refining operations, and 70 percent of all marketing operations in the early post-World War II years. Oil-producing states either sold their oil to these companies at the proffered price or did not sell their oil at all.

Encouraged by government incentives, in the aftermath of World War II, several smaller U.S. oil companies began to explore for oil abroad. More joined the pack when one of the first, Getty Oil, struck it rich by finding oil in Kuwait. Unlike the Sisters, these individual companies had little bargaining power. Gradually they cut into the share of the market controlled by the Sisters (whose control over production outside the United States and the Soviet Union dropped to 70 percent by 1970). More important, their individual operations were usually in one country only, and they either bought their oil from that state on its terms or did not acquire foreign oil at all. As the international oil market grew ever tighter during the 1968-1973 era of Western economic expansion, their host governments demanded—and received from these companies—much better financial payoffs than those offered by the Sisters to their producing states. Given the increasingly tight energy market, the Sisters had to extend the same deals to their host governments. Consequently, even before the October, 1973, Yom Kippur War led to the Arab oil embargo and OPEC's rise to prominence, the Seven Sisters' hold on the global industry was already eroding rapidly.

OPEC and the Global Economy

The 1973 Arab oil embargo on countries friendly to Israel created panic in the marketplace, as Western states bid against one another for oil that, in some instances, they did not have the storage facilities to accommodate. The price of oil on the spot market jumped from under three dollars per barrel (the Sisters' last posted price on the eve of the Yom Kippur War) to the twenty-dollar-per-barrel range. In turn, this hysteria allowed OPEC to buy out the Seven Sisters and other Western oil companies and establish itself as the new international cartel in charge of setting the price of oil. Subsequently, both the fortunes of the international petroleum industry and those of the international economy fluctuated with OPEC's fortunes and its ability to keep the price of oil stable and in a price range affordable enough to allow for overall economic growth and the economic development of Third World countries.

In general, OPEC's record has been spotty. The OPEC-endorsed price hikes in the 1970's—to twelve dollars per barrel in 1973 and to more than thirty dollars per barrel in 1979 (following the fall of the shah of Iran and resultant drop in the availability of Iranian oil in the market)—led to a prolonged recession in the oil-importing, economically advanced Western world throughout much of the 1980's, which depressed the price of OPEC oil significantly. As a result, oil was relatively cheap in the 1990's, which not only led to a renewed expansion of the global economy but also enabled both India and China to mount significant development plans fueled by low-cost, imported petroleum. With the tightening of the market at the turn of the twenty-first century and the uncertain

market conditions during the first decade thereof, OPEC again allowed the price of oil to soar to recession-inducing levels, slowing the base of globalization and, in many instances, encouraging countries to adopt protectionist policies antithetical to the ideals of a globalized economy.

Meanwhile, partly in order to survive the eras of depressed oil prices, the global petroleum industry reshaped itself, from the dominant Seven Sisters cartel of private oil companies into a complex mixture of private and state oil companies, complicated by the fact that not all of the state-owned oil companies in the world were the economic creatures of OPEC members.

Diversification, Mergers, and the Western Oil Corporations

When OPEC took over the international oil market in 1973, the possibility remained of discussing Western oil companies in the terminology that had been used for half a century. There were the "majors," the Seven Sisters, and then there were the "independents," that is, the comparatively small producers that included enterprises from family operations like Krumme Oil in Oklahoma to large, multinational oil companies like Getty Oil and Atlantic Richfield Company (ARCO). During the high-price energy era of the 1970's and early years of the following decade, all these companies reaped large profits, and the majors and many of the larger independents reinvested those profits in the pursuit of alternative energy schemes (Exxon, for example, invested five billion dollars in its oil-from-shale project in Colony, Colorado) and the acquisition of holdings in such other energy sectors as coal and uranium. These companies would do much the same vis-à-vis biofuels in the high-cost petroleum era that followed the U.S. invasion of Iraq in 2003.

The downturn in the price of oil in the 1980's caused the majors to shut down most of their alternative energy projects and to seek means for surviving in the lean years of the global recession. As these companies would do in even more publicized ventures during the 1990's, when the price of oil remained in the twenty-five-dollar-per-barrel range, many turned to mergers in order to economize, beginning with the 1984 merger between SoCal and

Gulf Oil that enabled Gulf to sell off many of its subsidiaries and service stations and led SoCal to change its name to Chevron. A decade later, what had been exceptional in 1984 became, momentarily, almost commonplace. The process began in December, 1998, when British Petroleum (BP) acquired Amoca (formerly the American Oil Company or Standard Oil of Indiana) for more than fifty billion dollars. The following year the combined corporation purchased ARCO (one of the major players in the discovery of Alaskan oil). Combined with its subsequent acquisition of Burmah Castrol, a lubricant manufacturing company, BP was able to pare approximately twenty thousand jobs worldwide and become temporarily the world's largest oil company.

BP held that distinction for only a few months until Exxon and Mobil merged in November, 1999, into the largest corporation in the world. The transaction enabled the combined corporation to sell off more than seventeen hundred service stations (to Tosco) and trim its payroll by nearly ten thousand employees. The merger mania did not end there, or on the U.S. side of the Atlantic Ocean. In 1999, two giant European petroleum firms, France's Total and Belgium's Petrofina, merged into TotalFina and then acquired France's other major petroleum company, Elf, to make TotalFina the fourth largest petroleum company in the world. Then, in 2001, another Sister-Sister marriage occurred, this time involving Chevron and Texaco.

In short, by the time oil prices began to rise appreciably shortly after the terrorist attacks on New York and Washington, D.C., on September 11, 2001, and especially following the U.S. invasion of Iraq in 2003, a new set of six "supermajors" had emerged in the world of private oil companies: Exxon-Mobil, BP-Amoco, Chevron-Texaco, TotalFina, Royal Dutch Shell, and Conoco-Phillips (whose two units completed their merger in August, 2002). All are vertically integrated and, compared to the "independents," have a commanding share of the market. Unlike that of the Seven Sisters, though, their power is rooted in sales, not in the production of oil: They accounted for only about 10 percent of the oil produced in the early years of the twenty-first century and their combined ownership of oil and gas accounts for less than 5 percent of the world's know oil and gas reserves.

State-Owned Petroleum Industries

Comparatively speaking, the real "supermajors" are not these private oil companies but the seven largest state-owned petroleum companies in the contemporary world, led by those of Saudi Arabia, Iran, Russia, and Venezuela, but also including the state-owned companies of China, Brazil, and Malaysia. Collectively, these seven account for nearly one-third of the world's gas and oil production and the majority of its know oil reserves. More important, when combined with the production capacity and reserve holdings of other state-owned oil companies, these actors account for the overwhelming majority of the world's oil and gas production and reserves. In that sense they are more analogous to the Seven Sisters than today's "supermajor" private oil corporations. However, unlike the Seven Sisters, they do not collaborate with one another. Quite to the contrary, they sometimes compete with one another for influence inside OPEC (where only Saudi Arabia, Iran, and Venezuela are represented) and for profits in the world's petroleum marketplace.

Oil Depletion and "Peak Oil"

Because petroleum is a nonrenewable natural resource, the industry will someday face the inevitable depletion of the world's oil supply. In 2007, the BP *Statistical Review of World Energy* predicted that the reserves in the Middle East would last 79.5 more years; those in Latin America, 41.2 years; and those in North America, 12 years. At 2008 production levels, the world's oil reserves would be depleted in 40.5 years.

A theory developed by geologists Marion King Hubbert in the 1940's, known as the Hubbert's peak theory (also known as "peak oil") or the "Hubbert curve," identified patterns of resource use that let Hubbert to predict that U.S. oil production would peak in the early 1970's. The extension of this analysis to world energy production to the prediction that a worldwide energy crisis would occur after production of oil and gas peaked. Research conducted by IBISWorld suggested that biofuels such as ethanol and biodiesel would continue to supplement petroleum but that, because output levels were low, these fuels would not replace local oil production and would play only a minor role in reducing dependence on imported crude oil.

Peak Oil and Climate Change

Perhaps the most damaging effect of petrochemicals is global warming. The use of fossil fuels releases carbon dioxide (CO_2) and methane, two greenhouse gases (GHGs) that trap the Earth's heat within the atmosphere and lead to warmer oceans, warmer climates, unstable weather, rising sea levels, and concomitant phenomena such as floods, drought, shifts in water flow, forest fires, famines, and extinctions. Such changes have already been noted but do not take place simultaneously or immediately; still they can occur over as short a time as decades, and their ecological and climate impacts are extremely complex and difficult to predict.

There is some consensus, however, that such changes, and their climate impacts, will increase in occurrence and severity, requiring ongoing mitigation and response. The projected effects of climate change are sufficient reason by themselves to seek to reduce fossil fuel dependency and replace energy needs with more sustainable resources. Peak oil will challenge humankind's ability to deal with and adapt to global warming: a decline in global oil supply, paired with an increase in the cost of goods and services, reduces the ability to make the investments in infrastructure that are required in order to respond to global warming in a proactive and coordinated fashion. Therefore, it is imperative that policy makers consider proposed responses to peak oil and global warming in tandem.

Joseph R. Rudolph, Jr.

Further Reading

Deffeyes, Kenneth S. *Hubbert's Peak: The Impending World Oil Shortage.* Rev. ed. Princeton, N.J.: Princeton University Press, 2003. Argues that oil production is peaking and will soon enter a terminal decline as existing oil fields run dry and large new finds become increasingly scarce.

Falola, Toyin, and Ann Genova. *The Politics of the Global Oil Industry: An Introduction.* Westport, Conn.: Praeger, 2005. Details connections throughout the world between business and political interests in the production and distribution of oil.

Heinberg, Richard. *The Party's Over: Oil, War, and the Fate of Industrial Societies.* 2d ed. Gabriola Island, B.C.: New Society, 2003. Argues that cheap energy sources, such as oil, are about to become more expensive, bringing about the end of the industrial era, which depends upon cheap energy. Recommends actions that Western nations and local communities can undertake to manage this transition.

International Energy Agency. *Oil Crises and Climate Challenges: Thirty Years of Energy Use in IEA Countries.* Paris: Author, 2004. Study of energy use during the last three decades of the twentieth century, detailing national and international responses to energy shortages and climatic shifts.

Mitchell, John V., and Beth Mitchell. "Structural crisis in the oil and gas industry." *Energy Policy* 64 (2014): 36-42. A number of converging developments are creating a crisis in the oil industry, as new technology increases supply and other factors reduce demand.

Paul, Bill. *Future Energy: How the New Oil Industry Will Change People, Politics, and Portfolios.* Hoboken, N.J.: John Wiley and Sons, 2007. Argues that a coming revolution in energy technology will transform the global economy, as well as sociopolitical structures. Recommends steps to benefit from the change.

Raymond, Martin S., and William L. Leffler. *Oil and Gas Production in Nontechnical Language.* Tulsa, Okla.: PennWell, 2006. Good overview of the oil and gas industries. Contains illustrative graphs, charts, and drawings that make the complex processes involved in oil and gas production understandable.

Sampson, Anthony. *The Seven Sisters: The Great Oil Companies and the World They Shaped.* New York: Bantam Books, 1976. History of the oil industry, its development into a major force, and its effects on the world.

Simmons, Matthew R. *Twilight in the Desert: The Coming Saudi Oil Shock and the World Economy.* Hoboken, N.J.: John Wiley and Sons, 2005. Focuses on Middle Eastern peak oil production and its threat to global economic stability.

Skeet, Ian. *OPEC: Twenty-five Years of Prices and Politics.* New York: Cambridge University Press, 1988. Focused study of the Organization of Petroleum Exporting Countries and the political motives and consequences of its decisions to raise and lower oil prices.

Solberg, Carl. *Oil Power: The Rise and Imminent Fall of an American Empire.* New York: Mason/Charter, 1976. Explication of the role of petroleum (both domestic and imported) in driving the United States' rise as a global superpower and the implications of peak oil for its future geopolitical status.

Yergin, Daniel. *The Prize: The Epic Quest for Oil, Money, and Power.* 1991. New ed. New York: The Free Press, 2008. Comprehensive, Pulitzer Prize-winning history of oil as a commodity and of the oil industry.

See also: Fossil fuel reserves; Fossil fuels and global warming; Fuels, alternative; Gasoline prices; Hubbert's peak; Organization of Petroleum Exporting Countries.

Oregon Institute of Science and Medicine

Category: Organizations and agencies
Date: Established 1980
Web address: http://www.oism.org

Mission

The Oregon Institute of Science and Medicine (OISM) figures in the debate on global climate change mostly because of controversial mailings sent out in 1998 and 2007, and its sponsorship of the Petition Project or Oregon Petition, a petition by scientists opposed to the idea of global warming that as of January, 2009, had more than thirty-one thousand signers. The OISM is a small, nonprofit private institution in Cave Junction, Oregon, a small community of about fourteen hundred residents in southwestern Oregon. The Institute was founded in 1980 and as of 2016 had eight faculty (though it does not teach classes) as well as a number of volunteers. Most of its stated research efforts involve research on medical subjects, but it also lists as research interests "improvement in precollege education curricula, especially in the sciences; and improved civilian emergency preparedness." In the climate debate it is prominent as sponsor of the Oregon Petition and distributor of papers rejecting the idea of global warming. The late, eminent scientist Frederick Seitz served as spokesperson for some of these efforts.

Significance for Climate Change

In 1998, the OISM first attracted widespread attention among climate researchers by distributing a mass mailing of a paper casting doubt on global warming. The paper was accompanied by a letter signed by Seitz, who had an illustrious career in solid state physics at the University of Illinois and Rockefeller University that included chairing the United States National Academy of Sciences from 1962 to 1969 and receiving the National Medal of Science in 1973. In 1979, Seitz became a paid consultant to the tobacco industry and led its scientific research program to discredit the evidence linking smoking to cancer and other illnesses. After his connections with the tobacco industry ended, Seitz became a critic of various environmental causes, including climate change.

The article mailed by the OISM looked like a genuine article reprinted by a scientific journal. In fact, the article almost exactly duplicated the format of the *Journal of the National Academy of Sciences.* The academy took the unusual step of issuing a public statement disclaiming any connection between itself and the paper, and many scientists expressed anger that Seitz had used his former connections to the academy to promote a paper that seemed purposely designed to resemble an academy publication. Seitz later claimed to have urged the authors of the paper to submit it for peer review.

In 2007, the OISM sent out another mass-mailed paper, "Environmental Effects of Increased Atmospheric Carbon Dioxide," by Arthur B. Robinson, Noah E. Robinson, and Willie Soon, which had been published in the *Journal of American Physicians and Surgeons.* The *Journal of American Physicians and*

Surgeons is published online by a politically conservative medical organization, the Association of American Physicians and Surgeons. The journal specializes in articles critical of government health care regulations, but it also frequently publishes nonmedical articles appealing to political conservatives. The paper by Robinson, Robinson, and Soon included much of the material from the 1998 paper and was also accompanied by a letter from Seitz.

The other major effort by the OISM to influence the climate debate is the Petition Project or Oregon Petition. The petition reads in part: We urge the United States government to reject the global warming agreement that was written in Kyoto, Japan in December, 1997, and any other similar proposals. The proposed limits on greenhouse gases would harm the environment, hinder the advance of science and technology, and damage the health and welfare of mankind.

The Petition Project provides lists of signatories and data on their academic credentials. As of January 23, 2009, the Petition Project claimed: The current list of 31,072 petition signers includes 9,021 PhD; 6,961 MS; 2,240 MD [medical] and DVM [veterinary]; and 12,850 BS or equivalent academic degrees. Most of the MD and DVM signers also have underlying degrees in basic science.

One significant problem with the Oregon Petition is that it does not list the institutional affiliations of signers. It also does not verify the credentials of signers. The words "credential" and "credibility" both come from the same Latin word meaning "believe." Credentials are evidence that someone is credible when talking about a subject. A patient may not have the technical training to understand or identify a medical problem, but a doctor's credentials are a strong indication that the doctor's opinion is credible. The doctor may also be well informed about many other nonmedical subjects (like climate change), but without credentials in those areas, there is no evidence one way or the other to judge his or her credibility.

Only 29 percent of the signers of the Oregon Petition have doctoral degrees in science, whereas 41 percent have only a bachelor's degree. A breakdown by specialty claimed 3,697 signers were scientists in the Earth, atmospheric, and environmental sciences, including 578 atmospheric scientists. Only 12 percent of the signers have degrees in Earth, atmospheric, and environmental science, and fewer than 2 percent are atmospheric scientists. The Oregon Petition Web site claims: "All of the listed signers have formal educations in fields of specialization that suitably qualify them to evaluate the research data related to the petition statement." Most of the signers, however, are in fields that have no real bearing on climate change, and most do not have the level of training needed to evaluate the research on climate change. They may be perfectly qualified in their own specialties, but they do not have credentials, or credibility, in climate research.

Although there are many uncertainties in predicting future climate changes and in formulating policy on climate change, campaigns such as the Oregon Petition and the Seitz mailings damage, rather than advance, the skeptic cause. Real experts on climate change are not persuaded by the Oregon Petition or the OISM mailings. The use of deceptive mailings and irrelevant credentials may persuade people without expertise in science, but those tactics serve only to erode the credibility of skeptics in the eyes of scientists.

As of 2016, the Oregon Petition site appears mostly inactive and still claims 31,000 signers, the same as in 2009. The OISM mentions the petition but does not link to it. The Institute web site mentions mostly its medical research. Institute founder Arthur B. Robinson ran for Congress in 2010, 2012, and 2014, but was not elected. He was on the ballot again in 2016.

Steven I. Dutch

Further Reading

Dessler, Andrew E., and Edward A. Parson. The Science and Politics of Global Climate Change: A Guide to the Debate. New York: Cambridge University Press, 2006. Overview of the controversy over global warming that includes analysis of the philosophical and practical issues raised in general by the intersection of politics and science.

Dunlap, Riley E., and Peter J. Jacques. "Climate change denial books and conservative think tanks: exploring the connection." American Behavioral Scientist (2013): 0002764213477096.

Most earlier books skeptical of climate change were produced by authors having ties to conservative think tanks but in recent years, self-published books have risen sharply in number. The vast majority of climate denial books have not undergone peer review.

Michaels, David. Doubt Is Their Product: How Industry's Assault on Science Threatens Your Health. New York: Oxford University Press, 2008. Examination of tactics used to exaggerate doubt in environmental health controversies. Many of these techniques were first used in the campaign to downplay the health risks of tobacco and are being applied to climate change.

Pearce, Fred. With Speed and Violence: Why Scientists Fear Tipping Points in Climate Change. Boston: Beacon Press, 2007. The chapter "Turning Up the Heat: A Skeptic's Guide to Climate Change" lays out a careful response to global warming skeptics that addresses their concerns.

Pittock, A. Barrie. Climate Change: Turning Up the Heat. Sterling, Va.: Earthscan, 2005. This book includes a chapter on the function of skeptics and uncertainty in the global warming debate.

Robinson, Arthur B., Noah E. Robinson, and Willie Soon. "Environmental Effects of Increased Atmospheric Carbon Dioxide." *Journal of American Physicians and Surgeons* 12, no. 3 (Fall, 2007). The 2007 paper of the OISM. Contains much of the information published in the 1998 paper but with some further discussion, figures, and references.

See also: Pseudoscience and junk science; Scientific credentials; Scientific proof; Skeptics.

Organization of Petroleum Exporting Countries

Categories: Organizations and agencies; fossil fuels; energy
Date: Established 1960
Web address: http://www.opec.org/

OPEC promotes the use of petroleum and attempts to make oil production profitable for its member countries. Petroleum is one of the three fossil fuels contributing to an increase in atmospheric CO_2, which is a key factor in global warming.

Key concepts

Arab oil embargo: a 1973 action by Arab members of OPEC to stop exporting petroleum to the United States and other supporters of Israel

Arab states: countries in the Middle East in which Arabs are the primary ethnic group, often incorrectly assumed to include Iran

cartel: an organization that limits competition in a particular industry to increase profitability for its members

fossil fuels: carbon-based fuels derived from decayed plants and animals under geologic pressure over millions of years

founding members: an official designation given to the five countries that formed OPEC

OPEC Conference: the chief authority within OPEC; composed of representatives of all the member states

secretariat: the administrative section of OPEC, which carries out the day-to-day functions of the organization

Mission

Meeting in Baghdad, Iraq, September 10-14, 1960, Iran, Iraq, Kuwait, Saudi Arabia, and Venezuela formed the Organization of Petroleum Exporting Countries (OPEC). The organization is often referred to as a cartel, with the aim of coordinating petroleum production and policies. Through these actions, the members hope to achieve maximum economic return, without causing a decrease in demand for petroleum. There are seven other active members (Algeria, Angola, Ecuador, Libya, Nigeria, Qatar, and the United Arab Emirates).

OPEC Organization

OPEC is an intergovernmental organization with its headquarters in Vienna, Austria. Any oil-exporting country may apply for membership, which is granted if the application is accepted by three-fourths of the current members, including all five of the founding members. At the OPEC Conference all members have one vote, and the conference alone has the power to make decisions on all policy issues of substance, by unanimous vote. At its semiannual sessions, the conference also appoints a secretary general who oversees the daily operation of the organization. All sections of the organization are charged with developing an acceptable policy that will match supply and demand for petroleum, allowing for stable prices and supplies. OPEC must take into consideration the effect these policies will have on other oil exporting countries and upon the nations that consume petroleum.

For the first decade of its existence, OPEC had limited opportunities. Large American and European petroleum companies dominated all aspects of oil production and marketing. Around the end of the decade, most members nationalized oil production, giving themselves greater power. This also gave the countries a greater stake in having a stable marketplace for trading petroleum. OPEC set production quotas for each country that insured relative stability, since slightly more than half the traded oil originates in OPEC. OPEC prices all of its oil in U.S. dollars.

The 1973 Arab Oil Embargo

In 1973, American economic problems had caused a decrease in the value of the dollar. Thus, OPEC leaders desired to increase the price of oil. On October 6, the Yom Kippur War was launched by Egypt and Syria against Israel, giving OPEC an excuse to raise prices and make a political statement. Led by the Persian Gulf oil producers, other Middle Eastern states sought to weaken the ties between the United States and Israel. On October 17, OPEC decided to end oil exports to those countries supporting Israel. OPEC cannot force its members to follow its decisions. However, within two days all the Arab states that were OPEC members joined in this action, as did Iran. In just over two weeks, OPEC cut production by one fourth. The actual military conflict was only twenty days in length, but the oil embargo lasted five months. The rising oil prices were a strong incentive for other oil producing countries to increase their production. This caused the resolve of some participating in the embargo to weaken, so with the military conflict ended and the

price of oil greatly increased, OPEC could declare victory and stop boycotting the American market.

Even though the Arab oil embargo only affected 7 percent of the oil available to the United States, it sent tremors throughout the United States. The psychological impact of people worrying about the gasoline supply resulted in long lines forming at gas stations across the country. Many areas of the country suffered gasoline shortages. By the end of the embargo, the price of gasoline had increased by about 45 percent nationwide. The threat of another oil embargo led to support for more domestic oil production and exploration and for non-OPEC suppliers. In the United States, the long term effect of the embargo was energy conservation, including mandated automobile efficiency standards. By decreasing dependence upon OPEC oil, the United States was able to decrease OPEC's economic and political power.

The Post-Embargo Era

The focus of OPEC's work since the mid-1970's has been to stabilize oil prices in a range that is advantageous for its members. About twice a decade an unexpected political or economic situation has affected the oil markets, causing rapid price increases or decreases. OPEC desires stability or orderly petroleum price increases. Since the 1990's a new challenge has arisen: the strong international movement to reduce dependence upon fossil fuels to combat global warming. OPEC fears this could significantly reduce the world's demand for oil. During the 2000's, OPEC produces about 45 percent of the world's petroleum and anticipates this increasing to over 50 percent in the next decade. While it has passed resolutions in favor of environmental responsibility and efficient automobiles, it also lobbies hard against any special taxation of oil, or petroleum products. OPEC believes that in the foreseeable future there is a continuing major role for oil. To offset any contribution oil makes to global warming, it advocates carbon capture and storage technologies, which captures carbon dioxide pollution and stores it underground. It is an advocate for investment in exploration for new oil fields and development of current fields.

Context

The balancing act that OPEC must undertake between optimal profitability and continued demand for oil products means that the secretariat is very active. The secretariat is constantly trying to forecast the global economic future, while at the same time watching for political activity that might be seen as a threat to the stability of the oil supply. OPEC was formed because there was a strong desire for the governments and citizens of the countries with the oil resources to get a larger share of the profits, rather than having them go to foreign firms. As such, it is only to be expected that OPEC will push hard to retain its markets for oil. With its central focus on economic interests, it should be expected that OPEC will fight hard to stop any movement to greatly reduce the use of oil, even if these changes are part of a process toward ending global warming. To get OPEC support on climate change issues, any proposals must allow continued use of petroleum products.

Donald A. Watt

Further Reading

Leggett, Jeremy. *Carbon War: Global Warming and the End of the Oil Era.* London: Routledge, 2001. Advocate of alternative forms of energy argues that lobbyists for OPEC and the major oil companies have caused continuing climate change by blocking the development of nonfossil fuel resources.

Mills, Robin. *The Myth of the Oil Crisis: Overcoming the Challenges of Depletion, Geopolitics, and Global Warming.* Santa Barbara, Calif.: Praeger, 2008. Asserts that oil resources will continue to be available; discusses the need to develop policies that will ensure political stability and keep climate change to a minimum.

Parra, Francisco. *Oil Politics: A Modern History of Petroleum.* London: I. B. Tauris, 2004. A former Secretary General of OPEC, the author focuses on the interactions between the producers and consumers of oil since the middle of the twentieth century.

Ross, Michael. *The oil curse: how petroleum wealth shapes the development of nations.* Princeton University Press, 2012. Oil wealth poses a paradox. Many poor countries with abundant oil do not benefit from it. In the last few decades they have become less democratic and more likely to have civil wars.

See also: Fossil fuel reserves; Fossil fuels and global warming; Gasoline prices; Hubbert's peak; Iran; Motor vehicles; Oil industry; Saudi Arabia; World Trade Organization.